T0178143

Lecture Notes in Computer Science   14554

Founding Editors

Gerhard Goos
Juris Hartmanis

Editorial Board Members

Elisa Bertino, *Purdue University, West Lafayette, IN, USA*
Wen Gao, *Peking University, Beijing, China*
Bernhard Steffen, *TU Dortmund University, Dortmund, Germany*
Moti Yung, *Columbia University, New York, NY, USA*

The series Lecture Notes in Computer Science (LNCS), including its subseries Lecture Notes in Artificial Intelligence (LNAI) and Lecture Notes in Bioinformatics (LNBI), has established itself as a medium for the publication of new developments in computer science and information technology research, teaching, and education.

LNCS enjoys close cooperation with the computer science R & D community, the series counts many renowned academics among its volume editors and paper authors, and collaborates with prestigious societies. Its mission is to serve this international community by providing an invaluable service, mainly focused on the publication of conference and workshop proceedings and postproceedings. LNCS commenced publication in 1973.

Stevan Rudinac · Alan Hanjalic · Cynthia Liem ·
Marcel Worring · Björn Þór Jónsson · Bei Liu ·
Yoko Yamakata
Editors

MultiMedia Modeling

30th International Conference, MMM 2024
Amsterdam, The Netherlands, January 29 – February 2, 2024
Proceedings, Part I

Springer

Editors
Stevan Rudinac 🆔
University of Amsterdam
Amsterdam, The Netherlands

Alan Hanjalic 🆔
Delft University of Technology
Delft, The Netherlands

Cynthia Liem 🆔
Delft University of Technology
Delft, The Netherlands

Marcel Worring 🆔
University of Amsterdam
Amsterdam, The Netherlands

Björn Þór Jónsson 🆔
Reykjavik University
Reykjavik, Iceland

Bei Liu 🆔
Microsoft Research Lab – Asia
Beijing, China

Yoko Yamakata 🆔
The University of Tokyo
Tokyo, Japan

ISSN 0302-9743   ISSN 1611-3349 (electronic)
Lecture Notes in Computer Science
ISBN 978-3-031-53304-4   ISBN 978-3-031-53305-1 (eBook)
https://doi.org/10.1007/978-3-031-53305-1

© The Editor(s) (if applicable) and The Author(s), under exclusive license
to Springer Nature Switzerland AG 2024

This work is subject to copyright. All rights are reserved by the Publisher, whether the whole or part of the material is concerned, specifically the rights of translation, reprinting, reuse of illustrations, recitation, broadcasting, reproduction on microfilms or in any other physical way, and transmission or information storage and retrieval, electronic adaptation, computer software, or by similar or dissimilar methodology now known or hereafter developed.
The use of general descriptive names, registered names, trademarks, service marks, etc. in this publication does not imply, even in the absence of a specific statement, that such names are exempt from the relevant protective laws and regulations and therefore free for general use.
The publisher, the authors, and the editors are safe to assume that the advice and information in this book are believed to be true and accurate at the date of publication. Neither the publisher nor the authors or the editors give a warranty, expressed or implied, with respect to the material contained herein or for any errors or omissions that may have been made. The publisher remains neutral with regard to jurisdictional claims in published maps and institutional affiliations.

This Springer imprint is published by the registered company Springer Nature Switzerland AG
The registered company address is: Gewerbestrasse 11, 6330 Cham, Switzerland

Paper in this product is recyclable.

Preface

These four proceedings volumes contain the papers presented at MMM 2024, the International Conference on Multimedia Modeling. This 30th anniversary edition of the conference was held in Amsterdam, The Netherlands, from 29 January to 2 February 2024. The event showcased recent research developments in a broad spectrum of topics related to multimedia modelling, particularly: audio, image, video processing, coding and compression, multimodal analysis for retrieval applications, and multimedia fusion methods.

We received 297 regular, special session, Brave New Ideas, demonstration and Video Browser Showdown paper submissions. Out of 238 submitted regular papers, 27 were selected for oral and 86 for poster presentation through a double-blind review process in which, on average, each paper was judged by at least three program committee members and reviewers. In addition, the conference featured 23 special session papers, 2 Brave New Ideas and 8 demonstrations. The following four special sessions were part of the MMM 2024 program:

- FMM: Special Session on Foundation Models for Multimedia
- MDRE: Special Session on Multimedia Datasets for Repeatable Experimentation
- ICDAR: Special Session on Intelligent Cross-Data Analysis and Retrieval
- XR-MACCI: Special Session on eXtended Reality and Multimedia - Advancing Content Creation and Interaction

The program further included four inspiring keynote talks by Anna Vilanova from the Eindhoven University of Technology, Cees Snoek from the University of Amsterdam, Fleur Zeldenrust from the Radboud University and Ioannis Kompatsiaris from CERTH-ITI.

In addition, the annual MediaEval workshop was organised in conjunction with the conference. The attractive and high-quality program was completed by the Video Browser Showdown, an annual live video retrieval competition, in which 13 teams participated.

We would like to thank the members of the organizing committee, special session and VBS organisers, steering and technical program committee members, reviewers, keynote speakers and authors for making MMM 2024 a success.

December 2023

Stevan Rudinac
Alan Hanjalic
Cynthia Liem
Marcel Worring
Björn Þór Jónsson
Bei Liu
Yoko Yamakata

Organization

General Chairs

Stevan Rudinac — University of Amsterdam, The Netherlands
Alan Hanjalic — Delft University of Technology, The Netherlands
Cynthia Liem — Delft University of Technology, The Netherlands
Marcel Worring — University of Amsterdam, The Netherlands

Technical Program Chairs

Björn Þór Jónsson — Reykjavik University, Iceland
Bei Liu — Microsoft Research, China
Yoko Yamakata — University of Tokyo, Japan

Community Direction Chairs

Lucia Vadicamo — ISTI-CNR, Italy
Ichiro Ide — Nagoya University, Japan
Vasileios Mezaris — Information Technologies Institute, Greece

Demo Chairs

Liting Zhou — Dublin City University, Ireland
Binh Nguyen — University of Science, Vietnam National University Ho Chi Minh City, Vietnam

Web Chairs

Nanne van Noord — University of Amsterdam, The Netherlands
Yen-Chia Hsu — University of Amsterdam, The Netherlands

Video Browser Showdown Organization Committee

Klaus Schoeffmann	Klagenfurt University, Austria
Werner Bailer	Joanneum Research, Austria
Jakub Lokoc	Charles University in Prague, Czech Republic
Cathal Gurrin	Dublin City University, Ireland
Luca Rossetto	University of Zurich, Switzerland

MediaEval Liaison

Martha Larson	Radboud University, The Netherlands

MMM Conference Liaison

Cathal Gurrin	Dublin City University, Ireland

Local Arrangements

Emily Gale	University of Amsterdam, The Netherlands

Steering Committee

Phoebe Chen	La Trobe University, Australia
Tat-Seng Chua	National University of Singapore, Singapore
Kiyoharu Aizawa	University of Tokyo, Japan
Cathal Gurrin	Dublin City University, Ireland
Benoit Huet	Eurecom, France
Klaus Schoeffmann	Klagenfurt University, Austria
Richang Hong	Hefei University of Technology, China
Björn Þór Jónsson	Reykjavik University, Iceland
Guo-Jun Qi	University of Central Florida, USA
Wen-Huang Cheng	National Chiao Tung University, Taiwan
Peng Cui	Tsinghua University, China
Duc-Tien Dang-Nguyen	University of Bergen, Norway

Special Session Organizers

FMM: Special Session on Foundation Models for Multimedia

Xirong Li	Renmin University of China, China
Zhineng Chen	Fudan University, China
Xing Xu	University of Electronic Science and Technology of China, China
Symeon (Akis) Papadopoulos	Centre for Research and Technology Hellas, Greece
Jing Liu	Chinese Academy of Sciences, China

MDRE: Special Session on Multimedia Datasets for Repeatable Experimentation

Klaus Schöffmann	Klagenfurt University, Austria
Björn Þór Jónsson	Reykjavik University, Iceland
Cathal Gurrin	Dublin City University, Ireland
Duc-Tien Dang-Nguyen	University of Bergen, Norway
Liting Zhou	Dublin City University, Ireland

ICDAR: Special Session on Intelligent Cross-Data Analysis and Retrieval

Minh-Son Dao	National Institute of Information and Communications Technology, Japan
Michael Alexander Riegler	Simula Metropolitan Center for Digital Engineering, Norway
Duc-Tien Dang-Nguyen	University of Bergen, Norway
Binh Nguyen	University of Science, Vietnam National University Ho Chi Minh City, Vietnam

XR-MACCI: Special Session on eXtended Reality and Multimedia - Advancing Content Creation and Interaction

Claudio Gennaro	Information Science and Technologies Institute, National Research Council, Italy
Sotiris Diplaris	Information Technologies Institute, Centre for Research and Technology Hellas, Greece
Stefanos Vrochidis	Information Technologies Institute, Centre for Research and Technology Hellas, Greece
Heiko Schuldt	University of Basel, Switzerland
Werner Bailer	Joanneum Research, Austria

Program Committee

Alan Smeaton	Dublin City University, Ireland
Anh-Khoa Tran	National Institute of Information and Communications Technology, Japan
Chih-Wei Lin	Fujian Agriculture and Forestry University, China
Chutisant Kerdvibulvech	National Institute of Development Administration, Thailand
Cong-Thang Truong	Aizu University, Japan
Fan Zhang	Macau University of Science and Technology/Communication University of Zhejiang, China
Hilmil Pradana	Sepuluh Nopember Institute of Technology, Indonesia
Huy Quang Ung	KDDI Research, Inc., Japan
Jakub Lokoc	Charles University, Czech Republic
Jiyi Li	University of Yamanashi, Japan
Koichi Shinoda	Tokyo Institute of Technology, Japan
Konstantinos Ioannidis	Centre for Research & Technology Hellas/Information Technologies Institute, Greece
Kyoung-Sook Kim	National Institute of Advanced Industrial Science and Technology, Japan
Ladislav Peska	Charles University, Czech Republic
Li Yu	Huazhong University of Science and Technology, China
Linlin Shen	Shenzhen University, China
Luca Rossetto	University of Zurich, Switzerland
Maarten Michiel Sukel	University of Amsterdam, The Netherlands
Martin Winter	Joanneum Research, Austria
Naoko Nitta	Mukogawa Women's University, Japan
Naye Ji	Communication University of Zhejiang, China
Nhat-Minh Pham-Quang	Aimesoft JSC, Vietnam
Pierre-Etienne Martin	Max Planck Institute for Evolutionary Anthropology, Germany
Shaodong Li	Guangxi University, China
Sheng Li	National Institute of Information and Communications Technology, Japan
Stefanie Onsori-Wechtitsch	Joanneum Research, Austria
Takayuki Nakatsuka	National Institute of Advanced Industrial Science and Technology, Japan
Tao Peng	UT Southwestern Medical Center, USA

Thitirat Siriborvornratanakul	National Institute of Development Administration, Thailand
Vajira Thambawita	SimulaMet, Norway
Wei-Ta Chu	National Cheng Kung University, Taiwan
Wenbin Gan	National Institute of Information and Communications Technology, Japan
Xiangling Ding	Hunan University of Science and Technology, China
Xiao Luo	University of California, Los Angeles, USA
Xiaoshan Yang	Institute of Automation, Chinese Academy of Sciences, China
Xiaozhou Ye	AsiaInfo, China
Xu Wang	Shanghai Institute of Microsystem and Information Technology, China
Yasutomo Kawanishi	RIKEN, Japan
Yijia Zhang	Dalian Maritime University, China
Yuan Lin	Kristiania University College, Norway
Zhenhua Yu	Ningxia University, China
Weifeng Liu	China University of Petroleum, China

Additional Reviewers

Alberto Valese
Alexander Shvets
Ali Abdari
Bei Liu
Ben Liang
Benno Weck
Bo Wang
Bowen Wang
Carlo Bretti
Carlos Cancino-Chacón
Chen-Hsiu Huang
Chengjie Bai
Chenlin Zhao
Chenyang Lyu
Chi-Yu Chen
Chinmaya Laxmikant Kaundanya
Christos Koutlis
Chunyin Sheng
Dennis Hoppe
Dexu Yao
Die Yu

Dimitris Karageorgiou
Dong Zhang
Duy Dong Le
Evlampios Apostolidis
Fahong Wang
Fang Yang
Fanran Sun
Fazhi He
Feng Chen
Fengfa Li
Florian Spiess
Fuyang Yu
Gang Yang
Gopi Krishna Erabati
Graham Healy
Guangjie Yang
Guangrui Liu
Guangyu Gao
Guanming Liu
Guohua Lv
Guowei Wang

Gylfi Þór Guðmundsson
Hai Yang Zhang
Hannes Fassold
Hao Li
Hao-Yuan Ma
Haochen He
Haotian Wu
Haoyang Ma
Haozheng Zhang
Herng-Hua Chang
Honglei Zhang
Honglei Zheng
Hu Lu
Hua Chen
Hua Li Du
Huang Lipeng
Huanyu Mei
Huishan Yang
Ilias Koulalis
Ioannis Paraskevopoulos
Ioannis Sarridis
Javier Huertas-Tato
Jiacheng Zhang
Jiahuan Wang
Jianbo Xiong
Jiancheng Huang
Jiang Deng
Jiaqi Qiu
Jiashuang Zhou
Jiaxin Bai
Jiaxin Li
Jiayu Bao
Jie Lei
Jing Zhang
Jingjing Xie
Jixuan Hong
Jun Li
Jun Sang
Jun Wu
Jun-Cheng Chen
Juntao Huang
Junzhou Chen
Kai Wang
Kai-Uwe Barthel
Kang Yi

Kangkang Feng
Katashi Nagao
Kedi Qiu
Kha-Luan Pham
Khawla Ben Salah
Konstantin Schall
Konstantinos Apostolidis
Konstantinos Triaridis
Kun Zhang
Lantao Wang
Lei Wang
Li Yan
Liang Zhu
Ling Shengrong
Ling Xiao
Linyi Qian
Linzi Xing
Liting Zhou
Liu Junpeng
Liyun Xu
Loris Sauter
Lu Zhang
Luca Ciampi
Luca Rossetto
Luotao Zhang
Ly-Duyen Tran
Mario Taschwer
Marta Micheli
Masatoshi Hamanaka
Meiling Ning
Meng Jie Zhang
Meng Lin
Mengying Xu
Minh-Van Nguyen
Muyuan Liu
Naomi Ubina
Naushad Alam
Nicola Messina
Nima Yazdani
Omar Shahbaz Khan
Panagiotis Kasnesis
Pantid Chantangphol
Peide Zhu
Pingping Cai
Qian Cao

Qian Qiao
Qiang Chen
Qiulin Li
Qiuxian Li
Quoc-Huy Trinh
Rahel Arnold
Ralph Gasser
Ricardo Rios M. Do Carmo
Rim Afdhal
Ruichen Li
Ruilin Yao
Sahar Nasirihaghighi
Sanyi Zhang
Shahram Ghandeharizadeh
Shan Cao
Shaomin Xie
Shengbin Meng
Shengjia Zhang
Shihichi Ka
Shilong Yu
Shize Wang
Shuai Wang
Shuaiwei Wang
Shukai Liu
Shuo Wang
Shuxiang Song
Sizheng Guo
Song-Lu Chen
Songkang Dai
Songwei Pei
Stefanos Iordanis Papadopoulos
Stuart James
Su Chang Quan
Sze An Peter Tan
Takafumi Nakanishi
Tanya Koohpayeh Araghi
Tao Zhang
Theodor Clemens Wulff
Thu Nguyen
Tianxiang Zhao
Tianyou Chang
Tiaobo Ji
Ting Liu
Ting Peng
Tongwei Ma

Trung-Nghia Le
Ujjwal Sharma
Van-Tien Nguyen
Van-Tu Ninh
Vasilis Sitokonstantinou
Viet-Tham Huynh
Wang Sicheng
Wang Zhou
Wei Liu
Weilong Zhang
Wenjie Deng
Wenjie Wu
Wenjie Xing
Wenjun Gan
Wenlong Lu
Wenzhu Yang
Xi Xiao
Xiang Li
Xiangzheng Li
Xiaochen Yuan
Xiaohai Zhang
Xiaohui Liang
Xiaoming Mao
Xiaopei Hu
Xiaopeng Hu
Xiaoting Li
Xiaotong Bu
Xin Chen
Xin Dong
Xin Zhi
Xinyu Li
Xiran Zhang
Xitie Zhang
Xu Chen
Xuan-Nam Cao
Xueyang Qin
Xutong Cui
Xuyang Luo
Yan Gao
Yan Ke
Yanyan Jiao
Yao Zhang
Yaoqin Luo
Yehong Pan
Yi Jiang

Yi Rong
Yi Zhang
Yihang Zhou
Yinqing Cheng
Yinzhou Zhang
Yiru Zhang
Yizhi Luo
Yonghao Wan
Yongkang Ding
Yongliang Xu
Yosuke Tsuchiya
Youkai Wang
Yu Boan
Yuan Zhou
Yuanjian He
Yuanyuan Liu
Yuanyuan Xu
Yufeng Chen
Yuhang Yang
Yulong Wang

Yunzhou Jiang
Yuqi Li
Yuxuan Zhang
Zebin Li
Zhangziyi Zhou
Zhanjie Jin
Zhao Liu
Zhe Kong
Zhen Wang
Zheng Zhong
Zhengye Shen
Zhenlei Cui
Zhibin Zhang
Zhongjie Hu
Zhongliang Wang
Zijian Lin
Zimi Lv
Zituo Li
Zixuan Hong

Contents – Part I

Where Are Biases? Adversarial Debiasing with Spurious Feature
Visualization 1
   Chi-Yu Chen, Pu Ching, Pei-Hsin Huang, and Min-Chun Hu

Cross-Modal Hash Retrieval with Category Semantics 15
   Mengying Xu, Hanjiang Lai, and Jian Yin

Spatiotemporal Representation Enhanced ViT for Video Recognition 28
   Min Li, Fengfa Li, Bo Meng, Ruwen Bai, Junxing Ren, Zihao Huang,
   and Chenghua Gao

SCFormer: A Vision Transformer with Split Channel in Sitting Posture
Recognition 41
   Kedi Qiu, Shoudong Shi, Tianxiang Zhao, and Yongfang Ye

Dive into Coarse-to-Fine Strategy in Single Image Deblurring 53
   Zebin Li and Jianping Luo

TICondition: Expanding Control Capabilities for Text-to-Image
Generation with Multi-Modal Conditions 66
   Yuhang Yang, Xiao Yan, and Sanyuan Zhang

Enhancing Generative Generalized Zero Shot Learning via Multi-Space
Constraints and Adaptive Integration 80
   Zhe Kong, Neng Gao, Yifei Zhang, and Yuhan Liu

Joint Image Data Hiding and Rate-Distortion Optimization in Neural
Compressed Latent Representations 94
   Chen-Hsiu Huang and Ja-Ling Wu

GSUNet: A Brain Tumor Segmentation Method Based on 3D Ghost
Shuffle U-Net 109
   JiXuan Hong, JingJing Xie, XueQin He, and ChenHui Yang

ACT: Action-assoCiated and Target-Related Representations for Object
Navigation 121
   Youkai Wang, Yue Hu, Wansen Wu, Ting Liu, and Yong Peng

Foreground Feature Enhancement and Peak & Background Suppression
for Fine-Grained Visual Classification 134
  Die Yu, Zhaoyan Fang, and Yong Jiang

YOLOv5-SRR: Enhancing YOLOv5 for Effective Underwater Target
Detection 147
  Jinyu Shi and Wenjie Wu

Image Clustering and Generation with HDGMVAE-I 159
  Yongqi Liu, Jiashuang Zhou, and Xiaoqin Du

"Car or Bus?" CLearSeg: CLIP-Enhanced Discrimination Among
Resembling Classes for Few-Shot Semantic Segmentation 172
  Anqi Zhang, Guangyu Gao, Zhuocheng Lv, and Yukun An

PANDA: Prompt-Based Context- and Indoor-Aware Pretraining for Vision
and Language Navigation 187
  Ting Liu, Yue Hu, Wansen Wu, Youkai Wang, Kai Xu, and Quanjun Yin

Cross-Modal Semantic Alignment Learning for Text-Based Person Search 201
  *Wenjun Gan, Jiawei Liu, Yangchun Zhu, Yong Wu, Guozhi Zhao,
  and Zheng-Jun Zha*

Point Cloud Classification via Learnable Memory Bank 216
  Lisa Liu, William Y. Wang, and Pingping Cai

Adversarially Regularized Low-Light Image Enhancement 230
  William Y. Wang, Lisa Liu, and Pingping Cai

Advancing Incremental Few-Shot Semantic Segmentation
via Semantic-Guided Relation Alignment and Adaptation 244
  Yuan Zhou, Xin Chen, Yanrong Guo, Jun Yu, Richang Hong, and Qi Tian

PMGCN:Preserving Measuring Mapping Prototype Graph Calibration
Network for Few-Shot Learning 258
  Zhengye Shen, Guangtong Lu, Qian Qiao, and Fanzhang Li

ARE-CAM: An Interpretable Approach to Quantitatively Evaluating
the Adversarial Robustness of Deep Models Based on CAM 273
  Zituo Li, Jianbin Sun, Yuqi Qin, Lunhao Ju, and Kewei Yang

SSK-Yolo: Global Feature-Driven Small Object Detection Network
for Images 286
  *Bei Liu, Jian Zhang, Tianwen Yuan, Peng Huang, Chengwei Feng,
  and Minghe Li*

MetaVSR: A Novel Approach to Video Super-Resolution for Arbitrary
Magnification 300
*Zixuan Hong, Weipeng Cao, Zhiwu Xu, Zhenru Chen, Xi Tao,
Zhong Ming, Chuqing Cao, and Liang Zheng*

From Skulls to Faces: A Deep Generative Framework for Realistic 3D
Craniofacial Reconstruction 314
*Yehong Pan, Jian Wang, Guihong Liu, Qiushuo Wu, Yazi Zheng,
Xin Lan, Weibo Liang, Jiancheng Lv, and Yuan Li*

Structure-Aware Adaptive Hybrid Interaction Modeling for Image-Text
Matching 327
Wei Liu, Jiahuan Wang, Chao Wang, Yan Peng, and Shaorong Xie

Using Saliency and Cropping to Improve Video Memorability 342
Vaibhav Mudgal, Qingyang Wang, Lorin Sweeney, and Alan F. Smeaton

Contextual Augmentation with Bias Adaptive for Few-Shot Video Object
Segmentation 356
*Shuaiwei Wang, Zhao Liu, Jie Lei, Zunlei Feng, Juan Xu, Xuan Li,
and Ronghua Liang*

A Lightweight Local Attention Network for Image Super-Resolution 370
Feng Chen, Xin Song, and Liang Zhu

Domain Adaptation for Speaker Verification Based on Self-supervised
Learning with Adversarial Training 385
Qiulin Li, Junhao Qiang, and Qun Yang

Quality Scalable Video Coding Based on Neural Representation 396
Qian Cao, Dongdong Zhang, and Chengyu Sun

Hierarchical Bi-directional Temporal Context Mining for Improved Video
Compression 410
Zijian Lin and Jianping Luo

MAMixer: Multivariate Time Series Forecasting via Multi-axis Mixing 422
Yongyu Liu, Guoliang Lin, Hanjiang Lai, and Yan Pan

A Custom GAN-Based Robust Algorithm for Medical Image Watermarking ... 436
Kun Zhang, Chunling Gao, and Shuangyuan Yang

A Detail-Guided Multi-source Fusion Network for Remote Sensing Object
Detection 448
Xiaoting Li, Shouhong Wan, Hantao Zhang, and Peiquan Jin

A Secure and Fair Federated Learning Protocol Under the Universal
Composability Framework 462
  Li Qiuxian, Zhou Quanxing, and Ding Hongfa

Bi-directional Interaction and Dense Aggregation Network for RGB-D
Salient Object Detection 475
  Kang Yi, Haoran Tang, Hongyu Bai, Yinjie Wang, Jing Xu, and Ping Li

Face Forgery Detection via Texture and Saliency Enhancement 490
  Sizheng Guo, Haozhe Yang, and Xianming Lin

Author Index 503

Where Are Biases? Adversarial Debiasing with Spurious Feature Visualization

Chi-Yu Chen[1]([✉]), Pu Ching[2], Pei-Hsin Huang[2], and Min-Chun Hu[2]

[1] National Yang Ming Chiao Tung University, Taipei, Taiwan
altis5526@gmail.com
[2] National Tsing Hua University, Hsinchu, Taiwan

Abstract. To avoid deep learning models utilizing shortcuts in a training dataset, many debiasing models have been developed to encourage models learning from accurate correlations. Some research constructs robust models via adversarial training. Although this series of methods shows promising debiasing performance, we do not know precisely what spurious features have been discarded during adversarial training. To address its lack of explainability especially in scenarios with low error tolerance, we design AdvExp, which not only visualizes the underlying spurious feature behind adversarial training but also maintains good debiasing performance with the assistance of a robust optimization algorithm. We show promising performance of AdvExp on BiasCheXpert, a subsampled dataset from CheXpert, and uncover potential regions in radiographs recognized by deep neural networks as gender or race-related features.

Keywords: Adversarial debiasing · Spurious feature · Explainable artificial intelligence · Fariness · Chest X-ray

1 Introduction

Deep learning methods have been taking the world by storm in recent years. Despite their promising potential, deep learning methods grapple with two pivotal challenges: fairness and explainability. Firstly, these methods heavily rely on the available training data, often resorting to shortcuts within the dataset to meet training objectives. These shortcuts, often referred to as spurious features, represent undesirable correlations misaligned with human concepts. For instance, demographic features are often imbalanced in the domain of radiology AI, where medical images necessitate expert annotations and are often challenging to collect. Larrazabal et al. [1] revealed that state-of-the-art classifiers exhibited poorer performance when confronted with patients of underrepresented genders within the training data. Moreover, Gichoya et al. [2] demonstrated that neural network models could discern race from chest X-ray images, a task even challenging for radiologists, thus also engendering risks for models trained on racially imbalanced datasets. Second, explainability plays an essential role especially in

© The Author(s), under exclusive license to Springer Nature Switzerland AG 2024
S. Rudinac et al. (Eds.): MMM 2024, LNCS 14554, pp. 1–14, 2024.
https://doi.org/10.1007/978-3-031-53305-1_1

scenarios with low error tolerance, such as emergency room, stock market, law court, etc. In such scenarios, an interpretable model becomes imperative, surpassing the merit of mere high-performance models alone. As for a successful debiasing model, it is appealing to *explain what bias the model disregards and how it was accomplished.*

To address dataset biases and achieve fairness, one line of research centers on gradient reversal techniques [3], which adversarially incentivizes the model to discard spurious features in intermediate layers [4–7]. Nevertheless, these methods lack explainability, failing to reveal the specific spurious features discarded during adversarial training. Wang et al. [8] introduced noise-based perturbations to mitigate spurious feature influence, yet their approach faces explainability challenges. Kehrenberg et al. [9] addressed explainability by null-sampling at intermediate layers to extract invariant representations. These high-dimensional representations are then transformed back into images via an invertible decoder. However, assessing and validating discarded spurious information in these transformed images may not be straightforward. Singla et al. [10] identified spurious features by generating neural activation heatmaps, but this approach necessitates human labor to distinguish between spurious and target features. Wang et al. [11] achieved explainability by utilizing gradient reversal techniques to mask spurious features from autoencoder outputs, but their target predicting performance drops significantly in a biased dataset.

Hence, we propose AdvExp, a framework achieving both explainability and fairness without human labor for identifying spurious features. Our contribution is twofold.

1. To pursue explainability, we extended the architecture proposed in [11] to leverage adversarial loss and train an autoencoder. This autoencoder creates a continuous-valued image mask outlining spurious feature areas. This empowers us to visualize the spurious features discarded by our model. Notably, our work mainly focuses on medical images, where error tolerance is often minimal and spurious features like race or gender are intricate to visualize and disentangle. We uncover potential regions in radiographs recognized by deep neural networks as gender or race-related features. Given the challenging nature of distinguishing race from a chest X-ray image even for human experts, the possible race-related image areas brings to light the potential misalignment between human cognition and computational processes.
2. For fairness, we trained the target classifier from the debiased outputs of the autoencoder and incorporate one of the distributionally robust optimization (DRO) methods, GroupDRO [12], which explicitly urges AdvExp to treat each subgroup equivalently and disregard shortcuts created by spurious features. To the best of our knowledge, we are the first to apply DRO to adversarial debiasing methods. In comparison with prior debiasing algorithms such as Learning Not to Learn (LnL) [5], Reweight [13], and baseline methods, our model significantly outperforms them in fairness metrics on the BiasCheXpert dataset, concurrently offering the advantage of explainable visualization of the discarded spurious features.

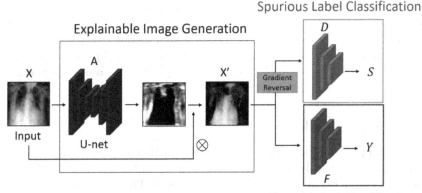

Fig. 1. The architecture of AdvExp.

2 Methods

2.1 Architecture

The goal of our problem is to predict the target label Y given the input image variable $X \in R^{H \times W \times C}$ while mitigating the influence of spurious features X^S. We assume that X is the combination of spurious features X^S and target features X^T. Additionally, each image is associated with a spurious label S, where S could represent gender, race, age, etc. For simplicity, both Y and S are considered as binary labels. Figure 1 illustrates the architecture of the proposed Adv-Exp model, which comprises three main components: Explainable Image Generation, Target Label Classification, and Spurious Label Classification. Within the Explainable Image Generation component, a U-Net-based autoencoder A generates a pixel-wise, single-channel spurious mask matching the size of the input image. Each pixel's value in the mask is continuous and ranges between 0 and 1, aiding in disentangling the impact of spurious features. The resultant explainable and debiased image X' is a Hadamard product of the input image with the spurious mask. The objective of the Explainable Image Generation is to minimize the mutual information $I(X'; S)$ while facilitating visualization of spurious features through X'. Proceeding on, the debiased image X' is sent to the Target Label Classifier F and the Spurious Label Classifier D. The output of the Target Label Classifier, denoted as $F(X')$, approximates $p(Y|X)$. On the other hand, the Spurious Label Classifier D plays a role in adversarial training and encourages the generation of the spurious mask, a process elaborated upon in Sect. 2.3.

2.2 GroupDRO

Shortcuts are often exploited by neural networks due to imbalanced data distributions. To ensure equal treatment of minority groups and enhance fairness within AdvExp, we implement the GroupDRO algorithm [12] in our Target Label Classifier F. GroupDRO assigns group weights to predefined groups and exponentially updates these weights based on their respective group losses. In our experimental setup, we consider spurious labels $S \in \{0, 1\}$ and target labels $Y \in \{0, 1\}$, resulting in a total of $|S| \times |Y|$ (i.e., 4) predefined groups. The detailed training process is elaborated in Sect. 2.3.

2.3   Training Procedure

Each training iteration consists of two steps. In the first step, we jointly train the autoencoder A and the Target Label Classifier F by minimizing the loss between the actual target label and the predicted target label. Therefore, the autoencoder A learns to preserve target features present in the input image, and the Spurious Label Classifier D is fixed throughout this step. In the second step, F is fixed and we optimize the Spurious Label Classifier D by minimizing the loss between the true and predicted spurious labels. Simultaneously, we update the autoencoder A through a reverse gradient from the Spurious Label Classifier to discard spurious features. This encapsulates the essence of adversarial debiasing within our approach, where the autoencoder A learns to confuse the Spurious Label Classifier while the latter endeavors to overcome this confusion. Combining step one and step two, the resulting image X' evolves into an explainable debiased image, which visualizes undesired spurious features and retains essential target features from the original input image. To be specific, during the first step, the autoencoder is trained by backpropagation of gradients from the Target Label Classifier F, along with a confusion loss L_{conf}. As mentioned earlier, the main goal of the autoencoder A is to minimize the mutual information $I(X'; S)$. By definition, $I(X'; S) = H(S) - H(S|X')$, where $H(\cdot)$ represents the marginal entropy, and $H(\cdot|\cdot)$ represents the conditional entropy. Since $H(S)$ is a constant, our focus shifts towards minimizing $-H(S|X') = E_{(x',s) \sim (X',S)}[\log p(s|x')]$, where $p(s|x')$ can be approximated by the output of the Spurious Label Classifier, denoted as $D(X')$. As a result, the confusion loss is defined as $L_{conf} = E[\log D(x')]$. The autoencoder A and the Target Label Classifier F are updated simultaneously. The loss function for the first step can be formulated as follows:

$$L_{first} = \min_{\theta_A, \theta_F} \{ E_{(x,y) \sim (X,Y)}[L_F(x', y)] + \lambda E[\log D(x')] \}, \tag{1}$$

where L_F is a cross entropy loss for Target Label Classification with GroupDRO implemented (will be explained later), λ is a tunable hyperparameter, and y is the target label of image x. In the second step, we optimize D and A based on the loss defined as follows. Notice that the minmax property is implemented as a gradient reversal layer [3].

$$L_{second} = \min_{\theta_A} \gamma \max_{\theta_D} \{ -\mu E_{(x,s) \sim (X,S)}[L_D(x', s)] \}, \tag{2}$$

where L_D is a cross entropy loss for spurious label classification, μ and γ are tunable hyperparameters, and s is the spurious label of image x. γ is implemented in the gradient reversal layer to control the propagating gradient from the Spurious Label Classifier.

GroupDRO is applied to Target Label Classification by assigning each training sample a group label g_i. Each group is also given a corresponding group weight W_{g_i}, where $W_g := [W_{g_1}, W_{g_2}, ..., W_{g_{n_g}}]$ forms a probability simplex, and n_g is the number of groups. In particular, given a learning rate η_{dro}, L_F is calculated based on Eq. 3,4,5 during iteration t.

$$W_{g_i}^t = W_{g_i}^{t-1} \exp \left(\eta_{dro} \, CrossEntropy(x_i, y_i) \right) \tag{3}$$

$$W_{g_i}^t = W_{g_i}^t \, / \sum_{i=1}^{n_g} W_{g_i}^t \tag{4}$$

$$L_F = \sum_{i=1}^{n_g} W_{g_i}^t \, CrossEntropy(x_i, y_i) \tag{5}$$

3 Datasets and Experimental Settings

3.1 Datasets

• **BiasCheXpert**: CheXpert-v1.0 [14] is a public dataset containing chest radiographs of 65240 patients. These chest X-rays were collected by Stanford Hospital between October 2002 and July 2017 in both outpatient and inpatient settings. In our experiment, we uniformly resized each image in the CheXpert-v1.0 Downsampled dataset to the resolution of 256×256 pixels. Each image in CheXpert-v1.0 is labeled with 14 observations (categories), and we selected four specific categories: *Cardiomegaly, Support Devices, Pneumonia, Pleural Effusion*, as the primary binary target labels. The four target labels are chosen because of their relatively distinct location and features. For the uncertain labels within each category, we adopted the U-ones method proposed in [14], which converts uncertain labels into positive labels as false-negatives are rather intolerable in clinical scenarios. To introduce gender and race as spurious attributes within our dataset, we created two extended datasets named BiasCheXpert-gender and BiasCheXpert-race. This was achieved by selecting images from CheXpert and ensuring that patients with positive labels are predominantly male or black, while patients with negative labels are predominantly female or white. For BiasCheXpert-gender, the dataset has the property of $p(Male|Pos) : p(Female|Pos) = p(Female|Neg) : p(Male|Neg) = \alpha : 1$. Similarly, for BiasCheXpert-race, it has the property of $p(Black|Pos) : p(White|Pos) = p(Black|Neg) : p(White|Neg) = \alpha : 1$. Note that α is a real number that stands for the bias ratio, and its value is larger than 1. For the gender bias experiment, we selected α to be 2, 10, and 100. In the race bias experiment, α values of 20, 40, and 100 are chosen. Further details regarding the numbers of samples in BiasCheXpert are revealed in Tables 1 and 2. Both the validation and testing sets of BiasCheXpert-gender and BiasCheXpert-race are composed of images randomly drawn from the original CheXpert-v1.0 Down-sampled dataset.

Table 1. Detailed description of BiasCheXpert-gender. (M: Male, F: Female, +: Positive label, -: negative label)

		Cardiomegaly		Support Devices		Pneumonia		Pleural Effusion	
		+	-	+	-	+	-	+	-
BiasCheXpert -gender (α=2)	M	7123	13417	M 25836	7085	M 4724	14070	M 20379	8548
	F	3561	26834	F 12918	14171	F 2362	28140	F 10189	17097
BiasCheXpert -gender (α=10)	M	7123	2683	M 25836	1417	M 4724	2814	M 20379	1709
	F	712	26834	F 2583	14171	F 472	28140	F 2037	17097
BiasCheXpert -gender (α=100)	M	7123	268	M 25836	141	M 4724	281	M 20379	170
	F	71	26834	F 258	14171	F 47	28140	F 203	17097
BiasCheXpert -gender Validation Set	M	920	4649	M 3224	2345	M 601	4968	M 2507	3062
	F	635	3373	F 2234	1774	F 423	3585	F 1837	2171
BiasCheXpert -gender Testing Set	M	947	4646	M 3160	2433	M 599	4994	M 2458	3135
	F	594	3390	F 2184	1800	F 432	3552	F 1816	2168

3.2 Metrics

We apply two measurements of fairness proposed in [15] to evaluate the performance of our debiasing model. First, equality of opportunity metric (EOM) calculates whether the true positive rate and true negative rate are similar in groups of different spurious attributes (i.e. groups of male and female for BiasCheXpert-gender, and groups of black and white for BiasCheXpert-race). Demographic disparity metric (DPM) evaluates whether the predicted positive rate and predicted negative rate are similar in groups of different spurious attributes. Let $\hat{y}$ be the target prediction, m be the number of target labels, and s be the spurious label. The definition of EOM and DPM can be written as:

$$EOM = \sum_{i=1}^{m} \frac{\min_{j \in S}[p(\hat{y} = i | y = i, s = j)]}{\max_{j \in S}[p(\hat{y} = i | y = i, s = j)]}, DPM = \sum_{i=1}^{m} \frac{\min_{j \in S}[p(\hat{y} = i | s = j)]}{\max_{j \in S}[p(\hat{y} = i | s = j)]} \quad (6)$$

The larger EOM and DPM are, the better our model achieves fairness and reduces the impact of spurious attribute. Moreover, since the training dataset is intentionally chosen to be biased, the model would perform poorly in testing accuracy if it learns from shortcut. Therefore, we also adopted area under ROC curve (AUC) to evaluate our debiasing models in testing performances.

3.3 Architecture Implementation

We adopted a U-net autoencoder containing 5 downsampling layers and 5 upsampling layers. The Spurious Label Classifier is a Resnet18 network, the Target

Table 2. Detailed description of BiasCheXpert-race. (B: Black, W: White, +: Positive label, -: negative label)

		Cardiomegaly			Support Devices			Pneumonia			Pleural Effusion	
		+	-		+	-		+	-		+	-
BiasCheXpert -race (α=20)	B	1972	3653	B	4168	1877	B	923	3833	B	3122	2314
	W	98	73077	W	208	37548	W	46	76673	W	156	46285
BiasCheXpert -race (α=40)	B	1972	1826	B	4168	938	B	923	1916	B	3122	1157
	W	49	73077	W	104	37548	W	23	76673	W	78	46285
BiasCheXpert -race (α=100)	B	1972	730	B	4168	375	B	923	766	B	3122	462
	W	19	73077	W	41	37548	W	9	76673	W	31	46285
BiasCheXpert -race Validation Set	B	253	773	B	533	493	B	141	885	B	412	614
	W	1675	9107	W	6101	4681	W	1177	9605	W	5009	5773
BiasCheXpert -race Testing Set	B	254	733	B	518	469	B	115	872	B	387	600
	W	1597	9149	W	6014	4732	W	1165	9581	W	4903	5843

Label Classifier is a Resnet50 network, and Adam is used as the optimizer. The learning rate for Spurious Label Classifier and the learning rate η_{dro} for Group-DRO is 0.00001; the learning rate for Autoencoder and Target Label Classifier is 0.000001. There is neither data augmentation nor learning scheduler applied to our model for experimental simplicity. Empirically, λ, μ, and γ are set as 1, 10, and 0.1 respectively for BiasCheXpert-gender. For BiasCheXpert-race, λ, μ, and γ are set as 10, 10, and 0.1 respectively. The learning rate for the Target Label Classifier is adjusted to 0.00001 in the ablation study of AdvExp without GroupDRO. All experiments are performed with fixed random seed. For evaluation, we picked the epoch with the best validation AUC. We conducted our experiments on a PC equipped with an Intel(R) Core(TM) i7-12700K CPU and an NVIDIA GeForce RTX 3090 GPU card.

4   Results

4.1   BiasCheXpert-Gender

Explainable Visualization Results: For BiasCheXpert-gender, gender of each patient is the existing bias in chest radiographs. Since breast is the most prominent feature for gender in a radiograph, we expected our generated explainable image would blur the contour of breast. Figure 2 (a)~(d) shows the examples that the contour of breast is obscured in our explainable debiased image.

Comparison with Other Methods: We compared the debiasing performance of our model with several different methods: Learning not to learn (LnL), Reweight, GradCAM masking, Random masking, and Baseline model. LnL is

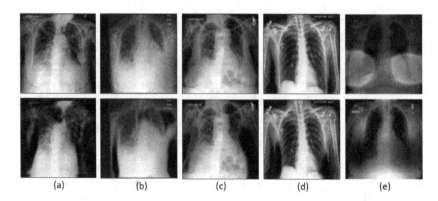

(a)   (b)   (c)   (d)   (e)

Fig. 2. (a)-(d) Visualization of gender debiasing ($\alpha = 10$). For each patient (a)-(d), the upper subfigure is the original BiasCheXpert image, where the contour of breast is manually sketched with red dashed lines. The lower subfigure is the explainable debiased image x', where the contour of breast is difficult to identify. (e) is an example of GradCAM masking result for one patient. The upper subfigure is the GradCAM activation map, and the lower one is the masked image. (Color figure online)

an existing adversarial debiasing method without spurious feature visualization [5]. For fair comparison, we changed the backbone of LnL as the U-net used in our model. Reweighting is another widely used simple algorithm dealing with fairness. It assigns each sample n a weight W_n, which is defined as

$$W_n = \frac{p(Y = y_n) \times p(S = s_n)}{p(Y = y_n, S = s_n)}, \tag{7}$$

where y_n and s_n stand for the target label and spurious label for sample n, respectively. The equation compares the difference between the expected probability and the observed joint probability. If the target and spurious labels are independent in the dataset, W_n equals to 1, otherwise W_n is less than 1. Therefore, the trained model focuses more on groups without spurious correlation. As for GradCAM masking, we pretrained a Spurious Label Classifier and built the spurious mask from its GradCAM activation map [16]. The original image was then masked by this spurious mask (cf. Fig. 2 (e)) and used for the training of the Target Label Classifier. In Random Masking, we randomly created spurious masks with value normalized between 0 and 1, and used the masked image for Target Label Classification. Baseline method simply used a regular Resnet50 network for Target Label Classification without any debiasing mechanism.

Table 3 shows that AdvExp outperforms the other methods among all metrics, except for pneumonia prediction, in which LnL [5] performs better in EOM and DPM. We can observe that if GroupDRO is not applied to our model, all metrics drop significantly, but still better than the Baseline in most cases except for pleural effusion prediction. Since the location of breast is in proximity with lower lung, where pleural effusion is often detected, the blurring of breast contour might also affect the prediction of pleural effusion. This shows

Table 3. BiasCheXpert-gender debiasing performances.

Method	Cardiomegaly			Support Devices			Pneumonia			Pleural Effusion		
	AUC	EOM	DPM	AUC	EOM	DPM	AUC	EOM	DPM	AUC	EOM	DPM
AdvExp	**0.773**	**1.514**	**1.290**	**0.789**	**1.634**	**1.473**	**0.669**	1.267	1.117	**0.798**	**1.760**	**1.677**
AdvExp w/o GroupDRO	0.701	0.841	0.670	0.725	0.906	0.674	0.583	0.734	0.676	0.689	0.711	0.519
LnL	0.724	1.017	0.790	0.617	1.230	1.155	0.558	**1.721**	**1.717**	0.707	1.193	1.030
Reweight	0.695	0.841	0.691	0.666	0.730	0.552	0.569	0.765	0.708	0.709	1.260	1.094
GradCAM masking	0.648	0.612	0.540	0.619	0.474	0.345	0.540	0.584	0.578	0.656	1.103	0.954
Random masking	0.692	0.844	0.696	0.653	0.667	0.498	0.501	0.637	0.636	0.681	0.889	0.489
Baseline	0.691	0.840	0.690	0.657	0.816	0.650	0.553	0.668	0.641	0.681	1.003	0.878

the benefit of implementing GroupDRO for debiasing. Reweight algorithm [13] treats each group in training dataset differently by their observed distribution and expected distribution, gaining a slight improvement on overall performance compared with the Baseline. GradCAM masking has poorest debiasing performance, showing that GradCAM mask cannot precisely locate breasts and results in loosing too much information. This urges the need of our model to provide both explainable and debiasing results at the same time. In Random masking method, the debiasing performance is either lower or similar to the Baseline model. Both GradCAM and Random masking shows that an effective spurious mask is essential for debiasing.

Comparison of Different Bias Ratio in the Dataset: We trained the proposed AdvExp on CheXpert-gender dataset with different bias ratio (including $\alpha = 2$, $\alpha = 10$, and $\alpha = 100$) and the corresponding performance on the testing dataset is shown in Table 4. When the bias ratio increases, both AdvExp and the Baseline performs worse in AUC, EOM, and DPM. This is expectable since the increase extent of bias makes deep learning models inclined to learn from shortcut spurious correlations. The good news is, AdvExp outperforms baseline ResNet50 model in even severe bias settings, which demonstrates its strength to reduce false correlations in the training dataset. From the debiased image visualization (cf. Fig. 3), we can observe that as the bias ratio increases, the masked area becomes narrower. This shows that AdvExp may not be not allowed to discard as many spurious features when there is a very strong correlation between target and spurious features, which makes the Autoencoder prone to retain both the target and spurious features. Luckily, such an extreme scenario doesn't happen often in real world.

4.2 BiasCheXpert-Race

As a kind of demographic feature, race is more abstract than gender, and even a radiologist can't distinguish a patient's race solely from chest x-ray images.

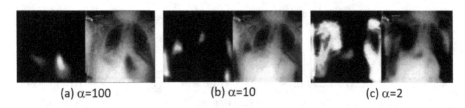

(a) α=100   (b) α=10   (c) α=2

Fig. 3. Explainable gender-debiased images of different α on label Cardiomegaly. (a-c) shows the debiased image results for α = 100, α = 10, and α = 2, respectively. Each subfigure consists of a spurious mask on the left, and a final debiased image result on the right. The masked area is the smallest when α = 100.

However, based on [13] and our experimental results, we found that race can be recognized by machines and potentially acts as a spurious feature. Most intriguingly, the visualization results of AdvExp (Fig. 4) indicates the possible locations where deep learning algorithms utilize to identify race.

Debiasing Performance: Table 5 demonstrates our race-debiasing performance under different bias ratio α across 4 oberservation labels. The results are less consistent compared with that in gender-debiasing, indicating that race-debiasing is a tougher task. As α in the training dataset increases, AdvExp performs worse, especially for Cardiomegaly. The phenomenon corresponds to what we've discussed in the gender-debiasing section that higher bias in the training dataset increases the difficulty of debiasing. Interestingly, we found that the performance of the Baseline model does not drop in accordance with the increased bias ratio. We inferred that since race features are more intricate in an medical image, the shortcut might not be learned as easy as in the gender bias case. Moreover, for Pneumonia, we found that several experimental results reach the maximum value of EOM/DPM (i.e., 2). Such high EOM/DPM is due to the collapse of deep learning models which predict every radiograph as negative target label. This could be attributed to the small size of positive pneumonia samples in BiasCheXpert-race (cf. Table 2) and the diverse lung patterns of pneumonia. Generally, our model performs better than the Baseline model, especially in the

Table 4. Gender-debiasing performance across datasets with different bias ratio.

	Setting	Cardiomegaly			Support Devices			Pneumonia			Pleural Effusion		
		AUC	EOM	DPM	AUC	EOM	DPM	AUC	EOM	DPM	AUC	EOM	DPM
α=2	AdvExp	0.817	1.902	1.825	0.826	1.928	1.863	0.703	1.964	1.911	0.832	1.922	1.912
	Baseline	0.794	1.414	1.265	0.764	1.321	1.126	0.635	1.244	1.183	0.807	1.651	1.524
α=10	AdvExp	0.773	1.514	1.290	0.789	1.634	1.473	0.669	1.267	1.117	0.798	1.760	1.677
	Baseline	0.691	0.840	0.690	0.657	0.816	0.649	0.561	0.636	0.595	0.694	0.835	0.634
α=100	AdvExp	0.661	0.710	0.537	0.698	1.043	0.850	0.559	0.374	0.323	0.648	0.706	0.536
	Baseline	0.619	0.483	0.388	0.595	0.335	0.247	0.530	0.313	0.294	0.589	0.356	0.252

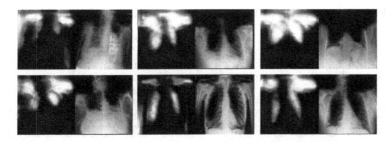

Fig. 4. Explainable race-debiased images on label Support Devices ($\alpha = 20$). Each subfigure consists of a spurious mask on the left, and a final debiased image result on the right. Most masked areas are distributed across shoulders and upper lungs.

fairness metrics, EOM and DPM, showing the debiasing strength of AdvExp in cases with more complicated spurious features like race.

Explainable Visualization Results: Since race can be hardly visualized in an X-ray image, to get the best explainable debiased image, we chose the best performance combination from Table 5 for demonstration: $\alpha = 20$ with label Support Devices. Figure 4 shows the explainble race-debiased image of six patients. Most masked area falls on shoulders and upper lungs. On the middle-lower radiograph in Fig. 4, the masked area particularly highlights the clavicles on both sides. Our visualization results indicate that a deep learning model might utilize these masked area to predict race. To further prove that the generated spurious mask (that indicates race features) is associated with upper body parts, we cropped the original image and trained a ResNet50 target classification model with the cropped image. To be more precise, the original image and the cropped image are in the size of $h \times w$, and $(h - 75) \times w$, respectively. We designed two cropping methods: Crop Lower and Crop Upper. In the Crop Lower method, the upper part of the image is removed, while in the Crop Upper method, the lower part of the image is removed (as illustrated in Fig. 5). The cropped images were further resized into (256, 256) for training the Target Label Classification model. We trained both models on BiasCheXpert-race ($\alpha = 20$) and tested them on the original testing set of BiasCheXpert-race.

Table 6 demonstrates the debiasing performance of the two cropping settings. Notice that we did not include pneumonia in this experiment since the trained model readily collapses in BiasCheXpert-race (as discussed in Sect. 4.2). One can find that except for AUC, Crop Lower has the highest EOM/DPM in almost all cases. Crop Upper has rather similar results compared with the Baseline. The drop in AUC might be due to partial loss in information since the removed part might also contain some target features. Nevertheless, the increase in EOM and DPM in Crop Lower experiment gives us a hint that the upper part of the image do include spurious features that have an impact on the debiasing results.

Table 5. Race-debiasing performance across datasets with different bias ratio.

	Setting	Cardiomegaly			Support Devices			Pneumonia			Pleural Effusion		
		AUC	EOM	DPM	AUC	EOM	DPM	AUC	EOM	DPM	AUC	EOM	DPM
α=20	AdvExp	**0.748**	**1.360**	**1.056**	**0.762**	**1.610**	**1.728**	**0.629**	**2**	**2**	**0.769**	**1.674**	**1.876**
	Baseline	0.746	1.175	1.021	0.708	1.173	1.042	0.572	1.107	1.106	0.739	1.153	1.110
α=40	AdvExp	0.730	**1.251**	0.977	**0.755**	**1.543**	**1.656**	**0.624**	1.098	1.082	**0.736**	**1.507**	**1.715**
	Baseline	**0.757**	1.147	**1.006**	0.697	1.098	0.988	0.562	**2**	**2**	0.704	1.064	0.990
α=100	AdvExp	0.710	1.117	0.879	**0.757**	**1.706**	**1.831**	**0.615**	1.078	1.062	0.713	**1.322**	**1.506**
	Baseline	**0.758**	**1.161**	**1.002**	0.697	1.117	1.007	0.575	**2**	**2**	**0.715**	1.122	1.073

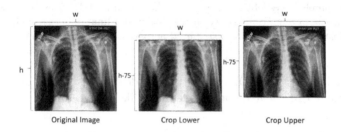

Fig. 5. Example of Crop Upper/Lower.

Table 6. Debiasing results of the cropping experiment.

Method	Cardiomegaly			Support Devices			Pleural Effusion		
	AUC	EOM	DPM	AUC	EOM	DPM	AUC	EOM	DPM
Crop Lower	**0.746**	**1.196**	1.072	0.707	**1.319**	**1.221**	**0.751**	**1.302**	**1.283**
Crop Upper	0.743	1.185	**1.074**	0.701	1.136	1.059	0.722	1.186	1.158
Baseline	**0.746**	1.175	1.021	**0.708**	1.173	1.042	0.739	1.153	1.110

5   Limitations

Although AdvExp seems promising, our work still encounters some limitations. First, in our experiments, we designed our target label and spurious label to be both binary for simplicity. Conceptually, AdvExp should be able to apply on multi-label tasks, and we may further validate our work in such scenario. Second, it is noted that AdvExp requires pre-defined spurious labels for debiasing. This prerequisite may overlook unknown spurious features that also correlates with target features.

6   Conclusion

For study fields where data collection is tedious, AdvExp not only reduces the impact of spurious attributes to achieve robustness but also visualizes the location of them in a given image to meet explainability. The discovery of spurious region even enables scientists to probe into what and how machines think and unravel the spurious information in traditional adversarial debiasing models.

References

1. Larrazabal, A.J., Nieto, N., Peterson, V., Milone, D.H., Ferrante, E.: Gender imbalance in medical imaging datasets produces biased classifiers for computer-aided diagnosis. In: Proceedings of the National Academy of Sciences, vol. 117, no. 23, pp. 12592–12594 (2020)
2. Gichoya, J.W., et al.: Ai recognition of patient race in medical imaging: a modelling study. Lancet Digit. Health 4(6), e406–e414 (2022)
3. Ganin, Y., Lempitsky, V.: Unsupervised domain adaptation by backpropagation. In: International Conference on Machine Learning, pp. 1180–1189. PMLR (2015)
4. Zhang, B.H., Lemoine, B., Mitchell, M.: Mitigating unwanted biases with adversarial learning. In: Proceedings of the 2018 AAAI/ACM Conference on AI, Ethics, and Society, pp. 335–340 (2018)
5. Kim, B., Kim, H., Kim, K., Kim, S., Kim, J.: Learning not to learn: training deep neural networks with biased data. In: Proceedings of the IEEE/CVF Conference on Computer Vision and Pattern Recognition, pp. 9012–9020 (2019)
6. Du, S., Hers, B., Bayasi, N., Hamarneh, G., Garbi, R.: FairDisCo: fairer AI in dermatology via disentanglement contrastive learning. In: Karlinsky, L., Michaeli, T., Nishino, K. (eds.) Computer Vision – ECCV 2022 Workshops. ECCV 2022. Lecture Notes in Computer Science, vol. 13804, pp. 185–202. Springer, Cham (2022). https://doi.org/10.1007/978-3-031-25069-9_13
7. Bahng, H., Chun, S., Yun, S., Choo, J., Oh, S.J.: Learning de-biased representations with biased representations. In: International Conference on Machine Learning, pp. 528–539. PMLR (2020)
8. Wang, Z., et al.: Fairness-aware adversarial perturbation towards bias mitigation for deployed deep models. In: Proceedings of the IEEE/CVF Conference on Computer Vision and Pattern Recognition, pp. 10379–10388 (2022)
9. Kehrenberg, T., Bartlett, M., Thomas, O., Quadrianto, N.: Null-sampling for interpretable and fair representations. In: Vedaldi, A., Bischof, H., Brox, T., Frahm, J.-M. (eds.) ECCV 2020. LNCS, vol. 12371, pp. 565–580. Springer, Cham (2020). https://doi.org/10.1007/978-3-030-58574-7_34
10. Singla, S., Feizi, S.: Salient ImageNet: how to discover spurious features in deep learning? arXiv preprint arXiv:2110.04301 (2021)
11. Wang, T., Zhao, J., Yatskar, M., Chang, K.-W., Ordonez, V.: Balanced datasets are not enough: estimating and mitigating gender bias in deep image representations. In: Proceedings of the IEEE/CVF International Conference on Computer Vision, pp. 5310–5319 (2019)
12. Sagawa, S., Koh, P.W., Hashimoto, T.B., Liang, P.: Distributionally robust neural networks for group shifts: on the importance of regularization for worst-case generalization. arXiv preprint arXiv:1911.08731 (2019)
13. Kamiran, F., Calders, T.: Data preprocessing techniques for classification without discrimination. Knowl. Inf. Syst. 33(1), 1–33 (2012)
14. Irvin, J., et al.: CheXpert: a large chest radiograph dataset with uncertainty labels and expert comparison. In: Proceedings of the AAAI Conference on Artificial Intelligence, vol. 33, no. 01, pp. 590–597 (2019)

15. Du, M., Yang, F., Zou, N., Hu, X.: Fairness in deep learning: a computational perspective. IEEE Intell. Syst. **36**(4), 25–34 (2020)
16. Selvaraju, R.R., Cogswell, M., Das, A., Vedantam, R., Parikh, D., Batra, D.: Grad-CAM: visual explanations from deep networks via gradient-based localization. In: Proceedings of the IEEE International Conference on Computer Vision, pp. 618–626 (2017)

Cross-Modal Hash Retrieval
with Category Semantics

Mengying Xu[1,2], Hanjiang Lai[1,2], and Jian Yin[1,2(✉)]

[1] Sun Yat-sen University, Guangzhou 510006, People's Republic of China
xumy55@mail2.sysu.edu.cn, {laihanj3,issjyin}@mail.sysu.edu.cn
[2] Guangdong Key Laboratory of Big Data Analysis and Processing, Guangzhou
510006, People's Republic of China

Abstract. With the multi-medal resource springing up, the field of cross-modal research has witnessed rapid advancement. Among these, Cross-Modal Hash retrieval has attracted attention due to its time-saving capabilities during the retrieval process. However, most existing cross-modal hash retrieval methods focus on capturing the relationship between modalities and often train in a one-to-one correspondence manner. Consequently, they often neglect the comprehensive distribution of the entire dataset. In response, we propose a novel approach that introduces semantic relationships between categories to capture class relationships and enable the measurement of the full distribution. First, the original image and text are encoded to obtain respective representations. Then, these are fused to generate a joint representation, which is passed through a classification head to obtain category semantics. The semantic information is used to allocate hash centers from a Hadamard matrix. Meanwhile, the image and text representations are fed into a hash layer to produce hash codes. We train the network with a modal matching loss for alignment, as well as center and class loss to capture the full data distribution. Extensive evaluations on two benchmark databases demonstrate that our approach achieves competitive performance compared to state-of-the-art methods.

Keywords: Cross-Modal · Hash Retrieval · Hash Center

1 Introduction

With the development of media resources, diverse multimodal data have emerged. The importance of cross-modal research [4,9,19,22] has become apparent, especially in cross-modal hash retrieval. Cross-modal hash retrieval [2,3, 8,13,20] has garnered considerable attention among researchers for its ability to quickly retrieve data from different modalities. In this paper, we primarily explore image-text hash retrieval.

Supported by the Key-Area Research and Development Program of Guangdong Province (2020B0101100001).

© The Author(s), under exclusive license to Springer Nature Switzerland AG 2024
S. Rudinac et al. (Eds.): MMM 2024, LNCS 14554, pp. 15–27, 2024.
https://doi.org/10.1007/978-3-031-53305-1_2

The key to successful hash retrieval is ensuring alignment between different modalities. Early research primarily focused on capturing modality relationships to realize alignment. For example, the DVSH model [3] uses deep neural networks for feature learning from dynamic data. The DCMH model [8] applies end-to-end deep cross-modal learning to alignment. However, these approaches lack consideration for intra-modal relationships. Subsequently, the PRDH model [16] integrated various pairwise constraints, considering intra-modal and inter-modal relationships. We observe that current cross-modal hash retrieval methods rely mainly on inter-modal and intra-modal relationships, where images and texts are paired for training models [14]. These methods solely focus on the similarities between pairwise data and relationships between modalities, as shown in Fig. 1(a) and 1(b). The prior methods fail to model the complete relationships within the data, leading to insufficient encoding of the underlying data distribution and category-level correlations.

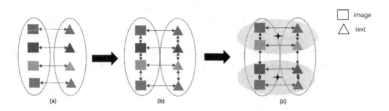

Fig. 1. Different capture forms of data distribution. Figure 1 (a) shows inter-modal relationships. Figure 1 (b) shows inter-modal and intra-modal relationships. Figure 1 (c)shows overall data distribution relationships.

To address the above issue, Fig. 1(c) demonstrates a more effective approach by considering category relationships and comprehensively capturing global relationships. For this reason, we leverage category relationships to better capture the overall data distribution. When it comes to categorizing relationships, simply treating all relationship categories equally fails to capture the rich semantic information that provides meaningful distinction between relationship types. As illustrated in Fig. 2, equally distributing class centers without considering semantic relationships fails to capture the underlying category correlations. This is another significant problem that needs to be addressed. Therefore, we consider the varying distances $c_i^T c_j$ between categories and allocate class centers based on category semantic information $w_i^T w_j$. This approach more comprehensively describes category relationships, improving cross-modal retrieval accuracy.

To tackle these problems, we propose a novel cross-modal hashing retrieval model that leverages semantic relationships between categories to capture class relationships. First, image and text inputs are encoded separately to obtain image and text representations. These representations are passed through a hash layer to produce corresponding hash codes. Next, the image and text representations are fused to generate a fused representation. We attach a classification

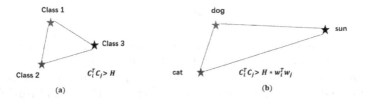

Fig. 2. The picture representation of category relationship. (a) shows the class relationships without semantic information. (b) shows the class relationships with semantic information. H represents the minimum distance between every two classes.

header and employ its weights as semantic information for different categories. We utilize to this semantic knowledge of categories to allocate hash centers generated by the Hadamard matrix. Then, we train an end-to-end network model using inter-modal and intra-modal paired loss to learn alignment between modalities. By leveraging the classification loss and center loss, we achieve class center similarity learning and increase distances between categories. The proposed method is evaluated on two benchmark datasets for image-text retrieval. Experiments show the generated hash centers better capture similarity relationships and achieve improved retrieval performance compared to existing methods.

Overall, our contributions are summarized as follows:

- This paper proposes a new method that addresses the issue of neglecting the overall data distribution by leveraging the relationships between categories.
- For category relationship construction, we consider the semantic relationships between categories reflecting the distances between different categories to achieve better performance.
- Compared with some state-of-the-art methods, extensive experiments are conducted on the MIRFLICKR-25K and NUS-WIDE benchmark datasets, demonstrating the effectiveness of the proposed method.

2   Related Work

In the 2010 s, hash-based retrieval research made remarkable strides with rapid advancements. Cross-modal hashing is widely used in multimedia retrieval due to notable advantages like reduced storage costs and faster query speeds. Moreover, image-text retrieval hashing grew increasingly popular.

Bronstein *et al.* presented the Cross-Modality Similarity-Sensitive Hashing (CMSSH) model [1] to learn hash functions that ensured binary codes were similar for relevant sample pairs and different for irrelevant pairs. Later research revealed leveraging semantic label information could improve performance. Consequently, the semantic correlation maximization (SCM) model [18] incorporated semantic labels to obtain a similarity matrix and perform binary encoding for reconstruction. Another study [10] introduced Semantics-Preserving Hashing (SePH), employing dual hashing and using training data semantic similarity as supervision. It approximated supervision to a probability distribution in Hamming space.

However, the above approaches relied on manual feature engineering, which may not be optimally compatible with hash code learning and require significant human effort for feature extraction. With the development of deep networks, more end-to-end designs emerged. Cao et al. [3] proposed the Deep Visual Semantic Hashing (DVSH) model, designing an end-to-end fusion network to learn joint image-text embeddings. However, DVSH was only applicable to specific cross-modal hashing scenarios where one modality must be temporally dynamic. For wider applicability, Jiang et al. [8] introduced the Deep Cross-Modal Hashing (DCMH) model as an end-to-end framework integrating feature learning and hash code learning. Subsequent research focused more on local relationships, such as CMHH [2] which optimized hash functions to better capture local similarities, and SCAHN [13] which utilized multi-layer network fusion to optimize representations. Overall, these image-text hashing (ITH) methods continuously improved performance. However, previous methods lacked consideration of intra-modal and inter-modal relationships.

To address the lack of an intra-modal model, Yang et al. [16] proposed the Pairwise Relationship Guided Deep Hashing (PRDH) model, which integrates various pairwise constraints to consider both intra-modal and inter-modal relationships. Further research has also explored different approaches to enhance semantic information and intra-modal links. For example, the Triplet-based Deep Hashing (TDH) model [6] leverages triplet labels to depict relative relationships between cross-modal instances. This enriches the relative semantic similarity of heterogeneous data to acquire more powerful hash codes. Additionally, Wang et al. [11] proposed the Fusion-Supervised Deep Cross-Modal Hashing (FDCH) model, which effectively enhances the correlation model of multi-modal data by learning unified binary codes through hash network fusion.

Overall, existing methods solely focused on inter-modality and intra-modality, overlooking comprehensive data distribution. In contrast, our approach considers the full distribution, using semantic relationships between categories to assign hash centers.

3   Methods

In this section, we first show the problem definition. Then, we introduce an overview of the proposed model architecture and each component in detail. Finally, we present the full loss functions including Pair loss, Center loss and Class loss.

3.1   Problem Definition

Suppose there are n training pairs, with each pair represented as an image x_i and text y_i. Let $X = \{x_i\}_{i=1}^n$ denote the images, where x_i is the i-th image modality.

Similarly, let $Y = \{y_i\}_{i=1}^n$ denote the text modality. Additionally, Each image-text pair (x_i, y_i) has an associated ground truth label l_i, which we denote as $L = \{l_i\}_{i=1}^n$.

3.2 Network Architecture

The overall architecture is shown in Fig. 3. We propose a new cross-modal hashing retrieval model that can comprehensively capture the overall data distribution. Our network can be divided into three parts: Encoder Module, Assign Module and Hash Module.

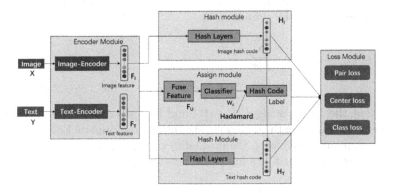

Fig. 3. Overview of the proposed architecture. The Encoder Module (left) encodes image and text inputs. The Assign Module (middle) uses fused semantics to assign hash centers. The Hash Module (middle) generates hash codes. The loss functions (right) comprise Pair loss, Center loss and Class loss.

Encoder Module. Given N (X, Y) pairs of images and texts, for the image modality we use a ResNet-34 as the Image-Encoder to obtain image representations F_I. For the text, we use bag-of-words (BoW) vector representations. We pass the BoW vectors through a Multi-Layer Perceptron (MLP) Text-Encoder to obtain high-level text representations F_T.

Assign Module. The random assignment of class centers would disregard the semantic information associated with different categories. Thereby, in this module, we need the incorporation of category semantic information to capture inter-class relationships. We fuse the image and text features F_I and F_T using attention to obtain a joint representation F_U. Then pass through a classification head and use its weights to obtain w representing the semantics of each category. We then use w to optimize the allocation of hash centers h_i:

Optimization Target: For m classes, we refer to [17], using the category features $w_1, w_2, ..., w_m$ to learn hash centers $h_1, h_2, ..., h_m$. Specifically, we formulate an optimization objective as:

$$\max_{h_1,h_2,...,h_m \in \{-1,1\}^q} \frac{1}{m(m-1)} \sum_i \sum_{j:j\neq i} \|h_i - h_j\|_H, \qquad (1)$$

where q is the hash code length and $\|.\|_H$ represents the Hamming distance

Since the Hamming distance $\|h_i - h_j\|_H$ is equivalent to $-h_i^T h_j$ [12] and $4\|h_i - h_j\|_H = 2q - 2h_i^T h_j$, we can add category semantics in an inner product form $w_i^T w_j$ to measure the distance between different categories. We define d to be the minimal Hamming distance $\|h_i - h_j\|_H \geq d$. With these facts, the objective in Eq. (1) can be simplified to:

$$\min_{h_1,h_2,...,h_m \in \{-1,1\}^q} \sum_{j:j\neq i} h_i^T h_j$$
$$\text{s.t. } h_i^T h_j \leq (q - 2d)w_i^T w_j (1 \leq i, j \leq m, i \neq j). \qquad (2)$$

We adopt the $\ell p - box$ algorithm [15] and fix one hash centres to solve the subproblem. We can drop the binary constraint and reformulate Eq.(2). By introducing an auxiliary variable z_3, the inequality constraints can be replaced by an equality constraint to the following equivalent form:

$$\min_{h_i,z_1,z_2,z_3} \sum_{j:j\neq i} h_i^T h_j$$
$$\text{s.t. } h_i^T H_{\sim i} + z_3 = (q - 2d)w_i^T w_j (1 \leq i, j \leq m, i \neq j) \qquad (3)$$
$$h_i = z_1, h_i = z_2, z_1 \in S_b, z_2 \in S_p.$$

The augmented Lagrange function w.r.t. Equation(3) is:

$$L(h_i, z_1, z_2, z_3, y_1, y_2, y_3) = \sum_{j\neq i} h_i^T h_j$$
$$+ y_1^T (h_i - z_1) + \frac{\mu}{2} \|h_i - z_1\|_2^2$$
$$+ y_2^T (h_i - z_2) + \frac{\mu}{2} \|h_i - z_2\|_2^2 \qquad (4)$$
$$+ y_3^T (h_i^T H_{\sim i} + z_3 - e) + \frac{\mu}{2} \|h_i^T H_{\sim i} + z_3 - e\|_2^2$$
$$\text{s.t.} z_1 \in S_b, z_2 \in S_p, z_3 \in R_+^{m-1}, e = (q - 2d)w_i^T w_j,$$

where $e = (q - 2d)w_i^T w_j$, and y1, y2, y3 are Lagrange multipliers.

Next, we will update each variable h_i, z_1, z_2, z_3.

Update h_i By setting this gradient to zero in Eq.(4), we can update h_i by

$$h_i \leftarrow (2\mu I_q + \mu H_{\sim i} H_{\sim i}^T)^{-1} (\mu(z_1 + z_2 + H_{\sim i}e - H_{\sim i}z_3)$$
$$- \sum_{j\neq i} h_j - y_1 - y_2 - H_{\sim i}y_3. \qquad (5)$$

Update z_1, z_2, z_3 Following [21], all of these projections have closed-form solutions:

$$
\begin{cases}
z_1 \leftarrow min(1, max(-1, h_i + \frac{1}{\mu}y_1)) \\
z_2 \leftarrow \sqrt{q} \frac{h_i + \frac{1}{\mu}y_2}{\left\| h_i + \frac{1}{\mu}y_2 \right\|_2} \\
z_3 \leftarrow max(0, e - h_i^T H_{\sim}i - \frac{1}{\mu}y_3)
\end{cases}
\tag{6}
$$

Update y_1, y_2, y_3 The Lagrange multipliers y1, y2 and y3 can be updated by

$$
\begin{cases}
y_1 \leftarrow y_1 + \mu(h_i - z_1) \\
y_2 \leftarrow y_2 + \mu(h_i - z_2) \\
y_3 \leftarrow y_3 + \mu(h_i^T H_{\sim}i - z_3)
\end{cases}
\tag{7}
$$

Finally, we can learn hash centers that contain semantic information of categories by maximizing the following optimization target.

Hash Module. We put the image F_I and text F_T from each modality encoder and use a fully connected layer and a tanh layer that restricts the values in the range $[-1, 1]$ as shown in Fig. 3. The output is the hash code for each modality.

3.3 Hash Objectives

Pair Loss. We use a traditional cross-modal hashing loss [8]. In summary, function as follows Eq.(8). We divide each row into three parts and detail each part below.

$$
L_p = - \sum_{i,j=1}^{n} \left(S_{ij}\Theta_{ij} - \log\left(1 + e^{\Theta_{ij}}\right) \right)
$$
$$
+ \gamma \left(\left\| \mathbf{B} - \mathbf{F_I} \right\|_F^2 + \left\| \mathbf{B} - \mathbf{F_T} \right\|_F^2 \right)
\tag{8}
$$
$$
+ \eta \left(\left\| \mathbf{F_I}1 \right\|_F^2 + \left\| \mathbf{F_T}1 \right\|_F^2 \right)
$$
$$
\text{s.t. } \mathbf{B} \in \{-1, +1\}^{c \times n},
$$

where, $\Theta_{ij} = \frac{1}{2}\mathbf{F_I}_{*i}^T \mathbf{F_T}_{*j}$. For the first part, we use S_{ij} to denote F_I and F_T similarity. $S_{ij} = 1$ means that they are similar. And conversely, $S_{ij} = 0$ means that they are not similar. The second part to keep the hash representation and hash code should be as similar as possible. The third part is used to maximize the information provided by each bit.

Center Loss. We use a loss function [12] that makes an output (approximate) hash code to be nearby to the hash center assigned to the corresponding class, and far away from other hash centers. The loss function towards hash centers is defined by:

$$
\boldsymbol{L_{ce}} = -\frac{1}{N} \sum_{j=1}^{N} \sum_{i=1}^{c} l_{j,i} log P_{j,i} + (1 - l_{j,i}) log(1 - P_{j,i})
\tag{9}
$$

with

$$P_{j,i} = \frac{e^{-S(b_j,h_i)}}{\sum_{k=1}^{c} e^{-S(b_j,h_k)}}. \tag{10}$$

Class Loss. We choose the cross-entropy loss function. By constructing the cross-entropy loss formula for classification scores and labels, the formula is calculated as follows:

$$L_{cl} = -\frac{1}{N}\sum_{i=1}^{N}\sum_{c=1}^{m} l_{i,c} log\widehat{l}_{i,c}. \tag{11}$$

Therefore, the overall loss is as follows:

$$L = L_p + L_{ce} + L_{cl}. \tag{12}$$

4 Experiment

In this section, we introduce our experiments. Firstly, we introduce the datasets, the evaluation measures and the setting of our model. And then, we show and analyze the results of the experiments.

4.1 Experimental Setting

Datasets. We validate the effectiveness of our proposed method and compare with other state-of-the-art methods on two widely used datasets:

1. MIRFLICKR-25K [7]: This dataset was obtained from the Flickr website with each image having a corresponding tag to form image-text pairs. We perform the same preprocessing as DCMH using the original images and generating 1,386-dimensional Bag-of-Words (BoW) vector representations for text tags.
2. NUS-WIDE [5]: The dataset consists of 269,648 web images divided into 81 classes. We perform the same preprocessing as for MIRFlickr-25K, using the original images and generating 1000-dimensional Bag-of-Words vector representations for the text tags.

Evaluation Measures. Mean Average Precision (mAP) measures the average precision across all classes. It is computed as a weighted mean of the Average Precision (AP) for each class. The formula is expressed as follows:

$$AP@K = \frac{\sum_{1}^{k} P(k) * rel(k)}{\sum rel(k)}. \tag{13}$$

In our experiments, we use mAP to measure overall performance. AP is calculated by the above equation. After the AP is acquired, it takes the average of all AP when calculating the mAP.

Implementation Details. We implemented our model in open-source PyTorch and configured the experiments as follows: For MIRFlickr-25K, we randomly selected 10,000 instances for training and 2,000 for querying. For NUS-WIDE, we used 10,500 for training and 2,100 for querying. The batch size was set to 128 and we trained using stochastic gradient descent. The learning rate was set to $10^{-1.1}$.

4.2   Experimental Results

Comparison to the State-of-the-Art Methods. In this section, we evaluate the performance of our methods and compare them with the state-of-the-art methods of cross-modal retrieval: DCMH [8], PRDH [16], DAH [21], CMHH [2], and SCAHN [13].

Table 1. Results compare with SOTA models in MIRFLICKR dataset

	I2T			T2I		
	16 bit	32 bit	64 bit	16 bit	32 bit	64 bit
DCMH	0.7410	0.7465	0.7485	0.7827	0.7900	0.7932
PRDH	0.7499	0.7546	0.7612	0.7890	0.7955	0.7964
DAH	0.7563	0.7719	0.7720	0.7922	0.8062	0.8074
CMHH	0.7830	0.8140	0.8210	0.758	0.782	0.793
SCAHN	0.7966	0.8173	0.8287	0.7857	0.8035	0.8151
Our	0.8101	0.8296	0.8352	0.7975	0.8112	0.8167

Table 1 shows the retrieval performance compared to baselines on the MIR-Flickr dataset. It can be observed that our method outperforms other baselines and achieves remarkable performance. Specifically, for 16-bit image-to-text retrieval, our MAP of 0.8101 is 1.69% higher than the best algorithm SCAHN. For 16-bit text-to-image, our performance improves over the best model by 1.50%. Overall, our method demonstrates effectiveness for cross-modal retrieval across different bit lengths.

Table 2 shows the retrieval performance compared to baselines on the NUS dataset. The baselines are the same as in Table 1. Our method achieves higher MAP than the best algorithms by 3.77%, 5.14%, and 4.27% on 16, 32, and 64 bits respectively for image-to-text retrieval. For text-to-image, our method improves over the best model by 4.53%, 4.12%, and 2.02% on 16, 32, and 64 bits. Our approach significantly outperforms traditional methods, demonstrating its effectiveness.

Comparison with Different Methods for Capturing Data Distributions. In this section, we validate the effectiveness of the proposed method

Table 2. Results compare with SOTA models in NUS-WIDE dataset

	I2T			T2I		
	16 bit	32 bit	64 bit	16 bit	32 bit	64 bit
DCMH	0.5903	0.6031	0.6093	0.6389	0.6511	0.6571
PRDH	0.6107	0.6302	0.6276	0.6527	0.6916	0.6720
DAH	0.6403	0.6294	0.6520	0.6789	0.6975	0.7039
CMHH	0.5530	0.5698	0.5924	0.5739	0.5786	0.5889
SCAHN	0.6342	0.6425	0.6515	0.6379	0.6454	0.6539
Our	0.6581	0.6755	0.6793	0.6668	0.6720	0.6671

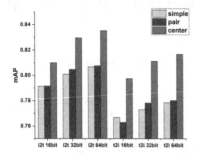 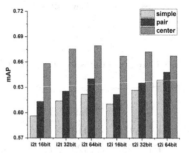

Fig. 4. Results in MIRFLICKR dataset **Fig. 5.** Results in NUS-WIDE dataset

to capture the overall data distribution and compare it with the simple and pair scheme.

To demonstrate that our method captures the overall data distribution and improves performance, we conducted a comparative study using the same network framework and experimental settings, but with different forms of hash objective functions. The "Simple" approach, inspired by DCMH, only considers the loss between modalities. The "Pair" approach, inspired by PRDH, considers both inter-modality and intra-modality losses. The "Center" approach, which is our method, incorporates relationships between categories.

The left graph shows the comparison of mAP results for the three methods at different bit lengths on the MIR dataset. Similarly, the right graph displays the results for the three methods on the NUS dataset. From Fig. 4 and Fig. 5, it is evident that our method significantly improves performance.

Comparison with Different Methods for Assigning Hash Centers As mentioned in Sect. 3.2, if we assign class centers randomly would overlook semantic information associated with different categories, so we capture inter-class relationships using category semantic information. This section aims to demonstrate the effects of our method to assign hash centres.

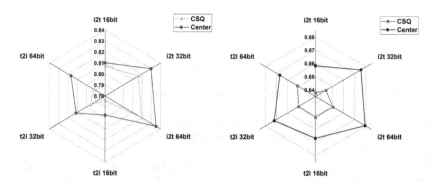

Fig. 6. Results in MIRFLICKR dataset **Fig. 7.** Results in NUS-WIDE dataset

To illustrate how utilizing multimodal information to allocate hash centers effectively enhances the discriminability of hash codes across different classes, we compare our "Center" approach with the traditional "CSQ" method. The "CSQ" method randomly selects hash centers for each class. In contrast, our approach leverages the semantic information of different classes to select distinct hash centers.

The left graph presents the comparison of results for the two methods, "CSQ" (represented by red) and "Center" (represented by blue), at different bit lengths on the MIR dataset. As shown in Fig. 6, the blue curve outperforms the red curve on all bits. Similarly, the right graph displays the comparison of results for the two methods on the NUS dataset. From Figs. 6 and 7, it is evident that utilizing multimodal information to allocate hash centers significantly improves performance.

5   Conclusion

We have proposed a novel cross-modal hash retrieval model that incorporates the semantic information of classes in assigning hash centers. This approach enables our model to capture the entirety of the data distribution, leading to better performance. In our experimental study, we observe that our method captures global relations, introduces richer semantic relations of categories, and improves overall performance. In the future, we plan to focus on creating specialized and efficient models that can make the best use of semantic information.

References

1. Bronstein, M.M., Bronstein, A.M., Michel, F., Paragios, N.: Data fusion through cross-modality metric learning using similarity-sensitive hashing. In: 2010 IEEE Computer Society Conference on Computer Vision and Pattern Recognition (2010)
2. Cao, Y., Liu, B., Long, M., Wang, J.: Cross-modal hamming hashing. In: European Conference on Computer Vision (2018)

3. Cao, Y., Long, M., Wang, J., Yang, Q., Yu, P.S.: Deep visual-semantic hashing for cross-modal retrieval. In: Proceeding of the 22nd ACM SIGKDD International Conference (2016)
4. Chen, H., Ding, G., Liu, X., Lin, Z., Liu, J., Han, J.: IMRAM: iterative matching with recurrent attention memory for cross-modal image-text retrieval. In: 2020 IEEE/CVF Conference on Computer Vision and Pattern Recognition (CVPR), pp. 12652–12660 (2020)
5. Chua, T.S., Tang, J., Hong, R., Li, H., Luo, Z., Zheng, Y.: Nus-wide: a real-world web image database from National University of Singapore. In: ACM International Conference on Image and Video Retrieval (2009)
6. Deng, C., Chen, Z., Liu, X., Gao, X., Tao, D.: Triplet-based deep hashing network for cross-modal retrieval. IEEE Trans. Image Process. **27**, 3893–3903 (2018)
7. Huiskes, M.J., Lew, M.S.: The MIR flickr retrieval evaluation. In: Proceedings of the 1st ACM International Conference on Multimedia Information Retrieval, pp. 39–43. Association for Computing Machinery, New York, NY, USA (2008)
8. Jiang, Q.Y., Li, W.J.: Deep cross-modal hashing. In: 2017 IEEE Conference on Computer Vision and Pattern Recognition (CVPR), pp. 3270–3278 (2016)
9. Kim, W., Son, B., Kim, I.: ViLT: vision-and-language transformer without convolution or region supervision. In: International Conference on Machine Learning, pp. 5583–5594. PMLR (2021)
10. Lin, Z., Ding, G., Hu, M., Wang, J.: Semantics-preserving hashing for cross-view retrieval. In: 2015 IEEE Conference on Computer Vision and Pattern Recognition (CVPR) (2015)
11. Wang, L., Zhu, L., Yu, E., Sun, J., Zhang, H.: Fusion-supervised deep cross-modal hashing. In: 2019 IEEE International Conference on Multimedia and Expo (ICME), pp. 37–42 (2019)
12. Wang, L., Pan, Y., Lai, H., Yin, J.: Image retrieval with well-separated semantic hash centers. In: Asian Conference on Computer Vision (2022)
13. Wang, X., Zou, X., Bakker, E.M., Wu, S.: Self-constraining and attention-based hashing network for bit-scalable cross-modal retrieval. Neurocomputing **400**, 255–271 (2020)
14. Williams-Lekuona, M., Cosma, G., Phillips, I.: A framework for enabling unpaired multi-modal learning for deep cross-modal hashing retrieval. J. Imaging **8**, 328 (2022)
15. Wu, B., Ghanem, B.: $\ell p - box$ admm: A versatile framework for integer programming. IEEE Trans. Pattern Anal. Mach. Intell. **41**(7), 1695–1708 (2016)
16. Yang, E., Deng, C., Liu, W., Liu, X., Tao, D., Gao, X.: Pairwise relationship guided deep hashing for cross-modal retrieval. In: AAAI Conference on Artificial Intelligence (2017)
17. Yuan, L., et al.: Central similarity quantization for efficient image and video retrieval. In: 2020 IEEE/CVF Conference on Computer Vision and Pattern Recognition (CVPR), pp. 3080–3089 (2019)
18. Zhang, D., Li, W.J.: Large-scale supervised multimodal hashing with semantic correlation maximization. In: AAAI Conference on Artificial Intelligence (2014)
19. Zhang, Q., Lei, Z., Zhang, Z., Li, S.Z.: Context-aware attention network for image-text retrieval. In: Proceedings of the IEEE/CVF Conference on Computer Vision and Pattern Recognition, pp. 3536–3545 (2020)
20. Zhang, X., Lai, H., Feng, J.: Attention-aware deep adversarial hashing for cross-modal retrieval. In: European Conference on Computer Vision (2017). https://api.semanticscholar.org/CorpusID:7770864

21. Zhang, X., Lai, H., Feng, J.: Attention-aware deep adversarial hashing for cross-modal retrieval. In: Ferrari, V., Hebert, M., Sminchisescu, C., Weiss, Y. (eds.) ECCV 2018. LNCS, vol. 11219, pp. 614–629. Springer, Cham (2018). https://doi.org/10.1007/978-3-030-01267-0_36
22. Zhen, L., Hu, P., Wang, X., Peng, D.: Deep supervised cross-modal retrieval. In: Proceedings of the IEEE/CVF Conference on Computer Vision and Pattern Recognition, pp. 10394–10403 (2019)

Spatiotemporal Representation Enhanced ViT for Video Recognition

Min Li[1,2(✉)], Fengfa Li[1,2], Bo Meng[3], Ruwen Bai[1,2], Junxing Ren[1,2],
Zihao Huang[1,2], and Chenghua Gao[1,2]

[1] Institute of Information Engineering, Chinese Academy of Sciences, Beijing, China
{limin,lifengfa,bairuwen,renjunxing,huangzihao,gaochenghua}@iie.ac.cn
[2] School of Cyber Security, University of Chinese Academy of Sciences,
Beijing, China
[3] First Research Institute of MPS, Beijing, China

Abstract. Vision Transformers (ViTs) are promising for solving video-related tasks, but often suffer from computational bottlenecks or insufficient temporal information. Recent advances in large-scale pre-training show great potential for high-quality video representation, providing new remedies to transformer limitations. Inspired by this, we propose a SpatioTemporal Representation Enhanced Vision Transformer (STRE-ViT), which follows a two-stream paradigm to fuse large-scale pre-training visual prior knowledge and video-level temporal biases in a simple and effective manner. Specifically, one stream employs a well-pretrained ViT with rich vision priors to alleviate data requirements and learning workload. Another stream is our designed spatiotemporal interaction stream, which first models video-level temporal dynamics and then extracts fine-grained and salient spatiotemporal representations by introducing appropriate temporal bias. Through this interaction stream, the model capacity of ViT is enhanced for video spatiotemporal representations. Moreover, we provide a fresh perspective to adapt well-pretrained ViT for video recognition. Experimental results show that STRE-ViT learns high-quality video representations and achieves competitive performance on two popular video benchmarks.

Keywords: Video recognition · Transformer · Spatiotemporal Representations

1 Introduction

In recent years, Vision Transformers (ViTs) [3,14,21] have exhibited strong representation capabilities for vision tasks. However, as a general architecture, transformers use minimal inductive bias, giving rise to several challenges that restrict the research progress of transformer architectures in the field of video recognition. Firstly, transformer-based video recognition often requires a large amount of supervised training data to learn sufficient spatiotemporal information, resulting in high training costs. Secondly, the transformer paradigm struggles to capture

© The Author(s), under exclusive license to Springer Nature Switzerland AG 2024
S. Rudinac et al. (Eds.): MMM 2024, LNCS 14554, pp. 28–40, 2024.
https://doi.org/10.1007/978-3-031-53305-1_3

local fine-grained spatiotemporal features. Additionally, due to limitations in hardware memory and computational resources, ViTs only allow sparse-sampled video frame inputs, making it difficult to capture temporal dynamic information. For the first problem, previous works [1, 2, 15] have demonstrated the benefits of pre-training on large-scale image datasets for video recognition, reducing the computational and time costs of training video models from scratch. Recently, several well-designed pretrained large models have emerged, such as CLIP [17] and ALIGN [8]. They leverage contrastive learning, masked image modeling, or traditional supervised learning approaches to learn high-quality and generic visual representations, exhibiting strong transferability across a range of downstream tasks.

For the second problem, recent efforts [10, 11, 22, 23] have attempted to enhance transformers by introducing an appropriate convolutional bias. The convolutional biases possessed by 3D CNNs allow for effective video processing but also impose limitations on the upper bound of model capacity. Structural combinations reduce the computational burden of transformers and accelerate convergence. However, the integration of CNN often compromises the inherent structure of transformers, hindering their ability to utilize the pretraining knowledge acquired from large-scale image datasets. This work introduces appropriate inductive bias through convolutions to address the limitations of transformers while preserving the original structure of ViT [3] to the greatest extent possible.

The third problem remains to be further studied. Extra temporal modeling often involves heavy computing costs. Some transformer-based works redesign the transformer to better handle video data, but they have to be trained from scratch. VTN [5] and ViViT [1] extend pre-trained image models to spatiotemporal dimensions, modeling global cross-frame dependencies on top of spatial representations. However, using long-distance self-attention makes VTN [5] and ViViT [1] lack the ability to model fine-grained spatiotemporal information. This work presents a simple but effective scheme to adapt the pre-trained model to video recognition and exploit fully spatiotemporal information in the video.

To address these challenges, we propose a SpatioTemporal Representation Enhanced Vision Transformer (STRE-ViT), which is a two-stream framework extended from ViT [3]. One stream of STRE-ViT utilizes a well-pretrained ViT backbone to extract spatial features while preserving the original ViT structure as much as possible. This ensures maximize reuse of prior knowledge from the ViT backbone [3] during supervised training. Another stream is designed for spatial-temporal information interaction. It first employs a lightweight 3D ResNet-like Temporal Dynamic Extractor (TDE) to efficiently model video-level temporal dynamics. Subsequently, a Temporal Feature Filter (TFF) and a transformer decoder are cross-stacked to find the interaction patterns between spatial and temporal representations, filtering salient information most relevant to video semantics. Overall, STRE-ViT with a two-stream paradigm preserves strong visual model capacity, while capturing video-level temporal dynamic information and fine-grained spatiotemporal semantic patterns. This work offers a fresh perspective for efficiently transferring well-pretrained ViT models for video recognition.

Due to computational constraints, we only train the baseline version of our model. On commonly used datasets such as Kinetics-400 [25] and Something-SomethingV2 [6], our approach outperforms other methods with similar FLOPs (Floating Point Operations). Empirically, as the network depth increases, the model capacity increases, resulting in higher accuracy. Therefore, our model has great potential as its network depth increases. Our main contributions are as follows:

1. We propose a novel architecture STRE-ViT with a two-stream framework that makes the best use of visual prior knowledge from the large-scale image pre-trained model, as well as effectively captures spatiotemporal semantic patterns in local and global contexts by introducing a spatiotemporal interaction stream.
2. We propose a fusion and filtering mechanism between pre-trained priors and ideal inductive bias. It's implemented by our designed spatiotemporal interaction stream, focusing on salient spatiotemporal information guided by an appropriate temporal bias, which is easily adaptable to other potential video tasks.
3. Experimental results on two standard video recognition benchmarks show the superiority of STRE-ViT. Our model effectively leverages existing large-scale pre-training knowledge, significantly reducing the training time and costs, while achieving competitive performance. This makes video recognition accessible to researchers with limited resources.

2 Related Work

Evolution of Video Recognition. For a long time, Convolutional Neural Networks (CNNs) have dominated various tasks in the computer vision field. Because of their locality and translational invariance, 3D CNNs and their variants have shown excellent performance in video recognition tasks. Recently, transformers [19] have shown great success handling long-range dependencies. In the case of large-scale pretrained models, pure transformer architectures like ViT [3] have proven to outperform CNN architectures in image classification tasks. These architectures have seen a quick adoption for solving video-related tasks.

TimeSformer [2] splits the temporal and spatial dimensions to achieve video understanding. ViViT [1] extends the ViT architecture [3] to 3D videos by transforming the 2D patch embedding into 3D embedding. Video Swin Transformer [15] is an extension of Swin Transformer [14] in the video domain, introducing the bias of locality in ViT. Works such as Uniformer [10] and UniformerV2 [11] propose relationship aggregators that unify convolution and self-attention in a transformer-like manner.

These approaches have attempted to model temporal information based on ViT [3], but they either modify the original structure of ViT [3], requiring training from scratch and incurring significant learning costs [10,15], or rely solely on long-distance self-attention, lacking the ability to model fine-grained spatiotemporal information [1,2,5,11,18]. Our approach aims to preserve the original ViT structure [3] as much as possible in one stream, allowing for the incorporation of rich pre-trained models. In the other stream, we introduce convolutional operations to introduce local biases and perform a series of feature fusion and selection processes, enabling better modeling of fine-grained features.

Efficient Transfer Learning. To address the challenges of high training costs and the difficulty of obtaining video data for transformer architectures, many studies are exploring methods to transfer the rich semantic knowledge from pretrained ViTs [3] to video tasks. Ju et al. [9] and Wang et al. [20] transform CLIP [8] into video recognition by learning semantics cues. Lin et al. [13] freezes CLIP features and incorporates temporal modeling for videos.

Inspired by this, our approach presents an efficient transfer learning architecture. Unlike the aforementioned methods, we design a flexible branch to adapt large-scale pre-training knowledge to video recognition, which effectively models temporal information with affordable computations.

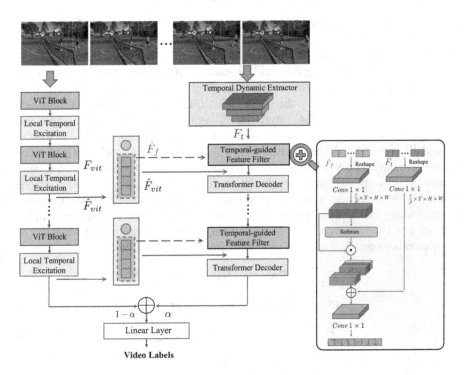

Fig. 1. Overall Framework of our STRE-ViT. STRE-ViT is a two-stream structure: the Spatial Feature Extraction Stream (**Left**) and the Spatiotemporal Interaction Stream (**Right**).

3 Method

3.1 Overview of STRE-ViT

As shown in Fig. 1, our STRE-ViT adopts a two-stream architecture. The spatial feature extraction stream extracts spatial features based on pre-trained ViT [3]. In this stream, we assemble the proposed Local Temporal Excitation (LTE) module, which excites the representation of spatial features within neighborhood temporal windows. Another spatiotemporal interaction stream consists of a Temporal Dynamic Extractor (TDE) and stacked Temporal-guided Feature Filters (TTF) and transformer decoders. TDE first models video-level temporal dynamics. Then, the stacked TTFs and decoders continuously filter the desired temporal information and guide the learning of fine-grained spatiotemporal representations that incorporate appropriate temporal biases. Finally, we utilized a learnable parameter α to fuse the features from two streams. We next present the specific details of each component in our model.

3.2 Spatial Feature Extraction Stream

Local Temporal Excitation. For video information, although ViT [3] has remarkable spatial feature extraction capabilities, it lacks interaction between adjacent frames in videos. Therefore, we introduce the Local Temporal Excitation (LTE) module to enable the interaction of features extracted by each ViT [3] block across adjacent frames. Consider the shape of the feature map $\mathbf{F}_{vit}^i$ obtained after encoding the input video frames through the ViT [3] block is $[N, L, C]$, where N is the batch size, C is the embedding dimension, and $\mathbf{L} = \mathbf{T} \times \mathbf{H} \times \mathbf{W} + 1$. First, we reshape $\mathbf{F}_{vit}^i$ into $[N, C, T, H, W]$, where T, H, and W represent the number of sampled frames, height, and width of video frames. We then apply a $1 \times 1 \times 1$ convolution to reduce the number of feature channels and improve efficiency, resulting in the feature $\mathbf{F}_{ch\_re}^i$ with reduced channels. Next, we use grouped convolution on $\mathbf{F}_{ch\_re}^i$ to fuse adjacent frame features along the temporal dimension channel by channel. Finally, we employ another $1 \times 1 \times 1$ convolution to expand the number of channels to match the input channel size. The specific process is formulated as follows:

$$\mathbf{F}_{ch\_re}^i = \mathrm{Conv}^{1 \times 1 \times 1}(C, C/r)\left(\mathbf{F}_{vit}^i\right) \tag{1}$$

$$\mathbf{F}_{te\_ex}^i = \mathrm{Conv}^{3 \times 1 \times 1}(C, C/r)\left(\mathbf{F}_{ch\_re}^i\right) \tag{2}$$

$$\hat{\mathbf{F}}_{vit}^i = \mathrm{Conv}^{1 \times 1 \times 1}(C/r, C)\left(\mathbf{F}_{te\_ex}^i\right) \tag{3}$$

where $\mathrm{Conv}^1(\mathbf{C}_{\mathrm{in}}, \mathbf{C}_{\mathrm{out}})(\cdot)$ represents a $1 \times 1 \times 1$ convolution with an input channel of $\mathbf{C}_{\mathrm{in}}$ and an output channel of $\mathbf{C}_{\mathrm{out}}$.

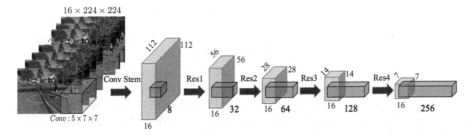

Fig. 2. TDE Architecture Diagram. We employ a stacking of ResBlocks to continuously downsample the spatial dimension while keeping the temporal dimension unchanged, aiming to extract temporal information between frames.

3.3 Spatiotemporal Interaction Stream

Temporal Dynamic Extractor. Inspired by SlowFast [4], we realize the effectiveness of motion information extracted by the fast pathway in video analysis. In this study, we employ a variant of the 3D Resnet convolutional network [7] to efficiently gather inter-frame motion information with lower computational cost. As shown in Fig. 2 (using 16 frames as an example), we perform spatial downsampling by stacking ResBlocks to reduce the spatial dimension while keeping the temporal dimension unchanged. To better exploit temporal information, we employ a linear layer to increase the dimensionality of the feature vectors extracted by the ResBlocks along the channel dimension. We then recombine the temporal information through convolutional layers and align it with the spatial features extracted by ViT [3].

Temporal-Guided Feature Filter. To address the spatial redundancy in video information, we employ a Temporal-guided Feature Filter (TFF), as shown in Fig. 1. The TFF denoises the input spatial features and then fuses them with the temporal information $\mathbf{F}_t$ extracted by the Temporal Dynamic Extractor (TDE). Specifically, we remove the class token from $\hat{\mathbf{F}}^i_{vit}$ to obtain $\hat{\mathbf{F}}_f$, which is reshaped into $[N, C, T, H, W]$. $\hat{\mathbf{F}}^i_{vit}$ is the output of LTE. Then, we apply a $1 \times 1 \times 1$ convolution to reduce the channel dimension and flatten it along the channel dimension, resulting in $\hat{\mathbf{F}}_{ch\_re}$. We set the value of β as 4. $\hat{\mathbf{F}}^i_{vit}$ is normalized by softmax and multiplied with itself, retaining the features with high self-responses. $\mathbf{F}_t$ undergoes a similar reshaping process and is added to $\mathbf{F}_{tff}$ after a $1 \times 1 \times 1$ convolution. We further apply another $1 \times 1 \times 1$ convolution to expand $\mathbf{F}_{tff}$ to the same channel dimension as the input.

$$F_{ch\_re} = \text{Conv}^{1 \times 1 \times 1}(C, C/\beta) \left(\text{Reshape}\left(\hat{F}_f \right) \right) \tag{4}$$

$$F_{ch\_re} = \text{Conv}^{1 \times 1 \times 1}(C, C/\beta) \left(\text{Reshape}\left(\hat{F}_f \right) \right) \tag{5}$$

$$F_{tff} = \text{Softmax}\left(\hat{F}_{ch\_re} \right) \cdot \hat{F}_{ch\_re} \tag{6}$$

$$\hat{F}_{tff} = \text{Conv}^{1\times1\times1}(C/\beta, C)\,(F_{tff} + F_t) \qquad (7)$$

The feature $\mathbf{F}_{tff}^{\hat{}}$, which contains both temporal information and local inductive bias, serves as the Q (query) in the transformer decoder, while the feature $\hat{\mathbf{F}}_{vit}^{i}$, which includes more comprehensive information, serves as the K (key) and V (value). Through the aforementioned process, we selectively filter the most relevant information related to the temporal features $\mathbf{F}_{tff}^{\hat{}}$ from the large-scale pre-trained ViT [3] model. This approach enriches final features used for classification with temporal information and inductive bias.

3.4   Feature Fusion Strategy

In scenarios with a large amount of data, ViT [3] has the ability to learn local relationships that surpass the predefined inductive bias in convolutions. Considering this, we introduce a learnable parameter α to balance the ViT stream and the spatiotemporal interaction stream. The parameter α allows the model to prioritize the features from the spatial feature extraction stream as the training dataset size increases.

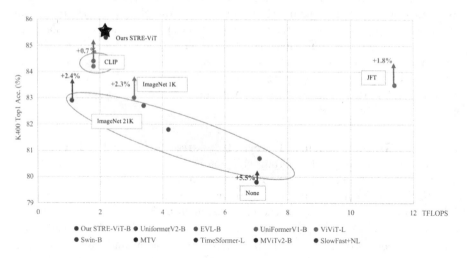

Fig. 3. Accuracy vs. previous SOTA on Kinetics-400 [25]. Compared with base models of previous SOTA, STRE-ViT achieves the best Top1 Accuracy.

4   Experiments

We evaluated our method on two popular video benchmarks: Kinetics-400 [25] and Sth-SthV2 [6]. Due to limited computational resources, we only tested the base version of our model. STRE-ViT-B consists of 12 ViT Blocks, labeled 0_{th} to

11_{th}. On Kinetics-400 [25], the $[8_{th}, 9_{th}, 10_{th}, 11_{th}]$ output of ViT Blocks interact with the spatiotemporal interaction stream. On Something-SomethingV2 [6], the interacting ViT Blocks are $[4_{th}, 5_{th}, 6_{th}, 7_{th}, 8_{th}, 9_{th}, 10_{th}, 11_{th}]$.

We adopt AdamW [16] optimizer with cosine learning rate schedule to train the entire network. We set the first 5 epochs for warm-up to overcome early optimization difficulty. The total epoch, stochastic depth rate, and weight decay are set to 55, 0.1, and 0.05 respectively for Kinetics-400 [25] and 5, 22, 0.1 and 0.05 respectively for Sth-SthV2 [6]. We linearly scale the base learning rates according to the batch size, which are $5e^{-6} \times \frac{\text{batchsize}}{32}$ and $4e^{-5} \times \frac{\text{batchsize}}{32}$ for Kinetics-400 [25] and Sth-SthV2 [6].

4.1   Main Results

Due to limited computational resources, we only compared our method with the base models of state-of-the-art video recognition methods with similar FLOPs on Kinetics-400 [25] and Sth-SthV2 [6]. Empirically, with deeper model layers, higher patch embedding dimensions, and more video frames, the model capacity tends to grow. Our model has the potential to achieve comparable or even better performance than the large models of these state-of-the-art video recognition methods. Experiments demonstrate that our model effectively leverages the pre-trained information from CLIP. As shown in Fig. 3, our model's performance surpasses models pre-trained on other datasets and stands out among models utilizing CLIP for pre-training.

Comparison with State-of-the-Art. Compared with state-of-the-art transformer-based models with comparable FLOPs, STRE-ViT achieves highly competitive performance on both Kinetics-400 [25] and Sth-SthV2 [6]. As shown in Table 1, STRE-ViT obtains a 85.3% Top-1 accuracy on Kinetics-400 [25] and achieves a 0.9% improvement in accuracy compared to Uniformerv2-B/16 [11], which applies the same large-scale image pre-training. It is exciting to note that by introducing the spatiotemporal interaction stream, our model achieves an accuracy improvement of 1.8% over the pure transformer-based ViViT [1] model, and 4.8% over TimeSformer-L [2] while utilizing fewer TFLOPs. Our model performs generally better in accuracy, showcasing its strong capacity in capturing spatiotemporal representations. The experimental results on Sth-SthV2 [6] are shown in Table 2. Compared to Kinetics-400 [25], the videos in this dataset exhibit stronger temporal correlations. STRE-ViT also shows excellent performance. This suggests that our model successfully integrates the desired temporal bias. After pre-training with the Kinetics-400 [25], our model performs even better.

Table 1. Comparison with state-of-the-arts on Kinetics-400.

Method	Pretraining	Acc. (%)	#Frames	TFLOPS
SlowFast+NL [4]	None	79.8	16×10	7.0
MViTv2-B [12]	IN-21K	82.9	32×3	1.1
TimeSformer-L [2]	IN-21K	80.7	96×1	7.1
MTV-B [18]	IN-21K	81.8	32×4	4.79
Swin-B [15]	IN-21K	82.7	32×4	3.4
UniFormerV1-B [10]	IN-1K	83.0	32×4	3.1
ViViT-L [1]	JFT	83.5	32×3	11.4
EVL-B [13]	CLIP	84.2	32×3	1.8
UniformerV2-B [11]	CLIP	84.4	8×4	1.8
Our STRE-ViT	CLIP	**85.3**	8×4	2.2

• Frame counts are reported as frames per view × number of views.
Note: Although UnitformerV2-B/16 achieves 85.6 Top-1 accuracy on Kinetics-400 after pretraining with K710, it uses extra video dataset which includes all categories of Kinetics-400. We do not list this version for fair comparison.

Table 2. Comparison with state-of-the-arts on Sth-SthV2.

Method	Pretraining	Acc. (%)	#Frames	TFLOPS
TimeSformer-L [2]	IN-21K	62.5	16×1	5.1
MTV-B [18]	IN-21K	68.5	32×4	11.2
Swin-B [15]	IN-21K	69.6	32×1	1.0
EVL-L [13]	CLIP	66.7	32×1	9.6
Uniformerv2-B [11]	CLIP	69.5	16×1	0.6
Our STRE-ViT	CLIP	70.3	16×1	0.6
MViTv2-B [12]	IN-21K+K400	72.1	32×3	1.1
Our STRE-ViT	CLIP+K400	**73.8**	16×1	0.6

Temporal Modeling Capability of Our Model. Through visual analysis in Fig. 4, we provide additional evidence that our model effectively represents spatiotemporal information. The selected video samples are from the Kinetics-400 dataset [25]. We use CAM [24] to show the most discriminative features located by the network. We find that our model is better able to capture action-related features in videos, which convey significant semantic information. For example, in the "chopping wood" sample, our model focuses more on the wood and the person, while UniformerV2 [11] pays more attention to the surrounding environment. For the sample "high jump", our model places more emphasis on the athlete, whereas UniformerV2 [11] fails to capture important information in several frames. These results show that our model, compared to UniformerV2 [11], captures fine-grained spatiotemporal features.

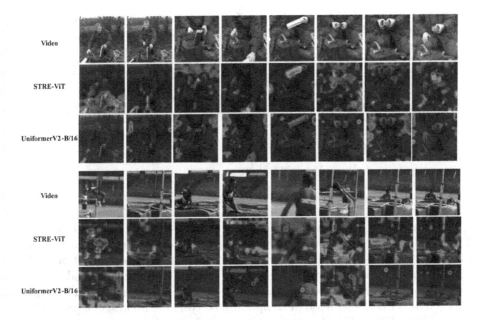

Fig. 4. Visualization Comparison between STRE-ViT and Uniformer-V2
[11]. We conducted visual comparisons of videos with strong temporal correlations.
The red regions indicate the areas that the model focuses on, while the blue regions
are ignored. (Color figure online)

4.2 Ablation Studies

We conducted comprehensive ablation studies to validate each design aspect of
STRE-ViT. Unless otherwise specified, results are obtained using ViT-B/16 [3]
backbone, 8 input frames, and 3 testing views on Kinetics-400 [25].

Spatiotemporal Interaction. We selectively varied the number of ViT Blocks
interacting with the spatiotemporal interaction stream, and the experimental
results are presented in Table 3(a). We observe that selecting more interacting
ViT blocks introduces stronger temporal biases, resulting in more significant
performance gains on Sth-SthV2 [6]. We struck a balance between FLOPs and
accuracy in our selection process.

Temporal Dynamic Extractor. To validate the effectiveness of TDE in temporal modeling, we conducted experiments as shown in Table 3(b). In these
experiments, we removed the TDE module and only fed $\hat{\mathbf{F}}_f$ (the features from
the spatial feature extraction stream, excluding the class token) into the TFF
module (denoted Method a. in Table 3(b)). The results demonstrated that our
TDE module improved accuracy by 0.5%.

Table 3. Results of ablation studies. (a) Effects of different selections of ViT blocks. (b) Effects of different designs.

Indices of selected ViT Blocks	k400		SSv2	
	Acc. (%)	TFLOPS	Acc. (%)	TFLOPS
$[10_{th}, 11_{th}]$	85.0	2.16	72.9	0.55
$[9_{th}, 10_{th}, 11_{th}]$	85.1	2.19	73.2	0.56
$[5_{th}, 7_{th}, 9_{th}, 11_{th}]$	85.0	2.24	73.2	0.57
$[8_{th}, 9_{th}, 10_{th}, 11_{th}]$	85.3	2.24	73.4	0.57
$[5_{th} - 11_{th}]$	85.3	2.36	73.7	0.60
$[4_{th} - 11_{th}]$	85.4	2.41	73.8	0.62
$[0_{th} - 11_{th}]$	85.4	2.58	73.8	0.66

(a)

Method	TDE	TFF	Acc. (%)	TFLOPS
a.	×	✓	84.8	2.09
b.	✓	×	85.1	2.23
c.	✓	✓	**85.3**	2.24

(b)

Temporal-Guided Feature Filter. We remove the TFF module and use only $\hat{F}^i_{vit}$ (the features from the spatial feature extraction stream for interaction, including the class token) as the K and V, while the features extracted by TDE (denoted as F_t) were used as the Q input to the transformer decoder. The experimental result (denoted as Method b. in Table 3(b)) indicates that with only a 0.01 TFLOPs increase in computational cost, the accuracy improved by 0.2%. This validates the effectiveness of the TFF module.

5 Conclusion

This paper presents a two-stream architecture called STRE-ViT. The Spatial Feature Extraction Stream aims to retain the prior knowledge from large-scale pre-trained ViT models as much as possible, while the Spatialtemporal Interaction Stream focuses on filtering spatial features from the Spatial Feature Extraction Stream and facilitating spatial-temporal feature interactions. STRE-ViT finally captures fine-grained and salient spatiotemporal representations. Due to computational resource limitations, we only tested the base version of the model, yet achieved competitive performance compared to state-of-the-art models with similar computational requirements. Importantly, our model exhibits strong scalability and potential for further advancements.

Acknowledgements. This work was supported by the National Key Research and Development Program of China under Grant 2021YFB2910109).

References

1. Arnab, A., Dehghani, M., Heigold, G., Sun, C., Lučić, M., Schmid, C.: ViViT: a video vision transformer. ArXiv, abs/2103.15691 (2021)
2. Bertasius, G., Wang, H., Torresani, L.: Is space-time attention all you need for video understanding? In: ICML, vol. 2, p. 4 (2021)
3. Dosovitskiy, A., et al.: An image is worth 16 × 16 words: transformers for image recognition at scale. ArXiv abs/2010.11929 (2020)

4. Feichtenhofer, C., Fan, H., Malik, J., He, K.: SlowFast networks for video recognition. In: Proceedings of the IEEE/CVF International Conference on Computer Vision, pp. 6202–6211 (2019)
5. Girdhar, R., Carreira, J., Doersch, C., Zisserman, A.: Video action transformer network. In: Proceedings of the IEEE/CVF Conference on Computer Vision and Pattern Recognition, pp. 244–253 (2019)
6. Goyal, R., et al.: The "something something" video database for learning and evaluating visual common sense. In: 2017 IEEE International Conference on Computer Vision (ICCV), pp. 5843–5851 (2017)
7. Hara, K., Kataoka, H., Satoh, Y.: Can spatiotemporal 3D CNNs retrace the history of 2D CNNs and ImageNet? In: 2018 IEEE/CVF Conference on Computer Vision and Pattern Recognition, pp. 6546–6555 (2017)
8. Jia, C., et al.: Scaling up visual and vision-language representation learning with noisy text supervision. In: International Conference on Machine Learning (2021)
9. Ju, C., Han, T., Zheng, K., Zhang, Y., Xie, W.: Prompting visual-language models for efficient video understanding. In: Avidan, S., Brostow, G., Cissé, M., Farinella, G.M., Hassner, T. (eds.) ECCV 2022. LNCS, vol. 13695, pp. 105–124. Springer, Cham (2022). https://doi.org/10.1007/978-3-031-19833-5_7
10. Li, K., et al.: UniFormer: unified transformer for efficient spatiotemporal representation learning. ArXiv abs/2201.04676 (2022)
11. Li, K., et al.: UniFormerV2: spatiotemporal learning by arming image ViTs with video UniFormer. arXiv preprint arXiv:2211.09552 (2022)
12. Li, Y., et al.: MViTv2: improved multiscale vision transformers for classification and detection. In: 2022 IEEE/CVF Conference on Computer Vision and Pattern Recognition (CVPR), pp. 4794–4804 (2021). https://api.semanticscholar.org/CorpusID:244799268
13. Lin, Z., et al.: Frozen clip models are efficient video learners. ArXiv abs/2208.03550 (2022)
14. Liu, Z., et al.: Swin transformer: hierarchical vision transformer using shifted windows. In: 2021 IEEE/CVF International Conference on Computer Vision (ICCV), pp. 9992–10002 (2021)
15. Liu, Z., et al.: Video swin transformer. In: 2022 IEEE/CVF Conference on Computer Vision and Pattern Recognition (CVPR), pp. 3192–3201 (2021). https://api.semanticscholar.org/CorpusID:235624247
16. Loshchilov, I., Hutter, F.: Fixing weight decay regularization in adam. ArXiv abs/1711.05101 (2017)
17. Radford, A., et al.: Learning transferable visual models from natural language supervision. In: International Conference on Machine Learning, pp. 8748–8763. PMLR (2021)
18. Yan, S., et al.: Multiview transformers for video recognition. In: CVPR (2022)
19. Vaswani, A., et al.: Attention is all you need. In: NIPS (2017)
20. Wang, M., Xing, J., Liu, Y.: ActionCLIP: a new paradigm for video action recognition. ArXiv abs/2109.08472 (2021)
21. Wang, W., et al.: Pyramid vision transformer: a versatile backbone for dense prediction without convolutions. In: 2021 IEEE/CVF International Conference on Computer Vision (ICCV), pp. 548–558 (2021)
22. Xiao, T., Singh, M., Mintun, E., Darrell, T., Dollár, P., Girshick, R.B.: Early convolutions help transformers see better. In: Neural Information Processing Systems (2021)

23. Yuan, K., Guo, S., Liu, Z., Zhou, A., Yu, F., Wu, W.: Incorporating convolution designs into visual transformers. In: 2021 IEEE/CVF International Conference on Computer Vision (ICCV), pp. 559–568 (2021)
24. Zhou, B., Khosla, A., Lapedriza, À., Oliva, A., Torralba, A.: Learning deep features for discriminative localization. In: 2016 IEEE Conference on Computer Vision and Pattern Recognition (CVPR), pp. 2921–2929 (2015)
25. Zisserman, A., et al.: The kinetics human action video dataset (2017)

SCFormer: A Vision Transformer with Split Channel in Sitting Posture Recognition

Kedi Qiu, Shoudong Shi$^{(\boxtimes)}$, Tianxiang Zhao, and Yongfang Ye

School of Computer Science and Technology, Ningbo University, Ningbo, China
{2211100274,shishongdong,2211100298,2111082399}@nbu.edu.cn

Abstract. Prolonged maintenance of poor sitting posture can have detrimental effects on human health. Thus, maintaining a healthy sitting posture is crucial for individuals who spend long durations sitting. The recent Vision Transformer (ViT) models have shown promising results in various computer vision tasks. However, it faces challenges such as limited receptive field and excessive parameter quantity. To tackle these issues, we propose SCFormer. To begin with, we utilize the Regular Split Channel (RSC) module to partition the feature map along the channel dimension using specific rules. This enables the flow of spatial information within the channel dimension while severing positional information between adjacent channels, ultimately improving the model's generalization. To extract local feature information and reduce computational complexity, we employ striped windows with parallel self-attention mechanisms over a subset of channels in the feature map. Lastly, we introduce Global Window Feedback (GWF), which exploits redundant information within the channel dimension through simple linear operations, enabling the extraction of inter-window global information and expanding the receptive field. By incorporating these design elements and employing a hierarchical structure, SCFormer demonstrates competitive performance in sitting posture recognition tasks. We achieve successful identification of 10 classes of sitting postures on our dataset, attaining an accuracy of 95%, surpassing current state-of-the-art models.

Keywords: sitting posture recognition · split channel · global window feedback

1 Introduction

With the popularization of modern office practices, many people spend long hours sitting at their desks, making it increasingly important to maintain proper sitting posture. Poor sitting posture can lead to various musculoskeletal issues such as back pain, neck tension and decreased productivity [1–3]. Therefore, the study of sitting posture has significant applications in human-computer interaction, healthcare, and traffic safety.

Supported by Innovation Challenge Project of China (Ningbo) (No. 2022T001).

© The Author(s), under exclusive license to Springer Nature Switzerland AG 2024
S. Rudinac et al. (Eds.): MMM 2024, LNCS 14554, pp. 41–52, 2024.
https://doi.org/10.1007/978-3-031-53305-1_4

There are two mainstream approaches to human sitting posture recognition: sensor-based and computer vision-based methods. Sensor-based methods for posture recognition involve using contact sensors that detect physical signals generated by changes in sitting posture [4]. These methods offer real-time performance but are costlier and can restrict free movement. In contrast, computer vision-based posture recognition methods are cost-effective and easier to deploy.

Currently, traditional computer vision-based posture recognition methods typically capture human body poses using cameras, analyze skeletal information through pose estimation algorithms, extract pose features using convolutional neural networks (CNN), and finally employ classifiers for classification [5]. This multi-step process increases computational complexity and slows down the model. To address this, we propose bypassing the pose estimation algorithm and directly classifying sitting posture. Thus, we have created a dataset consisting of 10 sitting posture classes and designed a network model for training.

Recently, various Transformer-based visual architectures have shown promising results in image classification tasks [6,7]. Leveraging the self-attention mechanism, Transformers inherently possess the crucial ability to capture long-range dependencies but suffer from a large number of parameters. One possible solution is to use local window self-attention [8], which confines the attention region of each token within a specified window size and obtains information between windows through a sliding window mechanism. However, this approach requires stacking numerous modules, resulting in limited global receptive field capability.

We propose SCFormer as an attempt to address these issues. Firstly, we employ the Regular Split Channel (RSC) module to divide feature maps according to specific rules, allowing spatial information to flow across channel dimensions while disconnecting positional information between adjacent channels, thereby enhancing model generalization. Secondly, we design the SCBlock to simultaneously acquire local window information and global receptive field. Within the SCBlock, we perform stripe window self-attention operations on a subset of the feature map to extract features within the window. Simultaneously, another subset undergoes Global Windows Feedback (GWF) operations to capture global receptive field, and these two operations occur in parallel. Notably, stripe window attention within the SCBlock is performed on a subset of the feature map, reducing significant computational costs. GWF utilizes simple linear operations to avoid additional computational overhead. The following are our contributions:

- RSC module that divides feature maps according to specific rules, enabling spatial information flow across channel dimensions and enhancing model generalization.
- We propose the GWF module that leverages redundant information between channels to efficiently capture global receptive field through simple linear operations.
- We construct SCFormer for sitting posture recognition and validate its effectiveness on our dataset.

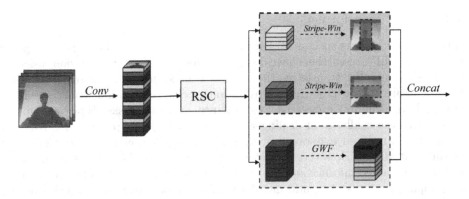

Fig. 1. The illustration of the SCFormer. It consists of RSC module, Stripe-Win Attention (red) and GWF (yellow) module, where Stripe-Win Attention and GWF modules are performed in parallel. (Color figure online)

2 Related Work

2.1 Sitting Posture Recognition

Sensor-Based Approach. Wan et al. [9] proposed a method that utilizes an array of pressure sensors to capture pressure information and devised a hip joint localization algorithm based on hip joint templates to classify four sitting postures. Hu et al. [10] developed a local sitting posture recognition system using six flexible sensors, an analog-to-digital converter, and a Spartan6 field-programmable gate array. However, these methods have limitations as they confine users to specific devices, restricting their mobility. Meyer et al. [11] designed a textile pressure sensor to measure pressure distribution on the human body. Electrodes made of conductive fabric are positioned on both sides of a compressible pad, creating a variable capacitor that is worn by the user to gather body posture information. They subsequently employed a Naive Bayes classifier to identify sitting posture categories. Nonetheless, this contact-based approach may potentially cause discomfort to users in the long term.

Vision-Based Approach. Li et al. [12] proposed a sitting posture recognition method based on multi-scale spatiotemporal features of skeletal graphs. This approach combines the spatial and temporal dimensions of human sitting postures along with the overall skeletal features to enhance recognition accuracy. However, due to the limited and fixed receptive field of CNNs, the model's ability to model long-range dependencies is constrained. Fang et al. [13] introduced Self-Attention modules to capture long-range dependency relationships, enabling the model to capture global features. Nevertheless, they still employ the traditional two-step sitting posture recognition method, which involves first extracting human skeletal keypoints and then using a classifier for classification, thereby increasing the computational overhead of the model.

2.2   Vision Transformer

After the tremendous success of Transformers in the field of natural language processing (NLP), researchers have designed visual Transformers for vision tasks. Dosovitskiy et al. [14] transferred the original Transformer architecture to visual tasks. Thanks to the powerful self-attention mechanism, Vision Transformers (ViTs) have achieved impressive performance in visual tasks. However, this comes with the challenge of dealing with a large number of parameters. Wang et al. [15] introduced a pyramid structure to merge tokens of key and query at different stages. However, consistent merging operations can result in the loss of fine-grained information. Liu et al. [8] divided the feature map into non-overlapping windows and performed local self-attention operations within each window, obtaining global receptive fields by moving the windows. However, even with a large number of stacked blocks, the expansion of the receptive field is still slow. Dong et al. [16] proposed Cross-Attention, designing equally sized rectangular windows that overlap and intersect with each other, to some extent expanding the receptive field. However, the problem of high computational cost remains unresolved. Therefore, we propose SCFormer, which reduces computational costs while maintaining accuracy. The GWF in SCFormer allows for fast modeling of global dependencies in images at a low cost.

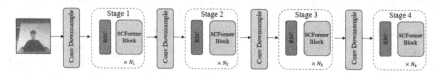

Fig. 2. (a) The overall architecture of the SCFormer. It is stacked by 4 stages, and the number of stacks $[N_1, N_2, N_3, N_4]$ can be changed according to actual needs. The typical number of stacks in this paper is $[2, 2, 18, 2]$; (b) stripe-win Attention internal structure.

3   Methodology

3.1   Overall Structure Design

The overall architecture of our proposed SCFormer is illustrated in Fig. 2. Following Liu et al. [8], we adopt a stacked block structure, dividing the model into four stages. In the first stage, we apply larger convolutional layers (stride = 4, kernel size = 7×7) to downsample the images. To reduce the number of tokens, we introduce convolutional layers (stride = 2, kernel size = 3×3) between two adjacent stages. To alleviate the computational cost of self-attention, we introduce the RSC module in each stage. The RSC module divides the feature map in the channel dimension according to specific rules, enabling spatial interactions in subsequent parallel operations. Details are described next.

3.2   Regular Split Channel

Self-Attention exhibits powerful capabilities in capturing long-range dependencies but comes with significant computational costs. The computational complexity of the original full-scale Self-Attention is quadratic in terms of the feature map size. It can be formulated as follow:

$$Q = XW_q, K = XW_k, V = XW_v; X = \{x_1, x_2, ..., x_n\} \tag{1}$$

$$Attention(Q, K, V) = \sigma((\frac{QK^T}{\sqrt{d_k}})V \tag{2}$$

where $\sigma(.)$ denotes the SoftMax function, d_k is the k-th channel dimension in X.

Recent works have proposed feasible solutions by suggesting the confinement of Self-Attention within a fixed-size rectangular window [8]. However, this approach limits the spatial dimension's global interaction, leading to slow expansion of the global receptive field.

Allowing systematic division and recombination of different local spatial windows in the channel dimension enables information interaction among these windows, facilitating the flow of spatial information in the channel dimension, as depicted in Fig. 3a. Islam et al. [17] provided evidence for the presence of positional information between channels, which follows a sequential distribution pattern. Channels that are closer in proximity exhibit greater similarity in positional information. By partitioning the feature maps along the channel dimension using a predefined rule (as depicted in Fig. 3b), the positional information between adjacent channels is severed, thereby enhancing the model's generalization capability.

We propose the RSC module. Assuming we have g windows, the output of the module will have $g \times n$ channels. The process involves reshaping the input channels into a shape of (g, n), followed by regular split and reorganization to create three new feature maps: F1, F2, and F3. The dimensions of the new feature maps are defined as follows:

$$F1 = (g, \frac{i}{k}), F2 = (g, \frac{i}{k+1}) \tag{3}$$

$$F3 = (g, \frac{i}{k+2} \cap \frac{i}{k+3}); \forall i \in \{1, 2, ..., n\} \tag{4}$$

In engineering practice, we set the parameter k to a value of 4. This means that the channels in feature maps F1 and F2 will be divided into groups of equal size, where each group contains one-fourth of the original channels. This division allows for effective information interaction among the windows and promotes the flow of spatial information in the channel dimension.

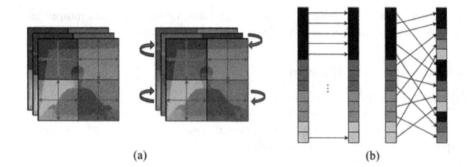

(a)   (b)

Fig. 3. (a) The illustration of the RSC module. The window is divided by specific rules to make the spatial information interact in the channel dimension; (b) After the RSC module, the position information between adjacent channels is split to increase the generalization ability of the model.

3.3   SCFormer Block

In this section, we will introduce the high-efficiency and low-computation SCFormer Block, as shown in Fig. 1, consisting of GWF and Stripe-Win Attention.

Global Window Feedback. The Local-Win Attention mechanism can lead to a lack of global receptive field. Existing approaches, such as shifting [8], expanding [18], or shuffling [19] operations, have been used to obtain global receptive fields, but these operations introduce additional computational costs. However, in well-trained deep neural networks, the channel dimension of the feature maps often contains rich and even redundant information [20], which can ensure comprehensive understanding of the input information. We propose leveraging this redundant information and performing simple operations to concurrently capture global receptive fields while modeling spatial information within the windows. To avoid introducing additional computational overhead, we propose a simple design called Global Window Feedback (GWF). As shown in Fig. 4, we suggest performing a linear operation on the feature maps to obtain feedback information from the global window. The feedback information is obtained based on the following transformation:

$$Fi = \phi i(Fi), \forall i = 1, ..., k \tag{5}$$

Here, Fi represents the i-th feature map, and ϕi denotes the linear operation applied to the i-th feature map. In practical engineering scenarios, we utilize a 3×3 depthwise convolution, followed by batch normalization (BN) and GELU activation, to obtain the feedback information from the global window. The depthwise convolution enables efficient modeling of the channel dimension, enhancing the ability to capture larger receptive fields.

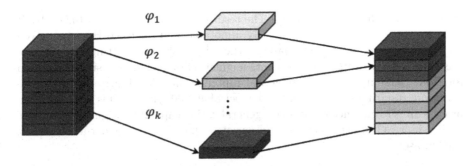

Fig. 4. The illustration of the GWF module. It operates by mapping the redundant information within the feature map to obtain global window feedback information in a simple linear fashion.

Stripe-Win Attention. To leverage the long-range dependency-building capabilities of Self-Attention (SA) more effectively, Dong et al. [16] proposed Cross Window Attention, which replaces equally sized rectangular windows with horizontally and vertically striped windows. However, the issue of high computational complexity remained unresolved. Therefore, based on CSwin, we introduce the Strip-Win Attention mechanism, which reduces computational overhead by applying horizontal and vertical striped windows to a partitioned portion of the feature maps, as highlighted in red in Fig. 1.

We divide the input sequence $F \in R^{h \times w \times \frac{c}{2}}$ into two equal parts along the channel dimension: $F_{hor} = F_{ver} \in R^{h \times w \times \frac{c}{4}}$. For the horizontal stripes, F_{hor} is evenly divided into non-overlapping horizontal stripes of equal width, denoted as $[F_{hor}^1, F_{hor}^2, ..., F_{hor}^M]$. Each token within a stripe is projected as Query, Key, and Value with a dimension of d_k. Thus, the definition of Stripe-Win Attention for F_{hor}^k is as follows:

$$F_{hor} = [F_{hor}^1, F_{hor}^2, ..., F_{hor}^M] \tag{6}$$

$$Attention(Q_k, K_k, V_k) = \sigma(\frac{F_{hor}^k W_k^Q (F_{hor}^k W_k^K)^T}{\sqrt{dk}}) F_{hor}^k W_k^V \tag{7}$$

where $\sigma(.)$ denotes the SoftMax function, $F_{hor}^k \in R^{(sw \times w) \times \frac{c}{4}}$, $M = H/sw$, and $k = 1, ..., M$. The projection matrices are denoted as $W_k^Q \in R^{\frac{c}{4} \times d_k}$, $W_k^K \in R^{\frac{c}{4} \times d_k}$, $W_k^V \in R^{\frac{c}{4} \times d_k}$, and d_k is set as $\frac{c}{4}$. The derivation for vertical striped self-attention follows a similar process.

4 Experiments

4.1 Experimental Dataset

In order to address the limited availability of a high-quality dataset for seated posture recognition, we have devised a solution in which we defined ten distinct seated postures and enlisted the participation of fifty volunteers for capturing images using a dual-camera setup. The resulting dataset comprises 10,000 RGB images, which have been meticulously annotated and selected. However, given the relatively small size of the dataset, there is a risk of overfitting. To mitigate this issue, we have applied various data augmentation techniques, such as random rotation, random cropping, horizontal flipping, Gaussian blur, and brightness variation, resulting in an expanded dataset of 40,000 images. This augmentation process aims to enhance the model's generalization capability. Figure 5 illustrates typical images from the dataset, and our experiments in this paper are based on this dataset.

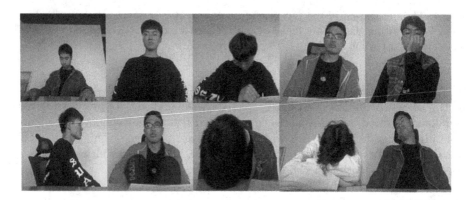

Fig. 5. The sitting posture dataset comprises typical images depicting various sitting positions, including "straight back and head down," "back up and head down," "back and head down," "leaning on the chair," "hand on the head," "body turned sideways," "feet on the chair," "head on the table," "sleeping on the table," and "leaning on the chair while sleeping." These ten distinct sitting postures provide comprehensive coverage of seated positions commonly observed in real-world scenarios.

4.2 Experimental Environment

All models were trained using the same hyperparameters on two NVIDIA RTX 3070Ti GPUs, running at Ubuntu 18.04 system, and employing the PyTorch version 1.8.0 framework. During the network model training stage, we utilized the Stochastic Gradient Descent with Momentum (SGDM) optimizer to update the network weights. The initial learning rate was set to 0.01, the optimizer momentum was set to 0.9, and a weight decay of $1e-4$ was applied. The batch size used for training was 32. We employed the cosine annealing method (linear warm-up of 4 epochs) to dynamically adjust the learning rate. And we set the total number of training epochs to be 50, 100, and 300 to show the convergence ability of the model. Throughout the training process, we measured the performance of the network model using Top-1 accuracy, parameter quantity, and Floating Point Operations (FLOPs).

4.3   Comparative Experiment on Different Network Models

We conducted experiments using the aforementioned experimental environment to train SCFormer and other mainstream Vision Transformers (ViT) models. The accuracy results obtained are presented in Table 1. On our dataset, models incorporating the local-win Attention mechanism demonstrate their advantages; however, due to the gradual expansion of the receptive field, a considerable number of epochs are required to fully exploit their benefits. In contrast, our proposed SCFormer model achieves commendable accuracy with a relatively small number of epochs. As outlined in Table 1, our model exhibits the lowest parameter count and FLOPs, while achieving the highest Top-1 accuracy.

Table 1. Comparison of different models on our datasets.

Models	Params	FLOPs	50 epoch Top-1 (%)	100 epoch Top-1 (%)	300 epoch Top-1 (%)
ViT-B/16 [14]	86M	55.4G	61.8	69.2	79.9
ViT-L/16 [14]	307M	190.7G	64.3	70.0	81.3
Swin-T [8]	29M	4.5G	81.9	87.3	92.0
Swin-S [8]	50M	8.7G	82.1	87.1	92.4
Cswin-T [16]	23M	4.3G	85.2	90.1	93.8
C-Former-T [21]	27.8M	2.9G	85.6	89.6	94.1
PVT-S [15]	25M	3.8G	82.3	88.9	92.3
Shunted-S [22]	22.4M	4.9G	83.2	89.6	93.4
SCFormer	**17M**	**3.2G**	**86.3**	**90.6**	**95.2**

In Fig. 6, we present the visualization of sitting pose recognition structures utilizing Class Activation Mapping (CAM). The results illustrate that SCFormer effectively focuses on crucial body parts while minimizing the influence of the background on the recognition capability.

4.4   Ablation Experiments

The experiments conducted demonstrate the superior Top-1 accuracy achieved by SCFormer on our dataset. To further showcase the effectiveness of our proposed approach, we conducted comparative evaluations with other architectures. Specifically, we examined the impact of removing the RSC and GWF modules, as well as compared SCFormer with the original CSwin model. These comparisons were conducted for 50 epochs, 100 epochs, and 300 epochs, respectively.

The experimental results, as presented in Table 2, confirm the crucial role played by the RSC module in improving accuracy. Joining this module led to an increase in accuracy by 2.1%. Additionally, the efficacy of the GWF module

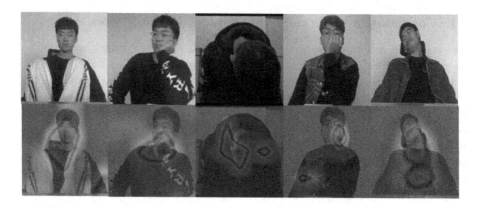

Fig. 6. The visualization results of SCFormer demonstrate its ability to effectively focus on important parts while minimizing the impact of the background on recognition accuracy.

Table 2. Ablation study on our datasets.

SCFormer	RSC	GWF	50 epoch Top-1 (%)	100 epoch Top-1 (%)	300 epoch Top-1 (%)
✓	–	–	79.3	85.5	92.4
✓	✓	–	81.1	87.9	94.5
✓	✓	✓	**86.3**	**90.6**	**95.2**

in rapidly capturing the global receptive field was demonstrated, enabling the model to converge quickly even with a limited number of epochs.

These findings highlight the significance of both the RSC and GWF modules in enhancing the performance of SCFormer. Their inclusion enables the model to effectively capture long-range dependencies and efficiently obtain global context information. The experimental results validate the effectiveness of our proposed scheme and underline the superiority of SCFormer over alternative architectures.

5   Conclusion

In this paper, we have presented SCFormer, an efficient vision Transformer model for sitting pose recognition. SCFormer achieves a balance between capturing local features and obtaining global receptive fields in parallel, resulting in significant computational cost reduction and faster model convergence. The introduced RSC module enables spatial information interaction in the channel dimension by dividing the feature map based on specific rules and effectively cutting off position information between adjacent channels, thereby enhancing the model's generalization capability. Our experimental results demonstrate that SCFormer outperforms other models on our dataset.

In future work, we plan to investigate the applicability of SCFormer to other computer vision tasks, such as semantic segmentation. We hypothesize that the efficient interaction of spatial information in the channel dimension, as demonstrated by SCFormer, may also benefit tasks requiring comprehensive understanding of visual scenes and precise object localization. By exploring these avenues, we aim to further validate the effectiveness and versatility of SCFormer in various computer vision applications.

Acknowledgements. Shoudong Shi is the corresponding author of this work. This work was supported by the Innovation Challenge Project of China (Ningbo) under Grant No. 2022T001.

References

1. Cagnie, B., Danneels, L., Van Tiggelen, D., De Loose, V., Cambier, D.: Individual and work related risk factors for neck pain among office workers: a cross sectional study. Eur. Spine J. **16**, 679–686 (2007). https://doi.org/10.1007/s00586-006-0269-7
2. Murphy, S., Buckle, P., Stubbs, D.: Classroom posture and self-reported back and neck pain in schoolchildren. Appl. Ergon. **35**(2), 113–120 (2004)
3. O'Sullivan, P.B., Mitchell, T., Bulich, P., Waller, R., Holte, J.: The relationship beween posture and back muscle endurance in industrial workers with flexion-related low back pain. Man. Ther. **11**(4), 264–271 (2006)
4. Ran, X., Wang, C., Xiao, Y., Gao, X., Zhu, Z., Chen, B.: A portable sitting posture monitoring system based on a pressure sensor array and machine learning. Sens. Actuators, A **331**, 112900 (2021)
5. Chen, K.: Sitting posture recognition based on OpenPose. In: IOP Conference Series: Materials Science and Engineering, vol. 677, p. 032057. IOP Publishing (2019)
6. Han, K., Xiao, A., Wu, E., Guo, J., Xu, C., Wang, Y.: Transformer in transformer. In: Advances in Neural Information Processing Systems, vol. 34, pp. 15908–15919 (2021)
7. He, J., et al.: TransFG: a transformer architecture for fine-grained recognition. In: Proceedings of the AAAI Conference on Artificial Intelligence, vol. 36, pp. 852–860 (2022)
8. Liu, Z., et al.: Swin transformer: hierarchical vision transformer using shifted windows. In: Proceedings of the IEEE/CVF International Conference on Computer Vision, pp. 10012–10022 (2021)
9. Wan, Q., Zhao, H., Li, J., Xu, P.: Hip positioning and sitting posture recognition based on human sitting pressure image. Sensors **21**(2), 426 (2021)
10. Hu, Q., Tang, X., Tang, W.: A smart chair sitting posture recognition system using flex sensors and FPGA implemented artificial neural network. IEEE Sens. J. **20**(14), 8007–8016 (2020)
11. Meyer, J., Arnrich, B., Schumm, J., Troster, G.: Design and modeling of a textile pressure sensor for sitting posture classification. IEEE Sens. J. **10**(8), 1391–1398 (2010)
12. Li, L., Yang, G., Li, Y., Zhu, D., He, L.: Abnormal sitting posture recognition based on multi-scale spatiotemporal features of skeleton graph. Eng. Appl. Artif. Intell. **123**, 106374 (2023)

13. Fang, Y., Shi, S., Fang, J., Yin, W.: SPRNet: sitting posture recognition using improved vision transformer. In: 2022 International Joint Conference on Neural Networks (IJCNN), pp. 1–6. IEEE (2022)
14. Dosovitskiy, A., et al.: An image is worth 16 × 16 words: transformers for image recognition at scale. arXiv preprint arXiv:2010.11929 (2020)
15. Wang, W., et al.: Pyramid vision transformer: a versatile backbone for dense prediction without convolutions. In: Proceedings of the IEEE/CVF International Conference on Computer Vision, pp. 568–578 (2021)
16. Dong, X., et al.: CSWin transformer: a general vision transformer backbone with cross-shaped windows. In: Proceedings of the IEEE/CVF Conference on Computer Vision and Pattern Recognition, pp. 12124–12134 (2022)
17. Islam, M.A., Kowal, M., Jia, S., Derpanis, K.G., Bruce, N.D.: Global pooling, more than meets the eye: position information is encoded channel-wise in CNNs. In: Proceedings of the IEEE/CVF International Conference on Computer Vision, pp. 793–801 (2021)
18. Vaswani, A., Ramachandran, P., Srinivas, A., Parmar, N., Hechtman, B., Shlens, J.: Scaling local self-attention for parameter efficient visual backbones. In: Proceedings of the IEEE/CVF Conference on Computer Vision and Pattern Recognition, pp. 12894–12904 (2021)
19. Xiao, J., Fu, X., Zhou, M., Liu, H., Zha, Z.J.: Random shuffle transformer for image restoration. In: International Conference on Machine Learning, pp. 38039–38058. PMLR (2023)
20. Han, K., Wang, Y., Tian, Q., Guo, J., Xu, C., Xu, C.: GhostNet: more features from cheap operations. In: Proceedings of the IEEE/CVF Conference on Computer Vision and Pattern Recognition, pp. 1580–1589 (2020)
21. Wang, W., et al.: CrossFormer: a versatile vision transformer hinging on cross-scale attention. arXiv preprint arXiv:2108.00154 (2021)
22. Ren, S., Zhou, D., He, S., Feng, J., Wang, X.: Shunted self-attention via multi-scale token aggregation. In: Proceedings of the IEEE/CVF Conference on Computer Vision and Pattern Recognition, pp. 10853–10862 (2022)

Dive into Coarse-to-Fine Strategy in Single Image Deblurring

Zebin Li[iD] and Jianping Luo[(⊠)][iD]

Guangdong Key Laboratory of Intelligent Information Processing, College of
Electronic and Information Engineering, Shenzhen University, Shenzhen, China
2210433027@email.szu.edu.cn, ljp@szu.edu.cn

Abstract. The coarse-to-fine approach has gained significant popularity in the design of networks for single image deblurring. Traditional methods used to employ U-shaped networks with a single encoder and decoder, which may not adequately capture complex motion blur patterns. Inspired by the concept of multi-task learning, we dive into the coarse-to-fine strategy and propose an all-direction, multi-input and multi-output network for image deblurring (ADMMDeblur). ADMMDeblur has two distinct features. Firstly, it employs four decoders, each generating a unique residual representing a specific motion direction. This enables the network to effectively address motion blur in all directions within a two-dimensional (2D) scene. Secondly, the decoders utilize kernel rotation and sharing, which ensures the decoders do not separate unnecessary components. Consequently, the network exhibits enhanced efficiency and deblurring performance while requiring fewer parameters. Extensive experiments conducted on the GoPro and HIDE datasets demonstrate that our proposed network achieves better performance in deblurring accuracy and model size compared to existing well-performing methods.

Keywords: coarse-to-fine · four decoders · kernel rotation and sharing

1 Introduction

Image deblurring, a low-level vision task in computer vision, aims to recover a sharp image from a blurry image caused by a camera shake or object motions. Notably, a blurry image not only has low perceptual image quality but also impacts the performance of various applications, such as objective detection [3], autonomous driving [4], and video monitoring [22]. Consequently, despite being a classical vision task, image deblurring remains a fruitful field of research.

Generally, the degradation model is formulated as follows:

$$B = I \otimes K + N \tag{1}$$

where $B, I, K, N, \otimes$ denote the blurry image, latent sharp image, blur kernel, additional random noise and 2D convolution operation, respectively. Solving Eq. (1) is a highly ill-posed problem because only all items but the blurry image B are

© The Author(s), under exclusive license to Springer Nature Switzerland AG 2024
S. Rudinac et al. (Eds.): MMM 2024, LNCS 14554, pp. 53–65, 2024.
https://doi.org/10.1007/978-3-031-53305-1_5

unknown. In conventional methods, researchers typically estimate the blur kernel and perform deconvolution or regularization using natural image priors [12,23] or convolutional neural networks (CNNs) [6,20] to address the inherent uncertainty of the deblurring problem. On the other hand, recent image deblurring methods [18,21] using end-to-end neural networks aim to directly learn the mapping relationship from a blurry image to its latent sharp image. Nah et al. [15] pioneered applying the classical coarse-to-fine optimization strategy to single-image deblurring and presented a deep network based on a multi-scale architecture for dynamic scene deblurring (DeepDeblur). DeepDeblur handles multi-scale degradation and progressively reconstructs a sharp image using stacked sub-networks. Motivated by the success of DeepDeblur, various CNN-based image deblurring methods [2,5,17] adopt the same strategy and similar architecture and attain remarkable performance improvements. In other words, deblurring in a coarse-to-fine manner has proven effective in single-image deblurring.

Although the methods perform image deblurring well in a coarse-to-fine manner, the network of the most is a U-shape network with a one-encoder-one-decoder architecture. Considering the complex nature of realistic motion blur, which arises from multiple intertwined motions with varying speeds and directions, we claim that the standard one-encoder-one-decoder architecture employed in a U-shaped network may not appropriately handle the problem. As a result, we, inspired by the previous works [9], decompose the deblurring problem into multiple sub-problems and present an all-direction, multi-input and multi-output network for image deblurring (ADMMDeblur). In terms of the network design of our proposed network, a one-encoder-four-decoder architecture is combined with a multi-input multi-output architecture. The proposed network in each scale contains one encoder and four decoders, as shown in Fig. 1. These decoders implicitly generate the residuals corresponding to different directions. We observe that the four decoders produce residuals along different directions, which collectively form the motion blur patterns of any directions within a 2D scene. Based on this observation, we apply the sharing and rotating operation to the convolutional kernels among the decoders. Unlike traditional parameter sharing in CNN that compromises performance [16], the parameter sharing in our proposed network reduces the number of model parameters while simultaneously improving network performance. The contribution of the proposed method can be summarized as follows:

1. We propose an all-direction, multi-input and multi-output network for image deblurring, which can decompose the deblurring problem into multiple sub-problems to handle the complexity of realistic motion blur. It is more efficient than the conventional image deblurring networks in a one-encoder-one-decoder architecture.
2. ADMMDeblur shares rotated convolution kernels from one decoder with the rest, leading to substantial improvement of the deblurring performance with less number of parameters.

2   Related Work

In this section, we review the conventional single-image deblurring methods adopting a coarse-to-fine strategy and the traditional networks with a one-encoder-multiple-decoder structure.

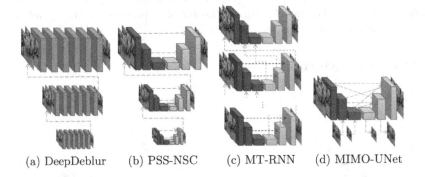

(a) DeepDeblur   (b) PSS-NSC   (c) MT-RNN   (d) MIMO-UNet

Fig. 1. Comparision of coarse-to-fine image deblurring network architectures.

2.1   Image Deblurring by Coarse-to-fine Approach

As a pioneering work, DeepDeblur [15] stacks sub-networks to build a multi-scale architecture for image deblurring, as shown in Fig. 1(a). The network effectively handles multi-scale motion blur in a coarse-to-fine manner. Inspired by the success of DeepDeblur, Gao et al. [5] presented a parameter selective sharing strategy and nested skip connections (PSS-NSC). The architecture of PSS-NSC illustrates in Fig. 1(b), in which each sub-network is structured as an Unet with one encoder, one decoder, and symmetric skip connections allowing direct transmission of feature maps from the encoder to the decoder. Furthermore, most network parameters are shared among these sub-networks since some modules in each sub-network perform a similar deblurring pattern. Park et al. [17] presented a multi-temporal recurrent neural network (MT-RNN), as depicted in Fig. 1(c). In MT-RNN, a single U-shape is repeated seven times, and the feature maps from its decoder in the current iteration will transfer to its encoder as inputs in the next iteration. This design reduces memory usage but leads to low runtime efficiency. Cho et al. [2] presented a multi-input multi-output U-net (MIMO-UNet), as depicted in Fig. 1(d). The MIMO-Unet has a single encoder to extract features from multi-scale blurry images, a single decoder to reconstruct multi-scale clear ones, and asymmetric feature fusion modules to merge all-scale features from the encoder in an efficient manner, which effectively improves the flexibility of the feature information flow.

2.2   One-Encoder-Multiple-Decoder Structure

In multi-task learning, a one-encoder-multiple-decoder structure has gained prominence. It typically features a single encoder responsible for extracting diverse information from the input data and contains multiple decoders for tackling individual tasks by the given labels. For example, Kendall et al. [11] proposed a multi-decoder architecture that leverages uncertainty estimates to weigh the losses for scene geometry and semantics tasks in computer vision. Furthermore, image decomposition, one of the multi-task problems, can be addressed by a one-encoder-multiple-decoder structure [7,13]. These networks typically necessitate distinct labels for each decoder during the training process or employ intricate loss functions based on prior knowledge [13].

3   Proposed Method

See Fig. 3.

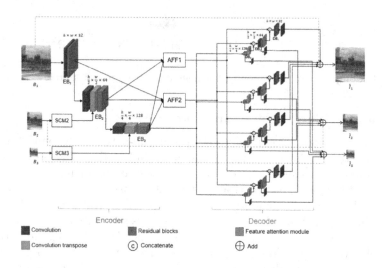

Fig. 2. The architecture of the proposed network.

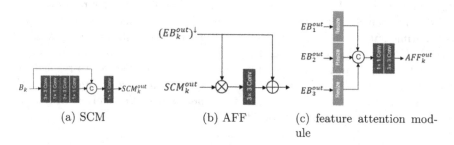

(a) SCM   (b) AFF   (c) feature attention module

Fig. 3. Comparison of coarse-to-fine image deblurring network architectures.

3.1 Architecture of the Proposed ADMMDeblur

Inspired by prior work [2,9], we propose an all-direction, multi-input and multi-output network for image deblurring (ADMMDeblur), combining a coarse-to-fine strategy with a one-encoder-multiple-decoder architecture, as depicted in Fig. 2. The network architecture shares similarities with the MIMO-UNet proposed in [2], comprising an encoder module, an asymmetric feature mixing module (AFF), and a decoder module. A blurry image and its down-sampled images serve as input to the network, which outputs restored images whose sizes correspond to the network inputs. The structures of the encoder module and AFF are the same as MIMO-UNet+. However, the decoder module in each scale includes four decoder blocks (DB) expected to generate different residuals. In each DB of all scales besides the smallest one, there is a convolutional block to merge the information from the decoder of the smaller scale and the AFF module. Each DB contains a residual block which consists of 20 modified residual blocks [21]. Each DB of all scales besides the largest one includes a transpose convolution layer. Since the output of DB is a feature map, not an image, we apply a single convolutional layer to it.

There are two notable differences between our proposed network and MIMO-UNet+. Firstly, the number of decoders increases, and secondly, the constructive manner of the decoders is unique. Each scale of our network incorporates one encoder accompanied by four distinct decoders. These decoders generate independent residuals along different directions, enabling comprehensive handling of image-blurring patterns across the entire 2D scene. Additionally, the four decoders are constructed by sharing rotated convolutional kernels.

3.2 Four Separated Decoders

Motivated by the previous work [9], we have decomposed the deblurring problem into multiple sub-problems in our proposed network. In other words, the network distributes the hybrid features, consisting of fused information from all scales combined with reconstructed features at smaller scales, into four distinct decoders. Leveraging the linear spanning theory [14], which suggests that a linear reconstruction involving multiple independent regression outputs can provide a larger solution space, each of the four decoders, ideally, generates independent regression outputs to maximize the solution space. In other words, it is expected that each decoder generates blurry residuals along four distinct directions, effectively encompassing the entire 2D scene. The generated residuals are complementary, working in synergy to facilitate image reconstruction.

The residuals displayed in Fig. 4 are estimated by each of the four decoders. Interestingly, among these residuals, two have more horizontal components, while the other two have more vertical components. Notably, the two decoders with more horizontal residuals provide complementary residuals in the vertical direction, while the two decoders with more vertical residuals offer complementary residuals in the horizontal direction. This observation underscores that although these four residuals may bear similarities in pairs, the two similar residuals represent opposing directions. Ultimately, the four residuals can be combined to represent any motion blur pattern in the entire 2D scene. Essentially, our proposed design allows the network to effectively decompose encoder features along four distinct directions, representing the motion blur pattern from any directions in a 2D scene without explicit constraints. This discovery has inspired us to devise a plan for decoder parameter sharing.

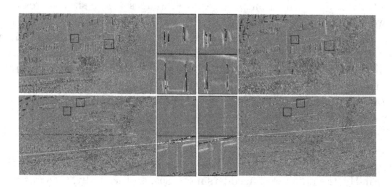

Fig. 4. Residual images generated by four decoders. For clarity, the magnified parts of residual images are displayed. From left-top to right-bottom: the residual images obtained by D_1, D_3, D_2, D_4.

3.3   Spatial Kernel Rotation for Parameter Sharing

Even though the separated decoders effectively decompose features into four distinct orientations, the degree of decoupling in the feature information falls short of our expectations due to the lack of explicit supervision. This undesirable partitioning of feature information restricts the size of the network solution space [14], subsequently impacting the deblurring performance of the network.

To address this issue, we can leverage the properties of the separated decoders discussed in Sect. 3.1. Given that these decoders extract information components along four different directions, we can share their parameters using the following approach:

$$\hat{r} = D_1(z; \theta_1) + D_2(z; \theta_2) + D_3(z; \theta_3) + D_4(z; \theta_4), \tag{2}$$

where $\hat{r}, D_1, D_2, D_3, D_4$ denote the generated residuals, input features, and four decoders with parameters $\theta_1, \theta_2, \theta_3$ and θ_4, respectively. Notably, the parameters θ_2, θ_3, and θ_4 are obtained by rotating parameter θ_1 by 90, 180, and 270°, respectively. The residuals generated by the decoders can be decomposed into two components along the horizontal (x-axis) and vertical (y-axis) axes. Assuming that the component with the greater signal energy corresponds to the main component, while the component with the lower signal capability represents the complementary component, it can be observed from Fig. 4 that the main components of D_1 and D_3 align with the y-axis, while the main components of D_2 and D_4 align with the x-axis. If we conceptualize the residuals as vectors, we can determine the direction of the main component as the direction of the vector. Upon observing the complementary nature of the components, we note that in terms of the residuals with approximated main components, their complementary components exhibit a complementary phenomenon, implying that their main component directions are opposite. Consequently, the residuals with similar main components have opposite directions, and with orthogonal main components are perpendicular to each other. In classical mechanics, two orthogonal vectors suffice to describe the entire motion within a 2D scene. In our proposed method, the residuals corresponding to the two orthogonal main components represent two orthogonal vectors, whereas the residuals with approximated main components signify two vectors with opposing directions. In other words, leveraging convolutional kernel rotation and parameter sharing operations, the four decoders execute the same feature extraction pattern for the four distinct directions, so that the generated residuals' directions cover the entire range of directions within a 2D scene. This approach aligns with the analysis process of motion in classical mechanics. As a result, ADMMDeblur achieves improved deblurring performance with a reduced number of model parameters.

3.4   Loss Function

Like other multi-scale deblurring networks, we use the multi-scale content loss function [15]. The content loss function is defined as follows:

$$L_c = \sum_{n=1}^{N} = \frac{1}{t_n} ||\hat{I}_n - I_n||_1, \tag{3}$$

where N is the number of levels. For normalization the loss is divided by the number of total elements tn.

Recent studies have demonstrated the effectiveness of incorporating perceptual loss for image recovery tasks [8,10]. Considering that the loss of detailed information in blurred images often corresponds to high-frequency information, it is crucial to minimize the discrepancy between blurred-clear image pairs in the frequency domain. To achieve this, we adopt the same perceptual loss function employed in [2]. The frequency loss function is defined as follows:

$$L_f = \sum_{n=1}^{N} = \frac{1}{t_n} ||F(\hat{I}_n) - F(I_n)||_1, \tag{4}$$

where F denotes the fast Fourier transform (FFT) that transfers the image signal to the frequency domain. The final loss function for training our network is determined as follows:

$$L = L_c + \lambda L_f, \tag{5}$$

where λ is experimentally set 0.1.

4   Experiment

4.1   Datasets and Implementation Details

For training of the networks, the GoPro training set [15] was used, which contains 2103 pairs of blurry and clear images. To assess the performance of various models, we utilize the GoPro test set and the HIDE test set [19], containing 1111 and 2025 pairs of blurry and clear images, respectively.

During every training iteration, we randomly select a batch of 4 images and apply random cropping to achieve a size of 256 × 256. In the data augmentation phase, horizontal flipping is applied to each batch with a probability of 0.5. For the GoPro dataset, we train our model for 4000 iterations until convergence is achieved. The initial learning rate is set to 10^{-4} and decayed by half every 500 iterations. Our experiments are conducted on an Intel Core i7-10700 and an NVIDIA RTX 2080Ti.

4.2   Performance Comparison

We conduct comparative analyses between ADMMDeblur and other deblurring networks for the performance in terms of model size and deblurring accuracy. More precisely, we conduct the following comparisons.

1. Comparison with the deblurring networks employing the coarse-to-fine strategy [2,5,15,17]. Quantitative results for ADMMDeblur, as well as the other networks trained on the GoPro train set, are evaluated on the GoPro test set and the HIDE test set, and recorded in Table 1. Additionally, Fig. 5 showcases the qualitative results of these networks on the GoPro test set.
2. Comparison with the recent well-performing deblurring networks. The above networks and our proposed network are trained on the GoPro train set and evaluated on the GoPro test set, the quantitative results of which are summarized in Table 2.

By conducting these comparisons, we aim to provide a comprehensive assessment of ADMMDeblur's performance relative to other deblurring networks.

As depicted in Table 1, ADMMDeblur exhibits superior performance compared to the other models utilizing the coarse-to-fine approach. On the GoPro test set, ADMMDeblur achieves an average PSNR of 33.01 dB and an average SSIM of 0.9614, surpassing the competing models. The same rank also occurs on the HIDE test set. Notably, ADMMDeblur, on the GoPro test set, outperforms the second-place MIMO-UNet+ by 0.56 dB in average PSNR, highlighting the effectiveness of our proposed multi-decoder structure. Analyzing the results presented in Table 2, it is evident that ADMMDeblur outperforms other methods for deblurring accuracy while maintaining less number of parameters. Notably, the number of parameters between MIMO-UNet+ and ADMMDeblur are equal, yet the latter consistently achieves higher average PSNR and SSIM scores. This improvement can be attributed to the effective utilization of shared rotation parameters in our proposed model.

The comprehensive comparison presented in Tables 1 and 2 implies the conclusion that ADMMDeblur outperforms other networks for model size and deblurring accuracy.

Table 1. Quantitative comparison of all image deblurring methods in a coarse-to-fine manner.

Model	GoPro		HIDE	
	PSNR	SSIM	PSNR	SSIM
DeepDeblur [15]	30.40	0.9162	25.73	0.8742
PSS-NSC [5]	30.96	0.9421	29.11	0.9133
MT-RNN [17]	31.14	0.9447	29.15	0.9177
MIMO-UNet+ [2]	32.45	0.9567	29.99	0.9304
ADMMDeblur	**33.01**	**0.9614**	**30.57**	**0.9368**

Fig. 5. Several examples on the GoPro test set. For clarity, the magnified parts of the resultant images are displayed. From left-top to right-bottom: the resultant images obtained by DeepDeblur, PSS-NSC, MT-RNN, MIMOUNet+, ADMMDeblur and ground-truth images, respectively.

Table 2. The average PSNR and SSIM on the GoPro test set.

Model	PSNR	SSIM	Params(M)
DMPHN [26]	31.20	0.9453	21.7
MIMO-UNet+ [2]	32.45	0.9567	**16.1**
MPRNet [25]	32.66	0.9589	20.1
HINet [1]	32.71	0.9593	88.7
Restormer [24]	32.92	0.9611	26.1
ADMMDeblur	**33.01**	**0.9614**	**16.1**

4.3 Ablation Study

In our proposed network, the decoder module holds significant importance. To substantiate the efficacy of our chosen number of decoders and rotating and sharing operations, we conducted the following two ablation experiments.

a. Increasing the number of decoders. We utilize MIMO-UNet+ as the baseline model, featuring a single decoder per scale. Building upon this baseline, we increase the number of decoders to 2 and 4 in each scale, resulting in MIMO-UNet2D and MIMO-UNet4D, respectively.
b. Rotating and sharing decoders. Based on the previous experiment, in each scale of MIMO-UNet2D, we rotate the convolution kernels of one decoder by 90° and then share them with the other decoder, leading to the model SP-MIMO-UNet2D. Similarly, we rotate the ones belonging to MIMO-UNet4D by 90, 180, and 270° and share them with the three remaining decoders, respectively, resulting in ADMMDeblur.

All models in the above ablation experiments are trained until convergence and evaluated on the GoPro set. The quantitative results of both experiments are documented in Table 3.

Table 3. Performance of various number of decoders

Model	PSNR	SSIM	Params (M)
MIMO-UNet+	32.45	0.9567	**16.1**
MIMO-UNet2D	32.36	0.9565	24.1
MIMO-UNet4D	32.50	0.9578	39.9
SP-MIMO-UNet2D	32.66	0.9586	**16.1**
ADMMDeblur	**33.01**	**0.9614**	**16.1**

As indicated in Table 3, MIMO-UNet2D exhibits inferior performance compared to MIMO-UNet+, while MIMO-UNet4D outperforms MIMO-UNet+, suggesting that the absence of explicit constraints limits the potential of a multi-decoder structure. Besides, the comparison of the quantitative results of MIMO-UNet2D and SP-MIMO-UNet2D, as well as MIMO-UNet4D and ADMMDeblur implies that our employed rotation of convolution kernels replaces the need for explicit constraints, which effectively partitions the solution space of the network and enhances the deblurring accuracy of the model.

Specifically, although MIMO-UNet2D and SP-MIMO-UNet2D, as well as MIMO-UNet4D and ADMMDeblur, share the same structure, rotated and shared convolution kernels yield superior performance with nearly half the number of parameters. Without explicit constraints, the residuals learned by the decoders with independent parameters exhibit more redundant components from a vector perspective. SP-MIMO-UNet2D incorporates rotated and shared parameters but it has only two decoders, thus capturing merely the range that can be represented by two non-reversible orthogonal components. It is only in our proposed ADMMDeblur, which includes four decoders utilizing rotation-sharing parameters, that the decoders generate residuals capable of representing the motion of any direction within the entire 2D scene. Consequently, ADMMDeblur demonstrates the best performance among the evaluated models.

5 Conclusion

In this work, we present a novel single-image deblurring network that performs optimally for model size and accuracy. Unlike the conventional one-encoder-one-decoder structure employed in U-shaped networks for coarse-to-fine deblurring, our proposed approach adopts a one-encoder-four-decoder structure. This design enables the network to effectively decompose the original deblurring problem into four sub-problems, assigning each sub-problem to a dedicated decoder. Through the rotation and sharing operations of decoder parameters, the features generated by these four decoders encompass the entirety of a 2D scene, representing motion in all directions. The experimental results demonstrate that our method outperforms other conventional methods regarding model size and deblurring accuracy.

Acknowledgements. This work was supported by the National Natural Science Foundation of China under Grant 62176161, and the Scientific Research and Development Foundations of Shenzhen under Grant JCYJ20220818100005011 and 20200813144831001.

References

1. Chen, L., Lu, X., Zhang, J., Chu, X., Chen, C.: HINet: half instance normalization network for image restoration. In: Proceedings of the IEEE/CVF Conference on Computer Vision and Pattern Recognition, pp. 182–192 (2021)
2. Cho, S.J., Ji, S.W., Hong, J.P., Jung, S.W., Ko, S.J.: Rethinking coarse-to-fine approach in single image deblurring. In: Proceedings of the IEEE/CVF International Conference on Computer Vision, pp. 4641–4650 (2021)

3. Dai, J., Li, Y., He, K., Sun, J.: R-FCN: object detection via region-based fully convolutional networks. In: Advances in Neural Information Processing Systems, vol. 29 (2016)
4. Franke, U., Joos, A.: Real-time stereo vision for urban traffic scene understanding. In: Proceedings of the IEEE Intelligent Vehicles Symposium 2000 (Cat. No. 00TH8511), pp. 273–278. IEEE (2000)
5. Gao, H., Tao, X., Shen, X., Jia, J.: Dynamic scene deblurring with parameter selective sharing and nested skip connections. In: Proceedings of the IEEE/CVF Conference on Computer Vision and Pattern Recognition, pp. 3848–3856 (2019)
6. Hradis, M., Kotera, J., Zemcík, P., Sroubek, F.: Convolutional neural networks for direct text deblurring, vol. 10, September 2015. https://doi.org/10.5244/C.29.6
7. Hui, Z., Chakrabarti, A., Sunkavalli, K., Sankaranarayanan, A.C.: Learning to separate multiple illuminants in a single image. In: Proceedings of the IEEE/CVF Conference on Computer Vision and Pattern Recognition, pp. 3780–3789 (2019)
8. Ignatov, A., Kobyshev, N., Timofte, R., Vanhoey, K., Van Gool, L.: DSLR-quality photos on mobile devices with deep convolutional networks. In: Proceedings of the IEEE International Conference on Computer Vision, pp. 3277–3285 (2017)
9. Ji, S.W., et al.: XYDeblur: divide and conquer for single image deblurring. In: Proceedings of the IEEE/CVF Conference on Computer Vision and Pattern Recognition, pp. 17421–17430 (2022)
10. Jiao, J., Cao, Y., Song, Y., Lau, R.: Look deeper into depth: monocular depth estimation with semantic booster and attention-driven loss. In: Ferrari, V., Hebert, M., Sminchisescu, C., Weiss, Y. (eds.) ECCV 2018. LNCS, vol. 11219, pp. 55–71. Springer, Cham (2018). https://doi.org/10.1007/978-3-030-01267-0_4
11. Kendall, A., Gal, Y., Cipolla, R.: Multi-task learning using uncertainty to weigh losses for scene geometry and semantics. In: Proceedings of the IEEE Conference on Computer Vision and Pattern Recognition, pp. 7482–7491 (2018)
12. Krishnan, D., Tay, T., Fergus, R.: Blind deconvolution using a normalized sparsity measure. In: CVPR 2011, pp. 233–240. IEEE (2011)
13. Li, Z., Snavely, N.: Learning intrinsic image decomposition from watching the world. In: Proceedings of the IEEE Conference on Computer Vision and Pattern Recognition, pp. 9039–9048 (2018)
14. Liu, C., Ke, W., Qin, F., Ye, Q.: Linear span network for object skeleton detection. In: Ferrari, V., Hebert, M., Sminchisescu, C., Weiss, Y. (eds.) ECCV 2018. LNCS, vol. 11206, pp. 136–151. Springer, Cham (2018). https://doi.org/10.1007/978-3-030-01216-8_9
15. Nah, S., Hyun Kim, T., Mu Lee, K.: Deep multi-scale convolutional neural network for dynamic scene deblurring. In: Proceedings of the IEEE Conference on Computer Vision and Pattern Recognition, pp. 3883–3891 (2017)
16. Ott, J., Linstead, E., LaHaye, N., Baldi, P.: Learning in the machine: to share or not to share? Neural Netw. 126, 235–249 (2020)
17. Park, D., Kang, D.U., Kim, J., Chun, S.Y.: Multi-temporal recurrent neural networks for progressive non-uniform single image deblurring with incremental temporal training. In: Vedaldi, A., Bischof, H., Brox, T., Frahm, J.-M. (eds.) ECCV 2020. LNCS, vol. 12351, pp. 327–343. Springer, Cham (2020). https://doi.org/10.1007/978-3-030-58539-6_20
18. Purohit, K., Rajagopalan, A.: Region-adaptive dense network for efficient motion deblurring. In: Proceedings of the AAAI Conference on Artificial Intelligence, vol. 34, pp. 11882–11889 (2020)
19. Shen, Z., et al.: Human-aware motion deblurring. In: Proceedings of the IEEE/CVF International Conference on Computer Vision, pp. 5572–5581 (2019)

20. Sun, J., Cao, W., Xu, Z., Ponce, J.: Learning a convolutional neural network for non-uniform motion blur removal. In: Proceedings of the IEEE Conference on Computer Vision and Pattern Recognition, pp. 769–777 (2015)
21. Tao, X., Gao, H., Shen, X., Wang, J., Jia, J.: Scale-recurrent network for deep image deblurring. In: Proceedings of the IEEE Conference on Computer Vision and Pattern Recognition, pp. 8174–8182 (2018)
22. Thorpe, C., Li, F., Li, Z., Yu, Z., Saunders, D., Yu, J.: A coprime blur scheme for data security in video surveillance. IEEE Trans. Pattern Anal. Mach. Intell. 35(12), 3066–3072 (2013)
23. Whyte, O., Sivic, J., Zisserman, A.: Deblurring shaken and partially saturated images. Int. J. Comput. Vision 110, 185–201 (2014)
24. Zamir, S.W., Arora, A., Khan, S., Hayat, M., Khan, F.S., Yang, M.H.: Restormer: Efficient transformer for high-resolution image restoration. In: Proceedings of the IEEE/CVF Conference on Computer Vision and Pattern Recognition, pp. 5728–5739 (2022)
25. Zamir, S.W., et al.: Multi-stage progressive image restoration. In: Proceedings of the IEEE/CVF Conference on Computer Vision and Pattern Recognition, pp. 14821–14831 (2021)
26. Zhang, H., Dai, Y., Li, H., Koniusz, P.: Deep stacked hierarchical multi-patch network for image deblurring. In: Proceedings of the IEEE/CVF Conference on Computer Vision and Pattern Recognition, pp. 5978–5986 (2019)

TICondition: Expanding Control Capabilities for Text-to-Image Generation with Multi-Modal Conditions

Yuhang Yang, Xiao Yan, and Sanyuan Zhang[✉]

College of Computer Science and Technology, Zhejiang University, Hangzhou, China
{cs_yhyang,syzhang}@zju.edu.cn

Abstract. Text-to-image generation models have achieved significant advancements, enabling the generation of high-quality and diverse images. However, solely relying on text prompts often leads to limited control over image attributes. In this paper, we propose a method for achieving multifaceted control in image generation via text prompts, reference images, and control tags. Our goal is to ensure that generated images align not only with the text prompts but also with attributes indicated by control tags in reference images. To achieve this, we leverage Grounded-SAM and data augmentation to construct a paired training dataset. Using the BLIP-VQA model, we extract multimodal features guided by control tags. With lightweight TICondition, we derive new features at textual and image levels. These features are then injected into the frozen Diffusion model, facilitating control over the image's background, structure, or subject matter during the generation process. Our experimental findings indicate that our approach demonstrates heightened multifaceted control capabilities and yields commendable generation outcomes compared to merely relying on text prompts for image generation.

Keywords: Text-to-Image generation · Controlled image generation · Subject-driven · Multimodal · Diffusion model

1 Introduction

In recent years, text-to-image technology has demonstrated significant potential [23,26,34], enabling the creation of high-quality images. The Stable Diffusion model [24] presently produces high-quality and diversified images based on text prompts. However, due to the inherent challenge of precisely conveying images through text prompts [16], the relationship between text prompts and generated images is not one-to-one. While the resulting images exhibit diversity, user controllability is relatively diminished.

To enhance text-to-image generation controllability [3,13,29], current methods include Textual Inversion [8] and DreamBooth [25], achieving subject consistency by binding the subjects of reference images to unique identifiers. Additionally, adhering to the adage "a picture is worth a thousand words," techniques like

© The Author(s), under exclusive license to Springer Nature Switzerland AG 2024
S. Rudinac et al. (Eds.): MMM 2024, LNCS 14554, pp. 66–79, 2024.
https://doi.org/10.1007/978-3-031-53305-1_6

T2I-Adapter [17] and ControlNet [36] implement structural control over image generation by utilizing reference images.

Methods like Textual Inversion and DreamBooth often rely on limited sample training, leading to significant overfittings, such as mistaking certain background features for subject attributes. Meanwhile, all the aforementioned image generation control methods merely address specific aspects of images. Textual Inversion achieves control over image subjects, while T2I-Adapter focuses on structural control.

Fig. 1. With the TICondition, given a reference image and corresponding control tags, we can generate diverse images guided by text prompts. The generated images demonstrate a high consistency with the reference image concerning the attributes defined by the control tags, such as background, structure, and subject.

Therefore, we propose a unified model to facilitate multifaceted image generation control. We adopt the T2I-Adapter approach, leveraging reference images for control. Given that images inherently encapsulate extensive and intricate information [27], their control is inherently more specific and distinctive compared to text prompts. However, this specificity can potentially lead to trivial solutions when directly using raw reference image features. Thus, feature extraction requires purposefulness and emphasis.

In this paper, we introduce reference images and control tags to guide the targeted extraction of corresponding features from the reference images. We employ techniques like Grounded-SAM (which combines Grounding DINO [15], Segment Anything [12], and Stable Diffusion [24]) and data augmentation to preprocess the training data, generating multiple paired datasets from a limited number of samples to alleviate overfitting. For multifaceted control in image generation, we leverage the pre-trained BLIP-VQA [14] model to extract specific multimodal features guided by control tags. After alignment via an adapter, these features fuse with textual features corresponding to text prompts, forming new textual features. Additionally, the aligned multimodal features engage in cross-attention with features extracted from the reference image, refined by an AutoEncoder and image adapter, resulting in supplementary image features. Subsequently, these processed attributes are injected into the frozen Stable Diffusion model. Notably, during training, parameter updates focus exclusively on the new components of TICondition, preserving the original model's integrity while significantly reducing training time compared to comprehensive fine-tuning. Figure 1 displays our generated results.

The main contributions of this paper are listed as follows:

- We devise a simple approach for generating multiple paired datasets based on Grounded-SAM and data augmentation, which requires only a reference image and its corresponding text prompt.
- We introduce a unified reasoning framework based on text prompts, reference images, and control tags, enabling multifaceted control in image generation.
- We propose a lightweight TICondition, achieving efficient injection of reference information into internal features at a lower training cost while maintaining the generalization capability of the pre-trained Stable Diffusion model.
- Experimental results demonstrate the effectiveness of our approach in both consistent text-to-image generation and multifaceted image generation control.

2   Related Work

2.1   Text-to-Image Generation

In recent years, diffusion models have garnered increasing attention in Text-to-Image Generation research. Diffusion models [21,24,26,28], employing iterative refinement, have demonstrated the capacity to produce highly realistic images. Building upon this foundation, some studies have introduced Text-to-Image models based on diffusion [16,22,24,26]. To enhance the performance of diffusion models, novel approaches have emerged. For instance, GLIDE [18] introduced a cascaded diffusion structure, utilizing Classifier-free guidance [11] for image generation and editing. DALL-E2 [22] uses one prior model and one decoder model to generate images from CLIP [20] latent embeddings. Imagen [26] discovered that scaling up the size of the text encoder [5,6,20] improves both sample fidelity and text-image alignment. In addition to diffusion models, the Latent Diffusion Model, also known as Stable Diffusion Model [24], has also attracted significant attention. In this paper, we utilize the Stable Diffusion (SD) model as our foundational framework to explore the realization of multifaceted control in text-to-image generation.

2.2 Subject-Driven Text-to-Image Generation

The task of subject-driven image generation from text involves generating images based on a given subject, guided by text prompts to generate the subject in a new context. Several GAN-based models [4,7,19] have been dedicated to achieving this objective. Subsequently, Textual Inversion [8] introduced an approach that binds placeholder neologisms to subject entities and optimizes word embeddings. Similarly, DreamBooth [25] adopted a comparable strategy but fine-tuned the entire diffusion model, resulting in improved expressiveness and subject fidelity. However, as widely recognized, limited data often lead to overfitting [31,33], both methods mentioned above are susceptible to overfitting due to limited sample training. In contrast, our approach mitigates the overfitting issue effectively by generating paired data and designing the TICondition network.

2.3 Controllable Diffusion Models

A controllable diffusion model enables user guidance for generating results. Currently, methods often rely on example-based generation [1,32,35,37] to transform structural inputs into realistic images based on example content. Among them, Controlnet [36] and T2I-Adapter [17] propose introducing adapters into publicly available SD models. Fine-tuning the adapters while keeping the original SD model parameters fixed reduces training costs and improves image structure control. In contrast, our TICondition introduces not only reference images as supplementary conditions but also control tags, thereby achieving multifaceted control over aspects such as background, structure, and subject in text-to-image generation.

3 Method

Given a set of data pairs (I_r, T, P), our objective is to generate images I that meet specific conditions based on the data pairs. Here, I_r represents a reference image, T denotes the corresponding control tags for the reference image I_r, and P is a global text prompt. The generated image I should satisfy both the global text prompt P and the reference conditions C_r, where C_r is a multimodal representation of the T portion of I_r, encompassing aspects such as the background, structure, or subject of I_r.

To achieve these requirements, in this paper, we introduce the TICondition approach for achieving more controllable image generation. The schematic diagram of our method is illustrated in Fig. 2, comprising components such as BLIP-VQA, Image and Text Adapters, Feature Fusion Modules, and a pre-trained Stable Diffusion model.

3.1 Paired Data Preparation

The purpose of this paper is to generate the desired target image I that satisfies both the global text prompt P and the reference condition C_r. To achieve this, we require a training dataset composed of (I_s, P, I_r, T, I), where I_s represents the original image, P represents the global text prompt, I_r denotes the reference image, T represents the control tag, and I signifies the desired target image.

To construct this data, we initially prepare a base dataset of text and image pairing examples, from which we obtain I_s and its corresponding P. We set the target image I as I_s for ease of data construction. In this paper, the control tag T takes on three forms: "subject," "structure," and "background." As an illustration, we consider the case where T corresponds to "subject." We need to ensure consistency between I_r and I_s in terms of the "subject" while allowing variability in other aspects such as the "background" and "structure." For instance, we use mask-based random background replacement to control the background and employ randomized image deformations to facilitate structural variations.

When a change in the subject is required, we utilize a pre-trained Grounded-SAM model to generate a subject-specific mask, M, and a novel subject image, I_{new}, based on text prompts. By applying suitable augmentations, we obtain the corresponding I_r and M_r. Notably, M_r is a supervisory signal for model optimization.

3.2 Model Designs

To fulfill the requirements above, we introduce TICondition to achieve a more controllable image generation. Our approach is illustrated in Fig. 2.

Conditionally Guided Multimodal Feature Extraction. To leverage the reference image, I_r, and control tag, T, for guiding image generation, precise feature extraction from the image is crucial. Inspired by BLIP-VQA [14], which generates high-quality answers based on user-provided images and questions, we adopt it as a means to extract specific multimodal features.

Table 1. Control tags and corresponding question templates.

Control Tags	Question templates
subject	What is the main object of this picture?
structure	What is the structure of main object in this picture?
background	What is the background of this picture?

Specifically, BLIP-VQA takes image and question inputs. Depending on control tag T, the corresponding template question is selected from the template library (Table 1). For example, for the "subject" tag, a template like "What

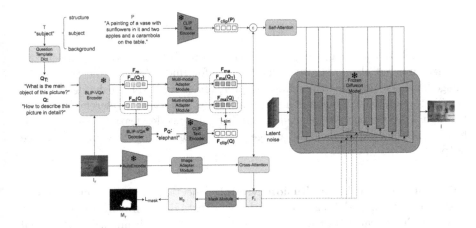

Fig. 2. The framework of our TICondition. The model accepts three inputs: text prompt P, control tag T, and reference image I_r. Through modules like multimodal feature extraction, adapters, self-attention, and cross-attention, text features and image features are injected separately into the Frozen Diffusion Model, enabling precise control generation of background, structure, and subject.

is the main object of this picture" is chosen. After input processing through a pre-trained BLIP-VQA model, the last encoder layer yields extracted features, forming the multimodal feature F_m.

Adapter Architecture. Directly injecting the multimodal feature F_m into the pre-trained SD model does not lead to improvements, as F_m is not aligned with the intermediate SD's features or CLIP features. Therefore, we introduce adapter modules for feature alignment.

The multimodal adapter module employs a lightweight adapter. The multimodal feature F_m is initially combined with a randomly shaped feature and processed through LayerNorm. This is followed by three residual attention modules [30] for feature extraction. Further LayerNorm operations are performed to generate the new feature.

To supervise feature alignment, we calculate a loss between the adapter-processed features and the CLIP features. As shown in Fig. 2, during training, we utilize a global template question Q, such as "How to describe this picture in detail" and feed it alongside the reference image I_r into the BLIP-VQA Encoder. This generates the corresponding multimodal feature $F_m(Q)$, further processed by the BLIP-VQA Decoder to obtain textual response P_Q. Employing a pre-trained CLIP model, P_Q yields the CLIP feature $F_{clip}(Q)$. Concurrently, the Multimodal Adapter processes F_m to produce $F_{ma}(Q)$. To align these features, we compute the cosine similarity between $F_{ma}(Q)$ and $F_{clip}(Q)$, expressed as follows:

$$L_{sim} = \frac{F_{ma}(Q) \cdot F_{clip}(Q)}{\|F_{ma}(Q)\| \cdot \|F_{clip}(Q)\|} \tag{1}$$

Inspired by the image adapter in T2I-Adapter [17], we not only inject textual features but also integrate image features for guided feature injection on the image level. We adjusted the image adapter, removing Pixel Unishuffle and converting the multi-layer outputs into a single layer. We pass the reference image I_r through a pre-trained AutoEncoder, acquiring its encoded feature. This encoded feature is then input into the image adapter, producing new image feature outputs.

Injection into Stable Diffusion Model. Injection in this paper occurs at both the textual and image feature levels. Textual feature injection involves merging CLIP features of the global text prompt P with aligned multimodal features $F_{ma}(Q_T)$. These concatenated features undergo self-attention and are injected into the SD module as new textual features, following the same injection methodology as in SD, with new features introduced at each layer.

For image-level feature injection, aligned multimodal features $F_{ma}(Q_T)$ and ImgAdapter-processed features are input into a cross-attention module, generating four distinct intermediate outputs F_I at varied resolutions. The cross-attention module here resembles the SD model's mechanism, aligning at the image level. Aligned multimodal features, F_{ma}, correspond to CLIP features, while ImgAdapter-processed features match SD's intermediate features. Inspired by [3,29], we opt to inject features during the middle layers of the decoding phase and at each distinctive feature scale for image-level feature injection.

For precise region-of-interest generation, we further pass the features obtained from the cross-attention module through a mask feature extraction module. The resulting mask feature, M_p, is then supervised against the ground truth mask, M_r, through binary cross-entropy (BCE) loss, as shown below:

$$L_{mask} = \sum_{i=1}^{n} bce_i, n = 4$$

$$bce_i = \frac{1}{N} \sum_{j=1}^{N} \left[-M_p^{ij} \cdot \log(M_r^{ij}) - (1 - M_p^{ij}) \cdot \log(1 - M_r^{ij}) \right] \tag{2}$$

Classifier-Free Guidance for Mul-Conditionings. Classifier-free diffusion guidance [11] is an approach that balances the quality and diversity of samples generated by diffusion models.

Inspired by InstructPix2Pix [2], we introduce three guiding scales, s_T, s_{NT}, and s_I, which can be adjusted to balance the correspondence between the generated image and the input image, their correspondence with the text prompt, and their correspondence with the reference image. Our corrected score estimate is formulated as follows:

$$\begin{aligned}
\tilde{e}_\theta\left(z_t, c_{NT}, c_I\right) = & e_\theta\left(z_t, \varnothing, \varnothing\right) \\
& + s_T \cdot \left(e_\theta\left(z_t, c_T, \varnothing\right) - e_\theta\left(z_t, \varnothing, \varnothing\right)\right) \\
& + s_{NT} \cdot \left(e_\theta\left(z_t, c_{NT}, \varnothing\right) - e_\theta\left(z_t, c_T, \varnothing\right)\right) \\
& + s_I \cdot \left(e_\theta\left(z_t, c_{NT}, c_I\right) - e_\theta\left(z_t, c_{NT}, \varnothing\right)\right)
\end{aligned} \quad (3)$$

Where c_T represents the features provided by the original text prompt P, c_{NT} is the newly injected text feature obtained through multimodal fusion, and c_I denotes the injected new image feature F_I.

Loss Function. During the optimization process, we keep the parameters of SD fixed and only optimize the newly added modules in TICondition. The overall loss is defined as follows:

$$L = \lambda_{sim} L_{sim} + \lambda_{mask} L_{mask} + \lambda_{ldm} L_{ldm} \quad (4)$$

λ_{sim}, λ_{mask}, and λ_{ldm} represent the weights for each loss term. Similar to SD, L_{ldm} is optimized based on image-condition pairs, new text features, and image features using the following procedure:

$$L_{ldm} = \mathbb{E}_{\mathcal{E}(x), c_I, c_{NT}, \epsilon \sim \mathcal{N}(0,1), t}\left[\|\epsilon - \epsilon_\theta\left(z_t, t, c_I, c_{NT}\right)\|_2^2\right] \quad (5)$$

L_{sim} (Eq. 1) aims to align the adapter-processed features with the CLIP features. L_{mask} (Eq. 2) ensures that the image injection focuses on the relevant regions of interest.

4   Experiment

4.1   Implementation and Experiment Details

This section outlines the implementation details of our approach. We utilize the Adam optimizer with a learning rate of 10^{-5}. The training was conducted on an RTX3090, employing a batch size of 1. In Eq. 4, λ_{sim} was set to 1, λ_{mask} to 0.1, and λ_{ldm} to 1.

We compiled a dataset from [8,13,25] and supplemented it with internet-sourced content, covering 15 categories and 45 subjects, including dogs, cats, perfumes, bicycles, and others. The dataset underwent annotation and manual BLIP curation to ensure precise image-text alignment.

We evaluated our method using the following metrics: (1) Image-alignment [8], assessing the visual similarity between generated images and target concepts using similarity metrics within the CLIP image feature space; (2) Text-alignment [9], measuring the alignment between generated images and provided text prompts using text-image similarity within the CLIP feature space; (3) FID [10], which quantifies the similarity between generated and real images.

4.2   Experimental Results

Main Qualitative Results. Our method covers multiple control aspects: background, structure, and subject. Figure 3 showcases image generation under distinct controls, yielding favorable outcomes and demonstrating strong compatibility and generalization.

Comparison Studies. In this section, we compare our approach with several state-of-the-art methods. Given our method's ability to achieve controlled generation across three distinct aspects, we conduct comparisons from three different perspectives accordingly.

In subject control, we compare our approach with Textual Inversion [8] and DreamBooth [25], employing Image-alignment and Text-alignment as quantitative metrics. Table. 2 outlines specific metrics, showing our method's superior Text-alignment performance while slightly trailing Textual Inversion in Image-alignment. Figure 4(a) visually contrasts the approaches, highlighting our solution's effectiveness in addressing overfitting.

In terms of structural control, we benchmark our method against T2I-Adapter [17] and ControlNet [36], utilizing Image-alignment, Text-alignment, and FID as quantitative assessment metrics. Table 3 highlights that our approach excels in FID and Text-alignment, albeit slightly trailing behind the others in Image-alignment. Figure 4(b) visually demonstrates our method's impressive structural control and generation prowess on the training dataset. Remarkably, our model achieves reasonable structural control without an extra structural extraction network even in a few-shot scenario.

For background control, we contrast our method with Paint-by-example [32]. As shown in Table 4, without needing extra masks for precise localization, our model slightly lags behind Paint-by-example in FID and Image-alignment. Nevertheless, our model enjoys a marginal Text-alignment advantage over Paint-by-example. Figure 4(c) demonstrates results under background control, revealing our method's superior coherence and harmony compared to Paint-by-example.

Ablation Studies. To gain a deeper understanding of TICondition, we conducted ablation experiments to evaluate our crucial design, which involves leveraging joint injection at both image and text levels and utilizing mask-guided supervision for achieving high-quality image-controlled generation.

The ablation experiment in Fig. 5(a) highlights the importance of joint injection at text and image levels. Solely relying on text-level injection affects background consistency, while exclusive image-level injection yields less harmonious results. This emphasizes the crucial role of simultaneous injection at both levels.

Figure 5(b) visually illustrates the impact of mask-guided supervision on structural generation. Guided generation with mask supervision leads to the generation of better images, capturing finer details.

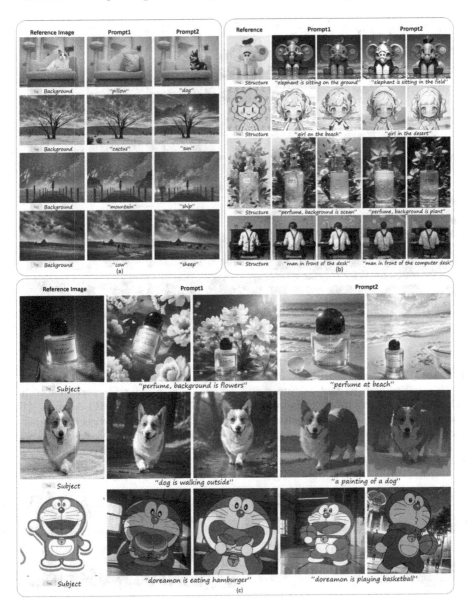

Fig. 3. (a) Qualitative results under background control. (b) Qualitative results under structure control. (c) Qualitative results under subject control.

Fig. 4. (a) Comparisons with Textual Inversion [8] and DreamBooth [25]. Our approach achieves high prompt relevance and demonstrates comparable performance in subject consistency compared to other models. (b) Comparison with T2I-Adapter [17] and ControlNet [36]. Our approach achieves high structure control and generation performance, and even in few-shot scenarios, we can achieve rudimentary structural control. (c) Comparison with Paint-by-Example. Our approach achieves favorable overall coherence and harmony.

Fig. 5. (a) Ablation study on Text-level and Image-level in TICondition. It shows the generated images with the guidance of Text-level Injection, Image-level Injection, and both, respectively. (b) Ablation study of mask loss supervision in TICondition.

Table 2. Quantitative analysis of our TICondition in subject control compared to existing methods. Best results are highlighted in bold.

Method	Image-alignment	Text-alignment
Textual Inversion	**0.742**	0.304
DreamBooth	0.727	0.346
Ours	0.739	**0.354**

Table 3. Quantitative analysis of our TICondition in structure control compared to existing methods. Best results are highlighted in bold.

Method	FID	Image-alignment	Text-alignment
T2I-Adapter	41.267	0.768	0.248
ControlNet	38.476	**0.781**	0.252
Ours	**37.192**	0.750	**0.257**

Table 4. Quantitative analysis of our TICondition in background control compared to existing methods. Best results are highlighted in bold.

Method	FID	Image-alignment	Text-alignment
Paint-by-example	**38.895**	**0.739**	0.315
Ours	43.419	0.709	**0.324**

5   Conclusions and Future Work

In this paper, we introduce TICondition to achieve multifaceted control, encompassing structure, background, and subject attributes during image generation. Using text prompts, reference images, and control tags as conditionings, TICondition harnesses BLIP-VQA for extracting multimodal information, enhancing precision in control. Employing Grounded-SAM and data augmentation, we create paired sample batches to alleviate overfitting. TICondition offers a cost-effective fine-tuning model by freezing pre-trained components during training. Experimental results demonstrate TICondition's robust control and image generation prowess, yielding high-fidelity outputs closely aligned with text prompts.

Currently, we have achieved multifaceted controlled generation. Our future focus is on extending the scope of control mechanisms. This extension may require not only reliance on a robust multimodal representation, but also further refinement of the model architecture.

References

1. Bhunia, A.K., et al.: Person image synthesis via denoising diffusion model. In: Proceedings of the IEEE/CVF Conference on Computer Vision and Pattern Recognition, pp. 5968–5976 (2023)

2. Brooks, T., Holynski, A., Efros, A.A.: InstructPix2Pix: learning to follow image editing instructions. In: Proceedings of the IEEE/CVF Conference on Computer Vision and Pattern Recognition, pp. 18392–18402 (2023)
3. Cao, M., Wang, X., Qi, Z., Shan, Y., Qie, X., Zheng, Y.: MasaCtrl: tuning-free mutual self-attention control for consistent image synthesis and editing. arXiv preprint arXiv:2304.08465 (2023)
4. Casanova, A., Careil, M., Verbeek, J., Drozdzal, M., Romero Soriano, A.: Instance-conditioned GAN. Adv. Neural. Inf. Process. Syst. 34, 27517–27529 (2021)
5. Devlin, J., Chang, M.W., Lee, K., Toutanova, K.: BERT: pre-training of deep bidirectional transformers for language understanding. arXiv preprint arXiv:1810.04805 (2018)
6. Dhariwal, P., Nichol, A.: Diffusion models beat GANs on image synthesis. Adv. Neural. Inf. Process. Syst. 34, 8780–8794 (2021)
7. Dong, Z., Wei, P., Lin, L.: DreamArtist: towards controllable one-shot text-to-image generation via contrastive prompt-tuning. arXiv preprint arXiv:2211.11337 (2022)
8. Gal, R., et al.: An image is worth one word: personalizing text-to-image generation using textual inversion. arXiv preprint arXiv:2208.01618 (2022)
9. Hessel, J., Holtzman, A., Forbes, M., Bras, R.L., Choi, Y.: CLIPScore: a reference-free evaluation metric for image captioning. arXiv preprint arXiv:2104.08718 (2021)
10. Heusel, M., Ramsauer, H., Unterthiner, T., Nessler, B., Hochreiter, S.: GANs trained by a two time-scale update rule converge to a local nash equilibrium. In: Advances in Neural Information Processing Systems, vol. 30 (2017)
11. Ho, J., Salimans, T.: Classifier-free diffusion guidance. arXiv preprint arXiv:2207.12598 (2022)
12. Kirillov, A., et al.: Segment anything. arXiv preprint arXiv:2304.02643 (2023)
13. Kumari, N., Zhang, B., Zhang, R., Shechtman, E., Zhu, J.Y.: Multi-concept customization of text-to-image diffusion. In: Proceedings of the IEEE/CVF Conference on Computer Vision and Pattern Recognition, pp. 1931–1941 (2023)
14. Li, J., Li, D., Xiong, C., Hoi, S.: BLIP: bootstrapping language-image pre-training for unified vision-language understanding and generation. In: International Conference on Machine Learning, pp. 12888–12900. PMLR (2022)
15. Liu, S., et al.: Grounding DINO: marrying DINO with grounded pre-training for open-set object detection. arXiv preprint arXiv:2303.05499 (2023)
16. Liu, X., et al.: More control for free! Image synthesis with semantic diffusion guidance. In: Proceedings of the IEEE/CVF Winter Conference on Applications of Computer Vision, pp. 289–299 (2023)
17. Mou, C., et al.: T2i-adapter: learning adapters to dig out more controllable ability for text-to-image diffusion models. arXiv preprint arXiv:2302.08453 (2023)
18. Nichol, A., et al.: Glide: towards photorealistic image generation and editing with text-guided diffusion models. arXiv preprint arXiv:2112.10741 (2021)
19. Nitzan, Y., et al.: MyStyle: a personalized generative prior. ACM Trans. Graph. (TOG) 41(6), 1–10 (2022)
20. Radford, A., et al.: Learning transferable visual models from natural language supervision. In: International Conference on Machine Learning, pp. 8748–8763. PMLR (2021)
21. Raffel, C., et al.: Exploring the limits of transfer learning with a unified text-to-text transformer. J. Mach. Learn. Res. 21(1), 5485–5551 (2020)
22. Ramesh, A., Dhariwal, P., Nichol, A., Chu, C., Chen, M.: Hierarchical text-conditional image generation with clip latents. arXiv preprint arXiv:2204.06125 (2022)

23. Ramesh, A., et al.: Zero-shot text-to-image generation. In: International Conference on Machine Learning, pp. 8821–8831. PMLR (2021)
24. Rombach, R., Blattmann, A., Lorenz, D., Esser, P., Ommer, B.: High-resolution image synthesis with latent diffusion models. In: Proceedings of the IEEE/CVF Conference on Computer Vision and Pattern Recognition, pp. 10684–10695 (2022)
25. Ruiz, N., Li, Y., Jampani, V., Pritch, Y., Rubinstein, M., Aberman, K.: Dreambooth: fine tuning text-to-image diffusion models for subject-driven generation. In: Proceedings of the IEEE/CVF Conference on Computer Vision and Pattern Recognition, pp. 22500–22510 (2023)
26. Saharia, C., et al.: Photorealistic text-to-image diffusion models with deep language understanding. Adv. Neural. Inf. Process. Syst. **35**, 36479–36494 (2022)
27. Sheikh, H.R., Bovik, A.C.: Image information and visual quality. IEEE Trans. Image Process. **15**(2), 430–444 (2006)
28. Tang, Z., Gu, S., Bao, J., Chen, D., Wen, F.: Improved vector quantized diffusion models. arXiv preprint arXiv:2205.16007 (2022)
29. Tumanyan, N., Geyer, M., Bagon, S., Dekel, T.: Plug-and-play diffusion features for text-driven image-to-image translation. In: Proceedings of the IEEE/CVF Conference on Computer Vision and Pattern Recognition, pp. 1921–1930 (2023)
30. Wang, F., et al.: Residual attention network for image classification. In: Proceedings of the IEEE Conference on Computer Vision and Pattern Recognition, pp. 3156–3164 (2017)
31. Wu, S., Yan, X., Liu, W., Xu, S., Zhang, S.: Self-driven dual-path learning for reference-based line art colorization under limited data. IEEE Trans. Circ. Syst. Video Technol. (2023)
32. Yang, B., et al.: Paint by example: exemplar-based image editing with diffusion models. In: Proceedings of the IEEE/CVF Conference on Computer Vision and Pattern Recognition, pp. 18381–18391 (2023)
33. Ying, X.: An overview of overfitting and its solutions. J. Phys. Conf. Ser. **1168**, 022022 (2019)
34. Yu, J., et al.: Scaling autoregressive models for content-rich text-to-image generation. arXiv preprint arXiv:2206.10789 2(3), 5 (2022)
35. Zhan, F., Yu, Y., Wu, R., Zhang, J., Lu, S., Zhang, C.: Marginal contrastive correspondence for guided image generation. In: Proceedings of the IEEE/CVF Conference on Computer Vision and Pattern Recognition, pp. 10663–10672 (2022)
36. Zhang, L., Agrawala, M.: Adding conditional control to text-to-image diffusion models. arXiv preprint arXiv:2302.05543 (2023)
37. Zhang, P., Zhang, B., Chen, D., Yuan, L., Wen, F.: Cross-domain correspondence learning for exemplar-based image translation. In: Proceedings of the IEEE/CVF Conference on Computer Vision and Pattern Recognition, pp. 5143–5153 (2020)

Enhancing Generative Generalized Zero Shot Learning via Multi-Space Constraints and Adaptive Integration

Zhe Kong[1,2], Neng Gao[1(✉)], Yifei Zhang[1], and Yuhan Liu[1,2]

[1] State Key Laboratory of Information Security, Institute of Information Engineering, CAS, Beijing, China
{kongzhe,gaoneng,zhangyifei,liuyuhan}@iie.ac.cn
[2] School of Cyber Security, University of Chinese Academy of Sciences, Beijing, China

Abstract. Generalized zero shot learning (GZSL) aims to recognize both seen classes and unseen classes without labeled data. Generative GZSL models strive to synthesize features of unseen classes given semantic embeddings as input. The key lies in retaining the semantic consistency and discriminative ability of generated features to approach the real feature distribution. In this paper, we tackle these challenges by introducing additional spaces apart from original image feature space to constrain the generation process and propose a GAN-based model called f-CLSWGAN-VAE2. Specifically, we incorporate cross-modal aligned variational auto encoders (VAE2) in our model to align generated features and corresponding semantic embeddings in semantic alignment space encouraging the generator to synthesize semantically-consistent features. Besides, we resort to contrastive learning in another image embedding space to alleviate confusion between generated instances. Moreover, we pioneer to propose a novel adaptive integration strategy to adaptively weight the classification results from multiple spaces by means of a binary classifier trained in semantic alignment space. Experiments on four popular GZSL datasets indicate that our model significantly outperformed baseline and achieved comparable results with other methods.

Keywords: Generalized Zero Shot Learning · Feature Generation · Multi-Space Constraints · Adaptive Integration

1 Introduction

Traditional image classification models [11] are trained and tested on data from a same set of classes that are independent and identically distributed. In generalized zero shot learning scenario, models are trained with labeled data of seen classes and tested on classes that may not be covered by the training data [15,24]. In other words, GZSL models are expected to recognize instances not only from seen classes but also from unseen classes. Serving as a bridge between seen classes and unseen classes, shared semantic space is introduced to

© The Author(s), under exclusive license to Springer Nature Switzerland AG 2024
S. Rudinac et al. (Eds.): MMM 2024, LNCS 14554, pp. 80–93, 2024.
https://doi.org/10.1007/978-3-031-53305-1_7

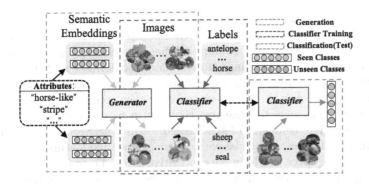

Fig. 1. The basic framework of generative GZSL models. A generator is trained on semantic embedding and image pairs of seen classes. Once trained, the generator takes semantic embeddings of unseen classes as input to complete images of them which are then combined with images of seen classes to train a supervised classifier. At test time, test images could be recognized by the trained classifier. Best viewed in color.

transfer knowledge between classes. Common class-specific semantic embeddings in semantic space include attributes [5], word vectors [7,19], etc.

Although challenging, tremendous works have made significant progress in improving the performance of GZSL models [18,33]. Early GZSL models are mainly embedding-learning-based [1,14,16] suffering from the so-called domain shift problem [8]. To mitigate the bias problem, recent works view GZSL as a missing data problem and propose generative GZSL models. The key insight to these works is to synthesize features of unseen classes by means of semantic embeddings. As depicted in Fig. 1, a generator is trained with pairs of semantic embeddings and image features from seen classes [21,23,31]. Subsequently, the generator takes as input semantic embeddings of unseen classes and output image features of unseen classes so that GZSL could be converted to a supervised image classification problem. In generative GZSL models, the generation process could be modeled using conditional variational autoencoders (cVAEs) [21] or conditional generative adversarial networks (cGANs) [20]. For example, Xian et al. [31] incorporated a discriminative classifier in the original GAN and proposed f-CLSWGAN.

The key to generative GZSL models lies in retaining the semantic consistency and discriminative ability of generated features to approach the real data distribution. On one hand, constraining the semantic consistency of generated features prevents the loss of semantic information during the generation process which is beneficial for generating semantically rich features for unseen classes. To guarantee generated features are consistent with corresponding semantic embeddings, cycle-WGAN [6] and TF-VAEGAN [23] introduced an additional constraint by ensuring generated features could be mapped back to corresponding semantic embeddings. However, as the mapping is trained with labeled data of seen classes, directly reducing the euclidean distance of mapped generated features of unseen classes and corresponding semantic embeddings may not be well suited for enhancing semantic consistency. In this paper, alternatively, we resort to cross-modal VAEs [26] to model the relationship between image features and semantic embeddings. On the assumption that the relationship between image features and semantic embeddings of seen classes conforms with that of unseen classes, we enforce the alignment of generated features and corresponding semantic embeddings in semantic alignment space. On the other hand, improving the discriminability of generated features contributes to subsequent classification process. To alleviate confusion of synthesized features between classes and facilitate training a discriminative classifier, we map generated features into another image embedding space and leverage contrastive loss [4,17] to facilitate the proximity of generated features among similar classes while keeping them distant from other classes. Finally, to take full advantage of features from various spaces, we train softmax classifiers in original image feature space, semantic alignment space and image embedding space respectively. Based on the observation that classification results in different spaces have different inclinations towards seen and unseen classes, we train an additional binary classifier in semantic alignment space to distinguish between seen classes and unseen classes. In this way, we could leverage different weighting strategies to adaptively integrate classification results from multiple spaces.

Our contributions are as follows: (1) We explore a novel approach to constrain semantic consistency of generative GZSL models. By incorporating cross-modal aligned VAEs in GAN-based architecture, the generator could generate semantically consistent features which is critical in generalized zero shot learning. (2) To address the issue that generated features may exhibit confusion with one another, we introduce another image embedding space to enhance the discriminability between generated instances. (3) We pioneer to propose a novel method to adaptively integrate classification results in various space. (4) Experiments on popular GZSL datasets indicate that our model significantly outperformed baseline and empirical analysis confirmed the rationality and effectiveness of multi-space constraints and adaptive integration.

2   Related Work

2.1   Embedding Learning Based GZSL Models

Early GZSL models focus on learning mappings between image feature space and semantic space. Depending on the space where embeddings are learned, these works are divided into (1) mapping from image feature space to semantic space [28] (2) mapping from semantic space to image feature space [3] and (3) mapping them into another common space [1,7]. Similarity measurement is performed in mapping space to align cross-modal features. For instance, semantic autoencoder [14] first mapped image features into semantic space and then mapped semantic embeddings back into image feature space, distance measurement could be performed in both space. CADA-VAE [26] encoded image features and attributes into a common latent space and aligned them there. Apart from works mentioned above that only utilize global features of images, attention-based methods [13,35] learn mappings between local features and semantic embeddings. For instance, Zhu et al. [35] weighted features of different channels obtaining attention maps to crop original images. DAZLE [13] regarded word vectors of attribute names as prototypes and measured the similarity between weighted image features and these prototypes. Local image features utilized in attention-based models are more discriminative and achieve better semantic alignment.

2.2 Generative GZSL Models

As primitive methods to solve GZSL problem, embedding learning based methods suffer from domain shift problem [8]. Projection learned from seen classes will introduce a bias when utilized on unseen classes. Alternatively, generative GZSL models leverage VAEs or GANs to synthesize image features of unseen classes from semantic embeddings. Once completed, missing image features of unseen classes could be merged with existing image features of seen classes to train a final supervised classifier. [21] sought to train a cVAE whose decoder is conditioned on semantic embeddings to generate samples of unseen classes. Since Wasserstein GAN (WGAN) [9] helps stable training of GANs, tremendous works based on WGAN have emerged. Xian proposed f-CLSWGAN [31] and f-VEAGAN [32] successively. The former incorporated a discriminative classifier in WGAN to guarantee generated features are well suited for training a discriminative classifier. The latter combined VAE with GAN to prompt a more stable generation process. Cycle-WGAN [6] and TF-VAEGAN [23] introduced a regressor as a new regularization for WGAN enforcing generated image features to reconstruct corresponding semantic embeddings. [2] equipped GAN with traditional ZSL models with closed-form solutions as probe model to improve the performance of the generator. [10] mapped generated features to embedding space and utilized class-level and instance-level supervision to obtain contrastive embeddings. [12] utilized attention-based embedding module to guide the generation process in order to boost the discrimination and to enrich semantics of generated features.

2.3 Other Methods

In addition to models mentioned above, [27] integrated GAN with embedding network and promoted mutual learning. [22,34] leveraged graph structures to output classifier parameters.

3 Method

3.1 Problem Formulation

As mentioned earlier, generalized zero shot learning (GZSL) strives to recognize objects from both seen and unseen classes whose labeled data are absent at training time. Given $\mathcal{T} = \{(x_i, y_i) \mid x \in \mathcal{X}, y \in \mathcal{Y}_s\}_{i=1}^{N}$ at training time where

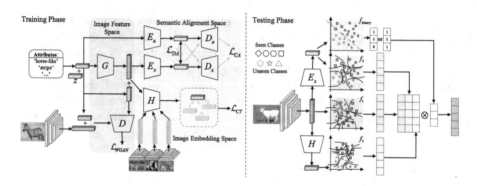

Fig. 2. Training f-CLSWGAN-VAE2 with multi-space constraints and testing by adaptively integrating classification results from mutiple spaces. Best viewed in color.

(x_i, y_i) stands for image feature and label pairs from training set $\mathcal{T}$, $\mathcal{X}$ denotes image feature space and $\mathcal{Y}_s$ is the label set of seen classes. The goal of GZSL models is to learn a classifier: $f : \mathcal{X} \rightarrow \mathcal{Y}_s \bigcup \mathcal{Y}_u$ where $\mathcal{Y}_u$ stands for the label set of unseen classes and $\mathcal{Y}_s \bigcap \mathcal{Y}_u = \emptyset$. Also, a semantic embedding set $A = \{a(y)|\forall y \in \mathcal{Y}_s \bigcup \mathcal{Y}_u\}$ is shared by all classes and utilized by GZSL models to transfer knowledge between seen classes and unseen classes.

3.2 Base Architecture: F-CLSWGAN

As the base architecture of our model, f-CLSWGAN [31] is a basic generation model extending conditional Wasserstein GAN (cWGAN) with a pretrained classifier. In f-CLSWGAN, the generator $G(z, a(y))$ takes a random noise $z \sim \mathcal{N}(0, 1)$ and semantic embedding $a(y)$ as input, and output an image feature $\tilde{x} \in \mathcal{X}$ of class y. The discriminator $D(x, a(y))$ scores the input feature x to determine its realness degree and matching degree with $a(y)$. Besides, to guarantee the generated features well suited for training a discriminative classifier, a classification loss is minimized over generated features. The whole model optimizes:

$$\mathcal{L}_{WGAN} = \min_G \max_D \left(\mathbb{E}[D(x, a(y))] - \mathbb{E}[D(\tilde{x}, a(y))] \right)$$
$$- \lambda \mathbb{E} \left[(\|\nabla_{\hat{x}} D(\hat{x}, a(y))\|_2 - 1)^2 \right] \qquad (1)$$
$$- \beta(\mathbb{E}[\log P(y \mid \tilde{x})] + \mathbb{E}[\log P(y \mid x)]),$$

where $\tilde{x} = G(z, a(y))$. The second term is the gradient penalty term [9] with $\hat{x} = \alpha x + (1 - \alpha)\tilde{x}$ and $\alpha \sim U(0, 1)$. The last two terms denote the classification loss. λ and β are hyperparameters weighting different losses.

Albeit the powerful generative capability, f-CLSWGAN could not guarantee the generated features are semantically-consistent. Moreover, whereas f-CLSWGAN encouraged G to generate discriminative features by virtue of a pretrained classifier, it failed to constrain the relationship between generated instances.

3.3   Cross-Modal Alignment in Semantic Alignment Space

To guarantee the generated features could well match corresponding semantic embeddings so that the generation process could be in a semantically-consistent manner, we introduce two variational autoencoders VAE_x for image features and VAE_a for semantic embeddings. Instead of enforcing the synthesized feature $\tilde{x}$ could be mapped back as corresponding semantic embedding $a(y)$ like Cycle-WGAN [32] did, we resort to these two VAEs to explicitly model the relationship between them through cross-modal interaction. Once aligned, VAEs could offer an additional regularization for the generation process encouraging G to synthesize semantically-consistent features. It is worth noting that directly training a mapping network on image feature and semantic embedding pairs from seen classes to constrain the generation process of unseen classes may introduce additional errors further aggravating the domain shift problem.

VAE_x strives to approximate the posterior distribution over the latent variables given image features and is comprised of an image feature encoder E_x and a decoder D_x. Analogously, VAE_a consists of a semantic embedding encoder E_a and a decoder D_a. The basic VAE loss could be formulated as:

$$
\begin{aligned}
\mathcal{L}_{VAE} = \; & \mathbb{E}_{q_\phi(z|x)}\left[\log p_\theta(x \mid z)\right] - \delta D_{KL}\left(q_\phi(z \mid x) \| p_\theta(z)\right) \\
& + \mathbb{E}_{q_\phi(z|a)}\left[\log p_\theta(a \mid z)\right] - \delta D_{KL}\left(q_\phi(z \mid a) \| p_\theta(z)\right)
\end{aligned}
\tag{2}
$$

where the encoder is denoted as $q_\phi(z|x)$ or $q_\phi(z|a)$ and the decoder is denoted as $p_\theta(x|z)$ or $p_\theta(a|z)$. D_{KL} is Kullback-Leibler divergence.

$\mathcal{L}_{VAE}$ enables VAE_x and VAE_a to learn meaningful representations in respective latent space. Furthermore, to align VAEs from different modalities, following [26], we add two additional regularization terms: cross-reconstruction loss and distribution-alignment loss. The former enforces latent embeddings from E_x could be reconstructed into semantic embeddings of the same class by D_a and vice versa. The latter minimizes the Wasserstein distance of latent multivariate Gaussian distributions from these two modalities. The cross-reconstruction loss could be written as:

$$
\mathcal{L}_{CA} = |a - D_a\left(E_x\left(x\right)\right)| + |x - D_x\left(E_a\left(a\right)\right)|
\tag{3}
$$

The distribution-alignment loss can be formulated as:

$$
\mathcal{L}_{DA} = \left(\|\mu_x - \mu_a\|_2^2 + \left\| \Sigma_x^{\frac{1}{2}} - \Sigma_a^{\frac{1}{2}} \right\|_{\text{Frobenius}}^2 \right)^{\frac{1}{2}}
\tag{4}
$$

where μ and Σ are predicted by E. The overall loss to train cross-modal aligned VAEs is:

$$
\mathcal{L}_{CA-VAE} = \mathcal{L}_{VAE} + \eta \mathcal{L}_{CA} + \gamma \mathcal{L}_{DA}
\tag{5}
$$

Once trained, VAE_x and VAE_a could serve as an evaluator to judge how well the generated features align with their corresponding semantic embeddings. We utilize them to constrain the generation process. Specifically, the latent embeddings of generated features $E_x(G(z, a(y)))$ should be reconstructed as corresponding semantic embeddings $a(y)$ by D_a. Also, the generated features should be able to be reconstructed from corresponding semantic embeddings. The loss function that caters to the above requirements is:

$$\mathcal{L}_{CA}(G) = |a - D_a\left(E_x\left(G(z,a)\right)\right)| + |G(z,a) - D_x\left(E_a\left(a\right)\right)| \tag{6}$$

Moreover, the latent embeddings of generated features should approach those of corresponding semantic embeddings. Their Wasserstein distance should be minimized:

$$\mathcal{L}_{DA}(G) = \left(\left\| \mu_{\tilde{x}} - \mu_a \right\|_2^2 + \left\| \Sigma_{\tilde{x}}^{\frac{1}{2}} - \Sigma_a^{\frac{1}{2}} \right\|_{\text{Frobenius}}^2 \right)^{\frac{1}{2}} \tag{7}$$

With the above constraints in semantic alignment space, the generator is expected to synthesize semantically-consistent features.

3.4   Contrastive Learning in Image Embedding Space

After restricting the relationship between generated features and corresponding semantic embeddings, we incorporate an image embedding net H in our framework to constrain relationships between generated features of the same class and those of different classes so as to alleviate confusion between instances. Image embedding net maps an image feature x_i to embedding space and obtain $h_i = H(x_i)$. In image embedding space, we define a contrastive loss [10,17] to compress inter-class similarities and increase intra-class similarities. Specifically, for an image feature from real data x_i or $\tilde{x}_i$ from G with label y_i, we set $h_i = H(x_i)$ or $h_i = H(\tilde{x}_i)$ as an anchor in embedding space. Then we randomly select another image feature x^+ with the same label y_i and project it into embedding space obtaining $h^+ = H(x^+)$ as the positive sample. Finally, K image features $\{x_1^-, x_2^-, \ldots, x_K^-\}$ with different labels are selected and then be mapped as negative samples $\{h_1^-, h_2^-, \ldots, h_K^-\}$. The objective is to increase the similarity between the anchor h_i and the positive sample while reducing the similarity between the anchor and any negative samples. This equals to minimize:

$$\mathcal{L}_{CT} = \mathbb{E}_{(h_i, h^+)} \left[-\log \frac{\exp\left(h_i h^+ / \tau_e\right)}{\exp\left(h_i h^+ / \tau_e\right) + \sum_{k=1}^{K} \exp\left(h_i h_k^- / \tau_e\right)} \right] \tag{8}$$

where $h_i = H(x_i)$ or $h_i = H(G(z,a))$ and $\tau_e > 0$ is the temperature parameter.

3.5   Generalized Zero Shot Classification in Various Spaces

The overall loss for training the generator is:

$$\mathcal{L} = \mathcal{L}_{WGAN} + \xi_{CA}\mathcal{L}_{CA}(G) + \xi_{DA}\mathcal{L}_{DA}(G) + \xi_{CT}\mathcal{L}_{CT} \tag{9}$$

where ξ are weights on different losses. By optimizing Eq. (9), the generator could generate an arbitrary number of features of unseen classes which are semantically-consistent and discriminative. The generation process could be formulated as: $\tilde{x} = G(z, a(y))$ where $y \in \mathcal{Y}_u$. After obtaining image features of all unseen classes, we integrate them with existing features of seen classes to train a supervised classifier: $f_1 : \mathcal{X} \rightarrow \mathcal{Y}_s \cup \mathcal{Y}_u$.

Also, we could map all features into semantic alignment space by cross-modal aligned VAEs and train another supervised classifier $f_2 : E_x(\mathcal{X}) \rightarrow \mathcal{Y}_s \cup \mathcal{Y}_u$ there. Likewise, we could train a supervised classifier $f_3 : H(\mathcal{X}) \rightarrow \mathcal{Y}_s \cup \mathcal{Y}_u$ in image embedding space. Given test features from seen classes or unseen classes, f-CLSWGAN-VAE2 could output classification results in three spaces.

3.6 Adaptively Integrating Classification Results from Multiple Spaces

An intuitive idea is to integrate the classification results from various spaces. Empirically, we find classifiers trained in different spaces exhibit different inclinations towards seen and unseen classes. Utilizing different strategies to weight classification results from various spaces based on whether the test feature belongs to seen classes or unseen classes will bring forth performance improvement. Nevertheless, it is a long-standing problem far from being solved to accurately determine whether a test feature belongs to seen classes or not. We propose to train a binary classifier f_{binary} in semantic alignment space and employ it as a reasonable discriminator to distinguish whether a test feature belongs to seen classes or not. The accuracy trend on test set of the binary classifier over training epochs is reported in Fig. 5.

Then we could fully exploit classification results from various spaces by adaptively weighting them and obtain the final classification score for a test feature x through:

$$S(x) = S_1(x) + \omega \times S_2(x) + \rho \times mask(x) \times S_3(x) \tag{10}$$

where S_1, S_2 and S_3 are the classification score of f_1, f_2 and f_3 respectively. ω and ρ are hyperparameters. $mask$ is generated by f_{binary} with binary value. If f_{binary} classifies x as seen class, the value of $mask$ is 1. Otherwise, the value of $mask$ is 0. In other words, considering the lack of injection of semantic information in image embedding space, we expect the final classification of unseen classes will not take classification results in image embedding space into account. The detailed generalized zero shot classification process by f-CLSWGAN-VAE2 is summarized in Algorithm 1 and is depicted in Fig. 2.

4 Experiment

4.1 Experimental Setup

Datasets and Evaluation Protocol. We evaluate our f-CLSWGAN-VAE2 on four benchmark datasets for GZSL: Animals with Attributes1(AWA1) [5], Animals with Attributes2(AWA2) [30], Caltech-UCSD-Birds (CUB) [29] and

---

Algorithm 1. Generalized zero shot classification using f-CLSWAGN-VAE2.

Input:

The training set $\mathcal{T} = \{(x_i, y_i, a_i) \mid x \in \mathcal{X}_s, y \in \mathcal{Y}_s, a \in \mathcal{A}_s\}_{i=1}^{N}$; semantic embedding set of unseen classes $A_u = \{a(y) | \forall y \in \mathcal{Y}_u\}$; a test feature x_j that may belongs to seen classes or unseen classes;

Output:

Label y_j of x_j;

1: Train E_x, E_a, D_x, D_a in cross-modal VAEs using Eq.(5);
2: Train G, D, H using Eq.(9);
3: Generate features of all unseen classes $\tilde{\mathcal{X}}_u = \{G(z, a) | \forall a \in A_u\}$;
4: Train supervised classifier f_1 on $\{(x_m, y_m) | x_m \in \mathcal{X}_s \cup \tilde{\mathcal{X}}_u, y_m \in \mathcal{Y}_s \cup \mathcal{Y}_u\}$;
    Train f_2, f_{binary} on $\{(E_x(x_m), y_m) | x_m \in \mathcal{X}_s \cup \tilde{\mathcal{X}}_u, y_m \in \mathcal{Y}_s \cup \mathcal{Y}_u\}$ and $\{E_a(A_u)\}$;
    Train f_3 on $\{(H(x_m), y_m) | x_m \in \mathcal{X}_s \cup \tilde{\mathcal{X}}_u, y_m \in \mathcal{Y}_s \cup \mathcal{Y}_u\}$;
5: Compute the final classification score $S(x_j)$ using Eq.(10);
6: **return** $y_j \leftarrow \text{argmax}_y s(x_j)$;

SUN Attribute (SUN) [25]. For data splits, We follow the data splits in [30] for fair comparisons. For image features, we extract 2048-dim features from Resnet101 [11] pretrained on Imagenet-1k without fine-tuning. We adopt class-specific human-annotated attribute vectors provided in [30] as semantic embeddings. As for evaluation protocols, following [30], We compute the average per-class top-1 accuracy on seen classes acc_s and unseen classes acc_u and compute the harmonic mean $H = \frac{2*acc_s*acc_u}{acc_s+acc_u}$ of them.

Implementation Details. In our model, the generator G, the discriminator D, encoders E_x and E_a, decoders D_x and D_a, embedding net H are all multi-layer perceptron with one hidden layer and Relu activation. Hidden units number of them are 4096, 1024, 1560, 1450, 1660, 660 and 2048 respectively. The size of latent embedding in latent space of cross-modal aligned VAEs is 64. Output feature size of H is 512. For GAN loss, we set λ is 10, β is 0.01 on all datasets. We follow parameter settings and training strategy in [26] to train cross-modal VAEs. We set the temperature parameter in Eq. (8) as 0.1 for all datasets. ξ_{CA} and ξ_{DA} are set to 0.0001. ξ_{CT} are set to 0.001. When integrating classification results in different spaces, we set ω equals to 8 on all datasets except AWA1 where it is set to 4. ρ is set to 2, 10, 3 and 4 on AWA1, AWA2, CUB and SUN respectively. Classifier f_1, f_2, f_3, f_{binary} are all softmax classifiers.

4.2 Comparison on Benchmark Datasets

In this section, we will compare f-CLSWGAN-VAE2 with baseline and other GZSL models. The comparison results are reported in Table 1. As could be seen from the table, our model achieves the best H on AWA1 and AWA2. It should be noted that the performance of our model significantly surpasses that of the baseline model f-CLSWGAN. This provides evidence that the introduced additional constraints and adaptive integration help to boost performance. Compared to

Table 1. Performance of different GZSL models measured by acc_s (%): top-1 accuracy on seen classes, acc_u (%): top-1 accuracy on unseen classes, H: harmonic mean of acc_s and acc_u.

Method	AWA1			AWA2			CUB			SUN		
	acc_s	acc_u	H	acc_s	acc_u	H	acc_s	acc_u	H	acc_s	acc_u	H
DAP	**84.7**	0.0	0.0	67.9	1.7	3.3	78.3	4.8	9.0	25.1	4.2	7.2
SAE	82.2	1.1	2.2	54.0	7.8	13.6	**80.9**	0.4	0.8	18.0	8.8	11.8
f-CLSWGAN	61.4	57.9	59.6	-	-	-	57.7	43.7	49.7	36.6	42.6	39.4
cycle-WGAN	64.0	56.9	60.2	-	-	-	61.0	45.7	52.3	33.6	49.4	40.0
CADA-VAE	72.8	57.3	64.1	75.0	55.8	63.9	53.5	51.6	52.4	35.7	47.2	40.6
f-VAEGAN-D2	-	-	-	70.6	57.6	63.5	60.1	48.4	53.6	38.0	45.1	41.3
TF-VAEGAN	-	-	-	75.1	**59.8**	66.6	64.7	**52.8**	**58.1**	40.7	45.6	**43.0**
ours	76.5	**63.6**	**69.4**	**80.4**	**59.8**	**68.6**	56.4	51.2	53.7	37.2	**49.6**	42.5

Table 2. Classification results in various spaces and their integrated results measured by acc_s (%): top-1 accuracy on seen classes, acc_u (%): top-1 accuracy on unseen classes, H: harmonic mean of acc_s and acc_u.

Classifier	AWA1			AWA2			CUB			SUN		
	acc_s	acc_u	H	acc_s	acc_u	H	acc_s	acc_u	H	acc_s	acc_u	H
f_1	67.7	55.6	61.1	78.9	30.3	43.8	43.4	33.7	37.9	31.5	30.3	30.9
f_2	70.7	63.4	66.9	76.4	55.8	64.4	53.3	51.0	52.1	35.0	46.5	40.0
f_3	**83.7**	43.4	57.2	**88.0**	35.4	50.5	54.3	29.3	38.1	28.8	27.5	28.2
Integrated	76.5	**63.6**	**69.4**	80.4	**59.8**	**68.6**	**56.4**	**51.2**	**53.7**	**37.2**	**49.6**	**42.5**

cycle-WGAN which maintains semantic consistency by ensuring generated features could be mapped back to corresponding semantic embeddings, our model outperforms it with 4.4% improvement on average in term of H. One major difference between them lies in the fact that our model explicitly constrains generated features should well match corresponding semantic embeddings by cross-modal aligned VAEs. The performance discrepancy implies that our model is more effective in maintaining semantic consistency. Also, our model exhibits superior performance on all dataset in comparison to f-VAEGAN-D2 even though f-VAEGAN-D2 utilizes VAEGAN to help stable training. On CUB and SUN, our model lags behind TF-VAEGAN the improvement of which may stem from the feedback module.

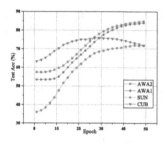

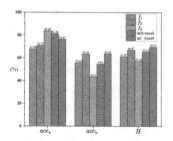

Fig. 3. Measuring the accuracy of the binary classifier on test set w.r.t. the training epoch which is trained on features of seen and unseen classes in semantic alignment space.

Fig. 4. The individual test accuracy of each classifier and the accuracy after fusing their respective results with and without using the *mask* after training our model for 24 epochs on AWA1.

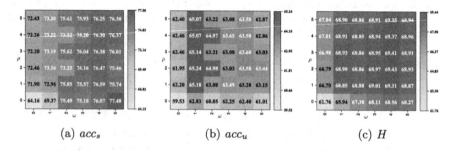

(a) acc_s (b) acc_u (c) H

Fig. 5. Accuracy on seen classes acc_s, accuracy on unseen classes acc_u and harmonic mean of them H on AWA1 w.r.t. different combinations of ω and ρ.

4.3 Ablation Study

In Table 2, we report the accuracy of different classifiers trained in various spaces and their integrated results. As indicated in the table, adaptively integrating classification results from various spaces leads to the best performance. On AWA1, f_3 achieves the best acc_s of 83.7%. However, integration leads to the best acc_u of 63.6% and the best H of 69.4%. This demonstrates that our model could effectively utilize the classification results of each classifier, combining the strengths of each classifier through adaptive integration. This is more pronounced on CUB and SUN where the final model achieves the best acc_s, acc_u and H. This suggests adaptive integration could increase the classification accuracy of unseen classes without undermining the classification accuracy of seen classes.

4.4 Model Component Analysis

Discussion on Adaptive Integration. In this section, we examine the justification for employing the binary classifier to adaptively integrate classification results from various spaces. Firstly, we measure the accuracy on test set of the binary classifier. The results are shown in Fig. 3. It could be observed that as the training progresses, the accuracy continues to rise. After 50 epochs, the binary classifier achieves accuracy of 84%, 85%, 72% and 72% on AWA1, AWA2, CUB and SUN. Accordingly, it is reasonable to employ it as a discriminator to distinguish whether a test feature belongs to seen class or not.

Then we evaluate the performance difference of f-CLSWGAN-VAE2 between adaptive integration of classification results using $mask$ generated by f_{binary} and direct integration of classification results (which means $mask(x)$ is 1 for all x in Eq.(10)). Figure 4 displays the comparison results on AWA1 (ω and ρ are set to 4 and 2 respectively for both). From the graph, it is evident that utilizing $mask$ to adaptively fuse the classification results of f_1, f_2 and f_3 achieved the best H of 69.43% which is 4.11% higher than fusion results without using $mask$. The H of direct integration is even lower than f_2 which illustrates the necessity of utilizing $mask$ to adaptively integrate classification results from various spaces. The graph also reveals that our model could balance the classification results from various spaces to achieve the optimal H.

Parameters. The acc_s, acc_u and H of our model on AWA1 under different combinations of ω and ρ are given in Fig. 5. As depicted in the figure, with the incorporation of classification results of f_2 or f_3, our model attains better performance in all three metrics. The peak performance is attained when results from all three classifier are adaptively integrated. Similar conclusion could be drawn that adaptive integration contributes in improving the performance of our model.

5   Conclusion

In this paper, we proposed f-CLSWGAN-VAE2, a GAN-based generative GZSL model. To encourage the model to generate semantically-consistent features, we introduced cross-modal aligned VAEs to ensure alignment between generated features and corresponding semantic embeddings in semantic alignment space. To alleviate confusion between generated instances, we incorporate a contrastive loss in another image embedding space encouraging generated features to be more discriminative. Moreover, to fully exploit features in various spaces, we pioneer to propose an integration strategy to adaptively weight the classification results from multiple spaces. Experiment results on four benchmark datasets show that our model attains decent performance and proposed modules are effective in boosting GZSL performance.

References

1. Akata, Z., Perronnin, F., Harchaoui, Z., Schmid, C.: Label-embedding for attribute-based classification. In: Proceedings of the IEEE Conference on Computer Vision and Pattern Recognition, pp. 819–826 (2013)

2. Cetin, S.: Closed-form sample probing for training generative models in zero-shot learning. Master's thesis, Middle East Technical University (2022)
3. Changpinyo, S., Chao, W.L., Sha, F.: Predicting visual exemplars of unseen classes for zero-shot learning. In: Proceedings of the IEEE International Conference on Computer Vision, pp. 3476–3485 (2017)
4. Chen, T., Kornblith, S., Norouzi, M., Hinton, G.: A simple framework for contrastive learning of visual representations. In: International Conference on Machine Learning, pp. 1597–1607. PMLR (2020)
5. Farhadi, A., Endres, I., Hoiem, D., Forsyth, D.: Describing objects by their attributes. In: 2009 IEEE Conference on Computer Vision and Pattern Recognition, pp. 1778–1785. IEEE (2009)
6. Felix, R., Reid, I., Carneiro, G., et al.: Multi-modal cycle-consistent generalized zero-shot learning. In: Proceedings of the European Conference on Computer Vision (ECCV), pp. 21–37 (2018)
7. Frome, A., Corrado, G.S., Shlens, J., Bengio, S., Dean, J., Ranzato, M., Mikolov, T.: Devise: a deep visual-semantic embedding model. Advances in neural information processing systems 26 (2013)
8. Fu, Y., Hospedales, T.M., Xiang, T., Gong, S.: Transductive multi-view zero-shot learning. IEEE Trans. Pattern Anal. Mach. Intell. 37(11), 2332–2345 (2015)
9. Gulrajani, I., Ahmed, F., Arjovsky, M., Dumoulin, V., Courville, A.C.: Improved training of wasserstein gans. Advances in neural information processing systems 30 (2017)
10. Han, Z., Fu, Z., Chen, S., Yang, J.: Contrastive embedding for generalized zero-shot learning. In: Proceedings of the IEEE/CVF Conference on Computer Vision and Pattern Recognition, pp. 2371–2381 (2021)
11. He, K., Zhang, X., Ren, S., Sun, J.: Deep residual learning for image recognition. In: Proceedings of the IEEE Conference on Computer Vision and Pattern Recognition, pp. 770–778 (2016)
12. Hong, Z., et al.: Semantic compression embedding for generative zero-shot learning. IJCAI, Vienna, Austria 7, 956–963 (2022)
13. Huynh, D., Elhamifar, E.: Fine-grained generalized zero-shot learning via dense attribute-based attention. In: Proceedings of the IEEE/CVF Conference on Computer Vision and Pattern Recognition, pp. 4483–4493 (2020)
14. Kodirov, E., Xiang, T., Gong, S.: Semantic autoencoder for zero-shot learning. In: Proceedings of the IEEE Conference on Computer Vision and Pattern Recognition, pp. 3174–3183 (2017)
15. Larochelle, H., Erhan, D., Bengio, Y.: Zero-data learning of new tasks. In: AAAI, vol. 1, p. 3 (2008)
16. Lei Ba, J., Swersky, K., Fidler, S., et al.: Predicting deep zero-shot convolutional neural networks using textual descriptions. In: Proceedings of the IEEE International Conference on Computer Vision, pp. 4247–4255 (2015)
17. Li, X., Yang, X., Wei, K., Deng, C., Yang, M.: Siamese contrastive embedding network for compositional zero-shot learning. In: Proceedings of the IEEE/CVF Conference on Computer Vision and Pattern Recognition, pp. 9326–9335 (2022)
18. Liu, B., Hu, L., Dong, Q., Hu, Z.: An iterative co-training transductive framework for zero shot learning. IEEE Trans. Image Process. 30, 6943–6956 (2021)
19. Mikolov, T., Chen, K., Corrado, G., Dean, J.: Efficient estimation of word representations in vector space. arXiv preprint arXiv:1301.3781 (2013)
20. Mirza, M., Osindero, S.: Conditional generative adversarial nets. arXiv preprint arXiv:1411.1784 (2014)

21. Mishra, A., Krishna Reddy, S., Mittal, A., Murthy, H.A.: A generative model for zero shot learning using conditional variational autoencoders. In: Proceedings of the IEEE Conference on Computer Vision and Pattern Recognition Workshops, pp. 2188–2196 (2018)
22. Naeem, M.F., Xian, Y., Tombari, F., Akata, Z.: Learning graph embeddings for compositional zero-shot learning. In: Proceedings of the IEEE/CVF Conference on Computer Vision and Pattern Recognition, pp. 953–962 (2021)
23. Narayan, S., Gupta, A., Khan, F.S., Snoek, C.G., Shao, L.: Latent embedding feedback and discriminative features for zero-shot classification. In: Computer Vision-ECCV 2020: 16th European Conference, Glasgow, UK, August 23–28, 2020, Proceedings, Part XXII 16. pp. 479–495. Springer (2020)
24. Palatucci, M., Pomerleau, D., Hinton, G.E., Mitchell, T.M.: Zero-shot learning with semantic output codes. Advances in neural information processing systems 22 (2009)
25. Patterson, G., Hays, J.: Sun attribute database: Discovering, annotating, and recognizing scene attributes. In: 2012 IEEE Conference on Computer Vision and Pattern Recognition, pp. 2751–2758. IEEE (2012)
26. Schonfeld, E., Ebrahimi, S., Sinha, S., Darrell, T., Akata, Z.: Generalized zero- and few-shot learning via aligned variational autoencoders. In: Proceedings of the IEEE/CVF Conference on Computer Vision and Pattern Recognition, pp. 8247–8255 (2019)
27. Shermin, T., Teng, S.W., Sohel, F., Murshed, M., Lu, G.: Integrated generalized zero-shot learning for fine-grained classification. Pattern Recogn. **122**, 108246 (2022)
28. Socher, R., Ganjoo, M., Manning, C.D., Ng, A.: Zero-shot learning through cross-modal transfer. Advances in neural information processing systems 26 (2013)
29. Welinder, P., Branson, S., Mita, T., Wah, C., Schroff, F., Belongie, S., Perona, P.: Caltech-ucsd birds 200 (2010)
30. Xian, Y., Lampert, C.H., Schiele, B., Akata, Z.: Zero-shot learning-a comprehensive evaluation of the good, the bad and the ugly. IEEE Trans. Pattern Anal. Mach. Intell. **41**(9), 2251–2265 (2018)
31. Xian, Y., Lorenz, T., Schiele, B., Akata, Z.: Feature generating networks for zero-shot learning. In: Proceedings of the IEEE Conference on Computer Vision and Pattern Recognition, pp. 5542–5551 (2018)
32. Xian, Y., Sharma, S., Schiele, B., Akata, Z.: f-vaegan-d2: A feature generating framework for any-shot learning. In: Proceedings of the IEEE/CVF Conference on Computer Vision and Pattern Recognition, pp. 10275–10284 (2019)
33. Xu, J., Le, H.: Generating representative samples for few-shot classification. In: Proceedings of the IEEE/CVF Conference on Computer Vision and Pattern Recognition, pp. 9003–9013 (2022)
34. Yi, K., Shen, X., Gou, Y., Elhoseiny, M.: Exploring hierarchical graph representation for large-scale zero-shot image classification. In: European Conference on Computer Vision, pp. 116–132. Springer (2022)
35. Zhu, Y., Xie, J., Tang, Z., Peng, X., Elgammal, A.: Semantic-guided multi-attention localization for zero-shot learning. Advances in Neural Information Processing Systems 32 (2019)

Joint Image Data Hiding and Rate-Distortion Optimization in Neural Compressed Latent Representations

Chen-Hsiu Huang$^{(\boxtimes)}$ and Ja-Ling Wu

National Taiwan University, Taipei 106319, Taiwan
{chenhsiu48,wjl}@cmlab.csie.ntu.edu.tw

Abstract. We present an end-to-end learned image data hiding framework that embeds and extracts secrets in the latent representations of a neural compressor. The message encoder/decoder design is flexible to support multiples of 64-bit message lengths. We jointly optimize with the rate-distortion function of the neural codec, which reduces the embedded size overhead significantly. By leveraging a perceptual loss function, our approach simultaneously achieves high image quality and low bit-error rate. Our approach offers superior image secrecy in steganography and watermarking scenarios than existing techniques. Processing messages in the compressed domain has much lower complexity, and our method can achieve about 30 times acceleration. Furthermore, with the prevalence of IoT smart devices, machines can extract hidden data directly from the compressed domain without decoding. Our framework can benefit both secure communication and the coding-for-machines concept.

Keywords: Image steganography · Watermarking · Neural compression · End-to-end learned data hiding

1 Introduction

Neural compression [34], the end-to-end learned image compression method [2,9, 10,24], has been actively developed in recent years and outperforms traditional expert-designed image codecs. Traditionally, most image processing algorithms cannot be directly applied to hand-crafted image codecs like JPEG. As a result, the first step before image processing or analysis is decompressing the image into raw pixels. With the evolution of neural compression, there is a growing trend to apply CNN-based methods directly to the compressed latent space [12, 31], leveraging the advantages of joint compression optimization and eliminating decompression.

Image steganography or watermarking hides secrets in a cover image to form a container image for communication or proof of ownership. Traditional methods worked on hiding and extracting secrets in either the spatial domain [14,27] or the frequency domain [5,13]. Modern techniques [3,30,35,37] use

© The Author(s), under exclusive license to Springer Nature Switzerland AG 2024
S. Rudinac et al. (Eds.): MMM 2024, LNCS 14554, pp. 94–108, 2024.
https://doi.org/10.1007/978-3-031-53305-1_8

deep neural networks (DNNs) and adversarial training to end-to-end learn an encoder/decoder pair that embeds and extracts the secrets with robustness against noise attacks. These data hiding techniques are highly relevant to image compression codecs and DNN-transformed latent representations.

In this work, to the best of our knowledge, we are the first to propose a learned image data hiding framework that embeds and extracts secrets in the latent representations of a neural compressor. Except for secret communication, the hidden message in the compressed latents can also be the metadata for machines to extract. With the increasing prevalence of IoT devices, drones, and self-driving cars, machines rather than humans are processing a more significant portion of captured visual content. Consequently, pursuing an efficient compressed representation that caters to human vision and machine tasks is crucial.

The possible contributions of our work include the following:

- We propose a generic data hiding framework for image steganography and watermarking. Our message encoder/decoder design is flexible to support multiples of 64-bit messages and up to 512 bits, which is higher than most existing benchmarked works.
- We use a perceptual loss function that can be incorporated into the messaging hiding task to ensure high image quality and low bit-error rate.
- We jointly optimize data hiding and the rate-distortion function of the neural codec to reduce the file size after message embedding significantly.

2   Related Works

2.1   Learned Image Compression

The field of learned image compression has witnessed significant advancements with the introduction of convolutional neural networks (CNNs). Several approaches have been proposed in the literature, starting with Ballé et al. [2], that surpassed traditional codecs like JPEG and JPEG 2000 in terms of PSNR and SSIM metrics. Minnen et al. [24] further improved coding efficiency by employing a joint autoregressive and hierarchical prior model, surpassing the performance of the HEVC [17] codec. More recently, Cheng et al. [10] developed techniques that achieved comparable performance to the latest coding standard VVC [26]. Several comprehensive survey and introduction papers [23,25,34] have summarized these advancements in end-to-end learned compression.

2.2   DNN-Based Steganography and Watermarking

Many traditional image steganography techniques laid their bases on the least significant bit (LSB) techniques, custom-designed algorithms [14,27], or changing mid-frequency components in the frequency domain [5,13]. Deep steganography methods like DeepStega [3] and UDH [35] defined a new task to hide one or more images into a cover image with a DNN. Unlike traditional methods that

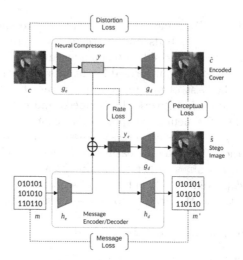

Fig. 1. The system block diagram of the proposed rate-distortion optimized joint data hiding and compression neural network.

require a perfect restoration of secret messages, Deep steganography methods minimize the distortion between the retrieved and the original secret images. Thus, the message is securely delivered because the authentic and recovered secret images are visually indistinguishable. Lu et al. [20] recently advanced deep steganography with a higher capacity of up to three or more secret images.

Zhu et al. [37] proposed the HiDDeN model that embeds the raw bits and extracts the secret message with a low bit error rate using a DNN. With the generative model adversarial trained against the noise layers, the HiDDeN model can achieve the purposes of steganography and digital watermarking with the same network architecture. Perhaps inspired by HiDDeN, many researchers have proposed similar adversarial network-based methods for steganography, such as StegaStamp [30], Hinet [15], and watermarking-related works such as [21,22,33].

3   Proposed Method

3.1   Problem Formulation

Figure 1 shows the overall architecture of the proposed neural data hiding framework. Our framework simultaneously fulfills steganography and watermarking scenarios and is jointly trained with the neural codec. We formulate our techniques as follows.

Steganography. A message $m \in \{0,1\}^n$ is hidden in a cover image c. A neural encoder/decoder pair g_e and g_d are used to obtain the compressed latent vector $y = g_e(c)$ and the encoded cover image $\hat{c} = g_d(y)$. A message encoder h_e is

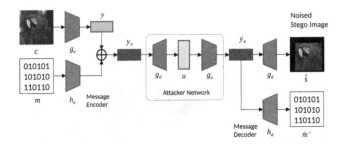

Fig. 2. The adopted schematic diagram for simulating the watermarking embedding-incurred noises.

trained to transform the message to the latent space with the same dimension as y. Then, the embedded latent vector y_e can be obtained as:

$$y_e = y \oplus h_e(m), \tag{1}$$

where $\oplus$ denotes the element-wise addition. The embedded latent vector y_e is then entropy-coded and transmitted. The stego image $\hat{s} = g_d(y_e)$ can be decoded by anyone accessing the publicly available g_e and g_d. However, only the specified receiver with a trained message decoder h_d can extract the secret message by:

$$m' = h_d(y_e). \tag{2}$$

In other words, we conduct the message embedding and extraction over the compressed latent spaces y and y_e, respectively.

Watermarking. We simulate the watermark embedding-induced noises by introducing the attacker network between the message encoder and decoder, as shown in Fig. 2. The noised stego image $\dot{s} = u(\hat{s})$ is derived using an attacker u to simulate noise attacks. The final noised stego image $\hat{\dot{s}}$ is obtained by re-encode and re-decode with the same neural codecs, that is

$$\hat{\dot{s}} = g_d(g_e(\dot{s})). \tag{3}$$

Then, the noised embedded latent vector $\dot{y}_e = g_e(\dot{s})$ is used to extract the embedded secrets with $\dot{m}' = h_d(\dot{y}_e)$.

3.2 Message Encoder/Decoder Network

We designed the message encoder h_e and the decoder h_d to be compatible with neural codecs from Balle [2], Minnen [24], and Cheng [10] using CompressAI [4] implementation. These neural codecs have compressed latent representations with 320 channels and a spatial resolution downsampled by 2^{-4}. For the 128×128-sized input images, the latent vector dimension is $320 \times 8 \times 8$. The secret message is designed to be a multiple of 64-bit so that we can reshape the message to 2D with size 8×8 and forward it to the message encoder. Both encoder and

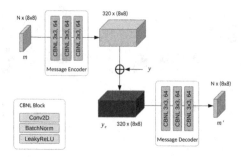

Fig. 3. The message encoder/decoder network architecture.

decoder use three CBNL (convolution, batch normalization, and leaky ReLU) blocks to transform vectors from N channels to 320 channels and vice versa. The decoder output is reshaped from $N \times (8 \times 8)$ tensor to a 1D vector as the decoded message m'. Figure 3 shows the detailed network architecture and the associated information flow.

It is worth noting that different neural codecs may have varying latent space dimensions. Still, our message encoder/decoder design is generic: We propose a simple and effective way to scale out message size from 64 to 512 bits. To improve the recovered bit accuracy, we can increase the encoder/decoder network capacity by altering the filter size of CBNL blocks.

3.3   Joint Training and Loss Functions

A typical end-to-end trained neural codec optimizes the rate-distortion efficiency by minimizing the corresponding codec loss $\mathcal{L}_C$, which is defined as:

$$\mathcal{L}_C = \mathsf{R} + \lambda \mathsf{D}. \tag{4}$$

For any input image $x \in \mathcal{X}$, the neural encoder g_e transforms x into a latent representation $y = g_e(x)$, which is later quantized to a discrete-valued vector $\hat{y}$. To reduce the notational burden, we refer to $\hat{y}$ simply as y in what follows. The discrete probability distribution P_y is estimated using a neural network and then encoded into a bitstream using an entropy coder. The *rate* of this discrete code, R, is lower-bounded by the entropy of the discrete probability distribution $H(P_y)$. On the decoder side, we decode y from the bitstream and reconstruct the image $\hat{x} = g_d(y)$ using the neural decoder. The *distortion*, D, is measured by a perceptual metric $d(x, \hat{x})$, where the mean square error (MSE) is commonly used.

Since the rate R is optimized for latents y in a standard neural codec, the distribution $H(P_y)$ cannot optimally model the message embedded latent y_e. We jointly optimize the message encoder/decoder and neural encoder/decoder interleaved as presented in Algorithm 1.

Algorithm 1: NeuralHidingTrain(c)

Data: Cover image c
Data: Neural encoder/decoder parameters θ_e, θ_d
Data: Message encoder/decoder parameters ϕ_e, ϕ_d
Result: Fine-tuned neural encoder/decoder g_e, g_d; Trained message
     encoder/decoder h_e, h_d
for $i \leftarrow 1$ **to** *num_epochs* **do**
  $\mathcal{L}_H \leftarrow \mathcal{L}_P + \alpha \mathcal{L}_M$;
  $\phi_e \leftarrow \phi_e + lr \times \text{Adam}(\mathcal{L}_H)$;
  $\phi_d \leftarrow \phi_d + lr \times \text{Adam}(\mathcal{L}_H)$;
  $\mathcal{L}_C \leftarrow \text{R} + \lambda\text{D}$;
  $\theta_e \leftarrow \theta_e + lr \times \text{Adam}(\mathcal{L}_C)$;
  $\theta_d \leftarrow \theta_d + lr \times \text{Adam}(\mathcal{L}_C)$;
end
return;

In each training epoch, we first update the message network parameters ϕ_e and ϕ_d by frozen neural codecs. We then fine-tune the standard neural codecs associated with both latents, y and y_e, in the rate function, that is

$$\text{R} = H(P_y) + H(P_{y_e}). \qquad (5)$$

We must model the coding rate on both latents y and y_e; otherwise, the fine-tuned neural codec will perform less optimally on raw cover images, making the jointly optimized neural codec less effective.

Message Hiding. We define our message-hiding loss function as a combination of perceptual loss $\mathcal{L}_P$ and message loss $\mathcal{L}_M$:

$$\mathcal{L}_H = \mathcal{L}_P + \alpha \mathcal{L}_M. \qquad (6)$$

Hyper-parameter α is used to control the relative weight of the two losses. The perceptual loss is measured by the DNN-based perceptual loss LPIPS [36]:

$$\mathcal{L}_P = \text{LPIPS}(\hat{c}, \hat{s}). \qquad (7)$$

We avoid using MSE to minimize image distortion because experiments show that the widely used MSE loss performs poorly with our proposed message encoder/decoder. The recovered bit accuracy is good, but the stego image deteriorated visually.

For measuring the decoded message error, we use binary cross-entropy as the loss function, that is

$$\mathcal{L}_M = \text{BCE}(m, m') + \beta\text{BCE}(m, \dot{m}'). \qquad (8)$$

In our joint training experiments, we set $\alpha = 1.5$ and $\beta = 1.0$ as hyper-parameters. For the rate-distortion optimization, the hyper-parameter λ depends on the quality setting of the lossy codec, which defaults to 0.18 as quality 8 in CompressAI.

3.4   Noise Attacks

For the watermarking scenario, we defined four common noise-causing attacks: Cropout, Dropout, Gaussian noise, and JPEG compression. We randomly generate the noise-related parameters during training.

– **Cropout:** We randomly crop out 70%-90% of the stego image in a rectangle and keep the rest pixels in black.
– **Dropout:** We randomly replace 10%-30% of the stego image pixels with black pixels.
– **Gaussian noise:** We randomly add Gaussian noise with mean $\mu = 0.5$ and variance σ ranging from 0.1 to 0.5 on 10%-30% pixels of the stego image.
– **JPEG compression:** We re-compress the stego image with random JPEG quality settings from 70 to 95. The quantization is approximated during training by cubic rounding used in [29].

Existing DNN-based data hiding techniques trained against the noises from the spatial domain, i.e., the stego image needs to be stored losslessly; otherwise, it encounters another image compression attack. In our case, because we need to transmit the image in the neural compressed format, we always have a second compression after the stego image is noised.

4   Experimental Results

We implemented our works on neural codecs hyper [2], mbt [24], and cheng [10] from CompressAI. We denote our data hiding methods as "Ours-hyper," "Ours-mbt," and "Ours-cheng," respectively. To ensure high visual quality after encoding, we set the highest coding quality as 8 in both codecs.

For training, we randomly selected 12,000 and 1,200 images from the COCO dataset [28] as the training and validation set, respectively. We resized the cover images to 128×128 pixels and randomly embedded 64-bit binary messages during training. We evaluated our model on the Kodak [16], DIV2K [1], and CelebA [19] datasets.

We trained our model using the PyTorch built-in Adam optimizer with a learning rate of 0.0001 and a batch size of 32. We trained our model for 160 epochs. We compared our model with HiDDeN[1], DeepStega, UDH[2], and StegaStamp[3]. Unlike other DNN-based methods that calculate distortion between the cover c and stego image s, we measured the distortion between $\hat{c}$ and $\hat{s}$, as described in Sect. 3.1. We follow the same training settings from the HiDDeN

---

[1] https://github.com/ando-khachatryan/HiDDeN.
[2] https://github.com/ChaoningZhang/Universal-Deep-Hiding.
[3] https://github.com/tancik/StegaStamp.

paper using a 32-bit message on 128×128 images and train for 400 epochs. We use the official implementation of UDH and download the pre-trained model on their website. Since DeepStega and UDH are methods to hide secret images into cover images of the same size, we report the flip of most significant bit (MSB) as the decoding error. The StegaStamp embeds 100-bit messages into 400×400 images in the official release code.

4.1 Steganography Secrecy

Quantitatively, we present image quality metrics in PSNR, SSIM [32], MAE (mean absolute error), and bit error rate, as shown in Table 1. While it is well-known that DNN-based methods cannot achieve zero-bit error rates, there are established techniques, such as BCH codes [8] and learning-based channel noise modeling [11], to mitigate this issue.

Since we jointly optimize with the rate-distortion function of the neural codecs, our evaluation result accounts for the effect of quantization on latent vectors. Table 1 indicates that our proposed methods have less perceptual distortion than others. The superior stego image quality of the hyper codec stems from its lower coding efficiency than the mbt codec, allowing more room for data hiding in the latent space without affecting the visual quality. On the other hand, the mbt and Cheng's codecs have more densely compressed latents, which results in more quality degradation after message embedding. We mark the order of quality superiority in Table 1 in blue, and the order highly correlates to the neural codecs' coding efficiency.

Table 1. Quality Metrics vs. Bit Error Rate Comparison

Method	Kodak				DIV2K				CelebA			
	PSNR↑	SSIM↑	MAE↓	Error	PSNR↑	SSIM↑	MAE↓	Error	PSNR↑	SSIM↑	MAE↓	Error
Ours-hyper	[1] 52.24	[1] 0.9989	[1] 0.33	0.0000	[1] 51.87	[1] 0.9993	[1] 0.36	0.0008	[1] 53.82	[1] 0.9991	[1] 0.25	0.0000
Ours-mbt	[2] 44.11	[2] 0.9942	[2] 1.16	0.0000	[2] 43.16	[2] 0.9959	[3] 1.30	0.0000	[2] 44.83	[2] 0.9949	[2] 1.03	0.0000
Ours-cheng	[3] 43.79	[3] 0.9936	[3] 1.20	0.0000	[3] 43.14	[2] 0.9959	[2] 1.29	0.0002	[3] 44.74	[2] 0.9949	[3] 1.05	0.0000
HiDDeN	39.61	0.9813	1.91	0.0000	37.59	0.9733	2.44	0.0013	39.28	0.9806	1.85	0.0000
DeepStega	36.51	0.9374	2.81	0.0156	34.72	0.9283	3.58	0.0184	38.27	0.9410	2.34	0.0182
UDH	37.88	0.9184	2.63	0.0213	38.35	0.9414	2.51	0.0248	38.27	0.9147	2.54	0.0150
StegaStamp	31.60	0.9430	4.54	0.0050	30.50	0.9451	5.38	0.0089	35.40	0.9586	2.67	0.0046

Qualitatively, we present the cover images and the resulting stego images in Fig. 4(b) and compare our methods' stego image residual with other DNN-based methods in Fig. 4(a).

Modern DNN-based data hiding methods add perturbations to the low-level feature space to extract messages from the spatial domain with robustness. As a result, these methods modify all the low-level pixels of the cover image, as shown in Fig. 4(a). Our neural data hiding method learns to modify the compressed latents, so the pixel modifications are placed in high-level image features, as

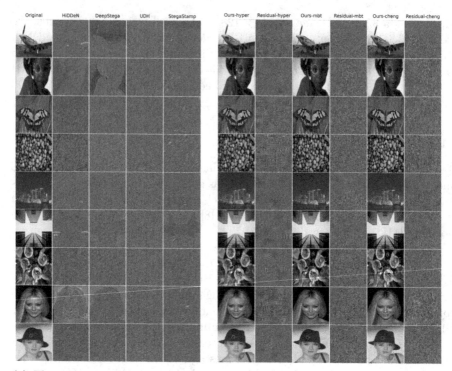

(a) The comparison of stego residual images (defined as the difference between the original and the message-embedded images).

(b) Quality comparisons of stego and stego residual images among our proposed neural data hiding methods.

Fig. 4. Qualitatively comparison of visual quality. Please zoom in to observe the modified pixel locations.

shown in Fig. 4(b). Our proposed method can generalize well on different neural codecs and has a less perceptual impact.

Although the receiver does not have access to the encoded cover image $\hat{c}$, in our scenario, we list the LPIPS [36] metrics between c and $\hat{s}$ in Table 2 and compare them with those of other methods. Our LPIPS metrics remain close to those of the HiDDeN and are superior to the other methods. The LPIPS is a learned perceptual metric based on a pre-trained DNN, which can be thought of as how effectively the stego image can be used as a proxy for the original cover image. Therefore, we believe our stego image $\hat{s}$ has not lost its general utility.

Table 2. LPIPS comparison of stego images on DIV2K

Method	$LPIPS(\hat{c}, \hat{s})$	$LPIPS(c, \hat{s})$	$LPIPS(c, s)$
Ours-hyper	0.00011	0.00353	-
Ours-mbt	0.00132	0.00341	-
Ours-cheng	0.00130	0.00603	-
HiDDeN	-		0.00375
DeepStega	-		0.07718
UDH	-		0.04261
StegaStamp	-		0.08039

4.2 Watermark Robustness

We evaluated the robustness of our method against trained noise attacks on the DIV2K dataset, as shown in Fig. 5(a). We varied the attack strength by increasing the noise parameter, which degrades image quality along the horizontal axis. Generally, we observe a clear trend that a less coding-efficient codec is more robust when the stego image is under noise attack. Figure 5(a) shows that our compress domain methods perform worse than HiDDeN and StegaStamp except for dropout attacks. In the Gaussian noise attack, the Ours-cheng bit-rate accuracy is close to 0.5, which means it cannot be robustly trained. We think the cause may be the attention module in Ours-cheng that makes it hard to model additive Gaussian noises, which requires further investigation.

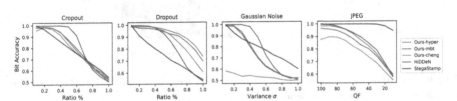

(a) As Ours-cheng is the most compact neural codec, its watermark robustness performs worst than Ours-mbt and Ours-hyper's.

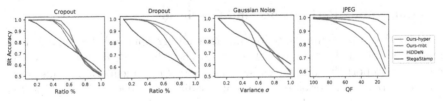

(b) The watermark robustness is improved without joint rate-distortion optimization.

Fig. 5. Watermark robustness against selected noise attacks.

Figure 5(b) shows another experiment in which we skip the joint rate-distortion optimization when trained with noise attacks. Our neural data hiding method performs equivalently to HiDDeN in the cropout attack and has superior robustness for the rest attacks. However, removing the rate-distortion optimization will increase the stego image's file size. In the watermarking scenario, the embedding size overhead is what we need to trade-off.

4.3   Steganalysis

We measured the ability of our model to resist steganalysis using publicly available steganalysis tools, including traditional statistical methods [6] and new DL-based approaches [7,18]. To assess our model's anti-steganalysis ability on the DIV2K dataset, we used the steganalysis tool StegExpose [6].

We varied the detection thresholds as input to StegExpose and plotted the ROC (receiver operating characteristic) curve. We then calculated the AUC (area under the curve) to indicate the classification effectiveness. Ideally, the AUC should be close to 0.5, indicating that the classifier performs no better than random guessing. Figure 6 shows the compared methods' ROC curve and AUC. Our proposed neural data hiding methods achieve better secrecy than HiDDeN and others, especially the Ours-cheng codec has the best AUC of 0.514.

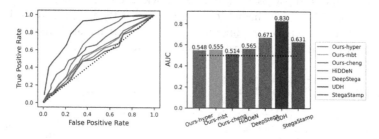

Fig. 6. The ROC curves and the AUCs associated with benchmarked works when the steganalysis tool StegExpose is applied.

4.4   Embedding Performance and Overhead

We show that the compressed domain data hiding approach has significantly lower computational complexity than approaches that hide data in the spatial domain. We report the message embedding/extraction time in Table 3, measured on an Intel i7-9700K workstation with an Nvidia GTX 3090 GPU. As expected, our compressed domain data hiding method is significantly faster than other spatial domain approaches.

Conceptually, data-hiding techniques convert secret messages to noise-liked signals and add them onto cover images as invisible perturbations. The embedding operation will increase the entropy of the stego image and increase file size

Table 3. Message embedding timing performance on DIV2K, in milli-seconds.

Method	Embed	Extract
Ours-hyper	**0.38**	**0.43**
Ours-mbt	**0.37**	**0.42**
Ours-cheng	**0.37**	**0.40**
HiDDeN	10.40	0.74
DeepStega	11.08	0.71
UDH	10.91	0.70
StegaStamp	58.93	34.12

Table 4. Message embedding overhead on DIV2K

Method	Embeded Size
Ours-hyper	**0.21%**
Ours-mbt	**0.47%**
Ours-cheng	**0.36%**
HiDDeN	5.03%
DeepStega	9.13%
UDH	15.67%
StegaStamp	14.51%

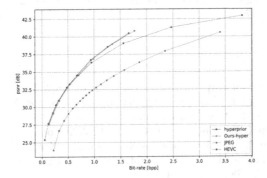

Fig. 7. The rate-distortion curve comparison of various image codecs.

after image compression. Table 4 shows the stego image file size overhead by percentage. For those spatial domain data hiding methods, the embedded stego images stored in PNG file format will cause 5–14% file size overhead. As mentioned in Sect. 3.3, the proposed compressed domain data hiding method jointly optimizes the data hiding process and coding efficiency. Thus, the produced stego image size has a negligible overhead of less than 1%.

Our fine-tuned and message capacity provisioned neural codec must remain optimal compared to vanilla ones. We compare the rate-distortion curve of the original hyperprior codec and Ours-hyper on the Kodak dataset. From Fig. 7, we demonstrate that the jointly optimized codec's coding efficiency is close to the pre-trained codec and superior to traditional human-designed codecs such as JPEG and HEVC.

5 Conclusion

In this work, we proposed a novel end-to-end learned framework for image data hiding that embeds secrets in the latent representations of a neural compressor. Our approach is generic and can be used with different neural compressors. The framework jointly optimizes with the rate-distortion efficiency to reduce message embedding overhead. We demonstrated its superior image secrecy and low bit-error rate in steganography and watermarking scenarios. Except for secret communication, the hidden message in the compressed latents can be the metadata for machines to extract. With the increasing prevalence of IoT devices, pursuing an efficient compressed representation for both human vision and machine understanding is crucial. Our proposed method offers about 30 times acceleration in this context. To summarize, the neural data hiding method in the neural compressor benefits both secure communication and the coding-for-machines concept.

Acknowledgements. The authors would like to thank the NSTC of Taiwan and CITI SINICA for supporting this research under the grant numbers 111-2221-E-002-134-MY3 and Sinica 3012-C3447.

References

1. Agustsson, E., Timofte, R.: Ntire 2017 challenge on single image super-resolution: Dataset and study. In: The IEEE Conference on Computer Vision and Pattern Recognition (CVPR) Workshops, June 2017
2. Ballé, J., Minnen, D., Singh, S., Hwang, S.J., Johnston, N.: Variational image compression with a scale hyperprior. arXiv preprint arXiv:1802.01436 (2018)
3. Baluja, S.: Hiding images in plain sight: deep steganography. Adv. Neural. Inf. Process. Syst. **30**, 2069–2079 (2017)
4. Bégaint, J., Racapé, F., Feltman, S., Pushparaja, A.: Compressai: a pytorch library and evaluation platform for end-to-end compression research. arXiv preprint arXiv:2011.03029 (2020)
5. Bi, N., Sun, Q., Huang, D., Yang, Z., Huang, J.: Robust image watermarking based on multiband wavelets and empirical mode decomposition. IEEE Trans. Image Process. **16**(8), 1956–1966 (2007)
6. Boehm, B.: Stegexpose-a tool for detecting lsb steganography. arXiv preprint arXiv:1410.6656 (2014)
7. Boroumand, M., Chen, M., Fridrich, J.: Deep residual network for steganalysis of digital images. IEEE Trans. Inf. Forensics Secur. **14**(5), 1181–1193 (2018)
8. Bose, R.C., Ray-Chaudhuri, D.K.: On a class of error correcting binary group codes. Inf. Control **3**(1), 68–79 (1960)
9. Chen, T., Liu, H., Ma, Z., Shen, Q., Cao, X., Wang, Y.: End-to-end learnt image compression via non-local attention optimization and improved context modeling. IEEE Trans. Image Process. **30**, 3179–3191 (2021)
10. Cheng, Z., Sun, H., Takeuchi, M., Katto, J.: Learned image compression with discretized gaussian mixture likelihoods and attention modules. In: CVPR, pp. 7939–7948 (2020)
11. Choi, K., Tatwawadi, K., Grover, A., Weissman, T., Ermon, S.: Neural joint source-channel coding. In: International Conference on Machine Learning, pp. 1182–1192. PMLR (2019)
12. Duan, Z., Ma, Z., Zhu, F.: Unified architecture adaptation for compressed domain semantic inference. IEEE Trans. Circuits Syst. Video Technol. (2023)

13. Fridrich, J., Pevny, T., Kodovsky, J.: Statistically undetectable jpeg steganography: dead ends challenges, and opportunities. In: Proceedings of the 9th Workshop on Multimedia & Security, pp. 3–14 (2007)
14. Holub, V., Fridrich, J., Denemark, T.: Universal distortion function for steganography in an arbitrary domain. EURASIP J. Inf. Secur. **2014**(1), 1–13 (2014)
15. Jing, J., Deng, X., Xu, M., Wang, J., Guan, Z.: Hinet: deep image hiding by invertible network. In: Proceedings of the IEEE/CVF International Conference on Computer Vision, pp. 4733–4742 (2021)
16. Kodak photocd dataset. https://r0k.us/graphics/kodak/
17. Lainema, J., Hannuksela, M.M., Vadakital, V.K.M., Aksu, E.B.: Hevc still image coding and high efficiency image file format. In: 2016 IEEE International Conference on Image Processing (ICIP), pp. 71–75. IEEE (2016)
18. Lerch-Hostalot, D., Megias, D.: Unsupervised steganalysis based on artificial training sets. Eng. Appl. Artif. Intell. **50**, 45–59 (2016)
19. Liu, Z., Luo, P., Wang, X., Tang, X.: Deep learning face attributes in the wild. In: Proceedings of the IEEE International Conference on Computer Vision, pp. 3730–3738 (2015)
20. Lu, S.P., Wang, R., Zhong, T., Rosin, P.L.: Large-capacity image steganography based on invertible neural networks. In: Proceedings of the IEEE/CVF Conference on Computer Vision and Pattern Recognition, pp. 10816–10825 (2021)
21. Luo, X., Li, Y., Chang, H., Liu, C., Milanfar, P., Yang, F.: Dvmark: A deep multiscale framework for video watermarking. arXiv preprint arXiv:2104.12734 (2021)
22. Luo, X., Zhan, R., Chang, H., Yang, F., Milanfar, P.: Distortion agnostic deep watermarking. In: Proceedings of the IEEE/CVF Conference on Computer Vision and Pattern Recognition, pp. 13548–13557 (2020)
23. Ma, S., Zhang, X., Jia, C., Zhao, Z., Wang, S., Wanga, S.: Image and video compression with neural networks: a review. IEEE Trans. Circuits Syst. Video Technol. (2019)
24. Minnen, D., Ballé, J., Toderici, G.D.: Joint autoregressive and hierarchical priors for learned image compression. Adv. Neural. Inf. Process. Syst. **31**, 10771–10780 (2018)
25. Mishra, D., Singh, S.K., Singh, R.K.: Deep architectures for image compression: a critical review. Signal Process. **191**, 108346 (2022)
26. Ohm, J.R., Sullivan, G.J.: Versatile video coding-towards the next generation of video compression. In: Picture Coding Symposium, vol. 2018 (2018)
27. Pevny, T., Filler, T., Bas, P.: Using high-dimensional image models to perform highly undetectable steganography. In: International Workshop on Information Hiding, pp. 161–177. Springer (2010)
28. Russakovsky, O., et al.: Imagenet large scale visual recognition challenge. Int. J. Comput. Vision **115**(3), 211–252 (2015)
29. Shin, R., Song, D.: Jpeg-resistant adversarial images. In: NIPS 2017 Workshop on Machine Learning and Computer Security, vol. 1 (2017)
30. Tancik, M., Mildenhall, B., Ng, R.: Stegastamp: invisible hyperlinks in physical photographs. In: Proceedings of the IEEE/CVF Conference on Computer Vision and Pattern Recognition, pp. 2117–2126 (2020)
31. Testolina, M., Upenik, E., Ebrahimi, T.: Towards image denoising in the latent space of learning-based compression. In: Applications of Digital Image Processing XLIV, vol. 11842, pp. 412–422. SPIE (2021)
32. Wang, Z., Bovik, A.C., Sheikh, H.R., Simoncelli, E.P.: Image quality assessment: from error visibility to structural similarity. IEEE Trans. Image Process. **13**(4), 600–612 (2004)

33. Wengrowski, E., Dana, K.: Light field messaging with deep photographic steganography. In: Proceedings of the IEEE/CVF Conference on Computer Vision and Pattern Recognition, pp. 1515–1524 (2019)
34. Yang, Y., Mandt, S., Theis, L.: An introduction to neural data compression. arXiv preprint arXiv:2202.06533 (2022)
35. Zhang, C., Benz, P., Karjauv, A., Sun, G., Kweon, I.S.: UDH: universal deep hiding for steganography, watermarking, and light field messaging. Adv. Neural. Inf. Process. Syst. **33**, 10223–10234 (2020)
36. Zhang, R., Isola, P., Efros, A.A., Shechtman, E., Wang, O.: The unreasonable effectiveness of deep features as a perceptual metric. In: CVPR (2018)
37. Zhu, J., Kaplan, R., Johnson, J., Fei-Fei, L.: Hidden: Hiding data with deep networks. In: Proceedings of the European Conference on Computer Vision (ECCV), pp. 657–672 (2018)

GSUNet: A Brain Tumor Segmentation Method Based on 3D Ghost Shuffle U-Net

JiXuan Hong[1], JingJing Xie[1], XueQin He[1(✉)], and ChenHui Yang[2(✉)]

[1] Department of Computer Science and Technology, Xiamen University, Xiamen, Fujian, China
xking_he@163.com
[2] Information College, Xiamen University, Xiamen, Fujian, China
chyang@xmu.edu.cn

Abstract. Research on MRI-based brain tumor segmentation methods has clinical significance and application value. The existing 3D brain tumor segmentation methods can make full use of the three-dimensional spatial information of MRI, but there are problems with large parameters and calculations. In view of the above problems, a brain tumor segmentation method based on 3D Ghost Shuffle U-Net(GSUNet) is proposed. In this paper, the 3D Ghost Module(3D GM) is utilized as the basic feature extractor of the network, which fundamentally solves the problem of the high complexity of the existing 3D brain tumor segmentation models. At the same time, the Ghost Shuffle Module(GSM) is designed, and GSM with stride 2 is utilized to realize down-sampling, optimize the feature extraction, and strengthen the information communication in the channel dimension. The GSM with stride 1 and the designed Dense Ghost Module(DGM) work together as a decoder to improve the representation ability of the network at a lower cost. Experimental results show that GSUNet can achieve segmentation performance comparable to mainstream brain tumor segmentation methods with extremely low model complexity.

Keywords: Magnetic Resonance Imaging (MRI) · Semantic Segmentation · 3D Ghost Shuffle · Lightweight Network

1 Introduction

Brain tumor is a common malignant brain disease that can occur anywhere in the brain and seriously threatens the life and health of patients. However, limited by existing medical technology, early detection of brain tumors and treatment are the only ways to minimize the impact of brain tumors. Magnetic resonance imaging (MRI), as a non-invasive and low-cost medical imaging technology, has become an important auxiliary tool in the diagnosis of brain tumors.

Since MRI is the three-dimensional data, the existing two-dimensional image-based brain tumor segmentation methods all need to slice the three-dimensional images. Compared with three-dimensional images, brain tumor segmentation on

© The Author(s), under exclusive license to Springer Nature Switzerland AG 2024
S. Rudinac et al. (Eds.): MMM 2024, LNCS 14554, pp. 109–120, 2024.
https://doi.org/10.1007/978-3-031-53305-1_9

two-dimensional slice images requires less calculation and memory. However, the tedious slicing operation takes a lot of time, and the data conversion from 3D to 2D will lead to the loss of spatial information of brain tumors. In addition, when the generated 2D segmentation results are restored to 3D space, problems such as slice tomography, aliasing, and jaggedness may appear, which affect the accuracy of brain tumor segmentation. In response to the above problems, more and more researchers tend to use 3D convolution to extract features from 3D MRI data to make full use of the spatial information in 3D data. However, segmentation models based on 3D convolution often have a large amount of parameters and calculations, occupy a large memory space, and take a long time to train. Therefore, in the case of limited computing resources and memory, inspired by GhostNet [8] and ShuffleNet [12], we design the Ghost Module in GhostNet as the basic unit and improve it to lighten the 3D brain tumor segmentation model and reduce the complexity of the model.

Based on the above ideas, we propose a 3D Ghost Shuffle U-Net (GSUNet). The main contributions of this work are as follows:

- In order to lighten the model, we devise a more lightweight 3D Ghost Module(3D GM) to extract as many detailed features as possible with as few parameters as possible.
- We design Dense Ghost Module(DGM), which acts as a local decoder to further enhance the feature expressive ability of the model, thereby improving the representational ability of the model.
- We propose Ghost Shuffle Module(GSM), applying GSM with stride 2 to reduce the loss of detailed feature information and strengthen the information between channels while avoiding an increase in computation. At the same time, GSM with stride 1 is employed as a local decoder to optimizes advanced semantic features.

2 Method

2.1 Architecture of GSUNet

The GSUNet is based on the encoder-decoder structure, which is mainly composed of 3D Ghost Module(3D GM), Dense Ghost Module(DGM) and Ghost Shuffle Module(GSM). As shown in Fig. 1, the overall architecture is as follows:

In the encoding path, GSUNet first expands the number of channels of the input image through 3D GM, from the original 4 channels to 16 channels. Second, use the 3D GM for feature extraction and keep the number of channels of input and output the same. When down-sampling, it is implemented by the GSM with stride 2 to prevent the increase in model parameters.

In the decoding path, $1 \times 1 \times 1$ convolution is adopted to adjust the number of channels of the high-level feature map, and transposed convolution is adopted to up-sampling and restore its resolution. After each up-sampling, the low-level feature map extracted from the same layer in the encoding path is concatenated with the high-level feature map obtained in the decoding path employing skip connections to make up for the lost details in the high-level semantic space. Second, the combined low-level and high-level feature information is refined and adjusted with the DGM to help the network better capture image features. Then, the GSM with stride 1 is employed to promote the information communication between low-level and high-level features, and obtain more compact deep semantic information. Finally, the 3D GM is employed to adjust the number of channels of the feature map to 4, and the segmentation result is obtained through the Softmax activation function.

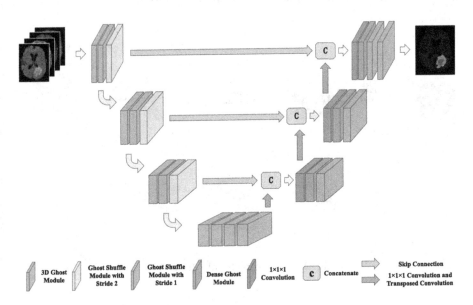

Fig. 1. Overall architecture of 3D Ghost Shuffle U-Net(GSUNet)

2.2 3D Ghost Module

Convolutional layers in deep convolutional neural networks extract feature maps that could include redundant features. In response to this problem, the Ghost Module [8] focuses on obtaining more feature information with fewer parameters. We expand the Ghost Module into 3D Ghost Module(3D GM), which as the basic feature extractor of GSUNet, so as to reduce the calculation of the 3D brain tumor segmentation model.

The 3D GM consists of two parts: pointwise convolution and cheap operation [8], as shown in Fig. 2. First, reduce channels of the input feature map through $1 \times 1 \times 1$ convolution and generate the ground-truth feature map. Second, the features are further enhanced through cheap linear operations, while increasing channels of feature map. All cheap operations use $3 \times 3 \times 3$ group convolutions, where the number of groups is set to half the output channels. By performing multiple linear operations on the same channels feature maps, obtain multiple new feature maps, which improves the computational efficiency. Finally, the original feature map and multiple feature maps obtained are concatenated in the channel dimension through the identity mapping to obtain the final output feature map. Therefore, compared with the conventional convolution layer, when the 3D GM obtains the same number of feature maps, it can effectively realize parameter compression and calculation optimization through simple linear operations. Batch normalization and ReLU activation are performed after all $1 \times 1 \times 1$ convolutions and $3 \times 3 \times 3$ group convolutions.

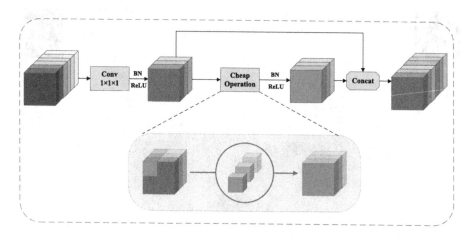

Fig. 2. 3D Ghost Module(3D GM)

2.3   Dense Ghost Module

Leveraging the advantages of 3D GM, we propose Dense Ghost Module(DGM) as a local decoder to enhance the feature representation ability of the global decoder. The structure of DGM is shown in Fig. 3. First, use $1 \times 1 \times 1$ convolution to reduce input channels. Secondly, two 3D GMs are adopted to refine the features, and the number of output channels is kept consistent with the input channels, thereby avoiding an increase in the calculation. In addition, short connections are added before and after each 3D GM, and the original feature map is fused with each refined feature map by element-by-element addition, which can preserve the feature information of different distributions and effectively alleviate the degradation problem of the model. It is worth noting that in the first 3D GM, both pointwise convolution and cheap operation are followed by batch normalization and ReLU activation, but in the second 3D GM, the ReLU activation function is not used.

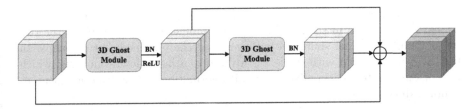

Fig. 3. Dense Ghost Module(DGM)

2.4 Ghost Shuffle Module

In group convolution, the convolution operation is divided into multiple paths, which effectively improves the computational efficiency, but it may also cause the output of each path to only depend on part of the input data. In addition, group convolutions are grouped in such a way that channels are sparsely connected, and the information exchange between channels may be affected. Therefore, in the proposed GSUNet model, we design the Ghost Shuffle Module (GSM) and apply it to optimize the feature extraction and strengthen the inter-channel feature information circulation.

Under different strides, the structure of the GSM is different. As shown in Fig. 4, the GSM with stride 1 contains two 3D GMs, and the number of input channels and output channels of each 3D GM are consistent. After the residual connection is adopted to add and fuse the output feature and the original feature, the channel shuffle operation is performed. The overall structure is similar to the DGM, batch normalization and ReLU nonlinear activation are performed in the first 3D GM, and only batch normalization is performed in the second 3D GM. The DGM together with the GSM with a stride of 1 constitutes the decoder in GSUNet, which is responsible for the refinement of deep features.

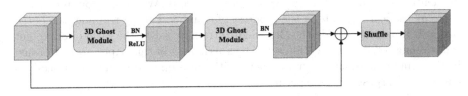

Fig. 4. Ghost Shuffle Module with stride 1(GSM)

As shown in Fig. 5, the GSM with stride 2 consists of two branches. In the first branch, first, utilize a 3D GM for channel expansion, increase output channels to twice the input channels, and improve the diversity of features. Second, a depthwise convolution with a stride 2 and a convolution kernel of $3 \times 3 \times 3$ are utilized to reduce the size of the feature. Then, another 3D GM is utilized to further optimize the feature, and keep the number of channels unchanged. In another branch, the size and number of channels of the original

feature are adjusted by depthwise separable convolution and then combined with
the feature extracted in the first branch to combine the information of the two
paths. Finally, channel shuffle is performed on the obtained feature maps. In
the GSUNet, the GSM with stride 2 is utilized to achieve down-sampling, which
can not only better retain effective detail information but also reduce memory
occupied during training.

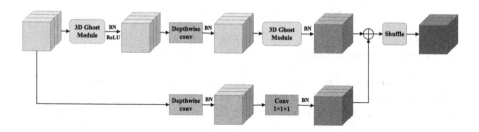

Fig. 5. Ghost Shuffle Module with stride 2(GSM)

The GSM is an independent module that can be applied to different network
architectures. It has the advantages of lightness, high efficiency, plug-and-play,
and portability, which improves the usability of GSM.

3   Experiments

The experiment is carried out on the BraTS [9] 2019 dataset, and the overall
process of the experiment is shown in Fig. 6. The proposed GSUNet is conducted
on the PyTorch framework with an NVIDIA GeForce RTX 3090. In this experiment, we use Adam [10] as the optimizer. The initial learning rate is 0.0003,
momentum is 0.90, and weight decay is set to 0.0001. We utilize poly police [14]
to decay the learning rate in the progress of training. During training, the batch
size is 2. Randomly divide the BraTS 2019 training set at the ratio of 8:2 to get
the training set and verification set. All experiments in this paper are performed
in the same experimental environment with the same experimental process to
ensure the fairness of the experiment.

3.1   Comparison Studies

This section compares the proposed GSUNet with other mainstream segmentation methods, mainly from three aspects of quantitative analysis, model complexity, and visualization results.

Quantitative Analysis. In the field of brain tumor segmentation, the Dice
score and Hausdorff Distance (HD) score are mainly used to evaluate the performance of the model. Therefore, when comparing with different methods, these
two quantitative indicators are mainly used for evaluation. We quantitatively

Table 1. Comparison of results between GSUNet and different 2D, 3D and lightweight segmentation methods on BraTS 2019 verification set

Type	Method	Dice(%)				HD(mm)		
		WT	TC	ET	Avg	WT	TC	ET
2D	Kotowski et al. [11]	77.4	72.5	68.1	72.7	22.16	16.47	11.33
	Baid et al. [4]	87.0	77.0	70.0	78.0	13.36	12.71	6.45
	Ben et al. [15]	85.0	76.0	74.0	78.3	–	–	–
	BrainSeg-Net [18]	86.9	77.5	70.8	78.4	–	–	–
	AGResU-Net [21]	87.0	**77.7**	70.9	78.5	–	–	–
	GSUNet (Ours)	**87.5**	77.3	**75.8**	**80.2**	11.08	12.29	4.44
3D	Bhalerao et al. [5]	85.2	70.9	66.6	74.2	8.07	9.57	7.27
	Ahmad et al. [2]	85.3	75.8	62.3	74.5	9.01	10.67	8.47
	Yan et al. [20]	86.0	73.0	66.0	75.0	40.31	10.40	18.53
	3D U-Net [6]	85.9	75.0	66.8	75.9	12.64	11.97	7.54
	Guo et al. [7]	87.2	72.8	67.7	75.9	**5.41**	**7.74**	6.21
	Amian et al. [3]	84.0	74.0	71.0	76.3	14.00	16.06	10.11
	V-Net [13]	86.9	76.7	66.9	76.8	10.42	8.88	9.26
	Attention U-Net [17]	86.8	76.4	67.6	76.9	16.71	13.80	8.35
	GSUNet (Ours)	**87.5**	**77.3**	**75.8**	**80.2**	11.08	12.29	**4.44**
Lightweight	MGF-Net [1]	86.0	71.1	63.2	73.4	–	–	–
	3D-ESPNet [16]	87.2	78.6	66.3	77.4	**7.42**	**9.74**	9.00
	Tai et al. [19]	84.3	**80.4**	68.4	77.7	–	–	–
	GSUNet (Ours)	**87.5**	77.3	**75.8**	**80.2**	11.08	12.29	**4.44**

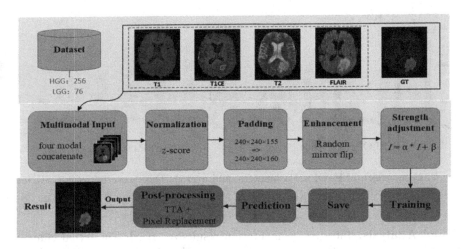

Fig. 6. Overall process of experiment

compare GSUNet with three types of different segmentation methods, namely 2D segmentation methods, 3D segmentation methods, and lightweight segmentation methods, as shown in Table 1. The segmentation results on the validation set are submitted to the CBICA platform for evaluation, and the quantitative scores of the three segmentation regions of intact tumor(WT), tumor core(TC), and enhanced tumor(ET) are obtained.

Complexity Analysis. The Floating-point Operations Per Second(FLOPs) and the total number of parameters that need to be trained in the model(Parameters) are utilized for complexity analysis. Among them, FLOPs is applied to measure the computational complexity while the Parameters is employed to measure the space complexity. We take histograms to display the FLOPs, Parameters, and average Dice scores of different methods on WT, TC, and ET, as shown in Fig. 7.

It can be seen from Fig. 7 that Attention U-Net [17] has fine segmentation performance, but its FLOPs is as high as 160.56G. 3D-ESPNet [16] uses efficient convolutional blocks instead of simple stacked convolutional layers to aggregate features, making the model more lightweight. Compared with 3D U-Net [6], FLOPs are reduced to 76.51G. The GSUNet proposed exploits the 3D GM, which reduces the FLOPs and parameters to about 12.5% and 3.7% of the 3D U-Net. It can also be seen that GSUNet has an extremely small amount of calculation and an extraordinarily low amount of parameters, and can perform well in WT, ET, and TC tumor segmentation, and achieve the highest average Dice score. Therefore, GSUNet has strong competitiveness among lightweight brain tumor segmentation methods.

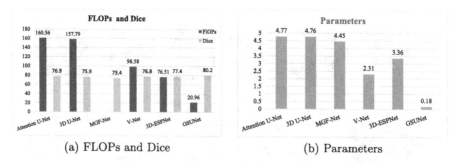

(a) FLOPs and Dice   (b) Parameters

Fig. 7. Comparison of FLOPs, Parameters, and average Dice score between GSUNet and different methods

Visual Analysis. Intuitive segmentation results are needed in actual clinical applications, so we further subjectively analyze the visualization results of different methods, as shown in Fig. 8. The leftmost column in the figure is the FLAIR sequence of the original brain MRI, and the second column is Image GT marked by experts, in which red represents gangrenous NCR and non-enhancing tumor NET, green represents edema ED, blue represents enhancing tumor ET, and black represents the background.

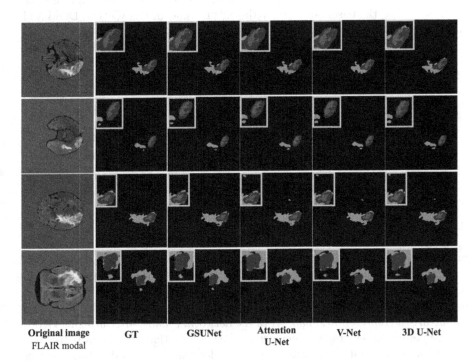

Original image GT GSUNet Attention V-Net 3D U-Net
FLAIR modal U-Net

Fig. 8. Comparison of visualization results of different brain tumor segmentation methods (the yellow box in the upper left corner of each segmentation result graph is the local enlarged result of the brain tumor) (Color figure online)

It can be seen from Fig. 8 that in the first sample and the third sample, the three methods of 3D U-Net [6], V-Net [13], and Attention U-Net [17] all have wrong predictions for edema. Observing the results of the four samples, it can be found that GSUNet not only has a higher segmentation accuracy but also retains more detailed information on the edge of brain tumors.

Based on the results of quantitative and complexity analysis, it can be found that compared with other 2D, 3D, and lightweight segmentation methods, GSUNet has extremely low model complexity, and is comparable to or even better than them in brain tumor segmentation accuracy.

3.2 Ablation Studies

This section verifies the effectiveness of GSUNet and discusses the impact of each module in GSUNet, including the 3D GM, DGM, and GSM, on segmentation performance. The ablation experiment is carried out on the BraTS 2019 dataset, and the evaluation is mainly based on the Dice coefficient acquired on the divided verification set. To eliminate the influence of other irrelevant factors, all experiments use the same experimental environment, hyperparameters, loss functions, and other experimental settings to ensure fairness. We capitalize on 3D U-Net [6] as the baseline to verify the effectiveness of the 3D GM, DGM and GSM. The experimental results are shown in Table 2.

Table 2. Ablation experiment of GSUNet

Method	Dice(%)			
	WT	TC	ET	Avg
Baseline	84.59	77.59	70.64	77.61
Baseline + 3D GM	83.98	80.46	71.98	78.81
Baseline + 3D GM + GSM	85.95	**81.58**	72.73	80.09
Baseline + 3D GM + GSM + DGM (GSUNet)	**86.16**	81.45	**73.43**	**80.35**

It can be found that by exploiting the 3D GM on 3D U-Net, the segmentation performance of the network is improved, which shows that the 3D GM has stronger feature extraction capabilities. Besides, GSUNet employs GSM to replace the convolution with stride 2 to compress the size of the feature map to achieve down-sampling, further improving the segmentation performance and enhancing the representation ability of the network. Finally, after utilizing DGM, the segmentation accuracy is increased, which shows that DGM can better utilize the low-level feature information in the decoder stage, and obtain a richer and more detailed feature representation.

4   Conclusion

This paper proposes a 3D ghost shuffle brain tumor segmentation method with low computational complexity and high accuracy, which is termed GSUNet. To evaluate the performance of the proposed model, the GSUNet is verified on public the BraTS 2019 dataset, the segmentation performance of GSUNet is evaluated quantitatively, and the model complexity and visualization results are compared and analyzed. The experimental results show that GSUNet achieves a high accuracy rate of brain tumor segmentation with extremely low computational complexity, and has strong competitiveness in the field of brain tumor segmentation.

References

1. Abraham, N., Khan, N.M.: Multimodal segmentation with MGF-Net and the focal Tversky loss function. In: Crimi, A., Bakas, S. (eds.) BrainLes 2019. LNCS, vol. 11993, pp. 191–198. Springer, Cham (2020). https://doi.org/10.1007/978-3-030-46643-5_18

2. Ahmad, P., Qamar, S., Hashemi, S.R., Shen, L.: Hybrid labels for brain tumor segmentation. In: Crimi, A., Bakas, S. (eds.) BrainLes 2019. LNCS, vol. 11993, pp. 158–166. Springer, Cham (2020). https://doi.org/10.1007/978-3-030-46643-5_15

3. Amian, M., Soltaninejad, M.: Multi-resolution 3D CNN for MRI brain tumor segmentation and survival prediction. In: Crimi, A., Bakas, S. (eds.) BrainLes 2019. LNCS, vol. 11992, pp. 221–230. Springer, Cham (2020). https://doi.org/10.1007/978-3-030-46640-4_21

4. Baid, U., Shah, N.A., Talbar, S.: Brain tumor segmentation with cascaded deep convolutional neural network. In: Crimi, A., Bakas, S. (eds.) BrainLes 2019. LNCS, vol. 11993, pp. 90–98. Springer, Cham (2020). https://doi.org/10.1007/978-3-030-46643-5_9

5. Bhalerao, M., Thakur, S.: Brain tumor segmentation based on 3D residual U-Net. In: Crimi, A., Bakas, S. (eds.) BrainLes 2019. LNCS, vol. 11993, pp. 218–225. Springer, Cham (2020). https://doi.org/10.1007/978-3-030-46643-5_21

6. Çiçek, Ö., Abdulkadir, A., Lienkamp, S.S., Brox, T., Ronneberger, O.: 3D U-Net: learning dense volumetric segmentation from sparse annotation. In: Ourselin, S., Joskowicz, L., Sabuncu, M.R., Unal, G., Wells, W. (eds.) MICCAI 2016. LNCS, vol. 9901, pp. 424–432. Springer, Cham (2016). https://doi.org/10.1007/978-3-319-46723-8_49

7. Guo, X., et al.: Brain tumor segmentation based on attention mechanism and multi-model fusion. In: Crimi, A., Bakas, S. (eds.) BrainLes 2019. LNCS, vol. 11993, pp. 50–60. Springer, Cham (2020). https://doi.org/10.1007/978-3-030-46643-5_5

8. Han, K., et al.: GhostNet: more features from cheap operations. In: 2020 IEEE/CVF Conference on Computer Vision and Pattern Recognition, CVPR 2020, Seattle, WA, USA, 13–19 June 2020, pp. 1577–1586. Computer Vision Foundation / IEEE (2020)

9. Jiang, Z., Ding, C., Liu, M., Tao, D.: Two-stage cascaded U-Net: 1st place solution to BraTS challenge 2019 segmentation task. In: Crimi, A., Bakas, S. (eds.) BrainLes 2019. LNCS, vol. 11992, pp. 231–241. Springer, Cham (2020). https://doi.org/10.1007/978-3-030-46640-4_22

10. Kingma, D.P., Ba, J.: Adam: a method for stochastic optimization. In: 3rd International Conference on Learning Representations, ICLR 2015, San Diego, CA, USA, 7–9 May 2015, Conference Track Proceedings (2015)

11. Kotowski, K., Nalepa, J., Dudzik, W.: Detection and segmentation of brain tumors from MRI using U-Nets. In: Crimi, A., Bakas, S. (eds.) BrainLes 2019. LNCS, vol. 11993, pp. 179–190. Springer, Cham (2020). https://doi.org/10.1007/978-3-030-46643-5_17

12. Ma, N., Zhang, X., Zheng, H.-T., Sun, J.: ShuffleNet V2: practical guidelines for efficient CNN architecture design. In: Ferrari, V., Hebert, M., Sminchisescu, C., Weiss, Y. (eds.) Computer Vision – ECCV 2018. LNCS, vol. 11218, pp. 122–138. Springer, Cham (2018). https://doi.org/10.1007/978-3-030-01264-9_8

13. Milletari, F., et al.: V-Net: fully convolutional neural networks for volumetric medical image segmentation. In: Fourth International Conference on 3D Vision, 3DV 2016, Stanford, CA, USA, 25–28 October 2016, pp. 565–571. IEEE Computer Society (2016)

14. Mou, L., et al.: CS-Net: channel and spatial attention network for curvilinear structure segmentation. In: Shen, D., et al. (eds.) MICCAI 2019. LNCS, vol. 11764, pp. 721–730. Springer, Cham (2019). https://doi.org/10.1007/978-3-030-32239-7_80

15. Ben Naceur, M., Akil, M., Saouli, R., Kachouri, R.: Deep convolutional neural networks for brain tumor segmentation: boosting performance using deep transfer

120   J. Hong et al.

learning: preliminary results. In: Crimi, A., Bakas, S. (eds.) BrainLes 2019. LNCS, vol. 11993, pp. 303–315. Springer, Cham (2020). https://doi.org/10.1007/978-3-030-46643-5_30

16. Nuechterlein, N., Mehta, S.: 3D-ESPNet with pyramidal refinement for volumetric brain tumor image segmentation. In: Crimi, A., Bakas, S., Kuijf, H., Keyvan, F., Reyes, M., van Walsum, T. (eds.) BrainLes 2018. LNCS, vol. 11384, pp. 245–253. Springer, Cham (2019). https://doi.org/10.1007/978-3-030-11726-9_22

17. Oktay, O., et al.: Attention U-Net: learning where to look for the pancreas. CoRR abs/1804.03999 (2018)

18. Rehman, M.U., et al.: BrainSeg-Net: brain tumor MR image segmentation via enhanced encoder-decoder network. Diagnostics (Basel, Switzerland) 11(2), 169 (2021)

19. Tai, Y., Huang, S., et al.: Computational complexity reduction of neural networks of brain tumor image segmentation by introducing fermi-dirac correction functions. Entropy 23(2), 223 (2021)

20. Yan, K., Sun, Q., Li, L., Li, Z.: 3D deep residual encoder-decoder CNNS with squeeze-and-excitation for brain tumor segmentation. In: Crimi, A., Bakas, S. (eds.) BrainLes 2019. LNCS, vol. 11993, pp. 234–243. Springer, Cham (2020). https://doi.org/10.1007/978-3-030-46643-5_23

21. Zhang, J., Jiang, Z., et al.: Attention gate reSU-Net for automatic MRI brain tumor segmentation. IEEE Access 8, 58533–58545 (2020)

ACT: Action-assoCiated and Target-Related Representations for Object Navigation

Youkai Wang, Yue Hu, Wansen Wu, Ting Liu, and Yong Peng[✉]

College of Systems Engineering, National University of Defense Technology,
Changsha, China
{wangyoukai,huyue11,wuwansen14,liuting20,yongpeng}@nudt.edu.cn

Abstract. Object navigation tasks require an agent to find a target in an unknown environment based on its observations. Researchers employ various techniques, such as extracting high-level semantic information and building a memory network, to enhance the perception and understanding of the environment. However, these methods neglect the correlation between the representations of the current scene and the target description, as well as the relationships between perception and actions. In this paper, we propose a model that uses semantic features of the visual observation as input for navigation, which are represented in the modality similar to that of the target embedding. On this basis, we fuse the visual features and spatial masks with an Encoder-Decoder transformer structure to reflect the association between perception and actions. Furthermore, in the memory module, this paper integrates the representations of explored scenes and the target information for more direct guiding of impending navigation direction selection. Our method enables the agent to perceive the position of the target more quickly and execute accurate actions to approach it. Our method outperforms the state-of-the-art (SOTA) models in the AI2Thor environment with higher navigation success rate and better learning efficiency.

Keywords: Object navigation · Semantic features · Spatial masks

1 Introduction

Object navigation [12,20] is a challenging task for embodied artificial intelligence. It requires an agent to reach instances of a given target category in an unknown environment based on the current visual observation. To complete the task, the agent need two abilities, i.e.: 1) to establish a state representation that reflects the structures of the local scene and holistic environment and the position of the agent in the scene and 2) to effectively use the state representation to plan

This research was supported partially by the National Natural Science Fund of China (Grant NO. 62306329) and the Natural Science Fund of Hunan Province (Grant NO. 2023JJ40676).

© The Author(s), under exclusive license to Springer Nature Switzerland AG 2024
S. Rudinac et al. (Eds.): MMM 2024, LNCS 14554, pp. 121–133, 2024.
https://doi.org/10.1007/978-3-031-53305-1_10

and navigate to the target. How to effectively build these two capabilities has become the focus of current research.

As the target in object navigation is usually given as an object category word, it has a significant difference in modalities from the visual inputs of environmental observations. If only image features are used, the agent usually suffer from poor image-text matching capabilities due to neglecting diverse and hierarchical semantics in the visual scenes. Thus, researchers [5,13,17] have extracted high-level semantic features from observed images to represent object-related information. These methods have greatly improved the success rate of navigation. However, there are limitations to current methods. First, although semantic features are extracted, image features are still directly used as one type of the key features for action prediction, thereby weakening the role of semantic features and increasing the difficulty of learning. Second, there is no emphasis on the features related to the target, which is not conducive to finding the target. Third, they focus on learning visual representations without considering the relationships between perception and actions, resulting in the agent being able to perceive the target location but unable to make more effective actions.

In order to address the aforementioned limitations, we propose a model based on Action-assoCiated and Target-related representations (ACT), which mainly consists of three modules: object semantic network (OSN), action associated features with a visual transformer (AF/VT), and historical memory network (HMN). In OSN, we employ an object detection model to extract high-level semantic features of objects and visual features of images from the visual observations, and use them as the basis for the other two modules. In AF/VT, we introduce spatial masks to cover the visual features obtained from processing image features in different regions based on the impact of action on perception. And we use the transformer structure [6] to treat the visual features as query, while using the semantic features as key and value to incorporate regional spatial information into the semantic features. In HMN, only the semantic features are retained in the memory node features, and we calculate the matching score between each memory node and the target embedding to retain the features that are more important for navigation. Finally, the output features from the three modules are concatenated and fed into the navigation module for action prediction. Our experiments show that the ACT outperforms the SOTA models.

Our contributions can be summarized as follows:

- We mainly use the high-level semantic features to guide the navigation of the agent. The memory features are composed of only semantic features of observation, and the visual features are fused with the semantic features using a Transformer network as query.
- We consider the correlation between the features and the target when applying them, and the agent can better recognize the spatial relationship of the target when learning feature representations.
- We introduce action associated visual features with the spatial masks to explicitly establish the relationships between perception and action.

– We validate the effectiveness of our model in the AI2Thor environment without using pre-training or expert experience. Our method improves the navigation success rates and the learning efficiency.

2   Relate Work

2.1   Visual Navigation

Visual navigation is a crucial task in robotics and AI. Traditional methods like SLAM [2] are inadequate for navigating in complicated unknown environments. Deep learning and reinforcement learning lead to more advanced visual navigation tasks, like point-goal navigation [16] and object-goal navigation [1,18,20]. Researchers [5,13,20] have applied deep reinforcement learning to solve visual navigation issues. Zhu et al. [20] employed an actor-critic model that has been widely used to improve the generalization capability of the agent. However, during training, the sparse reward received by the agent can hinder effective learning of the agent. Therefore, researchers [4–6] have employed methods such as imitation learning (IL) to provide expert experience to the agent. Du et al. [6] proposed a pre-training strategy where the agent learns to imitate the optimal navigation actions by IL. However, gathering expert policies is challenging. Comparatively, our method yields superior results compared to the SOTAs using IL.

2.2   Visual Representation in Visual Navigation

CNN-based navigation models are not sufficient to achieve optimal results, motivating researchers to introduce state representation techniques, such as object detection and semantic segmentation, to improve sample learning efficiency and strategy generalization. Anderson et al. [1] introduced an episodic semantic map and goal-oriented exploration strategies, while Yang et al. [15] utilized graph convolutional network (GCN) to learn prior knowledge of objects. There has been recent works [7,8] focusing on adding memory to navigation models. Fang et al. [7] employed memory-based Transformer to aggregate observation history for long time horizons. Fukushima el at. [8] improved navigation efficiency by incorporating semantic knowledge to Transformer. In contrast, our method mainly utilizes semantic features as the visual representations for the object navigation, which emphasizes the ability of the agent to learn spatial relationships between objects.

2.3   Attention in Visual Navigation

Attention mechanism allows models to focus its attention on input features that are relevant to the task in hand. Additionally, it has become an important approach for feature fusion. Du et al. [6] proposed VTNet, which fuses local object features and global image features to extract informative feature representations for visual navigation. Mayo et al. [11] mapped semantic and positional

spatial information of objects to improve navigation efficiency. And Chen *et al.*
[3] proposed an attention mechanism for combining perception with action to
overcome visual biases. Our approach combines these ideas by assigning visual
observations with spatial masks to pay attention to different regions according
to actions. It enables the agent to learn correlations between observations and
actions, choosing appropriate actions to achieve its navigation target.

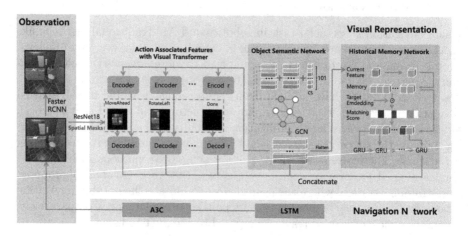

Fig. 1. Model overview. Our model consists of three modules: Object Semantic Network, Action Associated Features with Visual Transformer, Historical Memory Network. The object semantic features are obtained by GCN. The action associated features are obtained by spatial masks. The memory features are obtained by attention
with the target embedding and GRU. CS: the cosine similarity.

3   Task Definition

The object navigation task requires the agent to obtain visual observations from
a first-person perspective for navigating to a target location, without access to
prior global environmental information. The start location and pose of the agent
is set randomly at the beginning of each episode, and it is required to predict
the next action based on the current observation o_t and target class word T.
The agent can interact with the environment through six actions: MoveAhead,
RotateLeft, RotateRight, LookUp, LookDown, and Done. To be specific, the
forward step size is 0.25 m, and the angles of turning left/right and looking-
up/down are 45° and 30°, respectively. When the agent selects the Done, we
consider an episode a success if the target falls within the field of view of the
agent and is within a distance of 1.5m, otherwise it is considered a failure.

4   Proposed Method

Our goal is to guide the agent using high-level semantic features containing object spatial relationships, complemented by visual features that establish the correlation between perception and action. Our model contains three key modules, as illustrated in Fig. 1: 1) Object Semantic Network, which extracts high-level semantic information of objects from visual observations to construct an object semantic graph that represents object relationships, 2) Action Associated Features with a Visual Transformer, which employs spatial masks on the visual features corresponding to different actions, and integrates them with semantic features using VT fusion, and 3) Historical Memory Network, which obtains target-related memory features with the semantic features and the target emdedding, and extracts memory representations using GRU. Then we perform action prediction based on the concatenation of the three types of features.

4.1   Object Semantic Network

As the target location in the object navigation task is usually specified textually as a object category instead of a target image, it would be more modality-aligned to represent the streaming observations in a higher-level semantic form rather than mere visual features. Inspired by the work from Qiu *et al.* [13], we use Faster-RCNN [14] to detect objects and extract object categories and key attributes. The semantic feature of the object category o_i can be represented as:

$$v_i = [[b, x_i, y_i, bbox_i], MLP[bags, g_i], cs_{i,T}], o_i \in O, 1 \leq i \leq 101 \qquad (1)$$

where O is the list of all the 101 object categories in AI2Thor, and $[\cdot]$ denotes concatenation. The element b is a binary vector specifying whether the class o_i can be detected in the current frame. The elements (x_i, y_i) and $bbox_i$ correspond to the (x, y) coordinates of the center of the bounding box of the class o_i, and its covered area, respectively. If class o_i is not detected, the elements mentioned above will be set to 0. The element $bags \in R^{101 \times 1}$ is a bag-of-objects feature for object categories appearing in o_t. This element adds environment semantic information of the observation to the object semantic features. The element g_i is the word embedding of the class o_i. The element $cs_{i,T}$ is the cosine similarity between the word embeddings of the class o_i and the target object T, which can be calculated as:

$$cs_{i,T} = cosine(g_i, g_T) = \frac{g_i \cdot g_T}{\|g_i\| \cdot \|g_T\|}. \qquad (2)$$

Since the elements $bags$ and g_i have high dimensions, we use a multilayer perceptron (MLP) to transform them to lower dimensions. Then, we combine the seven elements to derive the object features $v_i \in R^{1 \times d}$, where d is the dimension of the embedding vector.

To integrate semantic knowledge into reinforcement learning, we represent it as a graph and apply GCN to calculate the relationships between nodes. We define a graph by $Graph = (V, A)$, where $V \in R^{101 \times d}$ and $A \in R^{101 \times 101}$ denote the nodes and the edges, respectively. The node features V consists of the semantic features v of 101 kinds of objects, and the adjacency matrix A is initialized using an external dataset VG [15], and it is continuously updated during training to reflect the relationships between objects. With the node features V and adjacency matrix A, our GCN outputs a graph representation $L \in R^{101 \times d}$ as the object semantic embedding:

$$L = ReLU(A \cdot V \cdot W) \tag{3}$$

where $W \in R^{d \times d}$ is a learnable matrix, and ReLU is a rectified linear unit.

4.2   Action Associated Features with Visual Transformer

Visual features contain crucial information for distinguishing specific environments. However, the direct concatenation of semantic knowledge and visual features can lead to excessive weight of visual features in the decision-making process, reducing efficiency and success rate. To address this, we first design a visual transformer structure utilized in many works [6–8] to integrate semantic features and visual features, exploiting the spatial relationship between detected objects and observed areas. Specifically, we enter the object semantic features L into the Encoder module to learn object associated representation $\hat{L}$, which utilizes a Transformer structure [9] consisting of multiple attention modules and feed-forward layers. Then the visual features G obtained by processing with ResNet18 is referred to as the location query and fed into the Decoder with $\hat{L}$ in order to embed region position information into the semantic features. And the attention function of the visual transformer decoder is expressed as:

$$Decoder(G, \hat{L}) = softmax(\frac{G\hat{L}^T}{\sqrt{d}})\hat{L}^T. \tag{4}$$

In visual navigation, the agent predicts actions based on its observations, which are in turn affected by its execution of those actions. We hope the agent can take effective actions so that the probability of new observations containing the target can be increased. For instance, when the target or objects related to

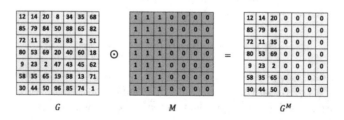

$$G \qquad\qquad M \qquad\qquad G^M$$

Fig. 2. Visualization of the spatial ask about RotateLeft. The masked features are computed through the inner product of visual features and the spatial mask.

the target appear on the left side of the visual field, the effective action of the agent is to turn left. If they appear in the center of the visual field, the effective action is to move forward. Therefore, if the agent is able to perceive how its actions affect observations and assess the correlation between different changes in observations and the target, it will be easier for the agent to choose effective actions that bring it closer to the target.

Inspired by the ANA method from Chen $et\ al.$ [3], we aim to explicitly couple visual observations with feasible actions. As shown in Fig. 1, we use a set of spatial masks that map the attention distribution to different candidate actions while highlighting the regions of interest according to different actions. The ANA employs learnable masks to focus on related objects in obtained visual attention, while we utilize fixed spatial masks on the visual features to highlight regions of interest. Our method enables a stronger correlation between perception and actions. For example, when the agent turns left, the method masks the right half of the visual features G, as shown in Fig. 2. Spatial masks can be expressed as:

$$G^M = G \cdot M \tag{5}$$

where $M \in R^{6 \times d \times d}$ represents the spatial masks, 6 is the number of candidate actions. After obtaining the 6 visual features corresponding to each action, we independently integrate 6 features with the object semantic features L by the VT, and obtain the action-associated features G_{action}:

$$G_{Action} = [Decoder(G_i^M, Encoder(L))]_{i=1}^6 \tag{6}$$

During the training, the agent acquires knowledge of the correlation between six features and the target, which in turn enhances its ability to select regions of interest that are proximal to the target and execute actions based on the corresponding features.

4.3   Historical Memory Network

The object navigation task is a partially observable Markov decision process where the agent needs to refer to historical observations and actions for current decision-making [10]. Most methods use a transformer network for memory encoding, but it can be computationally costly as the trajectory length increases. In addition, not all trajectory information is useful for the task at hand. Thus, we propose an external memory mechanism that encodes the object semantic features L of previous and current steps and use GRU to process the historical memory. Ours differs from the work of Du $et\ al.$ [5] in that we calculate attention weights separately for each memory slot with the target embedding, and directly use memory features for navigation. This avoids concentrating the attention of the agent on the explored areas unrelated to the target. We flatten the memory features $H = \{L_1, L_2, \cdots, L_t\}$ and use an MLP to keep them consistent with the dimension of the target emdedding g_T. Then we calculate the matching score between g_T and each memory slot of the memory features H by taking the inner product. And the embedded memory features is weighted by the matching score, and is passed into the GRU network:

$$H_M = MLP(Flatten(H)) \tag{7}$$

$$\hat{H} = softmax(H_M \cdot g_T) \cdot H_M \tag{8}$$

$$H_{GRU} = GRU(\hat{H}) \tag{9}$$

4.4 Navigation Network

The aforementioned object semantic features, action associated features, and memory features are concatenated as the input of the navigation module:

$$F = [L, G_{Action}, H_{GRU}]. \tag{10}$$

Following the previous literature [13,20], we approach this task as a reinforcement learning problem and employ the Asynchronous Advantage Actor-Critic (A3C) algorithm. As for the reward structure, we follow a similar setting to Qiu et al. [13]. When the agent successfully navigates to the target, it receives a reward of $R_s = +5$. If it identifies an object related to the target, it receives a partial reward $R_p = p \cdot R_s$, $p \in (0, 0.5)$, where p represents the proximity between an object and the target. The larger the value of p, the closer the object is to the target. Otherwise, a time penalty of -0.01 is given.

5 Experimental Results

5.1 Dataset and Metrics

Dataset. We use the AI2Thor dataset as the experimental platform for the target-driven navigation task. The AI2Thor dataset includes four types of rooms: kitchen, living room, bedroom, and bathroom. Each type consists of 30 rooms. The first 20 rooms of each type are used as the training set, and the remaining 10 rooms are used as the test set.

Metrics. We evaluate the navigation performance of the agent using the success rate (SR) and the success weighted path length (SPL). The SR estimates the effectiveness of navigation, and is formulated as $\frac{1}{N}\sum_{n=1}^{N} S_n$, where N stands for the total number of episodes, and S_n is the binary indicator of n-th episode. The SPL measures the efficiency of navigation, i.e. whether the agent can reach the goal via a shorter path. It is calculated as $\frac{1}{N}\sum_{n=1}^{N} \frac{Len_{opt}}{max(Len_n, Len_{opt})}$, where L_n and L_{opt} denote the length of the path of the agent and the optimal path for the n-th episode, respectively.

5.2   Training Details

Our model is implemented on top of the code provided by Qiu *et al.* [13]. We train the agent on 1.4 million episodes using the AI2Thor offline dataset. During the testing phase, we follow the same experimental setup as the previous literature [5,13], performing 250 test episodes for each of the four room types, resulting in a total of 1000 episodes. The VT contains two multi-head self-attention mechanism modules and two layers in the encoder and decoder. And we employ Adam optimizer to update the parameters of our networks with a learning rate 10^{-4}. All experiments are conducted on a server with 4 GeForce RTX 2080Ti GPUs.

5.3   Comparison Methods

In our experiments, we compare our proposed method with several baselines, including the following:

Random, randomly samples actions from uniform distribution. **SAVN** [15], uses meta-reinforcement learning to adapt to unseen environments. **MJOLNIR** [13], considers intrinsic relationships between target and objects. **OMT** [8], utilizes object-memory transformer to encode historical information. **ORG** [5], proposes object relation graph to learn relationships, external memory to detect and break deadlocks, and IL to provide guidance. **HOZ** [17], proposes hierarchical object-to-zone graph to capture prior knowledge of typical objects. **GVT** [19], proposes graph transformer viterbi inference network to discover new relationships and explore optimal actions. **VTNet** [6], uses visual transformer to encode observations and spatial-aware descriptors. **L-sTDE** [18], calculates the layout gap between the current environment and the prior knowledge to appropriately control the effect of the experience.

5.4   Quantitative Results

Table 1 shows the results of the comparative methods and our model. It demonstrates that our model outperforms prior works in metrics of SR and SPL under the same conditions without using pre-training or imitation learning (IL). Our model outperforms the SOAT method OMT by 6.3%/4.1%, 0.8%/4.6% and in SR and SPL (ALL/L $\geq$ 5). OMT embeds observed visual and semantic features as the memory nodes. Hence, our experimental results reveal that object semantic features play a critical role, and extracting memory features related to the target significantly aids in navigation.

We also compare the performance against the SOTA models with pre-training or IL. Their models have a higher SPL, because they benefit from prior experience or guidance from experts. However, our SR is higher than these methods. Considering that SR measures the effectiveness of navigation, which is the primary goal, it confirms the effectiveness of our model. ORG, HOZ, GVT, and L-sTDE emphasize the extraction and utilization of semantic features, but our SR is higher by 11.1%, 6.8%, 5.9%, and 2.4% than theirs, respectively, which indicates that the our method is more efficient and effective in constructing

Table 1. Comparison with state-of-the-art models in AI2Thor. Type I models don't employ pre-training or IL, while Type II models utilize these techniques.

Type	Method	ALL (%)		$\mathbb{L} \geq 5$ (%)	
		SR	SPL	SR	SPL
I	Random	11.2	2.1	1.1	0.5
	SAVN [15]	35.7	9.3	23.9	9.4
	MJNOIR [13]	65.3	21.1	50.0	20.9
	VTNet [6]	69.7	24.0	56.6	26.0
	OMT [8]	71.1	27.5	61.9	26.6
II	ORG [5]	66.3	38.4	57.4	37.4
	HOZ [17]	70.6	40.0	62.7	39.2
	GTV [19]	71.5	42.0	60.1	43.2
	VTNet [6]	72.2	44.9	63.4	44.0
	L-sTDE [18]	75.0	41.4	–	–
–	**Ours**	**77.4**	28.3	**66.0**	31.2

semantic features. Our method uses the VT structure like VTNet, but we use the spatial masks instead of positional embedding. Our SR improves by 5.2% compared to VTNet. The results show that identifying the relationship between perception and actions is important for visual navigation.

In addition, our model is only trained for 1.4 million episodes, while a minimum of 3 million episodes are required for training for the baselines. This suggests that our proposed semantic and memory features can help the agent perceive object position more quickly, while the explicit association between perception and actions helps the agent learn to make effective actions through perception more quickly.

Table 2. Ablation results on each module in the three sub-networks: object semantic features, action associated features and memory features. MS: the matching score.

ID	Object Semantic	Spatial Masks	Memory		ALL (%)		$\mathbb{L} \geq 5$ (%)	
			w/o MS	MS	SR	SPL	SR	SPL
1	✓				53.2	13.7	36.7	14.2
2		✓			72.2	27.1	58.9	28.5
3	✓	✓			74.7	27.5	61.6	29.5
4	✓	✓	✓		73.1	26.8	58.0	27.6
5	✓	✓		✓	77.4	28.3	66.0	31.2

5.5   Ablation Study

We verify the impact of each component of our model, including object semantic features, action associated features with VT, and memory feature, and we show the results of ablation experiments in Table 2. The baseline uses only object semantic feature, and achieves a SR of 53.2%. These results show the importance of high-level semantic information of object in navigating indoor environments.

Table 3. The ablation study of the learnable and fixed spatial masks.

ID	Masks	ALL (%)		$\mathbb{L} \geq 5$ (%)	
		SR	SPL	SR	SPL
1	learnable	76.0	27.9	64.6	30.8
2	fixed	77.4	28.3	66.0	31.2

Table 4. The ablation study of the memory features type. Vis: the viusal features, Sem: the semantic features.

ID	Memory	ALL (%)		$\mathbb{L} \geq 5$ (%)	
		SR	SPL	SR	SPL
1	Vis	69.0	25.7	54.2	26.9
2	Vis+Sem	72.3	26.4	58.2	27.7
3	Sem	77.4	28.3	66.0	31.2

Action Associated Features: When using only action-associated features with a visual transformer, we achieved a SR of 72.2%, indicating that using the spatial masks to mask the visual features can effectively inform the agent of the visual attention differences from different actions. Additionally, when adding object features in line 3, the SR and SPL are further improved, which shows explicit object feature representation is effective for guiding navigation.

Memory Features: After adding memory features, the SR and SPL achieves the best performance. And our target-related memory features have a positive effect on navigation in comparison with not using the matching score. The features help the agent understand the global information of the explored environment, and introduce additional operations to approach the target better.

Spatial Masks Are Learnable or Fixed? As shown in Table 3, fixed masks have a higher SR and SPL, indicating that they better reflect the correlation between perception and actions and help the agent choose effective actions.

Memory Slots are Represented by Semantic or Visual Features? Table 4 shows that the higher the proportion of visual features in the memory features, the lower the SR. The result demonstrates that high-level semantic features are more beneficial for object navigation.

5.6   Qualitative Study

Figure 3 presents the example top-down navigation trajectories maps of our approach and two competing methods. The starting point of the agent is located at the upper left of the room, and the target is a pillow located in the lower right of the room. The agent's initial observation lacked useful features due to facing the wall. When the agent first discovers the target, there are obstacles in the way.

Baseline   **VTNet**   **Ours**

Fig. 3. Qualitative results. The top-down map shows the trajectories of different models. The white triangle represents the starting point and pose of the agent. The red box depicts the target. The green curve represents navigation trajectory of the agent. (Color figure online)

After turning left, the target disappears from view once again. Due to the lack of memory features in the baseline and the VTNet, it is difficult for the agent to perceive information related to the target. As a result, the agent fails to reach the target. Our model can perceive the target's location thanks to its memory features. It also establishes correlations between perception and actions, which helps the agent make a timely right turn, enabling the target to reappear in the field of view. It allows the agent to continue moving forward towards the target until reaching it successfully.

6   Conclusion

In this work, we proposed the ACT model for the object navigation. Our motivation is to establish the association between perception and actions, as well as the correlation between feature representations and the target. Specifically, we utilized object semantic features, memory features based on target embedding, and action-associated features based on spatial masks for different actions to provide superior guidance for the planning. Experimental evaluation in the AI2Thor environment demonstrated the effectiveness of our model.

References

1. Anderson, P., Chang, A., Chaplot, D.S., et al.: On evaluation of embodied navigation agents. arXiv preprint arXiv:1807.06757 (2018)
2. Chaplot, D.S., Gandhi, D., Gupta, S., et al.: Learning to explore using active neural SLAM. In: 8th International Conference on Learning Representations (ICLR 2020), Addis Ababa, 26–30 April 2020. OpenReview.net (2020)
3. Chen, S., Zhao, Q.: Attention to action: Leveraging attention for object navigation. In: The British Machine Vision Conference (2021)
4. Dang, R., Shi, Z., Wang, L., et al.: Unbiased directed object attention graph for object navigation. In: Proceedings of the 30th ACM International Conference on Multimedia, pp. 3617–3627 (2022)

5. Du, H., Yu, X., Zheng, L.: Learning object relation graph and tentative policy for visual navigation. In: Vedaldi, A., Bischof, H., Brox, T., Frahm, J.-M. (eds.) ECCV 2020. LNCS, vol. 12352, pp. 19–34. Springer, Cham (2020). https://doi.org/10.1007/978-3-030-58571-6_2

6. Du, H., Yu, X., Zheng, L.: Vtnet: visual transformer network for object goal navigation. arXiv preprint arXiv:2105.09447 (2021)

7. Fang, K., Toshev, A., Fei-Fei, L., et al.: Scene memory transformer for embodied agents in long-horizon tasks. In: Proceedings of the IEEE/CVF Conference on Computer Vision and Pattern Recognition, pp. 538–547 (2019)

8. Fukushima, R., Ota, K., Kanezaki, A., et al.: Object memory transformer for object goal navigation. In: 2022 International Conference on Robotics and Automation (ICRA), pp. 11288–11294. IEEE (2022)

9. He, K., Zhang, X., Ren, S., et al.: Deep residual learning for image recognition. In: Proceedings of the IEEE Conference on Computer Vision and Pattern Recognition, pp. 770–778 (2016)

10. Kwon, O., Kim, N., Choi, Y., et al.: Visual graph memory with unsupervised representation for visual navigation. In: Proceedings of the IEEE/CVF International Conference on Computer Vision, pp. 15890–15899 (2021)

11. Mayo, B., Hazan, T., Tal, A.: Visual navigation with spatial attention. In: Proceedings of the IEEE/CVF Conference on Computer Vision and Pattern Recognition, pp. 16898–16907 (2021)

12. Mousavian, A., Toshev, A., Fišer, M., et al.: Visual representations for semantic target driven navigation. In: 2019 International Conference on Robotics and Automation (ICRA), pp. 8846–8852. IEEE (2019)

13. Pal, A., Qiu, Y., Christensen, H.: Learning hierarchical relationships for object-goal navigation. In: Conference on Robot Learning, pp. 517–528. PMLR (2021)

14. Ren, S., He, K., Girshick, R., et al.: Faster r-cnn: towards real-time object detection with region proposal networks. Adv. Neural Inf. Process. Syst. **28** (2015)

15. Yang, W., Wang, X., Farhadi, A., et al.: Visual semantic navigation using scene priors. arXiv preprint arXiv:1810.06543 (2018)

16. Ye, J., Batra, D., Wijmans, E., et al.: Auxiliary tasks speed up learning point goal navigation. In: Conference on Robot Learning, pp. 498–516. PMLR (2021)

17. Zhang, S., Song, X., Bai, Y., et al.: Hierarchical object-to-zone graph for object navigation. In: Proceedings of the IEEE/CVF International Conference on Computer Vision, pp. 15130–15140 (2021)

18. Zhang, S., Song, X., Li, W., et al.: Layout-based causal inference for object navigation. In: Proceedings of the IEEE/CVF Conference on Computer Vision and Pattern Recognition, pp. 10792–10802 (2023)

19. Zhou, K., Guo, C., Zhang, H., et al.: Optimal graph transformer viterbi knowledge inference network for more successful visual navigation. Adv. Eng. Inform. **55**, 101889 (2023)

20. Zhu, Y., Mottaghi, R., Kolve, E., et al.: Target-driven visual navigation in indoor scenes using deep reinforcement learning. In: 2017 IEEE International Conference on Robotics and Automation (ICRA), pp. 3357–3364. IEEE (2017)

Foreground Feature Enhancement and Peak & Background Suppression for Fine-Grained Visual Classification

Die Yu, Zhaoyan Fang, and Yong Jiang[✉]

School of Computer Science and Technology, Southwest University of Science and Technology, Mianyang, China
yudie@mails.swust.edu.cn, jiang_yong@swust.edu.cn

Abstract. Fine-grained visual classification task is challenging due to large variation in the appearance of the same subcategories and similarity in the appearance of different subcategories. To tackle the challenges, locating multiple discriminative regions plays a critical role. However, most previous works ignore the impact of background, which may provide negative clues that are not necessary or harmful for the network to classification. In this paper, we propose Foreground Feature Enhancement (FFE) module and Peak & Background Suppression (PBS) module, which are inserted in different layers of the CNN. The FFE module is designed to enhance and extract the most discriminative feature in the feature maps, and the PBS module is employed to suppress the peak features and background noise in the feature maps, forcing the network to mine other equally important features. Our proposed method can be trained end-to-end and does not require bounding boxes/part annotations. The experimental results achieve competitive performances on CUB200-2011, FGVC Aircraft, and Stanford Cars datasets.

Keywords: Fine-grained visual classification · Attention · feature enhancement · background suppression

1 Introduction

Fine-grained visual classification (FGVC) aims at delineating subcategories under a broad category, such as distinguishing different kinds of birds [1], cars [2], and aircrafts [3]. The challenges of fine-grained visual classification tasks can be segmented into two aspects: high intra-class and low inter-class visual variances, caused by illumination, different perspectives or poses of objects, and shooting angles, as shown in Fig. 1.

Previous approaches relied heavily on annotation information, which makes manual labelling is costly. To address above issues, recent works show that the key solution for fine-grained classification is to focus on extracting discriminative subtle features in multiple parts of object. However, it is hard for a solo CNN model to explain the discrepancies between subordinate classes of similar appearance. Consequently, the existing methods based on the CNN can be divided into the following two categories. The first type of

© The Author(s), under exclusive license to Springer Nature Switzerland AG 2024
S. Rudinac et al. (Eds.): MMM 2024, LNCS 14554, pp. 134–146, 2024.
https://doi.org/10.1007/978-3-031-53305-1_11

approach is localization which aims to generate local candidate regions with high distinguishability, such as NTS-Net [4], CCFR [5], S3N [6], and CP-CNN [7], but this type of method usually requires training a separate network to detect the target object, which may lead to high network complexity. The second type of approach is to strengthen the high-response areas among feature maps through the attention mechanism, such as MA-CNN [8], Cross-X [9], API-Net [10], and ACNet [11], but this type of method is affected by background, causing the network to notice features that are not favorable for classification. Here we refer to the areas except the target object itself as background.

Fig. 1. Two birds of similar appearance, American Crow, and Fish Crow. Each row denotes three distinct looks for the same bird. Each column represents two different birds, which look similar because of the lighting, the bird's pose and the angle of the shot, respectively.

Inspired by object enhancement and background suppression in [12, 13], we design an architecture interspersed with CNN middle layers for Foreground Feature Enhancement (FFE) and Peak & Background Suppression (PBS), where the FFE module boosts the most prominent regions of the attention maps obtained in the present stage to get region-specific representation, and the PBS module is applied to suppress the peak by the FFE module and background noise to force the network to mine useful features of other regions in the next stage. The two modules, FFE and PBS, complement each other, making the network much more capable of learning diverse features.

Our main contributions are as follows:

- Our PBS module not only masks out feature in peak regions to force the network to look for subtle differences among categories that look similar, but also suppresses background noise, allowing the network to focus more on the foreground itself.
- We use only two simple modules to achieve great performance, and the proposed method makes the inference time faster.

2 Related Work

In this part, we review two representative methods for finding discriminative regional features, including object-part-based approaches and attention-based approaches. The former aims at training a detection network to locate subregions used for performing classification. The latter focuses on improving feature learning and locating object details using attention mechanism.

2.1 Object-Part-Based Approaches

Due to the high cost of labor annotation, current works mainly depend on image-level labels to locate objects, then extract different specific parts. [14] propose to mark the maximally connected domains of different convolutional layers and obtain the bounding box of the object from their intersection, and then select distinguishable subtle features by sliding window. [5] propose to use the hierarchical structure of Feature Pyramid Network (FPN) and the non-maximum suppression (NMS) algorithm to locate local key regions, which are later fused to obtain embedded features to form a database for re-ranking the classification results. In addition, [7] propose to take advantage of the deep features to suppress the noise of large components in the shallow features, and using the shallow features to enrich the information of the deep features, as a way to improve the performance of locating locally discriminative regions. They all locate local parts of objects by generating part candidate boxes from the model.

2.2 Attention-Based Approaches

In fact, the local regions located by the object-part-based method still contain some redundant information, which can affect the classification performance. Therefore, many papers use attention mechanisms to capture prominent and potential features to resolve the ambiguity present in fine-grained visual classification, and so does this paper. [15, 16] propose to suppress the discriminative regions of the class activation maps and force the network to mine alternative features that are informative. Despite having performed well, they ignore the effect of background. Furthermore, [17] use attention maps to eliminate similar regions across classes, this is done by simply obtaining various attention maps based on different predictions of the same image, and then suppressing the similar high response regions in these attention maps. Nevertheless, the impact of background is still ignored. Unlike the former, [18] propose a module to incorporate background suppression knowledge in classification tasks, which adopts classification confidence scores generated by multiple backbone blocks to suppress eigenvalues in low confidence regions. In contrast to it, our PBS structure is easier than it does, and only one backbone block is used.

3 Method

In this section, we will describe our proposed method in detail. The overview is displayed in Fig. 2, which consists of two main modules: (1) A Foreground Feature Enhancement (FFE) module aims at learning a variety of different part-specific discriminative representations. (2) A Peak & Background Suppression (PBS) module not only suppresses most of the background, allowing the network to better distinguish between foreground regions, but also masks out features in peak regions, forcing the network to mine other features with subtle differences.

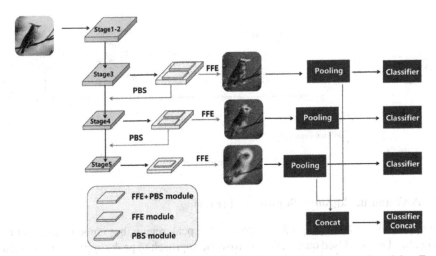

Fig. 2. An overview of our overall architecture. Our method mainly contains two modules: Foreground Feature Enhancement (FFE) and Peak & Background Suppression (PBS). The orange arrow points through the FFE module, and the green arrow points through the PBS module. The Classifier present two fully connected layers, ending with a softmax layer.

3.1 Foreground Feature Enhancement (FFE)

Let F be our backbone feature extractor, which has L stages. The output feature maps extracted from F are expressed as $F^i \in R^{c_i \times h_i \times w_i}$, where c_i, h_i, w_i refers to the number of channels, height and width of the feature map at i^{th} satge, and $i = \{1, 2, \cdots, L\}$. Inspired by CBAM [19], we feed the spatial attention of CBAM with the feature map output from the last S stages: $i = \{L - S + 1, \cdots, L - 1\}$. Note that for spatial attention, we concatenate the feature maps for GAP and GKMP, and then compress the channels using $7 \times 7, 5 \times 5, 3 \times 3$ convolution, respectively:

$$A_i = Conv(Concat[GAP\left(F^i\right), GKMP(F^i)]) \tag{1}$$

$$A_i' = sigmoid\,(A_i) \tag{2}$$

Here, the non-linear function Sigmoid is used to normalize, where GAP denotes global average pooling, and GKMP indicates global K-Max Pooling. The diagram of the FFE is shown in Fig. 3. Then, by enhancing the most outstanding part, we obtain the part with the boosted feature F_b^i:

$$F_b^i = F^i + \alpha * (A_i' \otimes F^i) \tag{3}$$

where α is a hyper-parameter, $\otimes$ means element-wise multiplication.

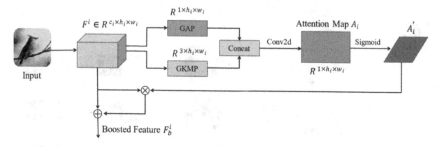

Fig. 3. The diagram of the FFE.

3.2 Peak and Background Suppression (PBS)

The approach of FBSD [16] only suppresses the peak region, but ignores the effect of background noise. Based on this, we improve the approach of peak suppression and then propose a new background suppression method. The diagram of the PBS is shown in Fig. 4.

Considering that the attention mechanism acts on the entire feature maps, either the target object itself or the regions containing the background noise can be activated. Furthermore, the high response areas generally tend to act on the target object. Therefore, we propose the peak suppression method which forces the network to mine other sub-response regions by suppressing the high response regions. Even if the high response areas are not on the target object but on the background, background noise can be suppressed indirectly through peak suppression. In addition, based on human experience, we speculate that most regions of homogeneous response are almost always situated on the background, so some of the background noise can be suppressed by suppressing regions smaller than the mean of the attention maps.

Based on the above analysis, here, we compute the maximum and the average of the obtained attention maps, then we consider values less than $mean(A_i)$ as background and values greater than $max(A_i) * \gamma$ as peak regions. So, by suppression of these parts, we can get the suppressed feature F_s^i:

$$M_s^i = \begin{cases} 0, \text{if } A_i < \text{mean}(A_i) \\ 0, \text{if } A_i > \text{max}(A_i) * \gamma \\ 1, \text{otherwise} \end{cases} \tag{4}$$

$$F_s^i = F^i + \beta * (F^i \otimes M_s^i) \tag{5}$$

where β, γ are hyper-parameters, γ is used for controlling the degree of suppression, and M_s^i is a mask.

3.3 Network Design

As depicted in Fig. 2, we use Resnet as an example. Resnet's feature extractor is divided into five stages, with the spatial dimension of the feature maps being reduced by half at the end of each stage. Bearing in mind that the semantic information at a deeper

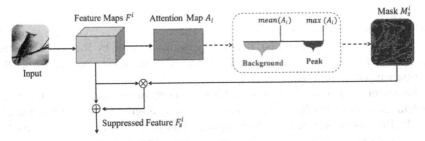

Fig. 4. The diagram of the PBS.

stage is richer and the receptive field is wider, we utilized the last S stages. The FFE module enhances several specific parts that depend on the PBS module's suppression of peaks of foreground objects and background regions. Our model is light and simple, it can gradually acquire multi-granularity feature representations at different stages of learning.

In the phase of the training, we use the augmented features F_b^i obtained in Eq. (3) for classification, and adopt cross entropy loss to compute each classification loss L_{cls}^i at i^{th} stage:

$$V_i = \varnothing\left(F_b^i\right), V^{concat} = \sum\nolimits_{i\in[L-S+1,\cdots,L]} V_i \tag{6}$$

$$p_i = cls_i(V_i), p_{concat} = cls_{concat}(V^{concat}) \tag{7}$$

$$L_{cls}^i = -log(p_i(c)), L_{cls}^{concat} = -log(p_{concat}(c)) \tag{8}$$

where V_i represents feature vector, $\varnothing$ represents a series of convolution and pooling operations, cls_i represents a classifier, p_i represents the probability of prediction, c represents the ground truth label of the input image. The total losses are as follows:

$$L_{total} = \sum\nolimits_{i\in[L-S+1,\cdots,L]} L_{cls}^i + L_{cls}^{concat} \tag{9}$$

where $S = 3$ denotes the last three stages of Resnet. For the inference, the final prediction is the average of the prediction values of the enhanced features at each stage.

4 Experiments

4.1 Datasets and Implementation Details

The datasets used to evaluate our model are the CUB-200-2011 [1], Stanford Cars [2], FGVC Aircraft [3]. Specifically, CUB-200-2011 contains 200 subcategories with 11788 images, in which 5994 ones as training set and 5794 ones as test set. Stanford Cars dataset owns 196 subcategories with 16185 images, in which 8144 ones as training set and 8041 ones as test set. FGVC Aircraft dataset has 100 subcategories with 10000 images, in which 6667 ones as training set and 3333 ones as test set. The details of each dataset are summarized in Table 1.

We validate the performance of our model on Resnet50 and Resnet101 [20], all of which are pretrained on the ImageNet dataset. The input image is preprocessed to 448 × 448. During the training period, data augmentation is done by Random Crop and Random Horizontal Flip, while Center Crop is used in the testing period. We set the hyper-parameters as follows: $\alpha = 0.5$, $\beta = 0.5$, $\gamma = 0.9$.

Our model is optimized by Stochastic Gradient Descent (SGD) with the momentum of 0.9, the number of epochs is set to 500, weight decay is set to 1e–5, mini-batch size is set to 20. Besides, we set the initial learning rate of the backbone layers to 0.002 and the newly added layers to 0.02. The learning rate is adjusted by the cosine anneal scheduler. Our experiments are implemented by Pytorch with two NVIDIA Titan V GPU.

Table 1. Three benchmark datasets with fine granularity.

Dataset	Name	#Class	#Train	#Test
CUB-200–2011 [1]	Bird	200	5994	5794
Stanford Cars [2]	Car	196	8144	8041
FGVC Aircraft [3]	Aircraft	100	6667	3333

4.2 Quantitative Results

In Table 2, we compare our method with some recent works on three of the most widely used benchmarks. On these datasets, our method delivers competitive results.

CUB-200–2011. The comparison of the results on the CUB-200–2011 are shown in the third column of Table 2, it can be seen that our proposed model achieves the best accuracy of 89.6% on Resnet50 backbone. Compared with the Resnet101 based methods, our model outperforms most of the comparable models, except that it is slightly lower than FBSD [16]. The possible reason of the performance drop is that images of bird dataset have a more complex background than those of car and aircraft, which may lead to overfitting when the network layers are deeper.

FGVC Aircraft. The fourth column shows the results for the FGVC Aircraft, on Resnet50, our model is 0.5% less than FDL [21], but compared with the FDL's two-stage approach, our one-stage approach is more lightweight. On Resnet101 backbone, our model outperforms all methods, we obtain an improvement of 2.1%, 1.9%, 0.7%, and 0.4% higher than ISQRT-COV [22], DBT-Net [23], CIN [24], and FBSD [16] respectively.

Stanford Cars. The fifth column show the results of the Stanford Cars, on this dataset, our method shows the best performance of 94.7% on Resnet50 and 95.0% on Resnet101 backbone, respectively. This phenomenon proves the validity of our model.

In Table 3, using FBSD [16] as the baseline model, we compare its parameters and inference time with our model on the CUB-200–2011 dataset. It can be concluded that although the number of parameters is the same for both, it is clear that the inference time of our model is 26% faster than its.

Table 2. Comparison of different methods on CUB-200–2011, FGVC Aircraft, and Stanford Cars. "-" indicates that the result was not reported in the related paper.

Methods	Backbone	CUB	AIR	CAR
RA-CNN [25]	VGG19	85.3	88.1	92.5
MA-CNN [8]	VGG19	86.5	89.9	92.8
MAMC [26]	Resnet50	86.2	-	92.8
NTS [4]	Resnet50	87.5	91.4	93.3
Cross-X [9]	Resnet50	87.7	92.6	94.5
DCL [27]	Resnet50	87.8	93.0	94.5
ACNet [11]	Resnet50	88.1	92.4	94.6
GCL [28]	Resnet50	88.3	93.2	94.0
FDL [21]	Resnet50	88.6	**93.4**	94.3
ELP [29]	Resnet50	88.8	92.7	94.2
MHEM [30]	Resnet50	88.2	92.9	94.2
FBSD [16]	Resnet50	89.3	92.7	94.4
Ours	Resnet50	**89.6**	92.9	**94.7**
MAMC [26]	Resnet101	86.5	-	93.0
DBT-Net [23]	Resnet101	88.1	91.6	94.5
CIN [24]	Resnet101	88.1	92.8	94.5
API-Net [10]	Resnet101	88.6	93.4	94.9
ISQRT-COV [22]	Resnet101	88.7	91.4	93.3
FBSD [16]	Resnet101	**89.5**	93.1	**95.0**
Ours	Resnet101	89.4	**93.5**	**95.0**

Table 3. Performance of FBSD and our model evaluated on CUB-200–2011 dataset. Resnet50 is used as the backbone.

	Methods	Parameters(M)	FPS
Resnet50 Backbone	FBSD	41.1	32.14
	Ours	41.1	39.35

4.3 Ablation Studies

In order to better understand our approach, we carry out ablation studies on the CUB-200-2011 dataset, using Resnet50 as the backbone.

Effect of FFE. As shown in Table 4, when the Foreground Feature Enhancement (FFE) is introduced into Resnet50, the accuracy achieves 88.9%, which is 2.4% higher than the baseline. As shown in Fig. 5, from the two heat maps of stage 3 and stage 4 with FFE, we can see that the prominent areas of the bird have been enhanced. This result demonstrates the validity of this component. But inevitably, there is still a lot of background noise.

Effect of PBS. As shown in Table 4, when the Peak & Background Suppression (PBS) is introduced, the accuracy of our model achieves 89.6%, which is 0.7% higher than the one based on Resnet50 with only Foreground Feature Enhancement (FFE). As shown in Fig. 5, we first discuss the effect of the peak suppression, take *Rufous_Hummingbird* for example, from the heat maps of stage 5 with PBS, we can see that the peak region including the head is suppressed, and so the network can search for another feature (the bird's beak) easily. As for the effect of the background suppression, from the heat maps of stage 3 and stage 4, comparing with the use of the FFE module alone, we can clearly see that the background noise is suppressed. The result demonstrates the validity of this component, and likewise demonstrates that the FFE module and the PBS module complement each other.

Effect of the threshold γ. As shown in Table 5, we evaluate the extent to which the suppression factor γ in Eq. (4) affects the accuracy from 0.7 to 1. We can observe that the best accuracy is achieved when $\gamma = 0.9$. When $\gamma < 0.9$, we infer that the possible reason for the performance degradation is that excessive strength of suppression resulting in the model suppressing most of the features that are favorable for classification, thus allowing the network to learn other useless features. When $\gamma = 1$, the possible reason of the performance drop is that the network only suppresses the most salient feature, while ignoring other features that are equally worthy of suppression.

Table 4. Contribution of each module.

Methods	FFE	PBS	ACC(%)
Resnet50			86.5
Resnet50	✓		88.9
Resnet50	✓	✓	89.6

Table 5. Ablation studies on the threshold γ.

γ	0.7	0.8	0.9	1
Accuracy	89.16	89.11	89.56	89.23

4.4 Visualization

To further validate the effectiveness of the model in this paper, we adopt the class activation visualization (Grad-CAM) approach to analyze the classification performances. As shown in Fig. 6, we take two raw images from each of the three datasets, the visualization

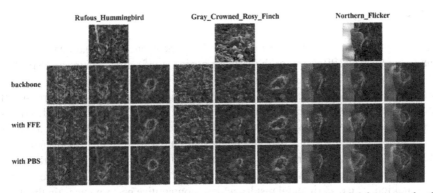

Fig. 5. Visualization of contrast activation maps. Take Rufous_Hummingbird for example, the first row represents the raw image. The second to fourth rows represent the heat maps with backbone, with FFE, and with PBS, respectively. The first to third columns represent stage 3, stage 4, and stage 5 of Resnet50.

results in the second to fourth rows correspond to the third to fifth stages on Resnet50, respectively. As a whole, based on peak suppression and background suppression with the PBS module, we can find that the FFE module achieves the effect of mining different features at different stages. Specifically, take the bird in the first column for example, whose background noise is gradually reduced from the stage 3 to the stage 5. Furthermore, the stage 3 of our model in the second row focuses on less granular features such as the bird's crown and neck, the stage 4 of our model in the third row excavates different parts of the bird's body such as the eye and wing, and the stage 5 of our model in the last row focuses on larger granular features such as the head and tail of the bird. The visualization demonstrates the ability of our model to mine different parts of the features and the suppression of peak and background noise.

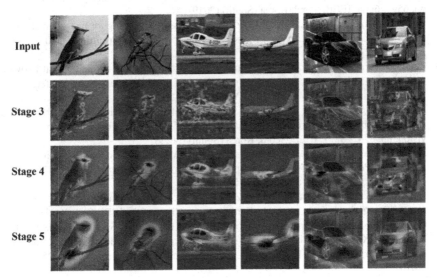

Fig. 6. Visualization of the attention maps with FFE and PBS modules on three datasets. The first line represents the original images of the input. The second to the fourth lines correspond to the last three stages of the Resnet50, respectively.

5   Conclusion

In this paper, we propose a Foreground Feature Enhancement (FFE) and Peak & Background Suppression (PBS) approach for fine-grained classification, which can be trained end-to-end and does not require bounding boxes/part annotations. Concretely, our network has two modules: One is the FFE module which enhances the most prominent features obtained from the feature maps in the current stage. The other is the PBS module which suppresses the most salient features obtained from the FFE module to better mine the sub-salient features at the subsequent stage and restraints the background noise in feature maps to learn more useful information at the next stage. Through the complementary role of the two modules, the network contributes to learning a more diverse representation. The experimental results achieve competitive performances on the three benchmark datasets.

References

1. Wah, C., Branson, S., Welinder, P., Perona, P.,Belongie, S.: The caltech-ucsd birds-200–2011 dataset. Tech. Rep. CNS-TR-2010–001, California Institute of Technology (2011)
2. Krause, J., Stark, M., Deng, J.,Fei-Fei, L.: 3d object representations for fine-grained categorization. In: Proceedings of the IEEE international conference on computer vision workshops, pp. 554–561. IEEE, Sydney, Australia (2013)
3. Maji, S., Rahtu, E., Kannala, J., Blaschko, M.,Vedaldi, A.: Fine-grained visual classification of aircraft. arXiv preprint arXiv:1306.5151 (2013)
4. Yang, Z., Luo, T., Wang, D., Hu, Z., Gao, J.,Wang, L.: Learning to navigate for fine-grained classification. In: Proceedings of the European conference on computer vision (ECCV), pp. 420–435 (2018). https://doi.org/10.1007/978-3-030-01264-9_26
5. Yang, S., Liu, S., Yang, C.,Wang, C.: Re-rank coarse classification with local region enhanced features for fine-grained image recognition. arXiv preprint arXiv:2102.09875 (2021)
6. Ding, Y., Zhou, Y., Zhu, Y., Ye, Q.,Jiao, J.: Selective sparse sampling for fine-grained image recognition. In: Proceedings of the IEEE/CVF International Conference on Computer Vision, pp. 6599–6608. IEEE, Seoul, Korea (South) (2019)
7. Liu, M., Zhang, C., Bai, H., Zhang, R., Zhao, Y.: Cross-part learning for fine-grained image classification. IEEE Trans. Image Process. **31**, 748–758 (2021). https://doi.org/10.1109/tip.2021.3135477
8. Zheng, H., Fu, J., Mei, T.,Luo, J.: Learning multi-attention convolutional neural network for fine-grained image recognition. In: Proceedings of the IEEE international conference on computer vision, pp. 5209–5217. IEEE, Venice, Italy (2017)
9. Luo, W., Yang, X., Mo, X., Lu, Y., Davis, L.S., Li, J., et al.: Cross-x learning for fine-grained visual categorization. In: Proceedings of the IEEE/CVF international conference on computer vision, pp. 8242–8251. IEEE, Seoul, Korea (South) (2019)
10. Zhuang, P., Wang, Y.,Qiao, Y.: Learning attentive pairwise interaction for fine-grained classification. In: Proceedings of the AAAI conference on artificial intelligence, pp. 13130–13137 (2020). https://doi.org/10.1609/aaai.v34i07.7016
11. Ji, R., Wen, L., Zhang, L., Du, D., Wu, Y., Zhao, C., et al.: Attention convolutional binary neural tree for fine-grained visual categorization. In: Proceedings of the IEEE/CVF Conference on Computer Vision and Pattern Recognition, pp. 10468–10477. IEEE, Seattle, WA, USA (2020)
12. Ai, D., Han, X., Ruan, X.,Chen, Y.: Color independent components based SIFT descriptors for object/scene classification. IEICE transactions on information and systems **93**(9), 2577–2586 (2010)

13. Kerdvibulvech, C.: Real-time augmented reality application using color analysis. In: 2010 IEEE Southwest Symposium on Image Analysis & Interpretation (SSIAI), pp. 29–32. IEEE, Austin, USA (2010)
14. Zhang, F., Li, M., Zhai, G., Liu, Y.: Multi-branch and multi-scale attention learning for fine-grained visual categorization. In: MultiMedia Modeling: 27th International Conference, MMM 2021, pp. 136–147. Springer, Prague, Czech Republic (2021)
15. Sun, G., Cholakkal, H., Khan, S., Khan, F.,Shao, L.: Fine-grained recognition: Accounting for subtle differences between similar classes. In: Proceedings of the AAAI Conference on Artificial Intelligence, pp. 12047–12054 (2020). https://doi.org/10.1609/aaai.v34i07.6882
16. Song, J.,Yang, R.: Feature boosting, suppression, and diversification for fine-grained visual classification. In: 2021 International Joint Conference on Neural Networks (IJCNN), pp. 1–8. IEEE, Shenzhen, China (2021). https://doi.org/10.1109/ijcnn52387.2021.9534004
17. Do, T., Tran, H., Tjiputra, E., Tran, Q.D.,Nguyen, A.: Fine-grained visual classification using self assessment classifier. arXiv preprint arXiv:2205.10529 (2022)
18. Chou, P.-Y., Kao, Y.-Y.,Lin, C.-H.: Fine-grained Visual Classification with High-temperature Refinement and Background Suppression. arXiv preprint arXiv:2303.06442 (2023)
19. Woo, S., Park, J., Lee, J.-Y.,Kweon, I.S.: Cbam: Convolutional block attention module. In: Proceedings of the European Conference on Computer Vision (ECCV), pp. 3–19 (2018) https://doi.org/10.1007/978-3-030-01234-2_1
20. He, K., Zhang, X., Ren, S.,Sun, J.: Deep residual learning for image recognition. In: Proceedings of the IEEE Conference on Computer Vision and Pattern Recognition, pp. 770–778 (2016)
21. Liu, C., Xie, H., Zha, Z.-J., Ma, L., Yu, L., Zhang, Y.: Filtration and distillation: Enhancing region attention for fine-grained visual categorization. In: Proceedings of the AAAI Conference on Artificial Intelligence, pp. 11555–11562 (2020). https://doi.org/10.1609/aaai.v34i07.6822
22. Lin, T.-Y., RoyChowdhury, A.,Maji, S.: Bilinear CNN models for fine-grained visual recognition. In: Proceedings of the IEEE International Conference on Computer Vision, pp. 1449–1457. IEEE, Santiago, Chile (2015)
23. Zheng, H., Fu, J., Zha, Z.-J., Luo, J.: Learning deep bilinear transformation for fine-grained image representation. Adv. Neural. Inf. Process. Syst. **32**, 4277–4286 (2019)
24. Gao, Y., Han, X., Wang, X., Huang, W.,Scott, M.: Channel interaction networks for fine-grained image categorization. In: Proceedings of the AAAI Conference on Artificial Intelligence, pp. 10818–10825 (2020)
25. Fu, J., Zheng, H.,Mei, T.: Look closer to see better: recurrent attention convolutional neural network for fine-grained image recognition. In: Proceedings of the IEEE Conference on Computer Vision and Pattern Recognition, pp. 4438–4446. IEEE, Honolulu, USA (2017)
26. Sun, M., Yuan, Y., Zhou, F.,Ding, E.: Multi-attention multi-class constraint for fine-grained image recognition. In: Proceedings of the European Conference on Computer Vision (ECCV), pp. 805–821 (2018). https://doi.org/10.1007/978-3-030-01270-0_49
27. Chen, Y., Bai, Y., Zhang, W.,Mei, T.: Destruction and construction learning for fine-grained image recognition. In: Proceedings of the IEEE/CVF Conference on Computer Vision and Pattern Recognition, pp. 5157–5166. IEEE, Long Beach, CA, USA (2019). https://doi.org/10.1109/CVPR.2019.00530
28. Wang, Z., Wang, S., Li, H., Dou, Z.,Li, J.: Graph-propagation based correlation learning for weakly supervised fine-grained image classification. In: Proceedings of the AAAI Conference on Artificial Intelligence, pp. 12289–12296 (2020) https://doi.org/10.1609/aaai.v34i07.6912

29. Liang, Y., Zhu, L., Wang, X.,Yang, Y.: A simple episodic linear probe improves visual recognition in the wild. In: Proceedings of the IEEE/CVF Conference on Computer Vision and Pattern Recognition, pp. 9559–9569, New Orleans, LA, USA (2022)
30. Liang, Y., Zhu, L., Wang, X.,Yang, Y.: Penalizing the hard example but not too much: a strong baseline for fine-grained visual classification. IEEE Trans. Neural Networks Learn. Syst., 1–12 (2022). https://doi.org/10.1109/tnnls.2022.3213563

YOLOv5-SRR: Enhancing YOLOv5 for Effective Underwater Target Detection

Jinyu Shi and Wenjie Wu[✉]

Dalian Maritime University, Dalian 116026, China
{sjy1967,greedy-hat}@dlmu.edu.cn

Abstract. Underwater target detection is a crucial aspect of ocean exploration, and advances in related technologies are of significant practical importance. When compared to other target detection tasks, such as those performed on land or in the air, underwater target detection presents a distinct set of challenges. These include poor image quality as a result of light attenuation and water turbidity, as well as the presence of small and densely packed targets that can be difficult to identify. Furthermore, the computational resources available within an underwater vehicle are often limited, further exacerbated the difficulty of detecting and identifying targets in this environment. Despite the excellent performance of existing underwater target detection algorithms with various underwater datasets, these methods often fail to achieve satisfactory results in complex underwater environments where small targets are prevalent. This paper proposes an improved version of the advanced target detection algorithm, YOLOv5, specifically tailored to address this issue. The proposed model incorporates a new SPD-Block module, Rep-Bottleneck structure, and RepBottleneck-ResNets structure, along with the Soft-NMS algorithm. The resulting model, YOLOv5-SRR, demonstrated remarkable effectiveness in underwater target detection, achieving an average accuracy (mAP) of 83.6% on the URPC2020 (Dalian) competition dataset. This result surpassed that of current general target detection models and proved to be more appropriate for use in complex underwater environments.

Keywords: Underwater Object Detection · SPD-Convolution module · RepVGG

1 Introduction

The ocean constitutes a significant portion of the earth's surface and contains abundant oil, gas, minerals, chemicals, and aquatic resources [25]. Unfortunately, due to the continuous expansion of human living space, land resources have been overexploited in recent years. As a result, many developed countries worldwide have shifted their focus toward marine resources, leading to an increased frequency of marine exploration. In support of these exploration efforts, a range of underwater tasks, including target location, biometrics, archaeology, object

© The Author(s), under exclusive license to Springer Nature Switzerland AG 2024
S. Rudinac et al. (Eds.): MMM 2024, LNCS 14554, pp. 147–158, 2024.
https://doi.org/10.1007/978-3-031-53305-1_12

search, environmental monitoring, and equipment maintenance, must be performed [11]. The underwater target detection techniques play a critical role in these efforts.

The detection of underwater objects is a challenging task compared to other computer vision tasks, due to issues such as poor image quality, the presence of small and dense targets that are difficult to detect, and limited computational power within underwater vehicles. Consequently, many researchers have made significant contributions in developing target detection algorithms [4,7,22]. For example, Huang et al. [6] integrated image enhancement techniques into an extended VGG16 feature extraction network and utilized a Faster R-CNN network [24] with feature mapping to detect and identify underwater biological targets using a URPC data set in 2019. In 2020, Chen et al. [2] introduced a sample distribution-based weighted loss function IMA (Invert Multi-Class AdaBoost) to mitigate the adverse effects of noise on detection performance. Furthermore, Qiao et al. [12] proposed a real-time accurate underwater target classifier in 2021 by combining LWAP (Local Wavelet Acoustic Pattern) and MLP (Multilayer Perceptron) neural networks to solve the challenge of underwater target classification.

Although various target detection methods have achieved success, research in such harsh underwater environments still has a long way to go. YOLOv5 is one of the most advanced algorithms in the YOLO family and is more suitable for industrial applications due to its accuracy and model size balance. In practical applications, models are typically equipped with mobile devices, such as underwater robots. Therefore, models must be robust and portable. Therefore, this paper proposes a YOLOv5-SRR model based on the latest version 7.0 YOLOv5s model with the corresponding improvements for underwater target detection. The proposed model is applied to underwater images, and experimental results on the URPC2020 (dalian) competition dataset demonstrate that the detection performance of our proposed yolov5-SRR surpasses current state-of-the-art methods.

The innovations in this paper are as follows:

(1) The integration of SPD-Block was proposed to address the issue of detecting numerous small targets in underwater environments. This mechanism effectively addresses the problems associated with losing a significant amount of fine information and inefficient feature learning due to small image resolution or object size.
(2) The RepBottleneck-ResNets module in YOLOv5-SRR has been designed to enhance the feature fusion capability of the neck network. This module not only accelerates the convergence of the network, but also improves the speed of the inference process.
(3) The Soft-NMS algorithm is implemented in the post-processing stage of YOLOv5-SRR to effectively address issues related to repetitive detections, improve noise immunity, and increase detection efficiency for the underwater target detection algorithm. By doing so, the accuracy and stability of the detection results are significantly improved.

The rest of this paper is organized as follows. Section 2 describes the proposed YOLOv5-SRR model and its theoretical basis. Section 3 evaluates and analyzes the performance of the YOLOv5-SRR model through experiments conducted on underwater image datasets. Finally, we give the conclusions of this work in Sect. 4.

2 Model

This section presents our design for an enhanced YOLOv5 architecture tailored specifically for underwater target detection. Our approach involved the development of the SPD-Block module by combining network structures such as Rep-Bottleneck and RepBottleneck-ResNets. Additionally, we incorporated the Soft-NMS algorithm during post-processing to produce the improved YOLOv5-SRR model, which demonstrated significantly better detection performance compared to the original model.

2.1 General Architecture

Glenn Jaucher released YOLOv5 [8] in 2020 and until 2023, released to v7.0, and this paper is based on the improved version of YOLOv5 v7.0. YOLOv5 extends the model structure of the previous YOLO family of algorithms. YOLOv5 includes four variants, namely YOLOv5s, YOLOv5m, YOLOv5l, and YOLOv5x, which have increasing network size and number of parameters in order.

Figure 1 shows the overall structure of the YOLOv5-SRR network with the addition of the SPD-Block module and the RepBottleneck-ResNets module to enhance the detection performance of the model for underwater target detection.

2.2 SPD-Block

Targets in underwater environments are commonly characterized by their small size, high density, and low image contrast. Consequently, effective underwater target detection algorithms should exhibit sensitivity to small targets, possess the ability to extract contour and detail information from complex backgrounds, and also demonstrate robustness to underwater images that may have low image quality and high levels of noise.

Previous underwater target detection studies [9] have generally overlooked the importance of small target features in underwater datasets, with most improvements focused on enhancing feature extraction within the backbone network. However, we recognize that this approach may present practical deployment challenges and difficult-to-resolve bottlenecks. As a result, we used SPD-Conv [13] within the network structure to better address small target detection in our proposed architecture.

The proposed SPD-Block module was constructed based on the previously introduced SPD-Conv. This SPD-Block module was then incorporated into the overall network structure of YOLOv5 7.0 to improve small target detection in

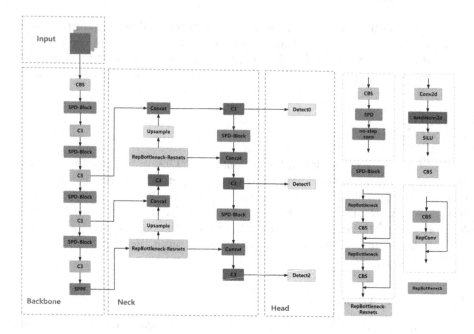

Fig. 1. The network structure of YOLOv5-SRR.

underwater environments. In this paper, we conducted several experiments to determine the optimal position of the SPD-Block module within the network architecture.

SPD-Conv. SPD-Conv consists of a space-to-depth (SPD) layer and a non-step convolution (Conv) layer. In the SPD layer, the feature map X is downsampled while preserving information in the channel dimension to prevent information loss. For any intermediate feature map X of size $S \times S \times C_1$, a series of sub-feature maps can be cut out according to the method shown in Eq. 1.

$$
\begin{aligned}
f_{0,0} &= X[0:S:scale, 0:S:scale], f_{1,0} = X[1:S:scale, 0:S:scale], \ldots, \\
f_{scale-1,0} &= X[scale-1:S:scale, 0:S:scale]; \\
f_{0,1} &= X[0:S:scale, 1:S:scale], f_{1,1}, \ldots, \\
f_{scale-1,1} &= X[scale-1:S:scale, 1:S:scale]; \\
&\vdots \\
f_{0,scale-1} &= X[0:S:scale, scale-1:S:scale], f_{1,scale-1}, \ldots, \\
f_{scale-1,scale-1} &= X[scale-1:S:scale, scale-1:S:scale].
\end{aligned}
\tag{1}
$$

For any feature map X, the subfeature map $f_{x,y}$ is made up of all terms X $(i + j)$ that are divisible by the scale of $i + x$ and $j + y$. Thus, each subfeature map is obtained by downsampling from X by a factor of scale. Figure 2(a)(b)(c) shows an example when scale = 2. It generates four subfeature maps, namely $f_{0,0}, f_{1,0}, f_{0,1}$ and $f_{1,1}$. Each submap has the shape $(\frac{S}{2}, \frac{S}{2}, C_1)$ and downsamples X by a factor of 2.

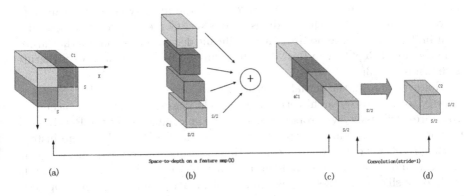

Fig. 2. Illustration of SPD-Conv when scale = 2 (see text for details).

Subsequently, these sub-feature maps are connected along the channel dimensions to obtain a new feature map X'. The spatial dimension is reduced by a factor of scale and the channel dimension is increased by a factor of $scale^2$. In other words, SPD converts the feature map X(S, S, C_1) into an intermediate feature map $X'(\frac{S}{scale}, \frac{S}{scale}, scale^2 C_1)$. Figure 2(d) gives an illustration using scale = 2.

After the SPD feature conversion layer, a non-step convolution layer with a C_2 filter is added (i.e., step = 1) where $C_2 < scale^2 C_1$ and is further transformed from $X'(\frac{S}{scale}, \frac{S}{scale}, scale^2 C_1)$ to $X''(\frac{S}{scale}, \frac{S}{scale}, C_2)$. Non-stepping convolution is used in order to retain all discriminative feature information as much as possible.Otherwise, for instance, using a 3 × 3 filer with stride=3, feature maps will get "shrunk" yet each pixel is sampled only once; if stride=2, asymmetric sampling will occur where even and odd rows/columns will be sampled different times.

Typically, a step size greater than 1 results in the loss of non-discriminatory information, although it appears that the feature map $X(S, S, C_1)$ is converted to $X''(\frac{S}{scale}, \frac{S}{scale}, C_2)$, but no information about X' is retained.

2.3   RepBottleneck-ResNets

The introduction of RepVGG has revived the use of the VGG [15] architecture in various applications. The use of convolutional RepConv [23] with the related RepVGG architecture has been implemented in YOLOv6 [10] and YOLOv7 [20] network structures, resulting in improved accuracy with a small increase in the number of parameters, due to the better balance between computational complexity and accuracy offered by RepVGG. However, in practical applications of target detection, the combination of RepVGG [3] and ResNet [17] architectures is not widely used due to a significant decrease in detection accuracy resulting from the direct use of RepConv in ResNet architectures.

After conducting several experiments to analyze the combination of RepConv and different architectures, we discovered that the identity connection in RepConv can interfere with the residuals in ResNet. Consequently, we proposed a solution for this issue to address the significant decrease in accuracy that arises from the use of RepConv in ResNet architectures. The specific network structure is shown in Fig. 3(a).

However, during the actual experiments, we observed that adding RepConv directly to the ResNet architecture led to slight degradation in model accuracy. To address this issue, we proposed increasing the complexity of the Rep-ResNet network and incorporating RepBottleneck into the ResNet architecture instead of using RepConv directly. After conducting experimental comparisons, we found that this modification improved the accuracy of the model by a small margin while only slightly increasing the number of model parameters. This approach presents a potential solution to the model size requirement imposed by underwater deployment problems.

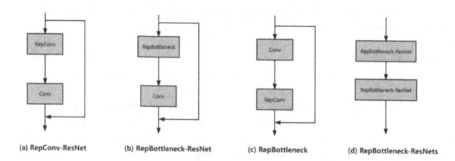

(a) RepConv-ResNet   (b) RepBottleneck-ResNet   (c) RepBottleneck   (d) RepBottleneck-ResNets

Fig. 3. (a) The solution structure of repconv used for the ResNet architecture; (b) the structure of RepBottleneck-ResNet; (c) the structure of RepBottleneck; (d) the structure of RepBottleneck-ResNets.

RepVGG. With the great achievements of the ResNet, VGG and Inception networks [14–16] in the field of images, models started to evolve toward depth and branching. While complex structures can often achieve higher accuracy than simple model structures, they also have significant drawbacks. On the one hand, complex branching structures tend to slow down model inference. On the other hand, certain components increase the memory overhead. RepVGG redesigned the network taking into account the above-mentioned drawbacks. Since the benefits of multi-branch architecture are all for training and the drawbacks are undesired for inference, it proposes to decouple the training-time multibranch and inference-time plain architecture via structural re-parameterization, which means converting the architecture from one to another via transforming its parameters.

2.4 Soft-NMS

Non-maximum suppression (NMS) is an important component of the target detection process. The traditional non-maximum suppression algorithm sorts all detection boxes on the basis of their scores. The detection box M with the maximum score is selected and all other detection boxes with a significant overlap (using a predefined threshold) with M are suppressed. This process is applied recursively to the remaining boxes. In this case, if a real object appears in the overlapping area, the detection of the object will fail and the average precision (AP) of the algorithm will be reduced.

The proposed Soft-NMS algorithm [1] effectively solves this problem by attenuating the detection fraction of nonmaximum detected frames instead of completely removing them, it requires only a simple modification of the regular NMS algorithm and does not add additional parameters.

The main improvement of Soft-NMS is the score reset function of NMS, the re-scoring function in Soft-NMS is

$$s_i = \begin{cases} s_i, iou(M, b_i) < N_t \\ s_i(1 - iou(M, b_i)), iou(M, b_i) \geq N_t \end{cases}, \qquad (2)$$

When the overlap between adjacent detection boxes and M exceeds the overlap threshold N_t, the detection fraction of the detection boxes decays linearly, and the detection boxes that are very close to M decay to a large extent, while the detection boxes far from M are not affected.

However, the above fraction reset function is not a continuous function, and this fraction reset function produces abrupt changes when overlap exceeds the overlap threshold N_t, which may lead to large changes in the sequence of detection results.

Therefore, we prefer to find a continuous score reset function. It produces no attenuation of the original detection scores for detection boxes that do not overlap, while producing large attenuation for highly overlapping detection boxes. Taking these factors into account, the score reset function is further improved in Soft-NMS by using a Gaussian penalty function.

$$s_i = s_i e^{-\frac{(iou(M, b_i))^2}{\sigma}}, \forall b_i \notin D \qquad (3)$$

This update rule is applied in each iteration and the scores of all remaining detection boxes are updated.

3   Experiments and Analysis

In this section, the configuration of the experimental environment, hyperparameters, and test data set are described. The experimental results show that the proposed YOLOv5-SRR model improves the accuracy and speed of underwater target detection, thus validating its effectiveness and superiority in challenging underwater detection environments.

3.1 Experimental Settings

The experimental platform is equipped with 14 vCPU Intel(R) Xeon(R) Gold 6330 CPU@2.00 GHz, NVIDIA GeForce RTX 3090 GPU with 24 G of video memory, and Ubuntu 20.04.4 LTS operating system. The software environment of the experimental platform is CUDA 11.3, compiler Python 3.8 .10 and the deep learning framework Pytorch 1.13.0+cu116.

3.2 Model Hyperparameter Setting

Training hyperparameter settings are shown in Table 1:

Table 1. Training hyperparameter settings.

Parameter Name	Parameter Settings
Weights	No
Image size	640 × 640
Epochs	300
Batch-size	64
Optimizer	SGD

The rest of the parameters are set as default parameters.

3.3 URPC Dataset

This dataset is the dataset of the 2020 National Underwater Robotics Professional Competition (Dalian) (URPC) and contains 6575 underwater optical images (including manually annotated real data). For more information, please visit https://www.flyai.com/d/312.The target groups to be evaluated in the experiment include four types of seafood, namely "holothurian", "echinus", "scallop", and "starfish".

3.4 Experimental Results and Analysis of the URPC Dataset

The detection performance of the proposed YOLOv5-SRR model was experimentally evaluated on the URPC dataset. As shown in Fig. 11, the results of the improved model showed that the detection efficiency was improved for various target categories, especially echinus, with an average accuracy (AP) value of 90.4%. The average accuracy (mAP) value of the model was calculated to

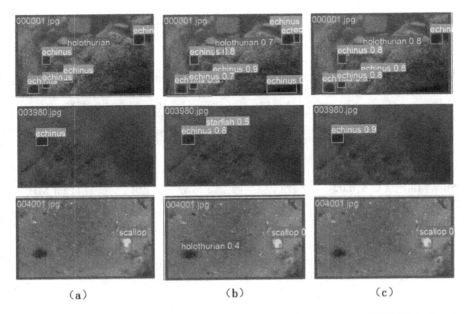

(a)   (b)   (c)

Fig. 4. Detection results of real label (a), YOLOv5 (b) and Yolov5-SRR (c) in harsh underwater scenes.

be 83.6%. As shown in Fig. 4, the proposed YOLOv5-SRR model outperforms the YOLOv5s model in terms of error detection and omission detection, and not only detects more targets accurately, but also has more accurate prediction boxes.

To demonstrate the superior performance of our proposed YOLOv5-SRR model, we compared it with other popular target detection models such as YOLOv6s, YOLOv7, YOLOv7-tiny and YOLOv8s [18]. We trained and tested these models on the URPC dataset and evaluated their mean average precision (mAP) scores. The comparison results are presented in Table 2.

As shown in the table, YOLOv7-tiny has the lowest number of parameters and smallest model size, but its mAP@0.5 and mAP@0.5:0.95 indexes are significantly lower than our proposed model. YOLOv7 has the highest recall, but its number of parameters and model size are much larger than our proposed model, and its mAP@0.5 and mAP@0.5-0.95 indexes are lower than our proposed model. YOLOv5s has the highest precision, but its recall, mAP@0.5, and mAP@0.5:0.95 indexes are lower than our proposed model.

In summary, our proposed YOLOv5-SRR model outperforms the other detection algorithms in the comparison test, demonstrating the practical advantages of our method in underwater target recognition. It is important to note that we used the original YOLOv5 hyperparameters without re-tuning, which suggests that our model may perform even better with specialized hyperparameter tuning.

In general, our experimental results suggest that our proposed YOLOv5-SRR model is a promising candidate for underwater target detection applications.

Table 2. Performance comparison of target detection model on the URPC dataset.

Model	Params (M)	Precision	Recall	mAP@0.5	mAP@0.5:0.95	Model size
Yolov4-csp [19]	52.51	76.7%	71.4%	76.2%	39.7%	105.5 MB
YoloR-csp [21]	52.51	81.6%	74.6%	81.9%	45.4%	105.5 MB
YoloX-s [5]	8.94	–	–	78.2%	42.2%	68.5 MB
Yolov6-s	18.5	–	–	78.4%	44.1%	38.7MB
Yolov7-tiny	**6.02**	79.1%	72.5%	78.5%	42%	**12.3 MB**
Yolov7	37.21	79.5%	**78.1%**	82.9%	46.5%	71.3 MB
Yolov8s	11.17	81.8%	74.6%	81.6%	47.8%	22.5 MB
Yolov5s	7.03	**84.3%**	72.8%	80.7%	44.8%	13.7 MB
Yolov5m	20.88	83.8%	76.4%	82.9%	47.6%	40 MB
Yolov5-SRR (ours)	8.75	83.4%	76.4%	**83.6%**	**50.9%**	17.9 MB

3.5 Ablation Experiment

The results of the ablation experiments are presented in Table 3. In order to validate the effectiveness of the proposed improvement strategies, we conducted a series of ablation experiments. These experiments aimed to compare and analyze the impact of different improvement strategies on the model's detection performance, thereby providing evidence of the efficacy of our proposed methods for underwater target detection.

Table 3. Ablation comparison of model performance improvement on the URPC dataset.

Model	SPD-Block	RepBottleneck-ResNets	Soft-NMS	Precision	Recall	mAP@0.5	mAP@0.5:0.95
Yolov5s				84.3%	72.8%	80.7%	44.8%
	✓			84.7%	74.8%	82.1%	46.6%
		✓		82.9%	75.1%	81.5%	45.2%
			✓	84.5%	73.1%	82.2%	50%
	✓	✓		83.7%	76.7%	82.6%	46.9%
	✓		✓	83.9%	75.6%	83.5%	51%
		✓	✓	82.8%	75.5%	82.7%	49.9%
	✓	✓	✓	83.4%	76.4%	83.6%	50.9%

4 Conclusions

In this paper, we present a novel underwater target detection model that integrates the SPD-Block module, the RepBottleneck-ResNets module, and the Soft-NMS algorithm. Previous models for underwater target detection have primarily focused on enhancing feature extraction capabilities without due consideration of small target features and dense target occlusion features inherent in underwater datasets, which may result in relative bottlenecks. In contrast, our proposed model addresses these issues by introducing the SPD-Block module to effectively mitigate information loss and improve feature learning efficiency for small image resolutions and target sizes. Furthermore, we improve the network's feature extraction capabilities by incorporating the RepBottleneck-ResNets module while striking a balance between the number of parameters and accuracy. Finally, the implementation of the Soft-NMS algorithm effectively mitigates the problem of dense target occlusions. Our experimental evaluations on the URPC dataset and ablation tests demonstrate the effectiveness of our proposed method, outperforming existing methods under complex underwater detection environments. These results further validate the efficacy of our proposed model in addressing unique challenges specific to underwater target detection.

It is important to note that our proposed method has not been applied to other advanced detection models. However, we believe that the characteristics of the dataset are similar, and our method can effectively improve other detection models as well. Further research could explore the applicability of our proposed method to other state-of-the-art detection models and assess its impact on their performance under challenging underwater detection conditions.

References

1. Bodla, N., Singh, B., Chellappa, R., Davis, L.S.: Soft-NMS-improving object detection with one line of code. In: Proceedings of the IEEE International Conference on Computer Vision, pp. 5561–5569 (2017)
2. Chen, L., et al.: Underwater object detection using invert multi-class AdaBoost with deep learning. In: 2020 International Joint Conference on Neural Networks (IJCNN), pp. 1–8. IEEE (2020)
3. Ding, X., Zhang, X., Ma, N., Han, J., Ding, G., Sun, J.: RepVGG: Making VGG-style convnets great again. In: Proceedings of the IEEE/CVF conference on computer vision and pattern recognition, pp. 13733–13742 (2021)
4. Dulhare, U.N., Ali, M.H.: Underwater human detection using faster R-CNN with data augmentation. Mater. Today: Proc. **80**, 1940–1945 (2023)
5. Ge, Z., Liu, S., Wang, F., Li, Z., Sun, J.: YOLOX: exceeding YOLO series in 2021. arXiv preprint arXiv:2107.08430 (2021)
6. Huang, H., Zhou, H., Yang, X., Zhang, L., Qi, L., Zang, A.Y.: Faster R-CNN for marine organisms detection and recognition using data augmentation. Neurocomputing **337**, 372–384 (2019)
7. Isa, I.S., Rosli, M.S.A., Yusof, U.K., Maruzuki, M.I.F., Sulaiman, S.N.: Optimizing the hyperparameter tuning of YOLOv5 for underwater detection. IEEE Access **10**, 52818–52831 (2022)
8. Jocher, G., et al.: ultralytics/yolov5: v7. 0-YOLOv5 SOTA realtime instance segmentation. Zenodo (2022)
9. Lei, F., Tang, F., Li, S.: Underwater target detection algorithm based on improved YOLOv5. J. Marine Sci. Eng. **10**(3), 310 (2022)
10. Li, C., et al.: YOLOv6: a single-stage object detection framework for industrial applications. arXiv preprint arXiv:2209.02976 (2022)

11. Liu, Y., Anderlini, E., Wang, S., Ma, S., Ding, Z.: Ocean explorations using auton-
    omy: technologies, strategies and applications. In: Su, S.-F., Wang, N. (eds.) Off-
    shore Robotics. OR, pp. 35–58. Springer, Singapore (2022). https://doi.org/10.
    1007/978-981-16-2078-2_2
12. Qiao, W., Khishe, M., Ravakhah, S.: Underwater targets classification using local
    wavelet acoustic pattern and multi-layer perceptron neural network optimized by
    modified whale optimization algorithm. Ocean Eng. **219**, 108415 (2021)
13. Sunkara, R., Luo, T.: No more strided convolutions or pooling: a new CNN build-
    ing block for low-resolution images and small objects. In: Amini, M.R., Canu, S.,
    Fischer, A., Guns, T., Kralj Novak, P., Tsoumakas, G. (eds.) ECML PKDD 2022,
    Part III. LNCS, vol. 13715, pp. 443–459. Springer, Cham (2023). https://doi.org/
    10.1007/978-3-031-26409-2_27
14. Szegedy, C., Ioffe, S., Vanhoucke, V., Alemi, A.: Inception-v4, inception-ResNet
    and the impact of residual connections on learning. In: Proceedings of the AAAI
    Conference on Artificial Intelligence, vol. 31 (2017)
15. Szegedy, C., et al.: Going deeper with convolutions. In: Proceedings of the IEEE
    Conference on Computer Vision and Pattern Recognition, pp. 1–9 (2015)
16. Szegedy, C., Vanhoucke, V., Ioffe, S., Shlens, J., Wojna, Z.: Rethinking the incep-
    tion architecture for computer vision. In: Proceedings of the IEEE Conference on
    Computer Vision and Pattern Recognition, pp. 2818–2826 (2016)
17. Targ, S., Almeida, D., Lyman, K.: ResNet in ResNet: generalizing residual archi-
    tectures. arXiv preprint arXiv:1603.08029 (2016)
18. Terven, J., Cordova-Esparza, D.: A comprehensive review of YOLO: from YOLOv1
    to YOLOv8 and beyond. arXiv preprint arXiv:2304.00501 (2023)
19. Wang, C.Y., Bochkovskiy, A., Liao, H.Y.M.: Scaled-YOLOv4: scaling cross stage
    partial network. In: Proceedings of the IEEE/CVF Conference on Computer Vision
    and Pattern Recognition, pp. 13029–13038 (2021)
20. Wang, C.Y., Bochkovskiy, A., Liao, H.Y.M.: YOLOv7: trainable bag-of-
    freebies sets new state-of-the-art for real-time object detectors. arXiv preprint
    arXiv:2207.02696 (2022)
21. Wang, C.Y., Yeh, I.H., Liao, H.Y.M.: You only learn one representation: unified
    network for multiple tasks. arXiv preprint arXiv:2105.04206 (2021)
22. Wang, H., Xiao, N.: Underwater object detection method based on improved faster
    RCNN. Appl. Sci. **13**(4), 2746 (2023)
23. Weng, K., Chu, X., Xu, X., Huang, J., Wei, X.: EfficientRep: an efficient RepVGG-
    style convnets with hardware-aware neural network design. arXiv preprint
    arXiv:2302.00386 (2023)
24. Xu, X., et al.: Crack detection and comparison study based on faster R-CNN and
    mask R-CNN. Sensors **22**(3), 1215 (2022)
25. Zhou, X., Ding, W., Jin, W.: Microwave-assisted extraction of lipids, carotenoids,
    and other compounds from marine resources. In: Innovative and Emerging Tech-
    nologies in the Bio-marine Food Sector, pp. 375–394. Elsevier (2022)

Image Clustering and Generation
with HDGMVAE-I

Yongqi Liu[1] , Jiashuang Zhou[1] , and Xiaoqin Du[1,2,3(✉)]

[1] School of Computer Science and Artificial Intelligence,
Wuhan Textile University, Wuhan, China
xiaoqindu@wtu.edu.cn
[2] Engineering Research Center of Hubei Province for Clothing Information,
Wuhan, Hubei, China
[3] Hubei Provincial Engineering Research Center for Intelligent Textile and Fashion,
Wuhan, Hubei, China

Abstract. Generating high-quality and realistic images has emerged as a popular research direction in the field of computer vision. We propose the Hierarchically Disentangled Gaussian Mixture Variational Autoencoder model with Importance sampling (HDGMVAE-I) for image generation and clustering. To enhance clustering performance, we introduce total correlation (TC) as a key to computing latent features between samples and cluster centroids and use slack variables to narrow down the feasible solution space. We also introduce Fisher discriminant as a regularization term to minimize within-class distance and maximize between-class distance, particularly effective for samples that are difficult to classify accurately. Moreover, we introduce importance sampling to reduce the information gap between the ELBO and the log-likelihood function, which leads to more realistic generated data. Experimental results demonstrate that our proposed method significantly outperforms the baseline in both clustering and generation tasks.

Keywords: VAE · Clustering · Disentanglement · Deep generative models · Gaussian mixture distribution

1 Introduction

Over the past decades, AI Generated Contents (AIGC) has garnered significant attention. Various generative models have been developed, including Variational Autoencoders (VAEs) [18,23], Generative Adversarial Networks (GANs) [10], Flow models, and diffusion models [13]. While GAN [6] has shown promising results, it faces challenges such as training instability and the lack of an exclusive inference model, limiting its applicability in transfer learning and downstream classification tasks. GANs and Flow models have also encountered difficulties in training, characterized by instability and mode collapse across different tasks. Diffusion models [5,14,15], on the other hand, demand extensive computational resources and long training times, rendering them less practical. In comparison,

© The Author(s), under exclusive license to Springer Nature Switzerland AG 2024
S. Rudinac et al. (Eds.): MMM 2024, LNCS 14554, pp. 159–171, 2024.
https://doi.org/10.1007/978-3-031-53305-1_13

VAE models offer advantages of easier training, lower computational requirements, and continued usage across various generation tasks.

To address these challenges, variational inference-based Variational Autoencoders (VAEs) provide a promising solution. While the vanilla VAE has shown satisfactory results on relatively simple datasets [18], it has been observed that the efficiency of disentanglement often significantly degrades when dealing with complex datasets [1,20]. Recently, Shao et al. proposes the Generalized Mixture Gaussian VAE (MCGM-VAE) [26], which introduces a Gaussian mixture model to the multi-modal VAE with multiple channels. These channels have different weight coefficients that follow channel-weight layers, resulting in a Gaussian mixture distribution. Jiang et al. [16] used a disentangled conditional-VAE to learn interpretable latent representations. Suekane et al. presents the Conditional and Hierarchical VAE (CHVAE), which constructs a fashion item recommendation system by extracting fashion-specific visual features for enabling conditional and hierarchical learning [27].

Disentanglement in image generation and clustering plays a crucial role in machine learning, computer vision, and pattern recognition. To address this, several researchers have incorporated information-theoretic disentanglement guidance into VAEs, proposing methods such as β-VAE [12], β-TCVAE [4], and FactorVAE [17]. These modifications have shown significant improvement in the degree of disentanglement in the learned representations compared to the vanilla VAE. In a recent study, Satheesh et al. [25] employ an intermediate non-linearly scaled time-frequency representation in conjunction with VAE to achieve better segregating performance on spectro-temporal features.

Zheng et al. [30] demonstrated the intrinsic relationship between log-likelihood estimation of Variational Autoencoders (VAEs) and Fisher information and Shannon entropy, highlighting their potential to enhance reconstruction accuracy. In a similar vein, Zacherl et al. [29] incorporated the Fisher information metric into the decoder of their proposed model. Motivated by the work of Dilokthanakul et al. [8], we integrate hierarchical factorization and importance sampling into the framework of Gaussian mixture variational auto-encoder. This integration aims to improve the reconstruction performance.

In this paper, we first construct the Evidence Lower Bound (ELBO) by a probabilistic graphical model, and decompose it into four meaningful items, referred to as HDGMVAE (Hierarchically disentangled Gaussian Mixture Variational AutoEncoder model) (Sect. 2.1) . Subsequently, we introduce slack variables to narrow down the feasible solution space (Sect. 2.2). To separate difficult-to-classify samples successfully, we incorporate Fisher discriminant as a regularization term (Sect. 2.3). Finally, we investigate the impact of importance sampling in the experiments of this paper (Sect. 2.4). In a result, we obtain the model HDGMVAE-I (Hierarchically disentangled Gaussian Mixture Variational AutoEncoder model with Importance sampling). Experimental results verify the effectiveness of HDGMVAE-I.

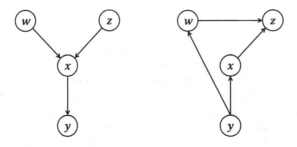

Fig. 1. Probabilistic graphic model in HDGMVAE. Left: the joint probability $p(y, x, w, z)$. Right: the conditional probability $q(x, w, z|y)$.

2 Methodology

2.1 Disentangled Representation Learning Using HDGMVAE

To enhance the representation of latent variables and tackle the difficulties in matching the ground truth posterior, we propose HDGMVAE, taking inspiration from the work of Dilokthanakul et al. [8]. We construct a probabilistic graphical model, as illustrated in Fig. 1. By leveraging this graphical model, we derive two equations:

$$p(y, x, w, z) = p(w)p(z)p(x|w, z)p(y|x).$$
$$q(x, w, z|y) = q(x|y)q(w|y)q(z|x, w) \tag{1}$$

In the equations, y represents the target observations, x represents latent attributes, z represents the mean sub-Gaussian distributions of the Gaussian Mixture Model (GMM), and w represents the mean coefficients of the distribution.

Thus the ELBO(Evidence Lower Bound) of variational process can be written as:

$$\log p(y) = \log \mathbb{E}_{q(x, w, z|y)} [p(y, x, w, z)]$$
$$\geq \mathbb{E}_{q(x, w, z|y)} \left[\log \frac{p(y, x, w, z)}{q(x, w, z|y)} \right] \tag{2}$$
$$= \mathcal{L}_{ELBO}$$

The maximization of Eq. 2 serves as the optimization objective of HDGMVAE and can be decomposed as follows:

$$\mathcal{L}_{ELBO} = \underbrace{\mathbb{E}_{q(x|y)}[\log p(y \mid x)]}_{\text{reconstruction term}} - \underbrace{KL(q(w \mid y)\|p(w))}_{w- \text{ prior}}$$
$$- \underbrace{\mathbb{E}_{q(x|y)q(w|y)}[KL(q(z \mid x, w)\|p(z))]}_{z- \text{ prior}} \tag{3}$$
$$- \underbrace{\mathbb{E}_{q(w|y)q(z|x, w)}[KL(q(x \mid y)\|p(x \mid w, z))]}_{\text{conditional prior}}$$

Subsequently, we can identify four terms within the ELBO: the w-prior, z-prior, conditional prior, and reconstruction term. The z-prior and w-prior terms impose constraints on the sub-Gaussian distribution and its coefficients, respectively. By encouraging the sub-Gaussian distribution to be as close as possible to the standard Gaussian distribution, the Gaussian mixture model becomes closer to the true distribution of the latent space. The conditional prior term ensures that the distribution obtained by sampling from the ground truth is as similar as possible to the distribution obtained from the latent space. This promotes consistency between the generated samples and the latent space representation. The reconstruction term measures the fidelity of the model by evaluating how closely the generated data matches the ground truth. It encourages the generated data to be as close as possible to the ground truth observations.

The total correlation (TC) can be calculated by the following equation:

$$TC(\boldsymbol{z}) = \mathbb{E}_{q_\phi(\boldsymbol{z})} \left[\log \frac{q_\phi(\boldsymbol{z})}{\prod_k q_\phi(\boldsymbol{z}_k)} \right] = \mathrm{KL} \left(q_\phi(\boldsymbol{z}) \| \prod_k q_\phi(\boldsymbol{z}_k) \right) \qquad (4)$$

Minimizing the total correlation encourages the distribution $q_\phi(\boldsymbol{z})$ to have statistically independent components $\boldsymbol{z}_k$, which can facilitate disentanglement of representations. This provides a potential mechanism for inducing disentanglement in the learned latent space. When our main objective is to achieve disentangled representations, we may choose to relax the constraint that the shape of the distribution matches the prior, and instead prioritize enforcing statistical independence among the latent variables. This idea is inspired by the hierarchically factorized VAE [9]. In the z-prior term, we decompose it into two components, denoted as A and B.

$$
\begin{aligned}
&- KL(q(\boldsymbol{z}|\boldsymbol{x},\boldsymbol{w}) \| p(\boldsymbol{z})) \\
&= \mathbb{E}_{q(\boldsymbol{z}|\boldsymbol{x},\boldsymbol{w})} \underbrace{\left[\log \frac{p(\boldsymbol{z})}{\prod_k p(\boldsymbol{z}_k)} - \log \frac{q(\boldsymbol{z}|\boldsymbol{x},\boldsymbol{w})}{\prod_k q(\boldsymbol{z}_k|\boldsymbol{x},\boldsymbol{w})} \right]}_{A} \\
&\quad - \underbrace{\sum_k KL\left(q(\boldsymbol{z}_k|\boldsymbol{x},\boldsymbol{w}) | p(\boldsymbol{z}_k) \right)}_{B}
\end{aligned}
\qquad (5)
$$

Term A matches the total correlation between variables in the inference model relative to the total correlation in the generative model. Term B minimizes the KL divergence between the inference marginal and prior marginal for each distribution of GMM $\boldsymbol{z}_k$, which is formally identical to Eq. 5. When the variable of distribution $\boldsymbol{z}_k$ contains sub-variables $\boldsymbol{z}_{k,d}$, we can recursively decompose the KL on the marginals $\boldsymbol{z}_k$.

$$- KL\left(q(\boldsymbol{z}_k|\boldsymbol{x}, \boldsymbol{w})|p(\boldsymbol{z}_k)\right)$$

$$= \mathbb{E}_{q(\boldsymbol{z}_k|\boldsymbol{x}, \boldsymbol{w}} \underbrace{\left[\log \frac{p(\boldsymbol{z}_k)}{\prod_k p(\boldsymbol{z}_{k,d})} - \log \frac{q(\boldsymbol{z}_k|\boldsymbol{x}, \boldsymbol{w})}{\prod_d q(\boldsymbol{z}_{k,d}|\boldsymbol{x}, \boldsymbol{w})}\right]}_{C}$$

$$- \underbrace{\sum_d KL\left(q(\boldsymbol{z}_{k,d}|\boldsymbol{x}, \boldsymbol{w})|p(\boldsymbol{z}_{k,d})\right)}_{D}$$

$$(6)$$

As a result, Eq. 5 incorporates a total correlation term to guarantee the statistical independence between individual sub-Gaussian distributions within the mixed Gaussian model. This term encourages disentanglement and ensures that each sub-Gaussian distribution captures distinct and independent factors of variation. On the other hand, Eq. 6 utilizes a total correlation term to enforce independence among the individual components of the sub-Gaussian distributions. It promotes disentanglement and encourages each component within a sub-Gaussian distribution to represent a separate and independent factor of variation. By minimizing the total correlation, the model is encouraged to learn disentangled representations at a finer level within each sub-Gaussian distribution.

2.2 Slack Variables

Inspired by β-VAE [12], we introduce three biases, denoted as r_1, r_2, r_3, which impose constraints on the terms of the ELBO. These biases serve to adjust the position of the total objective function within the feasible solution space.

$$KL(q(\boldsymbol{w} \mid \boldsymbol{y})\|p(\boldsymbol{w})) < r_1$$
$$\mathbb{E}_{q(x|y)q(w|y)}[KL(q(\boldsymbol{z} \mid \boldsymbol{y})\|p(\boldsymbol{z}))] < r_2 \qquad (7)$$
$$\mathbb{E}_{q(w|y)q(z|x,w)}[KL(q(\boldsymbol{x} \mid \boldsymbol{w}, \boldsymbol{z})\|p(\boldsymbol{x} \mid \boldsymbol{z}, \boldsymbol{w}))] < r_3$$

In the mean time, maximizing reconstruction term, our optimization objective can be expressed as follows:

$$\mathbb{E}_{q(x|y)}[\log p(\boldsymbol{y} \mid \boldsymbol{x})] + \alpha KL(q(\boldsymbol{w} \mid \boldsymbol{y})\|p(\boldsymbol{w}))$$
$$+ \beta \mathbb{E}_{q(x|y)q(w|y)}[KL(q(\boldsymbol{z} \mid \boldsymbol{y})\|p(\boldsymbol{z}))]$$
$$+ \gamma \mathbb{E}_{q(w|y)q(z|x,w)}[KL(q(\boldsymbol{x} \mid \boldsymbol{w}, \boldsymbol{z})\|p(\boldsymbol{x} \mid \boldsymbol{z}, \boldsymbol{w}))]$$
$$- (\alpha r_1 + \beta r_2 + \gamma r_3)$$

$$(8)$$

Due to the complexity of real data, it is often not feasible for the left-hand side of Eq. (7) to become exactly zero after optimization. This implies that Eq. (3) does not impose strict constraints on these terms. However, by introducing appropriate slack variables $\alpha r_1, \beta r_2, \gamma r_3$, we can ensure that the feasible solutions of the objective equation lie within the interior of the constraints defined by Eq. (7). The inclusion of slack variables $\alpha r_1, \beta r_2, \gamma r_3$ helps narrow down the feasible solution space. During practical computations, we treat the term $(\alpha r_1 + \beta r_2 + \gamma r_3)$

as a constant c and incorporate it into the objective function for optimization. This allows us to effectively balance the trade-offs between different components of the objective and guide the optimization process to find desirable solutions within the defined constraints.

2.3 Fisher Discriminant as Regularization

Our idea is to minimize the between-class distance and maximizing the within-class distance, which is consistent with the idea of the Fisher discriminant [30]. Inspired by this method, it is worth setting the latent space including K classes, which correspond to K sub-Gaussian distributions in this paper. Each sub-Gaussian distribution obeys $\mathcal{N}(w_i \boldsymbol{\mu}_i, w_i^2 \boldsymbol{\Sigma}_i)$, w_i denotes the mixture weight of each sub-Gaussian distribution, $\boldsymbol{\mu}_i$, $\boldsymbol{\Sigma}_i$ denotes the mean and variance of the sub-Gaussian distribution. The number of samples sampled in each sub-Gaussian distribution is n_i, and the sampled samples are denoted as $z_{i,j}$ for the i-th class and the j-th sample. The dataset $D = \{z_{i,j}\}$ is obtained, and the number of samples is $N = \sum_k n_i$. Next, we need to define the between-class covariance matrix $\boldsymbol{S}_B$ and the within-class covariance matrix $\boldsymbol{S}_W$.

First, within-class distance $\boldsymbol{S}_k \in \mathbb{R}$ is defined as:

$$\boldsymbol{S}_k = \sum_{n_i} (\boldsymbol{z}_i - w_i \boldsymbol{\mu}_i)(\boldsymbol{z}_i - w_i \boldsymbol{\mu}_i)^T = n_i w_i^2 \boldsymbol{\Sigma}_i \tag{9}$$

Within-class covariance matrix S_W is defined as the sum of the covariance matrices of each class:

$$\boldsymbol{S}_W = \sum_k \boldsymbol{S}_k \tag{10}$$

Thus the definition of the between-class covariance matrix $\boldsymbol{S}_B$ is obtained as:

$$\boldsymbol{S}_B = \sum_{k=1}^K n_k (w_k \boldsymbol{\mu}_k - \boldsymbol{\mu})(w_k \boldsymbol{\mu}_k - \boldsymbol{\mu})^T \tag{11}$$

In the training process, the global mean vector $\boldsymbol{\mu} = \boldsymbol{0}$. Then, the between-class covariance matrix $\boldsymbol{S}_B$ can be written as:

$$\boldsymbol{S}_B = \sum_{k=1}^K n_k w_k^2 \boldsymbol{\mu}_k \boldsymbol{\mu}_k^T \tag{12}$$

To maximize the between-class variance and minimize the within-class variance, we introduce the Fisher discriminant as regularization term F_{reg} as follows:

$$tr(\boldsymbol{S}_W^{-1} \boldsymbol{S}_B) = tr\left(\left(\sum_{k=1}^K n_k w_k^2 \boldsymbol{\Sigma}_i \right)^{-1} \left(\sum_{k=1}^K n_k w_k^2 \boldsymbol{\mu}_k \boldsymbol{\mu}_k^T \right) \right) \tag{13}$$

Assuming that the number of samples sampled from each sub-Gaussian distribution is the same, i.e., $n_1 = n_2 = ... = n_k$. The calculation of the Fisher regularization term could be simplified as followed:

$$F_{reg} = tr\left(\left(\sum_{k=1}^{K} w_k^2 \Sigma_i\right)^{-1} \left(\sum_{k=1}^{K} w_k^2 \mu_k \mu_k^T\right)\right) \tag{14}$$

So, the total loss can be written as:

$$\mathcal{L} = \mathcal{L}_{ELBO} + c + F_{reg} \tag{15}$$

2.4 Importance Sampling

To optimize the objective ELBO (Eq. 2), we apply the importance sampling technique.

$$\mathcal{L}'_{ELBO} = \mathbb{E}_{(x_i, w_i, z_i) \sim q(x,w,z|y)}\left[\log \frac{1}{n}\sum_{i=1}^{n} \frac{p(y, x_i, w_i, z_i)}{q(x_i, w_i, z_i|y)}\right] \tag{16}$$

For every sample (x_i, w_i, z_i), the probability of selecting samples within a batch is equal. Sampling n samples from the distribution of $q(x, w, z|y)$, and calculate the mean. When $n = 1$, $\mathcal{L}'_{ELBO} = \mathcal{L}_{ELBO}$, the objective degrades into GMVAE. If $n \to \infty$, $\mathcal{L}'_{ELBO}$ is infinitely close to ground truth $logp(y)$ [2].

To apply importance sampling, we set $v = \frac{p(y,x,w,z)}{q(x,w,z|y)}$, and calculate the gradient.

$$\nabla \mathbb{E}_{(x_i, w_i, z_i) \sim q(x,w,z|y)}\left[\log \frac{1}{n}\sum_{i=1}^{n} \frac{p(y, x_i, w_i, z_i)}{q(x_i, w_i, z_i|y)}\right]$$
$$= \mathbb{E}_{(x_i, w_i, z_i) \sim q(x,w,z|y)}\left[\frac{1}{\sum_{j=1}^{n} v_j}\sum_{i=1}^{n} v_i \nabla \log v_i\right] \tag{17}$$

Let $\tilde{v}_i = \frac{v_i}{\sum_{j=1}^{n} v_j}$,

$$\nabla \mathcal{L}'_{ELBO} = \mathbb{E}_{(x_i, w_i, z_i) \sim q(x,w,z|y)}\left[\sum_{i=1}^{n} \tilde{v}_i \nabla \log v_i\right]$$
$$= \left[\sum_{i=1}^{n} \tilde{v}_i \nabla \log v_i\right] \tag{18}$$

where $\tilde{v}_i$ can be seen as a importance factor to leverage different weights to different samples. It is worth noting that during the calculation of the gradient of Eq. (18), the gradient of $\nabla \log v_i$ is exactly the gradient of Eq. (2). The training process of the total neural network is depicted as Algorithm 1.

Algorithm 1. training process

$\theta, \phi \leftarrow$ Initialize parameters;
repeat
    $X \leftarrow$ Random minibatch of M datapoints (drawn from full dataset);
    Calculate the loss function $\mathcal{L}(\theta, \phi, X)$ by formula (15);
    Calculate $\triangledown \mathcal{L}$ as $\triangledown \log v_i$ by Neural Network;
    Calculate $\triangledown \mathcal{L}'_{ELBO}$ by formula (18);
    $\theta, \phi \leftarrow$ Update parameters using gradients $\triangledown \mathcal{L}'_{ELBO}$;
until convergence of parameters (θ, ϕ)
return (θ, ϕ)

Table 1. Clustering Result

Metrics	SC $\rightarrow$ 1	CH $\uparrow$	DB $\downarrow$
GMVAE	0.1716	1132.8412	4.0677
m = 2	0.3693	3081.6701	1.9343
m = 5	0.3697	3022.9172	1.3253
m = 8	**0.3805**	**3134.1266**	1.7893
m = 10	0.3798	3129.8682	**1.1473**

Clustering results. n represents the number of multiple samples. It can be observed that different values of m exhibit varying performance metrics compared to GMVAE. However, in general, all values of m demonstrate superior performance over GMVAE.

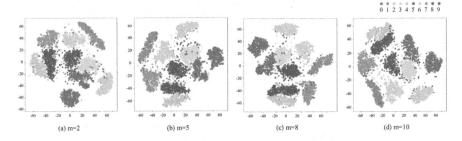

(a) m=2   (b) m=5   (c) m=8   (d) m=10

Fig. 2. 2-D visualization of the latent vector using t-SNE [22].

Fig. 3. Generation results on CelebA dataset.

3   Experimental Settings

In this section, we evaluate the effectiveness of our model on various downstream clustering and generation tasks. We adopt a simple neural network architecture consisting of 5-layer CNNs with a 3×3 kernel for both the encoder and decoder components. The latent space dimension z_{dim} is set to 128, and the mean squared error (MSE) reconstruction loss is utilized. Despite the simplicity of the model, it demonstrates excellent performance on the dataset used in this paper. To ensure reliable results, all experiments are conducted 10 times with 30 epochs each, using the aforementioned network structure. The final experimental results are obtained by averaging the outcomes of these repeated runs.

For clustering experiments, we use MNIST [19] dataset with 10 classes. Three metrics, namely Silhouette Coefficient (SC) [24], Calinski Harabasz Index (CH) [3], and Davies Bouldin Index (DB) [7], are used to verify the clustering performance of different models.

In generation experiments, we add Fashion MNIST [28], CelebA [21], dSprites datasets extraly. Three typical methods, including vanilla VAE [18], GMVAE [8], β-VAE [12], are performed for comparison. And four metrics, Fréchet Inception Distance (FID), Structural Similarity (SSIM), Multi-Scale Structural Similarity (MS-SSIM) and Learned Perceptual Image Patch Similarity (LPIPS), are used to verify the generation performance.

The experiments are conducted as follows. For each image, generations are executed based on HDGMVAE-I models with 4 different sampling numbers. In each experiment, the above four metrics of each model are calculated.

3.1   Ablation Experiments on Slack Variables

In the experiment, we set the number of samplings to 3. The experimental results are shown in Table 2. From the experiments, we can see that different data sets have different values for the slack variable. On the MNIST and Fashion MNIST data sets, the value of the slack variable is 10^{-2} is better, while on the CelebA and dSprites data sets, the value of the slack variable is 10^{-3}. It is worth noting that all optimal values are obtained when the slack variable is non-zero. This indicates that the slack variable positively affects the experimental index.

3.2   Ablation Experiments with Fisher Discriminant as Regularization

In the experiment, we set the number of samplings to 8. The experimental results are shown in Table 3. Through the experimental results, we can see that the Fisher Discriminant on regularization term has a huge impact on the clustering evaluation indicators, and the relative improvements in the three indicators on the MNIST dataset are 43.23%, 16.38%, and 32.20%.

Table 2. Ablation experiments of slack variables on MNIST, Fashion_MNIST, Celeba and dSprites datasets

c	MNIST				Fashion MNIST			
	FID ↓	SSIM ↑	MS-SSIM ↑	LPIPS ↓	FID ↓	SSIM ↑	MS-SSIM ↑	LPIPS↓
0	35.3861	0.9531	0.9897	0.0547	85.0373	0.8072	0.9259	0.0795
10^{-1}	**34.6363**	0.9575	0.9905	0.0491	**81.9173**	0.8027	0.9286	0.0844
10^{-2}	35.0791	**0.9582**	**0.9906**	**0.0483**	84.1146	**0.8083**	0.9298	**0.0779**
10^{-3}	35.3527	0.9549	0.9887	0.0502	85.5619	0.7973	0.9291	0.0860
10^{-4}	34.9833	0.9529	0.9894	0.0579	84.3786	0.7979	0.9282	0.0853
10^{-5}	35.7898	0.9539	0.9898	0.0520	84.4754	0.8041	**0.9316**	0.0813
c	CelebA				dSprites			
	FID ↓	SSIM ↑	MS-SSIM ↑	LPIPS↓	FID ↓	SSIM ↑	MS-SSIM↑	LPIPS↓
0	85.8112	0.7736	0.9541	0.0985	11.6752	0.9971	0.999506	0.0146
10^{-1}	86.1325	0.7705	0.9536	0.0969	11.8873	0.9969	0.999468	0.0151
10^{-2}	**84.8514**	0.7757	0.9532	0.0986	**11.0609**	0.9970	0.999481	**0.0143**
10^{-3}	85.4031	**0.7780**	**0.9560**	**0.0945**	12.5247	**0.9974**	0.999468	0.0203
10^{-4}	89.3177	0.7686	0.9536	0.1006	13.9871	0.9967	0.999425	0.0226
10^{-5}	89.0801	0.7681	0.9537	0.0985	11.4010	0.9971	**0.999549**	0.0211

Table 3. Ablation experiments of slack variables on MNIST

Metrics	SC ↑	CH ↑	DB ↓
HDGMVAE-I	**0.381**	**3134.127**	**1.789**
w/o Fisher Discriminant as Regularization	0.266	2693.011	2.639

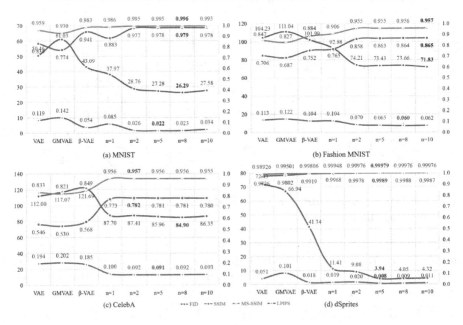

Fig. 4. Generation results on 8 models in 4 metrics for 4 datasets, where the bold indicates the best.

4   Results and Discussions

The resulting latent variables x for different sampling numbers is visualized using t-SNE [22], as shown in Fig. 2, and corresponding quantitative results is shown in Table 1. The results of various values of m on z is illustrated in Fig. 2 with $z_{dim} =$ 128. In general, using the number of samplings $m = 2, 5, 8, 10$, larger m brings better performance. But after reaching the best value, it started to deteriorate again. It can be described as a U-shaped curve. It is proved in the experiment of [11] that an excessively large batch size will cause a decrease in performance. On the other hand, increasing the number of samplings is equivalent to increasing the number of batch sizes. This also shows that although theoretically the larger the number of samplings, the better the result, the experimental results show that this is not the case.

Similarly, Fig. 3 shows the generation results on CelebA. In Fig. 4, these four sub-figures illustrates how these three different models and our methods with different sampling numbers effect the four datasets. For example, in MNIST dataset, our method has the best score in FID, which is 26.29. This result outperforms vanilla VAE by 24.17, GMVAE by 34.74 and β-VAE by 16.8. Other three measures are also better than other methods, so HDGMVAE-I produces better reconstruction results, which can be visually verified by Fig. 4(b)–(d).

5   Conclusion

In this paper, we explore the integration of factorization and importance sampling into the GMVAE framework for image clustering and generation tasks. Hierarchical factorization is employed to enhance the expressive power of the distributions and their individual components. By decomposing the latent space into hierarchical layers, we can capture more complex dependencies and interactions within the data. Furthermore, the introduction of importance sampling helps to reduce the bias between the model and the true underlying distribution. This leads to improved reconstruction performance, as the model can better approximate the true data distribution. To enhance the discriminative ability of the model, we incorporate Fisher discriminant as a regularization term. This term minimizes the within-class distance and maximizes the between-class distance, particularly for challenging samples that are difficult to accurately classify. This regularization encourages the model to generate samples that are more distinguishable and representative of different classes. To ensure that the model produces feasible solutions, we employ slack variables to narrow down the solution space. These slack variables help to enforce the constraints defined in the optimization problem, leading to more desirable and consistent solutions.

References

1. Aubry, M., Maturana, D., Efros, A.A., Russell, B.C., Sivic, J.: Seeing 3D chairs: exemplar part-based 2d–3d alignment using a large dataset of cad models. In: Proceedings of the IEEE Conference on Computer Vision and Pattern Recognition, pp. 3762–3769 (2014)
2. Burda, Y., Grosse, R., Salakhutdinov, R.: Importance weighted autoencoders. arXiv preprint arXiv:1509.00519 (2015)
3. Caliński, T., Harabasz, J.: A dendrite method for cluster analysis. Commun. Stat. Theory Methods 3(1), 1–27 (1974)
4. Chen, R.T., Li, X., Grosse, R.B., Duvenaud, D.K.: Isolating sources of disentanglement in variational autoencoders. In: Advances in Neural Information Processing Systems, vol. 31 (2018)
5. Chen, S., Huang, J.: Fec: three finetuning-free methods to enhance consistency for real image editing. arXiv preprint arXiv:2309.14934 (2023)
6. Chen, X., Duan, Y., Houthooft, R., Schulman, J., Sutskever, I., Abbeel, P.: InfoGAN: interpretable representation learning by information maximizing generative adversarial nets. In: Advances in Neural Information Processing Systems, vol. 29 (2016)
7. Davies, D.L., Bouldin, D.W.: A cluster separation measure. IEEE Trans. Pattern Anal. Mach. Intell. 2, 224–227 (1979)
8. Dilokthanakul, N., et al.: Deep unsupervised clustering with Gaussian mixture variational autoencoders. arXiv preprint arXiv:1611.02648 (2016)
9. Esmaeili, B., et al.: Structured disentangled representations. In: The 22nd International Conference on Artificial Intelligence and Statistics, pp. 2525–2534. PMLR (2019)
10. Goodfellow, I., et al.: Generative adversarial nets. In: Neural Information Processing Systems (2014)
11. Goyal, P., et al.: Accurate, large minibatch SGD: training imagenet in 1 hour. arXiv preprint arXiv:1706.02677 (2017)
12. Higgins, I., et al.: Beta-VAE: learning basic visual concepts with a constrained variational framework. In: International Conference on Learning Representations (2017)
13. Ho, J., Jain, A., Abbeel, P.: Denoising diffusion probabilistic models. In: Advances in Neural Information Processing Systems, vol. 33, pp. 6840–6851 (2020)
14. Huang, J., Liu, Y., Huang, Y., Chen, S.: Seal2real: prompt prior learning on diffusion model for unsupervised document seal data generation and realisation. arXiv preprint arXiv:2310.00546 (2023)
15. Huang, J., Liu, Y., Qin, J., Chen, S.: KV inversion: KV embeddings learning for text-conditioned real image action editing. arXiv preprint arXiv:2309.16608 (2023)
16. Jiang, J., Xia, G.G., Carlton, D.B., Anderson, C.N., Miyakawa, R.H.: Transformer VAE: a hierarchical model for structure-aware and interpretable music representation learning. In: ICASSP 2020–2020 IEEE International Conference on Acoustics, Speech and Signal Processing (ICASSP), pp. 516–520. IEEE (2020)
17. Kim, H., Mnih, A.: Disentangling by factorising. In: International Conference on Machine Learning, pp. 2649–2658. PMLR (2018)
18. Kingma, D.P., Welling, M.: Auto-encoding variational Bayes. arXiv preprint arXiv:1312.6114 (2013)
19. LeCun, Y., Bottou, L., Bengio, Y., Haffner, P.: Gradient-based learning applied to document recognition. Proc. IEEE 86(11), 2278–2324 (1998)

20. Liu, Z., Luo, P., Wang, X., Tang, X.: Deep learning face attributes in the wild. In: Proceedings of the IEEE International Conference on Computer Vision, pp. 3730–3738 (2015)
21. Liu, Z., Luo, P., Wang, X., Tang, X.: Large-scale celebfaces attributes (celeba) dataset. Retrieved August 15(2018), 11 (2018)
22. Van der Maaten, L., Hinton, G.: Visualizing data using t-SNE. J. Mach. Learn. Res. **9**(11) (2008)
23. Rezende, D.J., Mohamed, S., Wierstra, D.: Stochastic backpropagation and approximate inference in deep generative models. In: International Conference on Machine Learning, pp. 1278–1286. PMLR (2014)
24. Rousseeuw, P.J.: Silhouettes: a graphical aid to the interpretation and validation of cluster analysis. J. Comput. Appl. Math. **20**, 53–65 (1987)
25. Satheesh, C., Kamal, S., Mujeeb, A., Supriya, M.: Passive sonar target classification using deep generative *beta*-VAE. IEEE Sig. Process. Lett. **28**, 808–812 (2021)
26. Shao, J., Li, X.: Generalized zero-shot learning with multi-channel gaussian mixture VAE. IEEE Sig. Process. Lett. **27**, 456–460 (2020)
27. Suekane, K., et al.: Personalized fashion sequential recommendation with visual feature based on conditional hierarchical VAE. In: 2022 IEEE 5th International Conference on Multimedia Information Processing and Retrieval (MIPR), pp. 362–365. IEEE (2022)
28. Xiao, H., Rasul, K., Vollgraf, R.: Fashion-mnist: a novel image dataset for benchmarking machine learning algorithms. arXiv preprint arXiv:1708.07747 (2017)
29. Zacherl, J., Frank, P., Enßlin, T.A.: Probabilistic autoencoder using fisher information. Entropy **23**(12), 1640 (2021)
30. Zheng, H., Yao, J., Zhang, Y., Tsang, I.W., Wang, J.: Understanding VAEs in Fisher-Shannon plane. In: Proceedings of the AAAI Conference on Artificial Intelligence, vol. 33, pp. 5917–5924 (2019)

"Car or Bus?" ClearSeg: CLIP-Enhanced Discrimination Among Resembling Classes for Few-Shot Semantic Segmentation

Anqi Zhang[ID], Guangyu Gao[✉][ID], Zhuocheng Lv[ID], and Yukun An[ID]

School of Computer Science and Technology, Beijing Institute of Technology,
Beijing 100081, China
guangyugao@bit.edu.cn

Abstract. Few-shot semantic segmentation aims at learning to segment query images of unseen classes with the guidance of limited segmented support examples. However, existing models tend to confuse the resembling classes (*e.g.*, 'car' and 'bus') thus generating erroneous predictions. To address this, we propose the **CLIP-enhanced discrimination among** resembling classes for few-shot semantic **Segmentation (CLearSeg)**, which leverages information beyond support images, including the class name, through Contrastive Language-Image Pretraining (CLIP), to discriminate between resembling classes. Firstly, we modify the CLIP structure and design the Sliding Attention Pooling (SAP) to construct the *Text-Driven Activation (TDA) module*, learning the Class-Specific Activation (CSA) maps with class names. Since the semantic information is explicitly involved by the class name, the CSA maps exhibit clear distinctions among resembling classes. Meanwhile, to enrich fine-grained features ensuring distinguishability, the *Multi-Level Correlation (MLC) module* is designed to extract multi-level features of support and query images and generate various correlation maps. We further applied a decoder to fuse the CSA map and correlation maps with encoded features and obtain the final prediction. Experiments on the Pascal-5^i and COCO-20^i datasets have shown that CLearSeg outperforms previous methods, achieving the mIoU of 69.2% and 48.9% for 1-shot segmentation, particularly in distinguishing objects from resembling classes.

Keywords: Few-Shot Learning · Semantic Segmentation · CLIP · Resembling Class

1 Introduction

Few-shot Semantic Segmentation (FSS) allows for generalization to unseen domains with only a few annotated images, building upon the pioneering OSLSM [24]. Most FSS methods focus on improving feature representation extraction from labeled images (known as "support images") and enhancing

© The Author(s), under exclusive license to Springer Nature Switzerland AG 2024
S. Rudinac et al. (Eds.): MMM 2024, LNCS 14554, pp. 172–186, 2024.
https://doi.org/10.1007/978-3-031-53305-1_14

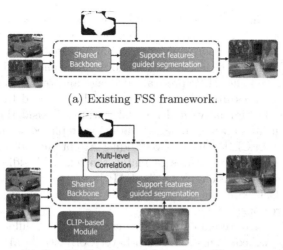

(a) Existing FSS framework.

(b) Our CLIP-enhanced framework *CLearSeg*.

Fig. 1. Comparison between the existing Few-shot Semantic Segmentation framework and our CLearSeg scheme. The previous methods in (a) have inadequate usage of features, leading to an unsatisfactory prediction on resemble classes. Our CLIP-enhanced CLearSeg utilized features from the category names and multi-grained levels, improving the discrimination among resembling classes.

similarity metrics between these feature representations and the "query image" to improve segmentation results. For example, while SG-One [35] proposed the Masked Average Pooling to extract guidance information from support images efficiently, PFENet [27] utilizes a learnable similarity metric with a multi-scale structure for precise predictions. HSNet [18] introduces the 4D Hypercorrelation to fully exploit the correlation of the support images and the query image, etc.

However, these models tend to confuse the resembling classes due to the lack of specific supervision. For example, objects resembling classes (*e.g.*, 'car' and 'bus', shown in Fig. 1) usually have similar features, leading to the feature maps of the backbone network that lack discriminability. To tackle this issue, several approaches have been proposed to enhance discrimination among different classes. For instance, PFENet [27] utilizes the high-level features from an ImageNet pre-trained backbone as prior knowledge to improve the identification of different classes. BAM [9] avoids model bias towards seen classes by explicitly identifying the regions that do not require segmentation by the base learner. Besides, IPRNet [22] mitigated the problem of resemblance between classes by the Relation Loss, which averages the cosine similarity between the prototypes in a batch. These methods may have enhanced the ability of the model to discriminate seen classes, yet the samples belonging to the unseen classes are still confused in such situations.

Moreover, the analogical ability of human beings heavily depends on the accumulated knowledge and the degree of understanding of object attributes. Namely, the more comprehensively people understand the attributes, the more accurate the recognition and differentiation of similar objects. Therefore, by leveraging the comprehensive knowledge embedded in CLIP [23], we propose the *CLIP*-enhanced Discrimination *a*mong *r*esembling classes for few-shot semantic *seg*mentation (CLearSeg) to better use the multi-modal and multi-level features.

CLearSeg utilized the CLIP pre-trained image and text encoders for excavating the text features from the class names and integrated them into image features as more discriminative multi-modal features. Considering that spatial information is crucial for semantic segmentation, we replaced the original Attention Pooling layer in CLIP with a novel Sliding Attention Pooling (SAP) layer, which transfers the image features to match the text embedding while maintaining the spatial dimension. With the SAP layer, we can get a text-guided heatmap called Class-Specific Activation (CSA) map, which explicitly emphasizes the target regions.

Moreover, objects of resembling classes may still present differences in multi-level features, so we construct the Multi-Level Correlation (MLC) module to generate multi-grained prior maps. For feature maps from each block of the backbone network, the MLC module obtains prototypes from support features and generates correlation maps with query features via relevance metric, allowing for the incorporation of fine-grained details and coarse-grained semantics for precise segmentation. Additionally, middle-level features of query and support images are concatenated and merged respectively following [9,11,27,32,33]. Finally, the multimodal CSA map and multi-grained correlation maps are fed into the decoder for feature fusion and segmentation prediction.

In summary, our main contributions can be outlined as follows:

- We introduced and modified the CLIP pre-trained framework to construct the Text-Driven Activation module with sliding attention pooling, which leverages text information beyond support images to explicitly discriminate resembling classes.
- We construct the well-crafted Multi-Level Correlation module that computes the correlation of multi-level query and support features, encompassing both coarse-grained and fine-grained features.
- Extensive experiments are conducted on Pascal-5^i and COCO-20^i, and the results show that our CLIP-enhanced method *CLearSeg* outperforms most previous work, especially for distinguishing among resembling classes.

2 Related Work

2.1 Few-Shot Semantic Segmentation

Emerging from the seminal OSLSM [24], Few-shot Semantic Segmentation (FSS) employs matched support image masks to guide query image semantic segmentation. A variety of FSS techniques with diverse network architectures and performance benchmarks have since emerged. The pivotal distinction among these methods lies in their approach to comparing support and query images, categorizing them into two main types based on their method of feature comparison.

Feature-Prototype Comparison: Pioneered by OSLSM [24], Prototype comparison involves transforming support features into global prototypes through pooling. These prototypes then interact with query features for segmentation. SGOne [35] designed the Masked Average Pooling operation, which reduces the influence of mask size. CANet [32] proposed a Dense Comparison Module, replacing the cosine similarity with 1×1 Convolution. PFENet [27] combined these and added a prior-generation operation and Feature Enrichment Module. Based on the previous methods, ASGNet [11] generated prototypes per MaskSLIC region and filtered them. BAM [9] designed a base-learner and utilized Gram matrices calculation to indicate which part of the image should not be segmented. NTRENet [15] excluded interference using a global background prototype.

Pixel-Wise Comparison: Unlike the feature-prototype comparison method, this method takes full advantage of each pixel in both query and support features. HSNet [18] introduced 4D hypercorrelation and a 4D convolutional pyramid encoder, fully utilizing each feature pixel. CyCTR [33] brought in self-alignment and cross-alignment transformers, retaining pixel-level data via deformable transformers. DCAMA [25] presented a cross-attention-like structure, fully capturing similarity from the query and masked support features. Subsequent work like [8,19,20,30] built upon these pixel-wise network structures, refining signal quality while amplifying effective information.

2.2 CLIP-Based Multi-modal Networks

Contrastive Language-Image Pre-Training (CLIP) [23] is a revolutionary multimodal pretraining network for open-vocabulary tasks. Trained on 400 million image-text pairs, CLIP excels in zero-shot tasks. Yet, its application to Few-shot Image Classification poses challenges, with weaker performance using just a few shots compared to zero-shot scenarios. To promote CLIP Few-shot capacity, CoOp [36] introduced Context Optimization to optimize the text prompts by learning the learnable context vectors. Adapter [5] adds linear layers and a residual connection to leverage knowledge from CLIP and support images. Tip-Adapter [34] presents a Training-Free CLIP-Adapter module using cached model weights. Meanwhile, Zero-shot Semantic Segmentation is blooming with magnificent CLIP-based structures. LSeg [10] uses cosine similarity between encoded and text features. ZegFormer [2], and OVSeg [12] combine MaskFormer and CLIP for a two-stage Zero-shot network.

3 Preliminaries

Task Description: In the standard FSS framework, datasets are divided into D_{train} and D_{test}, in which the corresponding object categories C_{train} and C_{test} are disjoint. Both sets are constructed as multiple *episodes*, which contain the query set Q and the support set S. The query set $Q = \{(I^q, M^q)\}$ contains a query image I^q and its corresponding ground truth segmentation mask M^q. The support set $S = \{(I_i^s, M_i^s)\}_{i=1}^{K}$ contains K pairs of the support image I_i^s and its mask M_i^s which belong to the same category as the query set. During each training episode, the learnable parameters of the model are optimized by the loss between the prediction of the query image I^q and the ground truth M^q. Following training episodes, model performance is assessed on D_{test} by making predictions for the query image I^q across all testing episodes.

Review on CLIP: Contrastive Language-Image Pre-Training (CLIP) consists of a ResNet-like image encoder $\mathcal{V}(\cdot)$ and a text encoder $\mathcal{L}(\cdot)$ based on Transformer. The image encoder transfers an input image $I \in \mathbb{R}^{H \times W \times 3}$ into a visual representation $X_v \in \mathbb{R}^C$, while the text encoder encodes n associated captions into a text representation $X_t \in \mathbb{R}^{n \times C}$. In training, matching image-text pairs are positives, while mismatched pairs are negatives. Using contrastive learning, the visual representation X_v approaches the corresponding text representation X_t and distances it from incompatible ones, aligning visual and textual information. Given the absence of specific category priors, current FSS models struggle to precisely differentiate instances from closely related classes. To address this, we extend our approach by incorporating class names. This involves generating a Class-Specific Activation map using CLIP, which enhances the model's discriminative capacity by introducing additional category information.

4   Approach

The CLearSeg, shown in Fig. 2, consists of two core modules: the Text-Driven Activation (TDA) module and the Multi-Level Correlation (MLC) module. The TDA module employs a pre-trained CLIP image encoder and text encoder to encode images and prompts with class names. The MLC module incorporates both shallow and deep features, facilitating the simultaneous extraction of coarse and fine details. While the TDA module generates the Class-Specific Activation (CSA) map, the MLC module produces diverse correlation maps. The final prediction is achieved by integrating the CSA map and correlation maps in the decoder, which empowers it to utilize both class-specific insights and multi-level features to distinguish objects from similar classes.

4.1   Text-Driven Activation Module

To make the most of the provided contents and distinguish objects from resembling classes, we introduce the text features from the class names by the CLIP framework. To suit the requirements of the FSS task, we made modifications to the original CLIP framework and developed a CLIP-based multi-modal module, known as the Text-Driven Activation module. The original CLIP can align the visual and textual representation in the embedding space and measure the correlation between them. However, employing this structure can only obtain features without any spatial information, which is not suitable for semantic segmentation. By employing a sliding-window approach across the entire image, we introduce Sliding Attention Pooling (SAP) to replace the original Attention Pooling layer in the CLIP structure. SAP effectively produces spatially dimensioned class-specific guidance information.

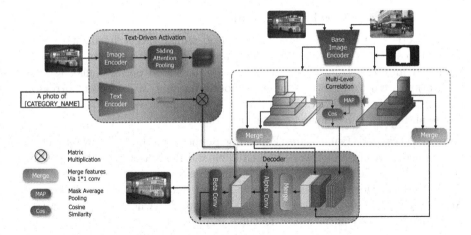

Fig. 2. Overall framework of CLearSeg, which contains the CLIP-based Text-Driven Activation (TDA) module, the Multi-Level Correlation (MLC) module, and feature integration in the decoder. Query images and class prompts are encoded by CLIP encoders, activating relevant regions within query features. Additionally, query and support images pass through the base encoder to create multi-level features for correlation computation. The decoder then refines these outputs to generate the final prediction. "Alpha conv" and "Beta conv" refer to the process in FEM [27].

Specifically, suppose the $X_v \in \mathbb{R}^{h \times w \times C_v}$ is the output of the CLIP image encoder without the Attention Pooling layer, We set a sliding window of size $p \times p$ to traverse the X_v in the stride of s. Once the operation is completed, we will get a group of feature representations $\tilde{X}_v = \{\tilde{x}_v^i \in \mathbb{R}^{p \times p \times C_v} \mid i \in [1, N_p]\}$ where $N_p = [(h - p)/s + 1] \times [(w - p)/s + 1]$. Then we flatten and transpose the $\tilde{X}_v$ to obtain the $\tilde{X}_v' \in \mathbb{R}^{p^2 \times N_p \times C_v}$. The SAP is computed as follows:

$$X_v^{sap} = W_t \cdot MHA(\mathbb{E}(\tilde{X}_v')W_q, \tilde{X}_v'W_k, \tilde{X}_v'W_v), \qquad (1)$$

where $W_q, W_k, W_v \in \mathbb{R}^{C_v \times C_v}$ and $W_t \in \mathbb{R}^{C_v \times C_t}$ represent the linear transform. $\mathbb{E}(\tilde{X}_v') \in \mathbb{R}^{1 \times N_p \times C_v}$ denotes the window-level representation obtained by averaging the first dimension of $\tilde{X}_v'$. $X_v^{sap} \in \mathbb{R}^{N_p \times C}$ is the output of the SAP operation. MHA denotes multi-head attention [28]:

$$MHA(q, k, v) = Softmax(\frac{q \cdot k^T}{\sqrt{d}} \cdot v), \qquad (2)$$

where q, k, v represents *query*, *key*, and *value* in the transformer, and d is the channels of them. Finally, we metric the similarity between X_v^{sap} and text representation $X_t \in \mathbb{R}^{1 \times C}$ to obtain the Class-Specific Activation (CSA) map M_{cls}:

$$M_{cls} = X_v^{sap} X_t^T. \qquad (3)$$

Benefiting from CLIP's powerful ability to align text and visual representations, the M_{cls} contains a wealth of class-specific information that can help the model distinguish resembling classes in the following steps.

4.2   Multi-level Correlation Module

While deep features carry semantic insights, aiding target localization, shallow features always hold spatial details. Thus, recent advancements [8,18,19,25] leveraged multi-level features to enhance segmentation performance. Drawing inspiration, we introduce the simple yet potent module, Multi-Level Correlation (MLC) module, which leverages all feature levels, concurrently introducing both coarse and fine features for segmentation.

Base Pretrained Backbone. Following previous methods [9], we select the pre-trained convolution network VGG-16 [26] and ResNet-50 [7] as the backbone network. We divide both the VGG-16 and ResNet-50 into five levels, which are named as $conv1\_x$, $conv2\_x$, $conv3\_x$, $conv4\_x$ and $conv5\_x$ respectively. Accordingly, the output features of each level are X_l^* where $l \in [0,4]$ and $* \in \{q,s\}$ denotes query and support features respectively.

Multi-level Correlation Computation. For each level l, we first achieve the support prototypes via Masked Average Pooling (MAP) [35] by

$$P_l = \frac{\sum_{x=0,y=0}^{h,w} X_{l;(x,y)}^s \cdot \mathcal{I}(M^s)_{(x,y)}}{\sum_{x=0,y=0}^{h,w} \mathcal{I}(M^s)_{(x,y)}}, \tag{4}$$

where $\mathcal{I}$ is an interpolation function to resize M^s to the same size as X_l^s. Then the correlation map $M_l^{co} \in \mathbb{R}^{h_l \times w_l}$ is calculated by channel-wise cosine similarity between the query features $X_l^q \in \mathbb{R}^{h_l \times w_l \times c_l}$ and the support prototype $P_l \in \mathbb{R}^{c_l}$:

$$M_{l;(x,y)}^{co} = \frac{x_l^q P_l}{\| x_l^q \|_2 \cdot \| P_l \|_2}, \quad x_l^q = X_{l;(x,y)}^q. \tag{5}$$

Notably, the MLC module achieves these correlation maps without introducing extra trainable parameters, ensuring both efficiency and optimization ease. Additionally, following [9,27,32], we concatenate and merge X_2^q, X_3^q, and X_2^s, X_3^s via 1×1 convolution for 256-dim features X^q and X^s.

4.3   Objective Function

Following FEM [27], we apply a 1×1 convolution to merge multi-level correlation map M^{co}, the fused query feature X^q and support prototype P^s. After the alpha convolution in FEM, we merge the class-specific activation map M_{cls} with the features and calculate through the beta convolution. Finally, we acquire the prediction Y, and employ the Cross-Entropy loss as the objective function by:

$$\mathcal{L}_f = -\frac{1}{HW} \sum_{i=1}^{HW} [M_i^q \log(\sigma(Y_i)) + (1 - M_i^q) \log(1 - \sigma(Y_i))], \tag{6}$$

where σ is the normalization function. We calculate n cross-entropy losses $\mathcal{L}_m^i$ ($i \in [1,n]$) of the intermediate prediction where n is the number of the spatial sizes of the FEM (set $n = 4$ here). The total loss $\mathcal{L}$ is computed as:

$$\mathcal{L} = \mathcal{L}_f + \frac{1}{n} \sum_i^n \mathcal{L}_m^i. \tag{7}$$

4.4 Extension to K-Shot Settings

Our network extends seamlessly to K-shot evaluation. Given K support images and masks $S = (I_i^s, M_i^s)_{i=1}^K$, alongside the query image I^q, we feed each image into pre-trained backbone. During middle-level merging and MLC calculation, support features and related masks turn into prototypes through Masked Average Pooling. To maintain network structure, we average support prototypes of 5 images into one. With averaged global features, the network tackles the 5-shot task using the predefined 1-shot solution.

5 Experiments

5.1 Dataset and Evaluation Metrics

Datasets. We evaluate CLearSeg on two widely used datasets, Pascal-5[i] [3] and COCO-20[i] [13]. Pascal-5[i] stems from PASCAL VOC 2012 [3] and SDS [6], comprising 20 distinct foreground categories. COCO-20[i] is built on the MS COCO [13] which has 80 categories. During training, both datasets' categories are split into four cross-validation folds, each containing a quarter of the categories. We divide the Pascal-5[i] dataset into groups of 5 in sequential order following [24]. Unlike the division setting in Pascal-5[i], for COCO-20[i] dataset, every 4 categories in sequential order are selected into the same fold according to [21] [29]. During the evaluation, Pascal-5[i] employs 1000 episodes, while COCO-20[i] uses 10000 episodes for the evaluation fold. Each validation episode involves segmenting 1 pair of query and support images. To ensure stable results, we use 5 random seeds and report the average outcome from 5 validation runs.

Evaluation Metrics. We use the mean Intersection over Union (mIoU) as the metric for evaluating our model's performance. The IoU for a particular class c is calculated as $IoU_c = \frac{TP_c}{TP_c + FP_c + FN_c}$, where $TP_c, FP_c,$ and FN_c represent the true positive, false positive, and false negative, respectively, of the predictions compared to the ground-truth mask. The mIoU is computed by taking the average of the IoU scores for all classes present in each fold, represented as $\frac{1}{N_c} \sum_c IoU_c$, where N_c is the number of classes. Additionally, we apply the FB-IoU metric, which only considers the IoU of foreground and background (i.e., $N_c = 2$) to compare our approach with other methods.

5.2 Implementation Details

We select VGG-16 [26] and ResNet-50 [7] as our backbones for comparisons with previous work. These backbone parameters are pre-trained following [9] and frozen during the training process. The whole framework is built on Pytorch. We employ an SGD optimizer with an initial learning rate of 0.005, momentum 0.9, and weight decay 10^{-4}. The batch size is set to 8 for Pascal-5[i] and 6 for COCO-20[i]. The input image size is 473 × 473 pixels for Pascal-5[i] and 641 × 641 pixels for COCO-20[i]. We train our network 200 epochs on Pascal-5[i] and 50 epochs on COCO-20^i. Moreover, we adopt the poly learning rate strategy [27] for dynamic learning rate as $1 - (iter/max\_iter)^{0.9}$. The whole training and evaluation process runs on the NVIDIA RTX 2080Ti GPU.

5.3 Comparison with State-of-the-Arts

We compare CLearSeg with other state-of-the-art methods on Pascal-5^i [3] and COCO-20^i [13], respectively, as shown in Table 1 and Table 2. Our CLearSeg surpasses previous methods in both the Pascal-5^i and COCO-20^i datasets, achieving a new state-of-the-art performance.

Table 1. Performance comparisons for 1-way 1-shot and 1-way 5-shot segmentation on Pascal-5^i in mIoU. The results in **Bold** refer to the best result among all the methods among all the methods, while the results underlined indicate the second-best result.

Method	1 shot					5 shot				
	Fold^0	Fold^1	Fold^2	Fold^3	Mean	Fold^0	Fold^1	Fold^2	Fold^3	Mean
Methods with Backbone of VGG-16										
PANet[ICCV19[29]]	42.3	58.0	51.1	41.2	48.1	51.8	64.6	59.8	46.5	55.7
PFENet[TPAMI20[27]]	56.9	68.2	54.4	52.4	58.0	59.0	69.1	54.8	52.9	59.0
HSNet[ICCV21[18]]	59.6	65.7	59.6	54.0	59.7	64.9	69.0	64.1	58.6	64.1
BAM[CVPR22[9]]	63.2	70.8	66.1	57.5	64.4	67.4	73.1	70.6	64.0	68.8
CLearSeg[Ours]	**67.9**	**74.4**	**69.6**	**65.3**	**69.3**	**68.9**	**74.8**	**71.0**	**66.3**	**70.3**
Methods with Backbone of ResNet-50										
PANet[ICCV19[29]]	44.0	57.5	50.8	44.0	49.1	55.3	67.2	61.3	53.2	59.3
CANet[CVPR19[32]]	52.5	65.9	51.3	51.9	55.4	55.5	67.8	51.9	53.2	57.1
PGNet[ICCV19[31]]	56.0	66.9	50.6	50.4	56.0	57.7	68.7	52.9	54.6	58.5
PFENet[TPAMI20[27]]	61.7	69.5	55.4	56.3	60.8	63.1	70.7	55.8	57.9	61.9
HSNet[ICCV21[18]]	64.3	70.7	60.3	60.5	64.0	70.3	73.2	67.4	67.1	69.5
SSP[ECCV22[4]]	60.5	67.8	66.4	51.0	61.4	68.0	72.0	**74.8**	60.2	68.8
CyCTR[NeurIPS21[33]]	65.7	71.0	59.5	59.7	64.0	69.3	73.5	63.8	63.5	67.5
DCAMA[ECCV22[25]]	67.5	72.3	59.6	59.0	64.6	70.5	73.9	63.7	65.8	68.5
BAM[CVPR22[9]]	69.0	73.6	67.6	61.1	67.8	70.6	75.1	70.8	67.2	70.9
CLearSeg[Ours]	**69.5**	**74.7**	67.2	**65.3**	**69.2**	70.5	**75.6**	72.0	70.1	**72.1**

Results on PASCAL-5^i. With a VGG-16 backbone, our CLearSeg shows improvements of 4.9% (1-shot) and 1.5% (5-shot) in mIoU compared to the previous SOTA method BAM [9]. Moreover, when we switched to a ResNet-50 backbone, our model still demonstrated significant improvements of 1.4% (1-shot) and 1.2% (5-shot) in mIoU, highlighting the effectiveness of our approach with two different backbones.

Results on COCO-20^i. Our method's performance on COCO-20^i still outperforms the current state-of-the-art. However, the previous state-of-the-art method BAM [9] has an inadequate validation process, as its evaluation is conducted over 1000 episodes instead of the 10000 episodes used in the vast majority of previous works [11,27,29,32], resulting in highly random mIoU score. By evaluating their provided weights on 10000 episodes, we got a more reliable result of BAM on COCO-20^i, achieving 45.2% (1-shot) and 48.5% (5-shot) with ResNet-50. In this context, our method achieved remarkable improvements of 3.7% (1-shot) and 5.8% (5-shot) over the previous state-of-the-art when using ResNet-50.

Table 2. Performance comparisons for 1-way 1-shot and 1-way 5-shot segmentation on COCO-20^i. The results in **Bold** refer to the best result, while the results underlined indicate the second-best result.

Method	1 shot					5 shot				
	Fold^0	Fold^1	Fold^2	Fold^3	Mean	Fold^0	Fold^1	Fold^2	Fold^3	Mean
Methods with Backbone of VGG-16										
PRNet[ECCV20[14]]	27.5	33.0	26.7	29.0	29.1	31.2	36.5	31.5	32.0	32.8
PFENet[TPAMI20[27]]	35.4	38.1	36.8	34.7	36.3	38.2	42.5	41.8	38.9	40.4
BAM[CVPR22[9]]	_39.0_	_47.0_	_46.4_	_41.6_	_43.5_	**47.0**	_52.6_	_48.6_	_49.1_	_49.3_
CLearSeg[Ours]	**42.2**	**52.0**	**48.0**	**46.5**	**47.2**	_46.1_	**57.6**	**53.7**	**51.4**	**52.2**
Methods with Backbone of ResNet-50										
PFENet[TPAMI20[27]]	36.5	38.6	34.5	33.8	35.8	36.5	43.3	37.8	38.4	39.0
RePRI[CVPR21[1]]	32.0	38.7	32.7	33.1	34.1	39.3	45.4	39.7	41.8	41.6
HSNet[ICCV21[18]]	36.3	43.1	38.7	38.7	39.2	43.3	51.3	48.2	45.0	46.9
CWT[ICCV21[17]]	32.2	36.0	31.6	31.6	32.9	40.1	43.8	39.0	42.4	41.3
SSP[ECCV22[4]]	35.5	39.6	37.9	36.7	37.4	40.6	47.0	45.1	43.9	44.1
CyCTR[NeurIPS21[33]]	38.9	43.0	39.6	39.8	40.3	41.1	48.9	45.2	47.0	45.6
IPMT[NeurIPS22[16]]	41.4	45.1	45.6	40.0	43.0	43.5	49.7	48.7	47.9	47.5
BAM[CVPR22[9]]	_43.4_	_50.6_	_47.5_	_43.4_	_46.2_	**49.3**	_54.2_	_51.6_	_49.6_	_51.2_
CLearSeg[Ours]	**43.7**	**55.7**	**48.0**	**48.0**	**48.9**	_47.3_	**61.8**	**54.3**	**53.8**	**54.3**

Given the quadrupled number of categories and the diversity in scale of the targets, the 80-category COCO-20^i dataset is undoubtedly more challenging than the 20-category Pascal-5^i dataset. Although some previous methods have shown high performance on the Pascal-5^i dataset, their effectiveness is limited when applied to the more comprehensive COCO-20^i dataset. By fully leveraging feature information, our method achieved a significantly greater improvement on the COCO-20^i dataset, demonstrating its impressive ability for generalization.

6 Ablation Study

We conducted various ablation studies to assess CLearSeg's components and the parameters in each component, which were performed using the ResNet-50 [7] backbone on the Pascal-5^i dataset in the 1-shot setting.

Various Components Integration. Table 3 shows that our CLearSeg model improves the baseline mIoU performance from 62.9% to 69.2%. Our MLC module is designed to exploit practical features at multiple levels, resulting in mIoU improvement of 3.9% compared to the baseline. Meanwhile, the TDA module is responsible for extracting effective text features based on category prompts, which contributes 3.7% improvement in mIoU. Furthermore, we find that the addition of the TDA module leads to a further 2.4% improvement in mIoU compared to CLearSeg without TDA, demonstrating the importance of incorporating text feature information of categories as the guidance of FSS tasks.

Table 3. Ablation studies for CLearSeg with various components integration.

Base	MLC	TDA	mIoU(%)	Δ
✓			62.9	0.0
✓	✓		66.8	+3.9
✓		✓	66.6	+3.7
✓	✓	✓	69.2	+6.3

Windows Size in SAP. The TDA module tackles the challenge of distinguishing similar features by utilizing category text features as a reference. In Sect. 4.1, we elaborate on the sliding-window approach to retain spatial information. Experiments, detailed in the upper part of Table 4, highlight the window size's impact on performance, performing best with a size of 3 or 4. Reducing the window size enhances the precision of the CSA map while weakening interaction among adjacent features. The time consumption simultaneously grows rapidly due to the necessary upsampling process before the Attention Pooling operation and higher computational cost. Hence, a window size of 4 strikes the right balance, offering peak performance for our TDA module.

Table 4. Ablation studies on the window size in SAP and layer selection of MLC.

Windows size in SAP				
Window size	3	4	5	7
mIoU(%)	69.1	**69.2**	68.7	67.7
Layer selection of MLC				
Selected layers	{2,3}	{1,2,3}	{0,1,2,3}	{1,2,3,4}	{0,1,2,3,4}
mIoU(%)	63.5	63.7	63.5	**66.8**	66.6

Layer Selection of MLC. The results presented in the lower part of Table 4 confirm the effectiveness of our selected levels of feature correlation. We set the middle-level feature correlation as the baseline and achieved the mIoU of 63.5%. Incorporating layer 1 feature correlation marginally enhances performance, while layer 0 feature correlation negatively affects it. Notably, layer 4 feature correlation elevates mIoU by 3.1% in comparison to structures lacking it. Given these experimental findings, we opted for correlations spanning layer 1 to layer 4.

Qualitative Results. We visualize some representative prediction results in PASCAL-5^i and COCO-20^i under the 1-shot setting in Fig. 3. We presented the result of baseline, baseline+MLC, and CLearSeg following the settings of the ablation study in Table 3. Additionally, we present a visualization of the CSA map to display its effectiveness. The first three columns of the segmentation results demonstrate that TDA activates the potential regions and suppresses the irrelevant categories, resulting in accurate object recognition. In the following three columns, MLC identifies the approximate object locations, and TDA refines the segmentation result towards the ground truth region by providing the CSA map as guidance. Our qualitative results showcase the effectiveness of our proposed model in distinguishing objects with similar features.

Fig. 3. Qualitive result of CLearSeg on 1-shot Pascal-5^i task. The settings of "Baseline", "Baseline+MLC" and "Ours" follows the settings of the ablation study in Table 3.

7   Conclusions

In this paper, we proposed a novel CLIP-enhanced Discrimination among Resembling Classes for Few-shot Semantic Segmentation (CLearSeg) that fully excavates the features for more precise performance. For one thing, we introduce the CLIP-based Text-Driven Activation module, which leverages text features from category prompts to generate an activation map for guidance. Especially, we consider that spatial information is crucial for semantic segmentation, and present a novel Sliding Attention Pooling to replace the original Attention Pooling layer in CLIP, which explicitly emphasizes the target regions. For another, we construct a simple yet efficient Multi-Level Correlation module. It fully exploits the correlation between query and support features, using both coarse-grained and fine-grained information. Our experiments on Pascal-5^i and COCO-20^i demonstrate that CLearSeg achieves state-of-the-art performance in these benchmarks.

Acknowledgment. This work was supported by the Industry-University-Institute Cooperation Foundation of the Eighth Research Institute of China Aerospace Science and Technology Corporation (No. SAST2022-049) and the National Natural Science Foundation of China under Grant No. 61972036.

References

1. Boudiaf, M., Kervadec, H., Masud, Z.I., Piantanida, P., Ben Ayed, I., Dolz, J.: Few-shot segmentation without meta-learning: a good transductive inference is all you need? In: Proceedings of IEEE Conference on Computer Vision and Pattern Recognition, pp. 13979–13988 (2021)
2. Ding, J., Xue, N., Xia, G.S., Dai, D.: Decoupling zero-shot semantic segmentation. In: Proceedings of IEEE Conference on Computer Vision and Pattern Recognition, pp. 11583–11592 (2022)
3. Everingham, M., Van Gool, L., Williams, C.K., Winn, J., Zisserman, A.: The pascal visual object classes (voc) challenge. Inter. J. Comput. Vis. **88**, 303–338 (2010)
4. Fan, Q., Pei, W., Tai, Y.W., Tang, C.K.: Self-support few-shot semantic segmentation. In: Proceedings of European Conference on Computer Vision, pp. 701–719 (2022)
5. Gao, P., Geng, S., Zhang, R., et al.: Clip-adapter: better vision-language models with feature adapters. arXiv preprint arXiv:2110.04544 (2021)
6. Hariharan, B., Arbeláez, P., Girshick, R., Malik, J.: Simultaneous detection and segmentation. In: Fleet, D., Pajdla, T., Schiele, B., Tuytelaars, T. (eds.) ECCV 2014. LNCS, vol. 8695, pp. 297–312. Springer, Cham (2014). https://doi.org/10.1007/978-3-319-10584-0_20
7. He, K., Zhang, X., Ren, S., Sun, J.: Deep residual learning for image recognition. In: Proceedings of IEEE Conference on Computer Vision and Pattern Recognition, pp. 770–778 (2016)
8. Hong, S., Cho, S., Nam, J., Lin, S., Kim, S.: Cost aggregation with 4d convolutional swin transformer for few-shot segmentation. In: Proceedings of European Conference on Computer Vision. pp. 108–126 (2022)
9. Lang, C., Cheng, G., Tu, B., Han, J.: Learning what not to segment: a new perspective on few-shot segmentation. In: Proceedings of IEEE Conference on Computer Vision and Pattern Recognition, pp. 8057–8067 (2022)
10. Li, B., Weinberger, K.Q., Belongie, S., et a.: Language-driven semantic segmentation. arXiv preprint arXiv:2201.03546 (2022)
11. Li, G., Jampani, V., Sevilla-Lara, L., Sun, D., Kim, J., Kim, J.: Adaptive prototype learning and allocation for few-shot segmentation. In: Proceedings of IEEE Conference on Computer Vision and Pattern Recognition, pp. 8334–8343 (2021)
12. Liang, F., Wu, B., Dai, X., et al.: Open-vocabulary semantic segmentation with mask-adapted clip. arXiv preprint arXiv:2210.04150 (2022)
13. Lin, T.-Y.: Microsoft COCO: common objects in context. In: Fleet, D., Pajdla, T., Schiele, B., Tuytelaars, T. (eds.) ECCV 2014. LNCS, vol. 8693, pp. 740–755. Springer, Cham (2014). https://doi.org/10.1007/978-3-319-10602-1_48

14. Liu, Y., Zhang, X., Zhang, S., He, X.: Part-aware prototype network for few-shot semantic segmentation. In: Vedaldi, A., Bischof, H., Brox, T., Frahm, J.-M. (eds.) ECCV 2020. LNCS, vol. 12354, pp. 142–158. Springer, Cham (2020). https://doi.org/10.1007/978-3-030-58545-7_9

15. Liu, Y., Liu, N., Cao, Q., et al.: Learning non-target knowledge for few-shot semantic segmentation. In: Proceedings of IEEE Conference on Computer Vision and Pattern Recognition, pp. 11573–11582 (2022)

16. Liu, Y., Liu, N., Yao, X., Han, J.: Intermediate prototype mining transformer for few-shot semantic segmentation. In: Proceedings of Neural Information Processing Systems (2022)

17. Lu, Z., He, S., Zhu, X., Zhang, L., Song, Y.Z., Xiang, T.: Simpler is better: few-shot semantic segmentation with classifier weight transformer. In: Proceedings of IEEE International Conference on Computer Vision, pp. 8741–8750 (2021)

18. Min, J., Kang, D., Cho, M.: Hypercorrelation squeeze for few-shot segmentation. In: Proceedings of IEEE International Conference on Computer Vision, pp. 6941–6952 (2021)

19. Moon, S., Sohn, S.S., Zhou, H., et al.: Hm: hybrid masking for few-shot segmentation. In: Proceedings of European Conference on Computer Vision, pp. 506–523 (2022)

20. Moon, S., Sohn, S.S., Zhou, H., et al.: Msi: maximize support-set information for few-shot segmentation. arXiv preprint arXiv:2212.04673 (2022)

21. Nguyen, K., Todorovic, S.: Feature weighting and boosting for few-shot segmentation. In: Proceedings of IEEE International Conference on Computer Vision, pp. 622–631 (2019)

22. Okazawa, A.: Interclass prototype relation for few-shot segmentation. In: Proceedings of European Conference on Computer Vision, pp. 362–378 (2022)

23. Radford, A., Kim, J.W., Hallacy, C., et al.: Learning transferable visual models from natural language supervision. In: Proceedings of International Conference on Machine Learning, pp. 8748–8763 (2021)

24. Shaban, A., Bansal, S., Liu, Z., Essa, I., Boots, B.: One-shot learning for semantic segmentation. In: Proceedings of British Machine Vision Conference, pp. 167.1-167.13 (2017)

25. Shi, X., et al.: Dense cross-query-and-support attention weighted mask aggregation for few-shot segmentation. In: Proceedings of European Conference on Computer Vision, pp. 151–168 (2022)

26. Simonyan, K., Zisserman, A.: Very deep convolutional networks for large-scale image recognition. In: International Conference on Learning Representations (2015)

27. Tian, Z., Zhao, H., Shu, M., Yang, Z., Li, R., Jia, J.: Prior guided feature enrichment network for few-shot segmentation. IEEE Trans. Pattern Recog. Mach. Intell. 44(2), 1050–1065 (2020)

28. Vaswani, A., Shazeer, N., Parmar, N., et al.: Attention is all you need. In: Proceedings of Neural Information Processing Systems 30 (2017)

29. Wang, K., Liew, J.H., Zou, Y., Zhou, D., Feng, J.: Panet: few-shot image semantic segmentation with prototype alignment. In: Proceedings of IEEE International Conference on Computer Vision, pp. 9197–9206 (2019)

30. Xiong, Z., Li, H., Zhu, X.X.: Doubly deformable aggregation of covariance matrices for few-shot segmentation. In: Proceedings of European Conference on Computer Vision, pp. 133–150 (2022)

31. Zhang, C., Lin, G., Liu, F., Guo, J., Wu, Q., Yao, R.: Pyramid graph networks with connection attentions for region-based one-shot semantic segmentation. In: Proceedings of IEEE International Conference on Computer Vision, pp. 9587–9595 (2019)
32. Zhang, C., Lin, G., Liu, F., Yao, R., Shen, C.: Canet: class-agnostic segmentation networks with iterative refinement and attentive few-shot learning. In: Proceedings of IEEE Conference on Computer Vision and Pattern Recognition, pp. 5217–5226 (2019)
33. Zhang, G., Kang, G., Yang, Y., Wei, Y.: Few-shot segmentation via cycle-consistent transformer. Proc. Neural Inform. Process. Syst. **34**, 21984–21996 (2021)
34. Zhang, R., Zhang, W., Fang, R., eg al.: Tip-adapter: training-free adaption of clip for few-shot classification. In: Proceedings of European Conference on Computer Vision, pp. 493–510 (2022)
35. Zhang, X., Wei, Y., Yang, Y., Huang, T.S.: Sg-one: similarity guidance network for one-shot semantic segmentation. IEEE Trans. Cybern. **50**(9), 3855–3865 (2020)
36. Zhou, K., Yang, J., Loy, C.C., Liu, Z.: Learning to prompt for vision-language models. Inter. J. Comput. Vis. **130**(9), 2337–2348 (2022)

PANDA: Prompt-Based Context- and Indoor-Aware Pretraining for Vision and Language Navigation

Ting Liu, Yue Hu$^{(\boxtimes)}$, Wansen Wu, Youkai Wang, Kai Xu, and Quanjun Yin

College of Systems Engineering, National University of Defense Technology, Changsha, China
{liuting20,huyue11,wuwansen14,wangyoukai,xukai09}@nudt.edu.cn

Abstract. Pretrained visual-language models have extensive world knowledge and are widely used in visual and language navigation (VLN). However, they are not sensitive to indoor scenarios for VLN tasks. Another challenge for VLN is how the agent understands the contextual relations between actions on a path and performs cross-modal alignment sequentially. In this paper, we propose a novel Prompt-bAsed coNtext- and inDoor-Aware (PANDA) pretraining framework to address these problems. It performs prompting in two stages. In the indoor-aware stage, we apply an efficient tuning paradigm to learn deep visual prompts from an indoor dataset, in order to augment pretrained models with inductive biases towards indoor environments. This can enable more sample-efficient adaptation for VLN agents. Furthermore, in the context-aware stage, we design a set of hard context prompts to capture the *sequence-level* semantics in the instruction. They enable further tuning of the pretrained models via contrastive learning. Experimental results on both R2R and REVERIE show the superiority of PANDA compared to existing state-of-the-art methods.

Keywords: visual and language · multimodal representation

1 Introduction

Creating intelligent agents that can follow human instructions continues to be a major challenge in embodied artificial intelligence [1]. In particular, Vision and Language Navigation (VLN) [2,3], which requires an agent to follow natural language instructions and make sequential decisions in a photo-realistic simulated environment.

In recent years, there has been a growing tendency for VLN methods [3–6] to build upon foundationally pretrained vision-and-language models. However, the distributional statistics of the large-scale web-scraped dataset conventionally employed for such pre-training often diverge substantially from the indoor domain of VLN environments. This domain gap leads to a disconnect in the ability to understand VLN scenarios. In the meantime, the nature of sequential decision-making under partially observable environments, which essentially differs VLN from other reasoning tasks, requires a VLN agent to align between the

© The Author(s), under exclusive license to Springer Nature Switzerland AG 2024
S. Rudinac et al. (Eds.): MMM 2024, LNCS 14554, pp. 187–200, 2024.
https://doi.org/10.1007/978-3-031-53305-1_15

sequences of action descriptions and viewpoints. With the above in mind, we argue that most existing works suffer from two significant shortcomings:

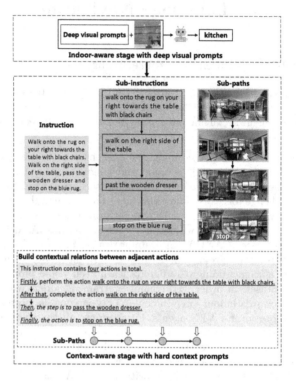

Fig. 1. A simple demonstration of PANDA. In the indoor-aware stage, we exploit deep visual prompts learned from an indoor dataset by an auxiliary supervised task to help the VLN agent to abstract the indoor image semantics, then align sub-instructions with sub-paths sequentially by understanding the contextual relations between adjacent actions using hard context prompts in the context-aware stage.

- While pretrained models have powerful knowledge, their distributions diverge starkly from indoor navigation domains. Insensitive to indoor semantics, these models struggle to recapitulate the understanding requisite for dynamic reasoning within unfamiliar indoor scenes.
- Existing works ignore the contextual semantics implicitly embedded in the given instruction and the sequential relationships between fine-grained instructions (e.g., the instruction *"Walk out of the bathroom and turn left"* can be divided into two ordinal sub-instructions, i.e., *"first walk out of the bathroom"*, and *"then turn left"*), and the agent may perform chaotic actions.

Prompt engineering has been extensively studied and achieved significant success in NLP [7,8] and CV [9,10]. Prompt engineering refers to the design of an input template that converts the expected output into a fill-in-the-blank format.

With hand-designed prompts, the pioneer pretrained language model GPT-3 [11] has shown strong potential in the few-shot or zero-shot settings. CLIP [9] embeds the text label of the object into the prompt template, which transforms the image recognition task into an image-text matching problem. CoOp [12] uses learnable vectors as text prompts to obtain the improvement of few-shot image classification. VPT [13] introduces only a few learnable parameters into the input space while keeping the parameters of the pretrained model frozen during training. Prompt learning has also been introduced into VLN tasks [6,14] in some latest works. For example, our previous work DAP [14] applies a low-cost prompt tuning paradigm to learn shallow visual prompts for extracting in-domain image semantics. However, we found the shallow prompt lacks deep scene understanding.

Inspired by these works, we attempt to exploit prompt learning to solve the above problems and propose a novel Prompt-bAsed coNtext- and inDoor-Aware (PANDA) pretraining framework. A simple demonstration is shown in Fig. 1. PANDA improves pretrained general vision-language models in VLN with two stages: (i) In the indoor-aware stage, PANDA make a pretrained vision-language model aware of the specific VLN domain by adding deep visual prompts learned from indoor scene knowledge; (ii) In the context-aware stage, we aim to manually design context prompts to make the pretrained model aware of the contexts between navigational actions and reasoning about the entire sequence.

Specifically, to narrow the domain gap between the pretrained model and VLN tasks, we first generate a set of indoor datasets. Then we introduce a set of deep visual prompts in the input space of the visual encoder in a pretrained model. The aim is to enable the agent to identify the objects and scenes in the indoor images. Only deep visual prompts and an MLP head are learnable during training with the indoor datasets, while the parameters of the pretrained model are kept frozen. With prompt learning, deep visual prompts learned from the indoor dataset can adapt the pretrained models to VLN scenes very efficiently. In the context-aware stage, we first divide the R2R dataset [1] into sub-instructions and sub-paths. Then, we use manually designed hard context prompts to explicitly align the predicted action sequence, which is characterized by a series of viewpoints, with the contextual information implicitly embedded in the given instruction, and instill both out-of-context and contextual knowledge in the instruction into cross-modal representations. Contrastive learning is introduced to further tune the pretrained models for *sequence-level* cross-modal alignment.

In summary, the contributions of this work are summarized as: (i) We present PANDA to pretrain a representation model for VLN tasks that captures the indoor scene semantics and context semantics along the action sequence; (ii) We introduce deep visual prompts to adapt pretrained models to VLN tasks. Contrastive learning is also introduced to achieve effective alignment between textual prompts and visual semantics; (iii) PANDA shows promising performances and generalization ability with the help of prompt-based learning, and outperforms existing state-of-the-art methods.

2 Related Works

Vision and Language Navigation. Many methods [3–5] have been proposed for VLN, a famous embodied artificial intelligence task. Recently, the powerful representation abilities of pretrained models have attracted great attention. While the VLN-BERT [3] model was pretrained on a large set of web-crawled datasets to improve image-text matching, the PREVALENT [15] model is trained on a large amount of image-text-action triplets to learn generic representations of visual environments and language instructions. Following these works, a recurrent function [16] was introduced into the BERT model that significantly improves sequential action prediction. However, current VLN methods ignore the contextual information implicitly embedded in the given instruction and the sequential relations between sub-instructions. In addition, most pretrained models are trained on web-crawled general-purpose datasets, which incurs a considerable domain gap when used for VLN tasks.

Large-Scale Vision-Language Pretraining. Motivated by the success of the BERT model [17] on NLP tasks, numerous vision-language pretrained (VLP) models [15,18] have been recently proposed. These VLP models have been applied to various vision-language tasks such as visual grounding [19] and vision and language navigation [15], etc., which have all made great performance breakthroughs. Despite their powerful visiolinguistic representation abilities, VLP models are not designed for tasks that entail sequential decision-making, such as VLN tasks. In this work, we aim to improve the VLP models to make them more suitable for VLN tasks.

Prompt Learning. The idea of prompt learning is to put the expected output as an unfilled blank into a prompt template that is incorporated into the input information, which has sparked significant interest in NLP [7,8]. There are two types of prompt templates: one is hard prompts designed manually, and the other is soft prompts learned automatically. For example, a cloze template [20] is designed manually to probe knowledge in pretrained language models can benefit many downstream tasks, and in P-tuning [21], deep prompt templates are learned in the continuous space by gradient descent without intricate design. Recent works [9,10,22,23] subsequently introduce prompt learning into the pretrained vision-language models. For example, CLIP [9] embeds the text label of an image into a discrete template such as *"A photo of a {object}"*, and the image recognition task can be transformed into an image-text matching problem.

3 Method

3.1 VLN Problem Setup

The VLN task can be expressed as follows: The VLN agent is put in a photorealistic environment such as Matterport3D [24] simulator [8], it is assigned a random initial position and given a language instruction I which contains l word tokens. The VLN agent is required to find a route from the initial position

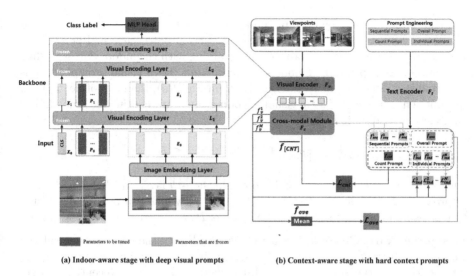

(a) Indoor-aware stage with deep visual prompts   (b) Context-aware stage with hard context prompts

Fig. 2. The overview of PANDA for VLN. We explore two forms of prompts: (a) deep visual prompts are inserted into the input space of the vision encoder, where only the parameters of deep visual prompts and the MLP head are updated during training; (b) we design hard context prompts to abstract the sub-instructions semantics and their sequential relations. The indoor-aware stage focuses on learning deep visual prompts to enhance the adaptation of backbone models to VLN tasks, while the context-aware stage aims to capture the contextual relations between actions on a trajectory and performs cross-modal alignment sequentially.

to the target position. At each time step t, the agent observes the environment and makes a current action decision that updates the agent state s_t to a new state s_{t+1}. The state includes historical information and the current spatial information consists of a viewpoint and orientation. All viewpoints are on the connectivity graph $G = \langle V, E \rangle$ of the environment [25], V and E represent navigable nodes and edges respectively. With the instructions, current state s_t and visual observations O_t, the agent needs to execute the next actions a_{t+1} one by one to navigate on the connectivity graph until stop at the target position.

3.2 Prompt Engineering

We introduce different forms of prompts in two stages for adapting pretrained vision-language models to VLN tasks in this subsection.

Learning Deep Visual Prompts Automatically. In the indoor-aware stage, to narrow the domain gap, we adopt the supervised learning method to learn deep visual prompts, taking indoor images as the input and the text of the corresponding object as labels. We introduce a set of continuous embeddings, i.e., deep visual prompts, in the input space after the input images are initially processed by the embedding layer. Deep visual prompts are automatically learned from an indoor dataset by prompt tuning, which helps the VLN agent to ground the object and scene descriptions in the instruction onto the visual perception. Deep visual prompts are learnable during training with the indoor dataset, while the parameters of pretrained models are kept frozen. Each visual prompt token is a learnable d-dimensional vector. As shown in Fig. 2(a), the form of deep visual prompts encoded by the i-th visual encoding layer are continuous embeddings that can be represented as:

$$P_i = \left\{ p_i^j \in \mathbb{R}^d \mid j \in \mathbb{N}, 1 \leq j \leq H \right\}, \tag{1}$$

where p_i^j represents the j-th visual prompt in the i-th layer and H is the number of prompts. Note that besides P_i ($1 \leq i \leq N$), P_0 which is a part of the inputs to the visual encoder, takes the same form as P_i.

Designing Hard Context Prompts Manually. In the context-aware stage, we aim to abstract the contextual semantics implicitly embedded in the given instruction and the semantics of sub-instructions in order. Recently, Bridge-Prompt [26], the latest work in activity recognition from videos, discovers that human language is a powerful tool to depict the ordinal semantics between correlated actions. Motivated by this, we intend to manually design text prompts to realize the above purpose. Supposing that a sub-instruction (such as *"walk into the hallway"*) constitutes a specific kind of agent action, we can use the prompt template as *"perform the action "* (refers to the action description for the i-th sub-instruction.) to abstract the semantics of the sub-instruction. However, if the sub-instruction is treated as a separate prompt instance, this out-of-context policy cannot describe the contextual semantics between adjacent ordinal sub-instructions. A more effective form of textual prompts should not only capture the semantics of individual sub-instructions, but also capture the contextual semantics and the sequential information between sub-instructions, and describe the overall semantics of the instruction. To this end, we manually design the hard context prompts consisting of four kinds of text prompts for VLN, as shown at the bottom of Fig. 1. We will empirically show the superiority of such a design over the exploitation of individual sub-instructions in Subsect. 4.3. Considering the instruction with M sub-instructions:

1) **A count prompt** abstracts the total number information of a sequence of actions contained in an instruction. We use the template as *"This instruction contains {num (M)} actions"* and denote the count prompt as Y_{cnt}.

2) **A sequential prompt** abstracts the ordinal information for every sub-instruction. We use the template as *"this is the {seq_i} action"* and denote the sequential prompt as y_{seq}^i. The set of sequential prompts is as follows:

$$Y_{seq} = \left[y_{seq}^1, \dots, y_{seq}^M \right]. \tag{2}$$

3) **An individual prompt** abstracts the semantic information of a sub-instruction. To integrate contextual information, we incorporate sequential information into the individual prompt and use the template as *"{seq_i},*

perform the action $\{a_i\}$ " for action a_i. We denote the individual prompt set as:

$$Y_{ind} = \left[y_{ind}^1, \ldots, y_{ind}^M \right]. \tag{3}$$

4) **An overall prompt** abstracts the overall information for the complete instruction. The overall prompt is made up of all individual prompts, which can be expressed as:

$$Y_{ove} = Concat \left(y_{ind}^1, \ldots, y_{ind}^M \right). \tag{4}$$

3.3 Indoor-Aware Stage with Deep Visual Prompts

Deep visual prompts can be widely used in vision-language pretrained models to better understand indoor image semantics. We apply the pretrained model PREVALENT [15] for demonstration. We inject indoor visual knowledge into the visual encoder F_v of the pretrained PREVALENT model by prompt tuning.

Text Generation with the CLIP Model. we take the powerful cross-modal pretrained model CLIP to automatically generates text labels corresponding to the indoor image from the Matterport3D [24] dataset. The method takes full advantage of the knowledge learned from the CLIP [9] and builds an indoor image-text dataset. We first encode the prompt template *"A photo of a {object}"* by the text encoder of the CLIP, where the label represents object classes or rooms. Then we encode the image by the image encoder and calculate the similarity of the text embedding and image embedding. Finally, we choose the text with the highest matching score for the image. Through the above methods, we can automatically generate the indoor image-text dataset.

Deep Visual Prompt Tuning. Deep visual prompts are inserted into the input space of the N-layer vision encoder in the PREVALENT. The output of the i-*th* visual encoding layer is formulated as:

$$[\mathbf{X}_i, \mathbf{P}_i, \mathbf{E}_i] = L_i \left(\mathbf{X}_{i-1}, \mathbf{P}_{i-1}, \mathbf{E}_{i-1} \right), \quad i = 1, 2, \ldots, N \tag{5}$$

where $\mathbf{X}_i$, $\mathbf{P}_i$, and $\mathbf{E}_i$ denote the $[CLS]$, prompts and image features respectively encoded by the i-*th* visual encoding layer. The output $\mathbf{X}_N$ at the N-th layer of the visual encoder is mapped by an MLP head to a predicted class probability distribution y.

The PREVALENT model is retrained on indoor image-text pairs that we have prepared, as shown in Fig. 2(a). Firstly, we freeze all parameters of the PREVALENT backbone model, which could not be updated during the training process. Then we add additional visual prompts on the first n layers ($n \leq N$), and an MLP head after the N-*th* layer and deep visual prompts are learnable during training. We apply a cross-entropy loss to only optimize deep visual prompts and the linear head via gradients during prompt tuning. With such a low-consumption auxiliary classification task, the visual prompts are expected to inject the knowledge of object-level and scene-level indoor image semantics into the PREVALENT model.

3.4   Context-Aware Stage with Hard Context Prompts

Sub-instructions and Sub-paths Generation. In order to learn the ordinal relations between sub-instructions, and match a sub-instruction with its corresponding sub-path, we generate a fine-grained training dataset. We apply the FGR2R [27] method to divide the instructions into sub-instructions and pair each sub-instruction with its corresponding sub-path. Instructions are divided by *"and"*, *"comma"*, and *"period"* delimiters. An illustrative example is provided here. We divide the given instruction *"Walk onto the rug on your right towards the table with black chairs. Walk on the right side of the table, past the wooden dresser and stop on the blue rug."* into *"Walk onto the rug on your right towards the table with black chairs"*, *"Walk on the right side of the table"*, *"past the wooden dresser"* and *"stop on the blue rug"*, as shown in Fig. 1.

Fine-Grained Alignment by Contrastive Learning. The viewpoints along a path are first passed through the visual encoder F_v updated in the indoor-aware pretraining stage to generate visual embeddings. We manually design the hard context prompts $(Y_{cnt}, Y_{seq}, Y_{ind}, Y_{ove})$ for the path. As shown in Fig. 2(b), the text encoder F_t abstracts the embeddings of the hard context prompts as $(f_{cnt}, f_{seq}, f_{ind}, f_{ove})$, respectively. The visual embeddings and the sequential prompts embeddings are then passed through the cross-modal encoder F_c to abstract the image features f_v^i of the i-th sub-path.

We input the i-th sequential prompt feature f_{seq}^i to the cross-modal module, which allows the cross-modal module to focus on the sequential information of each ordinal action. In addition, we add a learnable count token $\overline{f_{[CNT]}}$ in F_c to extract quantitative information to match the count prompt f_{cnt}.

Contrastive vision-text learning maximizes the similarity between encoded visual features and text features. We encode the sub-instruction x and its corresponding sub-path y with the text encoder and the visual encoder, respectively, generating text representation r_x and vision representation r_y. The cosine similarity between r_x and r_y can be calculated as follows:

$$s\left(r_x, r_y\right) = \frac{r_x \cdot r_y}{\mid r_x \mid\mid r_y \mid}. \tag{6}$$

For a batch of the text representation R_x and the vision representation R_y, the batch similarity matrix S can be denoted as:

$$S\left(R_x, R_y\right) = \begin{bmatrix} s\left(r_{x_1}, r_{y_1}\right) & \cdots & s\left(r_{x_1}, r_{y_M}\right) \\ \vdots & \ddots & \vdots \\ s\left(r_{x_M}, r_{y_1}\right) & \cdots & s\left(r_{x_M}, r_{y_M}\right) \end{bmatrix}, \tag{7}$$

where M is the number of sub-instructions (sub-paths). We respectively apply a normalized function to the rows and columns on $S\left(R_x, R_y\right)$ to get $S_V\left(R_x, R_y\right)$ and $S_T\left(R_x, R_y\right)$. We assign the similarity score of positive pairs to 1 while negative pairs to 0, thus obtaining the batch similarity matrix GT of ground truth.

Our training objective is to maximize the similarity between the matrix S and GT. We use the Kullback-Leibler divergence as the contrastive loss:

$$D_{KL}(P\|Q) = \frac{1}{N^2} \sum_{i=1}^{N} \sum_{j=1}^{N} P_{ij} \log \frac{P_{ij}}{Q_{ij}}, \tag{8}$$

where P and Q are $N \times N$ matrices. The contrastive loss for vision-text pairs can be defined as:

$$\mathcal{L} = \frac{1}{2} \left[D_{KL}(S_T\|GT) + D_{KL}(S_V\|GT) \right]. \tag{9}$$

The above formula is used to calculate the three parts of vision-text contrastive losses:

1) $\mathcal{L}_{ind}^i$ is the contrastive loss between f_{ind}^i and f_v^i, which allows the model to align each sub-instruction and the corresponding sub-path.
2) $\mathcal{L}_{ove}$ is the contrastive loss between the mean-pooled $\overline{f_{ove}}$ and overall prompt feature f_{ove}, where $\overline{f_{ove}}$ is the mean-pooled features of all image features. $\mathcal{L}_{ove}$ captures the relationship between the overall instruction and the whole path.
3) $\mathcal{L}_{cnt}$ is the contrastive loss between the learnable count token $\overline{f_{[CNT]}}$ and the count prompt feature f_{cnt}, which captures quantitative information of all actions.

The overall loss of the context prompt framework can be denoted as follows:

$$\mathcal{L} = \lambda_1 \mathcal{L}_{ove} + \lambda_2 \mathcal{L}_{cnt} + \sum_{i=1}^{M} \mathcal{L}_{ind}^i, \tag{10}$$

where λ_1 and λ_2 are balance coefficients.

4   Experiment

4.1   Experimental Setup

Implementation Details. Our training process includes updating the pretrained model and adapting it to downstream VLN tasks. Without loss of generality, our baseline agent follows the architecture of RecBERT [16], which initializes from the pretrained model OSCAR [18] learned from out-of domain datasets or PREVALENT [15] learned from VLN datasets. In the indoor-aware stage, we pretrain the PREVALENT model on our generated about 1000 indoor image-text pairs with prompt tuning for 20 epochs with batch size 10, and the number of deep visual prompts is 10. In the context-aware stage, we continue to train the PREVALENT model updated in the indoor-aware with pairs of sub-instruction and sub-paths for 20 epochs with batch size 20. After that, we adapt PANDA to the downstream generative VLN task with fine-tuning. Based on a simple parameter sweep, values of λ_1 and λ_2 are set to be 0.5 and 0.1, respectively. For R2R, we train the agent on the raw training data and the augmented data from PREVALENT for 300,000 iterations, and the batch size is 8. For REVERIE, we train the agent for 200,000 iterations with batch size 8. All experiments are conducted on a single NVIDIA 3090 GPU.

Table 1. Comparison with the SOTA methods on R2R dataset.

Agent	Val Seen				Val Unseen				Test Unseen			
	TL	NE↓	SR↑	SPL↑	TL	NE↓	SR↑	SPL↑	TL	NE↓	SR↑	SPL↑
Random	9.58	9.45	16	–	9.77	9.23	16	–	9.89	9.79	13	12
Human	–	–	–	–	–	–	–	–	11.85	1.61	86	76
PRESS [28]	10.57	4.39	58	55	10.36	5.28	49	45	10.77	5.49	49	45
EnvDrop [29]	11.00	3.99	62	59	10.70	5.22	52	48	11.66	5.23	51	47
PREVALENT [15]	10.32	3.67	69	65	10.19	4.71	58	53	10.51	5.30	54	51
EnvDrop+REM [30]	11.13	3.14	70	66	14.84	4.99	53	48	10.73	5.40	54	50
AuxRN [31]	–	3.33	70	67	–	5.28	55	50	–	5.15	55	51
ORIST [32]	–	–	–	–	10.90	4.72	57	51	11.31	5.10	57	52
NvEM [33]	11.09	3.44	69	65	11.83	4.27	60	55	12.98	4.37	58	54
EnvDrop+SEvol [34]	12.55	3.70	61	57	14.67	4.39	59	53	14.30	3.70	59	55
NvEM+SEvol [34]	11.97	3.56	67	63	12.26	3.99	62	57	13.40	4.13	62	57
ProbES [35]	10.75	2.95	73	69	11.58	4.03	61	55	12.43	4.20	62	56
ADAPT [6]	11.39	2.70	74	69	12.33	3.66	66	59	13.16	4.11	63	57
GRVLN-BERT [36]	11.08	2.58	75	71	12.49	3.81	62	56	12.78	3.96	63	57
RecBERT (init. OSCAR) [16]	10.79	3.11	71	67	11.86	4.29	59	53	12.34	4.59	57	53
RecBERT (init. PREVALENT) [16]	11.13	2.90	72	68	12.01	3.93	63	57	12.35	4.09	63	57
PANDA(Ours)	10.65	**2.54**	**75**	**72**	12.08	**3.50**	**66**	**60**	12.31	3.86	**64**	**59**

Table 2. Comparison with the state-of-the-art methods on REVERIE dataset.

Methods	REVERIE Validation Seen						REVERIE Validation Unseen						REVERIE Test Unseen					
	Navigation				RGS↑	RGSPL↑	Navigation				RGS↑	RGSPL↑	Navigation				RGS↑	RGSPL↑
	SR↑	OSR↑	SPL↑	TL			SR↑	OSR↑	SPL↑	TL			SR↑	OSR↑	SPL↑	TL		
Random	2.74	8.92	1.91	11.99	1.97	1.31	1.76	11.93	1.01	10.76	0.96	0.56	2.30	8.88	1.44	10.34	1.18	0.78
Human	–	–	–	–	–	–	–	–	–	–	–	–	81.51	86.83	53.66	21.18	77.84	51.44
Seq2Seq-SF [37]	29.59	35.70	24.01	12.88	18.97	14.96	4.20	8.07	2.84	11.07	2.16	1.63	3.99	6.88	3.09	10.89	2.00	1.58
RCM [38]	23.33	29.44	21.82	10.70	16.23	15.36	9.29	14.23	6.97	11.98	4.89	3.89	7.84	11.68	6.67	10.60	3.67	3.14
SMNA [39]	41.25	43.29	39.61	7.54	30.07	28.98	8.15	11.28	6.44	9.07	4.54	3.61	5.80	8.39	4.53	9.23	3.10	2.39
FAST-Short [40]	45.12	49.68	40.18	13.22	31.41	28.11	10.08	20.48	6.17	29.70	6.24	3.97	14.18	23.36	8.74	30.69	7.07	4.52
FAST-MATTN [2]	50.53	55.17	45.50	16.35	31.97	29.66	14.40	28.20	7.19	45.28	7.84	4.67	19.88	30.63	11.61	39.05	11.28	6.08
ProbES [35]	46.52	48.49	42.44	13.59	33.66	30.86	27.63	33.23	22.75	18.00	16.84	13.94	24.97	28.23	20.12	17.43	15.11	12.32
RecBERT (init. OSCAR) [16]	39.85	41.32	35.86	12.85	24.46	22.28	25.53	27.66	21.06	14.35	14.20	12.00	24.62	26.67	19.48	14.88	12.65	10.00
RecBERT (init. PREVALENT) [16]	51.79	53.90	47.96	13.44	38.23	35.61	30.07	35.02	24.90	16.78	18.77	15.27	**29.61**	32.91	**23.99**	15.86	16.05	**13.51**
PANDA (Ours)	**54.39**	**55.80**	**51.08**	13.04	**40.62**	**38.29**	**32.66**	**37.66**	**27.88**	15.74	**20.76**	**17.74**	27.81	**33.30**	20.89	18.30	**17.20**	12.95

4.2 Comparison to State-of-the-Art Methods

Results on R2R. Results in Table 1 compare the performance of different methods on the R2R. Compared to the baseline model RecBERT [16], PANDA improves the agent's performance, achieving 66% SR (+3%) and 60% SPL (+3%) on the validation unseen. On the test unseen split, we achieve 64% SR (+1%) and 59% SPL (+2%). The large performance improvement suggests that improving the indoor-aware and context-aware capacity for pretrained models benefits the learning of navigation for the VLN agent. Compared to existing state-of-the-art methods, we can see that only 2% performance gap on SR exists between the validation unseen and the test unseen splits, indicating that our agent improves the generalization ability to new environments. Among all the methods, PANDA has the best results across all metrics, even compared against some newest entries such as SEvol [34], ProbES [35] and ADAPT [6]. Notice that the counterpart methods, with data augmentation or better navigation inference, are orthogonal to the proposed PANDA, meaning that they can be integrated to yield even stronger solutions.

Results on REVERIE. We compare PANDA with existing state-of-the-art methods on the REVERIE dataset, as shown in Table 2. Compared to the baseline model RecBERT (init. PREVALNRT) [16], we achieve 1.99% improvement on RGS and 2.47% improvement on RGSPL on the validation unseen split. On the test unseen split, we achieve 1.15% improvement on RGS. Despite RecBERT having a higher SR and SPL, one of the possible reasons is that their agent is wandering in the process of finding the target object. This suggests that PANDA is better for locating target objects.

4.3 Ablation Study

In this subsection, we conduct experiments to further study the effectiveness of the prompting stages in PANDA, verify the effects of visual prompts based on different pretrained models, and subsequently investigate the effects of different design choices in the context-aware stage.

Table 3. Ablation study of different prompt forms on R2R.

Methods	Val seen			Val Unseen		
	NE ↓	SR ↑	SPL ↑	NE ↓	SR ↑	SPL ↑
RecBERT (init. PREVALENT) [16]	2.90	72.18	67.72	3.93	62.75	56.84
+ Deep visual prompts	**2.31**	**76.49**	71.70	3.72	65.05	59.33
+ Hard context prompts	2.65	74.14	70.04	3.78	64.84	59.41
PANDA	2.54	75.02	**71.84**	**3.50**	**65.60**	**59.71**

Overall Effectiveness of the Two Prompting Stages. Table 3 shows the comparison of using different prompt forms on the R2R dataset. Introducing deep visual prompts and hard context prompts can effectively improve the performance of the strong baseline model RecBERT (init. PREVALENT) [16]. By comparing the results between the baseline and only with deep visual prompts, we can find that deep visual prompts can effectively enhance navigation performance, demonstrating that deep visual prompts with additional knowledge for visual recognition are useful for understanding indoor image semantics. Comparing the results between the baseline and only with hard context prompts, we can see that the introduction of the hard context prompts improves the navigation performance, which shows that attending to the contextual relations between actions and the sequential cross-modal alignment is helpful for making corrective action decisions. By comparing the results between only with deep visual prompts and PANDA, we can find that introducing hard context prompts can further improve navigation performance.

Effects of Different Contrast Learning Objectives. Table 3 shows the effectiveness of hard prompts used in the context-aware stage where we incorporate three key components into the total loss function: count, individual, and overall losses. We have evaluated the efficacy of each loss component, and Table 4 presents the quantitative results, revealing the positive impact of all three losses on the final performance.

Table 4. Ablation study of the context-aware stage.

Methods	Val seen			Val Unseen		
	NE ↓	SR ↑	SPL ↑	NE ↓	SR ↑	SPL ↑
RecBERT (init. PREVALENT) [16]	2.90	72.18	67.72	3.93	62.75	56.84
+ $\mathcal{L}_{sub}$	2.73	**74.22**	68.51	3.88	63.14	57.23
+ $\mathcal{L}_{cnt}$	2.86	72.92	68.52	3.91	63.37	57.92
+ $\mathcal{L}_{cnt}+\mathcal{L}_{ind}$	2.71	73.57	69.43	3.83	64.26	58.64
+ $\mathcal{L}_{cnt}+\mathcal{L}_{ind}+\mathcal{L}_{ove}$	**2.65**	74.14	**70.04**	**3.78**	**64.84**	**59.41**

In order to identify whether it is the context prompting or merely further exposure to the target domain that actually provides an improvement, we remove all hard context prompts and just match sub-instructions with sub-paths by contrastive learning. This is simply represented as $\mathcal{L}_{sub}$. Through a comparison of the experimental results presented in Table 4, we observe that aligning only sub-instructions and sub-paths yields only a marginal improvement in navigation performance by merely considering the out-of-context cross-modal alignment, while incorporating hard context prompts is proved to be more effective at enhancing performance by explicitly capturing sequence-level semantics and performing a sequential cross-modal alignment. Obviously, $\mathcal{L}_{sub}$ has even lower performance than only $\mathcal{L}_{cnt}$ on the validation unseen split even though $\mathcal{L}_{cnt}$ only considers the number of navigation steps, which is a very general contextual hint. With $\mathcal{L}_{cnt}$ and $\mathcal{L}_{ind}$ combined, the gap becomes much larger. VLN is a task characterized by sequential decision-making, where not only out-of-context knowledge is important but also contextual relations. This also suggests that hard context prompts are capable of capturing higher-order relationships among navigational actions at the sequence level.

5 Conclusion

In this work, we propose a Prompt-bAsed coNtext- and inDoor-Aware (PANDA) pretraining framework, which prompts the VLN agent with the capability of recognizing objects and scenes in visual perceptions in the indoor-aware stage. In the context-domain stage, PANDA enables a sequence-level representation via hard context prompts that are aware of the semantics of individual image-text pairs and across navigational actions along the trajectory. The context prompts in this work dig into the potential of prompt-based learning approaches for understanding ordinal actions and contextual relations. We believe that PANDA also can benefit future studies in other vision and language tasks. We will leave this for future work.

Acknowledgement. This research was supported partially by the National Natural Science Fund of China (Grant Nos. 62306329 and 62103425, and the Natural Science Fund of Hunan Province (Grant Nos. 2023JJ40676 and 2022JJ40559).

References

1. Das, A., Datta, S., Gkioxari, G., Lee, S., Parikh, D., Batra, D.: Embodied question answering. In: Proceedings of CVPR, pp. 1–10 (2018)
2. Qi, Y., Wu, Q., Anderson, P., et al.: Reverie: remote embodied visual referring expression in real indoor environments. In: Proceedings of CVPR, pp. 9982–9991 (2020)
3. Majumdar, A., Shrivastava, A., Lee, S., Anderson, P., Parikh, D., Batra, D.: Improving vision-and-language navigation with image-text pairs from the web. In: Vedaldi, A., Bischof, H., Brox, T., Frahm, J.-M. (eds.) ECCV 2020. LNCS, vol. 12351, pp. 259–274. Springer, Cham (2020). https://doi.org/10.1007/978-3-030-58539-6_16
4. Hao, W., Li, C., Li, X., Carin, L., et al.: Towards learning a generic agent for vision-and-language navigation via pre-trainin. In: CVPR 2022, pp. 13134–13143. IEEE (2022)
5. Guhur, P.-L., Tapaswi, M., Chen, S., et al.: Airbert: in-domain pretraining for vision-and-language navigation. In: Proceedings of ICCV, pp. 1634–1643. IEEE (2021)
6. Lin, B., Zhu, Y., Chen, Z., et al.: ADAPT: vision-language navigation with modality-aligned action prompts. In: CVPR, pp. 15375–15385. IEEE (2022)
7. Liu, P., Yuan, W., Fu, J., Jiang, Z., Hayashi, H., Neubig, G.: Pre-train, prompt, and predict: a systematic survey of prompting methods in natural language processing, CoRR, vol. abs/ arXiv: 2107.13586 (2021)
8. Lester, B., Al-Rfou, R., Constant, N.: The power of scale for parameter-efficient prompt tuning. In: EMNLP (1), pp. 3045–3059. ACL (2021)
9. Radford, A., Kim, J.W., et al.: Learning transferable visual models from natural language supervision. In: ICML, pp. 8748–8763. PMLR (2021)
10. Yao, Y., Zhang, A., Liu, Z., et al.: CPT: colorful prompt tuning for pre-trained vision-language models, CoRR, vol. abs/ arXiv: 2109.11797 (2021)
11. Brown, T.B., Mann, B., et al.: Language models are few-shot learners. In: NeurIPS (2020)
12. Zhou, K., Yang, J., Loy, C.C., Liu, Z.: Learning to prompt for vision-language models. Int. J. Comput. Vis. 130(9), 2337–2348 (2022)
13. Jia, M., et al.: Visual prompt tuning. In: Avidan, S., Brostow, G.J., Cissé, M., Farinella, G.M., Hassner, T. (eds.) Computer Vision - ECCV 2022–17th European Conference, Tel Aviv, Israel, 23–27 October 2022, Proceedings, Part XXXIII, vol. 13693. LNCS, pp. 709–727, Springer (2022). https://doi.org/10.1007/978-3-031-19827-4_41
14. Liu, T., Hu, Y., Wu, W., Wang, Y., Xu, K., Yin, Q.: Dap: domain-aware prompt learning for vision-and-language navigation (2023)
15. Hao, W., Li, C., Li, X., Carin, L., et al.: Towards learning a generic agent for vision-and-language navigation via pre-training. In: Proceedings of CVPR, pp. 13137–13146 (2020)
16. Hong, Y., Wu, Q., et al.: VLN BERT: a recurrent vision-and-language BERT for navigation. In: CVPR, pp. 1643–1653. IEEE (2021)
17. Devlin, J., Chang, M., et al.: BERT: pre-training of deep bidirectional transformers for language understanding. In: NAACL-HLT (1), pp. 4171–4186 (2019)
18. Li, X., et al.: OSCAR: object-semantics aligned pre-training for vision-language tasks. In: Vedaldi, A., Bischof, H., Brox, T., Frahm, J.-M. (eds.) ECCV 2020. LNCS, vol. 12375, pp. 121–137. Springer, Cham (2020). https://doi.org/10.1007/978-3-030-58577-8_8

19. Liu, X., Huang, S., Kang, Y., Chen, H., Wang, D.: VGDiffZero: text-to-image diffusion models can be zero-shot visual grounders (2023)
20. Petroni, F., et al.: Language models as knowledge bases? EMNLP/IJCNLP (1), pp. 2463–2473. ACL (2019)
21. Liu, X.: GPT understands, too, CoRR, vol. abs/ arXiv: 2103.10385 (2021)
22. Zhou, K., Yang, J., Loy, C.C., Liu, Z.: Learning to prompt for vision-language models. Int. J. Comput. Vision **130**(9), 2337–2348 (2022)
23. Tsimpoukelli, M., Menick, J.L., Cabi, S., Eslami, S., Vinyals, O., Hill, F.: Multimodal few-shot learning with frozen language models. Adv. Neural. Inf. Process. Syst. **34**, 200–212 (2021)
24. Chang, A.X., Dai, A., Funkhouser, T.A., Halber, M., Nießner, M., et al.: Matterport3d: learning from RGB-D data in indoor environments. In: 3DV, 667–676. IEEE (2017)
25. Anderson, P., Chang, A.X., Chaplot, D.S., et al.: On evaluation of embodied navigation agents, CoRR, vol. abs/ arXiv: 1807.06757 (2018)
26. Li, M., et al.: Bridge-prompt: Towards ordinal action understanding in instructional videos. In: Proceedings of CVPR, pp. 19880–19889 (2022)
27. Hong, Y., Opazo, C.R., Wu, Q., Gould, S.: Sub-instruction aware vision-and-language navigation. In: EMNLP (1), pp. 3360–3376. Association for Computational Linguistics (2020)
28. Li, X., Li, C., Xia, Q., Bisk, Y., Celikyilmaz, A., et al.: Robust navigation with language pretraining and stochastic sampling. In: EMNLP/IJCNLP (1), pp. 1494–1499. ACL (2019)
29. Tan, H., Yu, L., et al.: Learning to navigate unseen environments: back translation with environmental dropout. In: NAACL, pp. 2610–2621. ACL (2019)
30. Liu, C., Zhu, F., Chang, X., et al.: Vision-language navigation with random environmental mixup. In: ICCV, pp. 1624–1634. IEEE (2021)
31. Zhu, F., Zhu, Y., Chang, X., et al.: Vision-language navigation with self-supervised auxiliary reasoning tasks. In: Proceedings of CVPR, pp. 10012–10022. IEEE (2020)
32. Qi, Y., Pan, Z., Hong, Y., Wu, Q., et al.: The road to know-where: an object-and-room informed sequential BERT for indoor vision-language navigation. In: ICCV, pp. 1635–1644. IEEE (2021)
33. An, D., Qi, Y., Wu, Q., et al.: Neighbor-view enhanced model for vision and language navigation. In: ACM MM, pp. 5101–5109. ACM (2021)
34. Chen, J., Gao, C., Meng, E., et al.: Reinforced structured state-evolution for vision-language navigation. In: Proceedings of CVPR, pp. 15429–15438. IEEE (2022)
35. Liang, X., Zhu, F., Li, L., Xu, H., Liang, X.: Visual-language navigation pretraining via prompt-based environmental self-exploration. In: ACL (1), pp. 4837–4851. ACL (2022)
36. Zhang, Z., Qi, S., Zhou, Z., et al.: Reinforced vision-and-language navigation based on historical BERT. In: ICSI, pp. 427–438 (2023)
37. Anderson, P., Wu, Q., Teney, D., Bruce, J., Johnson, M., et al.: "Vision-and-language navigation: interpreting visually-grounded navigation instructions in real environments. Proc. CVPR **22**, 3674–3683 (2018)
38. Wang, X., Huang, Q., Celikyilmaz, A., Gao, J., et al.: Reinforced cross-modal matching and self-supervised imitation learning for vision-language navigation. In: Proceedings of CVPR, pp. 6629–6638. IEEE (2019)
39. Ma, C., Lu, J., Wu, Z., AlRegib, G., Kira, Z., et al.: Self-monitoring navigation agent via auxiliary progress estimation. In: ICLR (2019)
40. Ke, L., Li, X., et al.: Tactical rewind: self-correction via backtracking in vision-and-language navigation. In: Proceedings of IEEE CVPR, pp. 6741–6749 (2019)

Cross-Modal Semantic Alignment Learning for Text-Based Person Search

Wenjun Gan[1], Jiawei Liu[1(✉)], Yangchun Zhu[1], Yong Wu[2], Guozhi Zhao[2], and Zheng-Jun Zha[1]

[1] University of Science and Technology of China, Hefei 230027, China
{ganwenjun,zhuychun}@mail.ustc.edu.cn, {jwliu6,zhazj}@ustc.edu.cn
[2] China Merchants Bank, Shenzhen 518040, China
{wuyong139,gzzhao}@cmbchina.com

Abstract. Text-based person search aims to retrieve pedestrian images corresponding to a specific identity based on a textual description. Existing methods primarily focus on either the alignment of global features through well-designed loss functions or the alignment of local features via attention mechanisms. However, these approaches overlook the extraction of crucial local cues and incur high computational costs associated with cross-modality similarity scores. To address these limitations, we propose a novel Cross-Modal Semantic Alignment Learning approach (SAL), which effectively facilitates the learning of discriminative representations with efficient and accurate cross-modal alignment. Specifically, we devise a Token Clustering Learning module that excavates crucial clues by clustering visual and textual token features extracted from the backbone into fine-grained compact part prototypes, each of which is corresponding to a specific identity-related discriminative semantic. Furthermore, we introduce the optimal transport strategy to explicitly encourage the fine-grained semantic alignment of image-text part prototypes, achieving efficient and accurate cross-modal matching while largely reducing computational costs. Extensive experiments on two public datasets demonstrate the effectiveness and superiority of SAL for text-based person search.

Keywords: Text-Based Person Search · Optimal Transport · Feature Alignment

1 Introduction

Person search is the task of identifying a specific pedestrian across multiple non-overlapping camera networks. It has received heightened attention in recent years due to its wide application, such as activity analysis, automated tracking, and smart retail *etc.* This task remains challenging since the existence of dramatic variations in illumination, camera viewpoint, image resolution, and human pose, as well as cluttered backgrounds and occlusions *etc.* In line with the type of query, person search can be divided into two categories: image-based and

© The Author(s), under exclusive license to Springer Nature Switzerland AG 2024
S. Rudinac et al. (Eds.): MMM 2024, LNCS 14554, pp. 201–215, 2024.
https://doi.org/10.1007/978-3-031-53305-1_16

text-based person search. Image-based person search (person re-identification) requires at least one image as the query, which is difficult to obtain in certain scenes. By contrast, text-based person search which utilizes free-form natural language descriptions is more friendly in practical applications. Text-based person search aims at searching for the person images of interest that best match natural language descriptions in a large-scale image gallery. Compared to image-based person search, text-based person search suffers from severe inter-modality gaps in addition to the troubles of image-based person search, thus making this task more challenging.

To handle the upper challenges, early text-based person search methods separately extract global visual and textual representations from two modalities and project them into a jointly shared space through well-designed loss functions or network models [37,40]. Although these global-matching methods preliminarily reduce modality gaps, they bring noises of irrelevant regions in pedestrian images into global representations and can not effectively excavate the discriminative local details from images and texts. Some recent works [1,16,21,22] explore more effective local-matching algorithms with cross-modality attention mechanism for handling the semantic misalignment issue brought by modality gaps. These local-matching methods employ pre-defined rules (e.g., stripe partition strategy [4] and word-level segmentation [7]) or introduce external models (e.g., human parsing [29], pose estimation [16] and Natural Language ToolKit [21]) to acquire local representations, and estimate cross-modal alignment (e.g., region-word or region-phrase correspondences [22]) to calculate the similarity score. Nevertheless, both external models and expensive pairwise interaction operations for cross-modality similarity score incur substantial computational costs.

Different from prior works, we propose a novel Cross-Modal Semantic Alignment Learning approach (SAL) for text-based person search. SAL is designed to effectively learn discriminative image-text representations with efficient and accurate cross-modal alignment to handle inter-modality gaps. Specifically, SAL consists of a backbone network, a Token Clustering Learning module (TCL), and a Feature Alignment Learning module (FAL). TCL maintains a set of visual and textual semantic prototypes to cluster diverse token features extracted from the backbone into fine-grained compact part prototypes for excavating pivotal cues. Each part prototype is corresponding to identity-related discriminative parts of pedestrians. FAL introduces the optimal transport strategy that explicitly aligns fine-grained semantics between the image-text part prototypes, towards accurate cross-modal matching with lower computation complexity. As for the backbone network, we initialize our model with the CLIP pre-trained parameters attributed to the robust feature capabilities of the Contrastive Language-Image Pre-training (CLIP) [23] model. Extensive experimental results on two benchmark datasets demonstrate the effectiveness and superiority of our SAL.

The main contributions can be summarized as follows: (1) We propose a novel Cross-Modal Semantic Alignment Learning approach for text-based person search to learn discriminative image-text representations with efficient and accurate cross-modal alignment, which significantly reduces inter-modality gaps.

(2) We devise a Token Clustering Learning module that maintains a set of visual and textual semantic prototypes to aggregate diverse token features extracted from both modalities into fine-grained compact part prototypes. (3) We design a Feature Alignment Learning module that introduces the optimal transport strategy to align such image-text part prototypes for effectively addressing semantic misalignment while mitigating computational costs. (4) Extensive experiments on the CUHK-PEDES and ICFG-PEDES datasets demonstrate that SAL outperforms previous methods in supervised text-based person search tasks.

2 Related Work

2.1 Image-Based Person Search

In the past decade, a variety of image-based person search (person re-identification) methods have sprung up, aiming at extracting discriminative and robust representations from pedestrian images. Early works primarily focus on designing hand-crafted features with body structures or robust distance metric learning [8,36]. For example, Farenzena et al. [8] extracted three complementary features that mimic human appearance: overall color content, spatial arrangement of colors in stable regions, and the presence of repeated local patterns with high entropy to achieve model robustness. Recently, deep learning technique has been widely adopted for person re-identification, towards learning discriminative deep representations of pedestrians. For example, Hou et al. [14] designed the Spatial Interaction-and-Aggregation (SIA) module for person ReID, which aggregates the correlated spatial features to consider the local relation and the multi-scale appearance relations. Shen et al. [26] proposed a Similarity-Guided Graph Neural Network (SGGNN) to model the pairwise relationships among different probe-gallery pairs so as to learn the probe-gallery relation features. Chen et al. [2] proposed a cascaded feature suppression mechanism, which mines all potential salient features stage-by-stage and integrates these discriminative salience features with the global feature, producing the representation.

2.2 Text-Based Person Search

Li et al. [20] first purpose the task of text-based person search with a challenging dataset CUHK-PEDES and exploit a recurrent neural network with gated neural attention (GNA-RNN) for this task. The existing methods for text-based person search can be divided into two categories: global-matching methods and local-matching methods. Global-matching methods extract global visual and textual representations respectively from two modalities and project them into a jointly shared space. For example, Zhang et al. [37] proposed a cross-modal projection matching loss (CMPM) and cross-modal projection classification (CMPC) loss for learning discriminative image-text embeddings. Zhang et al. [40] proposed an end-to-end dual-path convolutional network to learn the visual and textual representations with instance loss. These methods reduce the modality

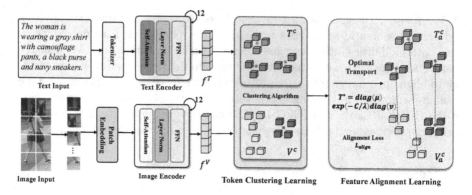

Fig. 1. The overall architecture of the proposed Cross-Modal Semantic Alignment Learning approach. It contains a backbone network, a Token Clustering Learning module (TCL) as well as a Feature Alignment Learning module (FAL).

gap with well-designed loss functions, which overlook noteworthy local details of vision and language. To exploit the potential of discriminative local features, some recent works explore more effective local-matching models to address the semantically inconsistent issue caused by the modality gap. These methods are usually divided into two steps, 1) extract effective local representations by pre-defined rules or external models, 2) and then perform cross-modal matching operations with attention mechanisms. For example, Chen et al. [1] utilized the attention mechanism to automatically associate related phrases and local visual features. Jing et al. [16] devised a pose-guided attention network to implement fine-grained alignment of noun phrases and visual regions. Gao et al. [10] adopted a contextual non-local attention mechanism that adaptively aligns the full-scale representations of images and texts. Zheng et al. [39] proposed a Gumbel attention module to alleviate the matching redundancy issue via choosing semantically strong relevant image regions for all the regions of images, and the corresponding words/phrases. Yan et al. [35] designed a novel transformer architecture which leverages the power of CLIP to achieve cross-modal fine-grained local alignment. In contrast to previous methods, we first cluster diverse token features from both modalities into fine-grained compact part prototypes and then introduce the optimal transport strategy to efficiently align such image-text part prototypes.

3 Methodology

In this section, we first introduce the overall architecture and then elaborate the two key modules of SAL in the following subsections.

3.1 Overall Architecture

Let $S = \{v_i, t_i\}_{i=1}^{N}$ denotes a dataset consists of N image-text pairs, where v_i represents a pedestrian image and t_i is the corresponding textual description.

The identities of the K pedestrians in the dataset are denoted as $Y = \{y_i\}_{i=1}^{N}$ with $y_i \in \{1, ..., K\}$. Given a textual description of a pedestrian, the goal is to find the images of the most relevant person from an image gallery. To address this task, we propose a novel Cross-Modal Semantic Alignment Learning approach for text-based person search, aiming at learning discriminative efficiently aligned image-text representations. As illustrated in Fig. 1, SAL consists of two key modules, $i.e.$, Token Clustering Learning module and Feature Alignment Learning module, as well as a backbone network.

For pedestrian images, the backbone network utilizes pre-trained CLIP Vision Transformer model as the image encoder to extract the visual token features $\boldsymbol{f}^V \in \mathbb{R}^{N \times d}$, where $N = HW/P^2$ is the number of patches, where (H, W) is the resolution of the original image, P is the size of each image patch, d denotes the dimension of the feature. For textual descriptions, the backbone network introduces the lower-cased byte pair encoding (BPE) [24] with a 49,408 vocabulary size as the tokenizer and utilizes CLIP Transformer model as the text encoder to extract textual token features $\boldsymbol{f}^T \in \mathbb{R}^{L \times d}$, where L denotes the length of sequences. After obtaining the initial visual and textual features from image and text modalities, we devise a Token Clustering Learning module that aggregates token features extracted from the backbone network into fine-grained compact part prototypes via clustering algorithm, each of which is corresponding to identity-related discriminative semantics. To align these part prototypes of the two modalities, FAL reformulates the semantic alignment task as an optimal transport problem and harnesses the Sinkhorn algorithm [6] to explicitly encourage fine-grained semantic alignment between vision and language, towards accurate cross-modal matching with lower computational complexity.

3.2   Token Clustering Learning Module

To harness the inherent expressiveness of fine-grained local features and avoid substantial costs from external models, we propose a Token Clustering Learning module that excavates crucial clues by clustering diverse token features extracted from the backbone network into compact part prototypes via the soft k-means based algorithm [15].

In this module, we define visual and textual semantic prototypes, denoted as $\boldsymbol{V} \in \mathbb{R}^{M \times d}$ and $\boldsymbol{T} \in \mathbb{R}^{M \times d}$, which can be updated iteratively. The clustering process involves several key steps. For the sake of clarity, we describe these steps using the visual modality as an example. We assume that each visual token feature $\boldsymbol{f}_i^V \in \mathbb{R}^d$ belongs to one of M visual semantic prototypes $\boldsymbol{V} \in \mathbb{R}^{M \times d}$, making it necessary to compute the association map between the visual token feature and visual semantic prototypes. At iteration t, this token-prototype association is formulated as follows:

$$\boldsymbol{Q}_{ij}^t = e^{-\left\| \boldsymbol{f}_i^V - \boldsymbol{V}_j^{t-1} \right\|^2} \tag{1}$$

Next, one of the visual semantic prototypes is updated using the following formulation:

$$V_i^t = \frac{1}{Z_i^t} \sum_{p=1}^{N} \left(Q_{pi}^t\right)^\mathsf{T} f_i^V \tag{2}$$

where Z_i^t represents a constant used for normalization, T denotes the matrix transpose operation. After several iterations, we capture visual and textual semantic prototypes, denoted as $V^c \in \mathbb{R}^{M \times d}$ and $T^c \in \mathbb{R}^{M \times d}$. These prototypes are regarded as discriminative part prototypes for their respective modalities. Each part prototype corresponds to a specific identity-related compact semantic, which is the crucial clue for distinguishing different persons.

3.3 Feature Alignment Learning Module

To explicitly encourage the fine-grained semantic alignment between the visual and textual modalities, we design a Feature Alignment Learning Module that introduces the optimal transport strategy to align the image-text part prototypes, towards efficient and accurate cross-modal matching while significantly reducing computational costs.

Considering that the learned part prototypes in both images and texts may exhibit semantic misalignment, we explicitly address this issue by promoting fine-grained local alignment across modalities through the optimal transport strategy. Optimal transport-based methods [3,33,38] are widely known for their capacity to minimize the cost of transporting from one distribution to another. In this work, we reformulate the cross-modal alignment as an optimal transport problem, which aims to generate an optimal plan for semantic alignment. Mathematically, the optimization objective for this problem is defined as follows:

$$\arg\min \langle T, C \rangle_F \tag{3}$$

subject to

$$T\mathbf{1} = \mu, T^\mathsf{T}\mathbf{1} = \nu, T \in \mathbb{R}_+^{M \times M} \tag{4}$$

where $\langle \cdot, \cdot \rangle_F$ denotes Frobenius dot product, $\mathbf{1}$ is a vector of all ones of the appropriate dimension, C is the cost matrix between the text-image part prototypes. T is the transport plan, interpreting the alignment between two modalities. As traditional methods to solve this problem are mostly complex and non-differentiable, we adopt the Sinkhorn algorithm [6] to utilize entropy regularization term constraint for fast optimization, which is formulated as follows:

$$\arg\min_{T \in \Pi(\mu,\nu)} \langle T, C \rangle_F - \lambda \mathrm{H}(T) \tag{5}$$

where $\Pi(\mu,\nu) = \left\{ T \in \mathbb{R}_+^{M \times M} \mid T\mathbf{1} = \mu, T^\mathsf{T}\mathbf{1} = \nu \right\}$, H is the entropy function, $\mathrm{H}(T) = -\sum_{i,j} \log(T_{ij}) T_{ij}$, and λ is a hyper-parameter that controls the smoothness. Then we can obtain a fast optimization solution T^*, which can be expressed as:

$$T^* = \mathrm{diag}(\mu)\exp(-C/\lambda)\,\mathrm{diag}(\nu) \tag{6}$$

In this way, we acquire an optimal plan T_{t2i}^*, which minimizes the cost of transporting each part prototype from the text modality to the corresponding one in the image modality. The optimal plan T_{t2i}^* thus serves as a propeller for fine-grained cross-modal alignment. We can capture the aligned visual and textual part prototypes via the optimal plan, defined as follows:

$$
\begin{aligned}
T_a^c &= T_{t2i}^* \cdot V^c \\
V_a^c &= (T_{t2i}^*)^{\mathsf{T}} \cdot T^c
\end{aligned}
\tag{7}
$$

To further enhance semantic alignment between these aligned textual and visual part prototypes, we employ the alignment loss function, defined as:

$$
\mathcal{L}_{\text{align}} = \mathcal{L}_{t2i} + \mathcal{L}_{i2t}
\tag{8}
$$

where

$$
\mathcal{L}_{t2i} = -\log \frac{\exp(S(T_a^c, V_a^c)/\tau)}{\sum \exp(S(T_a^c, V_a^c)/\tau)}, \mathcal{L}_{i2t} = -\log \frac{\exp(S(V_a^c, T_a^c)/\tau)}{\sum \exp(S(V_a^c, T_a^c)/\tau)}
\tag{9}
$$

where S denotes the cosine similarity function, and τ is a temperature hyperparameter which controls the probability distribution.

3.4 Model Optimization

We adopt the common identification loss [12] and ranking loss [13] to optimize the proposed model. The identification loss is formulated as follows:

$$
\mathcal{L}_{\text{id}} = -\frac{1}{B}\sum_{i=1}^{B} \log softmax\left(W_{y_i}^{\mathsf{T}} x_i + b\right)
\tag{10}
$$

where y_i represents the pedestrian identity of the i-th sample, B is batch size. x_i refers to the image feature or text feature. We compute the identification loss with the aligned part prototypes V_a^c and T_a^c, which is termed as $\mathcal{L}_{\text{id}}^c$. Besides, we calculate the ranking loss between the aligned visual prototypes V_a^c and textual prototypes T_a^c, which is described as follows:

$$
\begin{aligned}
\mathcal{L}_{\text{rank}}^c = &max\left(\alpha - S\left(V_a^c, T_a^{c,+}\right) + S\left(V_a^c, T_a^{c,-}\right), 0\right) + \\
&max\left(\alpha - S\left(T_a^c, V_a^{c,+}\right) + S\left(T_a^c, V_a^{c,-}\right), 0\right)
\end{aligned}
\tag{11}
$$

where $T_a^{c,+}$, $V_a^{c,+}$ belong to the corresponding positive sample in a mini-batch. $T_a^{c,-}$, $V_a^{c,-}$ belong to the corresponding hard negative sample in a mini-batch. α denotes a margin hyper-parameter, and S denotes the cosine similarity function. The overall loss of SAL is computed as follows:

$$
\mathcal{L} = \mathcal{L}_{\text{id}}^c + \mathcal{L}_{\text{align}} + \mathcal{L}_{\text{rank}}^c
\tag{12}
$$

During the inference stage, we extract semantically aligned representations from images and texts. We compute the cosine similarity between the aligned representations V_a^c, T_a^c using Sinkhorn algorithm to obtain the final similarity score.

Table 1. Performance comparison with the state-of-the-art methods on CUHK-PEDES dataset.

Method	Rank-1	Rank-5	Rank-10
A-GANet [21]	53.14	74.03	81.95
HGAN [39]	59.00	79.49	86.62
PMA [16]	59.94	79.86	86.70
NAFS [10]	59.98	80.41	87.56
MGEL [28]	60.27	80.01	86.74
SSAN [7]	61.37	80.15	86.73
CM-MoCo [11]	61.65	80.98	86.78
LapsCore [34]	63.40	–	87.80
TIPCB [4]	63.63	82.82	89.01
LBUL [31]	64.04	82.66	87.22
CAIBC [30]	64.43	82.87	88.37
AXM-Net [9]	64.44	80.52	86.77
LGUR [25]	65.25	83.12	89.00
IVT [27]	65.59	83.11	89.21
TFAF [19]	65.69	84.75	89.93
MPFD [5]	66.11	84.05	90.24
SAL	**69.14**	**85.90**	**90.81**

4   Experiment

4.1   Experimental Setup

Dataset. CUHK-PEDES [20] is a large-scale dataset for text-based person search, which includes 40,206 images and 80,440 textual descriptions of 13,003 identities. The training set contains 34,054 images and 68,126 text descriptions of 11,003 pedestrians. The validation set has 3,078 images, 1000 person IDs, and 6,158 textual descriptions. The testing set consists of 3,074 images and 6,156 textual descriptions of the rest 1,000 pedestrians. Each image comprises at least two sentence descriptions, each of which constitutes no less than 23 words on average. **ICFG-PEDES** [7] is a recently published dataset that contains 54,522 pedestrian images of 4,102 identities, of which all images are collected from the MSMT17 database [32]. ICFG-PEDES is fixedly divided into a training set and a testing set. The training set includes 34,674 images of 3,102 pedestrians and the testing set contains 19,848 images for the rest 1,000 pedestrians. The caption corresponding to each pedestrian image has at least 37 words.

Evaluation Metrics. We employ the standard metric in most person search literature, $i.e.$, Cumulative Matching Characteristic at Rank-1, Rank-5, Rank-10 to evaluate the recognition performances of different person ReID models. Rank-k denotes the probability that when given a textual description of a person as a query, we can find at least one matching person image in the top-k candidate list ranked by their similarities.

Table 2. Performance comparison with the state-of-the-art methods on ICFG-PEDES dataset.

Method	Rank-1	Rank-5	Rank-10
Dual Path [40]	38.99	59.44	68.41
CMPM/C [37]	43.51	65.44	74.26
MIA [22]	46.49	67.14	75.18
SCAN [18]	50.05	50.98	77.21
ViTAA [29]	50.98	68.79	75.78
SSAN [7]	54.23	72.63	79.53
IVT [27]	56.04	73.60	80.22
MPFD [5]	57.29	75.84	82.35
SAL	**62.77**	**78.64**	**84.21**

Table 3. Performance comparisons in terms of computational complexity on CUHK-PEDES dataset.

Method	Inference Time	Rank-1
MIA [22]	42 ms	53.10
NAFS [10]	42 ms	59.94
SCAN [18]	46 ms	55.86
SAL	**6 ms**	**69.14**

Implementation Details. We adopt the CLIP-Transformer pre-trained model parameters as the initialization of the image encoder and text encoder in the backbone network. We resize all the images to the size of 384×128. The data augmentation strategies during training include random horizontal flipping with 50% probability, padding 10 pixels, random cropping, and random erasing. The length of the input text is set to at most 76 words. The patch size is 16. In terms of parameter setting, the margin α in the ranking loss is set to 0.3, the hyper-parameter λ of Sinkhorn algorithm is set to 0.1. The iteration in TCL is set to 1. The temperature hyper-parameter τ is set to 0.02. Besides, the dimension d is 512 and M is 6. During training, the batch size of the model is set to 64. We use Adam optimizer [17] to train the overall framework. The framework is in total trained for 60 epochs with the initial learning rate of 1e−5 and the lr is decayed by 0.1 at the 20th, 40th, and 50th epochs. We conduct all the experiments on one 3090ti GPU with Pytorch framework.

4.2 Comparisons with State-of-the-Art Methods

Results on CUHK-PEDES. We report the comparison of our proposed SAL with the state-of-the-art methods including A-GANet [21], HGAN [39], PMA [16], NAFS [10], MGEL [28], SSAN [7], CM-MoCo [11], LapsCore [34], TIPCB [4], LBUL [31], CAIBC [30], AXM-Net [9], LGUR [25], IVT [27], MPFD [5] on

Table 4. Evaluation of the effectiveness of each component of SAL on ICFG-PEDES dataset.

Model	Rank-1	Rank-5	Rank-10
Baseline	52.84	72.74	79.93
Baseline+TCL	60.42	77.23	82.94
Baseline+TCL+FAL	62.77	78.64	84.21

the CUHK-PEDES dataset. From Table 1, we can observe that our SAL achieves the performance of 69.14%, 85.90%, and 90.81% in Rank-1, Rank-5, and Rank-10 metrics, respectively. Notably, SAL outperforms previous methods across all three metrics. Particularly, when compared to the local matching method like MPFD [5], SAL obtains a performance improvement of 3.03% in Rank-1 accuracy. This is because our method effectively learns aligned, identity-related discriminative representations for text-based person search tasks, which are crucial clues for distinguishing different persons.

Comparisons on ICFG-PEDES. The comparison with the state-of-the-art approaches of Dual Path [40], CMPM/C [37], MIA [22], SCAN [18], ViTAA [29], SSAN [7], IVT [27], MPFD [5] on ICFG-PEDES dataset is presented in Table 2. SAL consistently demonstrates superior performance, achieving remarkable results of 62.77%, 78.64%, and 84.21% in Rank-1, Rank-5, and Rank-10 metrics, respectively. Notably, when compared to attention-based methods such as IVT [27], our SAL achieves significant performance improvement of 6.73% in Rank-1 accuracy. This is because that our SAL achieves fine-grained accurate alignment of image-text part prototypes for text-based person search tasks, which is more efficient and accurate than attention-based methods.

Comparisons on Computational Complexity. As discussed before, our SAL presents significant advantages in terms of computational complexity. To this end, we present the comparisons of the inference time and Rank-1 accuracy between SAL and three methods employing cross-attention mechanisms, namely MIA [22], SCAN [18], and NAFS [10] in Table 3. The inference time refers to the average time that includes the feature extraction and similarity computation for a given query. For a fair comparison, we resize all images to the size of 384×128 and all experiments are conducted on one Titan X GPU. Notably, SAL demonstrates significantly reduced inference times while simultaneously achieving remarkable Rank-1 accuracy compared to all other methods that rely on cross-modal attention mechanisms. This efficiency advantage is attributed to the employment of optimal transport strategy in SAL.

4.3 Ablation Studies

Effectiveness of Components. To further verify the impact of the proposed components, we conduct a series of ablation studies on the ICFG-PEDES

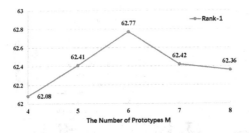

Fig. 2. Effect of the number of prototypes M in terms of Rank-1 accuracy on ICFG-PEDES dataset.

dataset. The results are shown in Table 4. We utilize the CLIP Transformer model [23] with contrastive loss and identification loss as the *Baseline* to extract the global visual and textual representations. *Baseline+TCL* indicates that we employ the backbone and TCL to learn the fine-grained discriminative visual and textual part prototypes. *Baseline+TCL+FAL* represents the whole SAL, which effectively learns discriminative image-text representations with efficient and accurate cross-modal alignment for text-based person search, addressing inter-modality gaps. Compared to the *Baseline*, *Baseline+TCL* achieves 7.58% performance improvement in Rank-1 accuracy. The improvement demonstrates the effectiveness and superiority of TCL in capturing identity-related discriminative part prototypes from the two modalities. Moreover, *Baseline+TCL+FAL* surpasses *Baseline+TCL* by 2.35% in Rank-1 accuracy. This performance comparison highlights that the optimal transport strategy in FAL indeed efficiently promotes the fine-grained semantic alignment of image-text part prototypes, which contributes to easing the misalignment between images and texts while largely reducing computational complexity.

Analysis of the Number of Prototypes in TCL. We investigate the effect of the number of prototypes M on performance. As illustrated in Fig. 2, when adjusting the number of prototypes M from 4 to 6, SAL obtains performance improvement of 0.69% in Rank-1 accuracy. This indicates that a larger number of prototypes is beneficial to capture more diverse semantic patterns for better feature representation. When increasing M from 6 to 8, SAL achieves slight performance degradation, indicating that an excessive number of prototypes incorporates redundant information and background noise into the learned part prototypes, damaging their discrimination. Thus, the parameter M is set to 6 in our experiments.

4.4   Visualization Results

We present the visualization results of the aligned visual and textual representations from two matched image-text pairs in Fig. 3 (a). It can be observed that the highlighted image regions correspond to the highlighted words or phrases

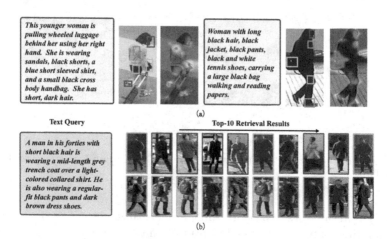

Text Query   Top-10 Retrieval Results

(b)

Fig. 3. (a) Visualization results of the aligned visual and textual representations from two matched image-text pairs. (b) Top-10 retrieval results on ICFG-PEDES dataset between Baseline (the first row) and our SAL (the second row) for the same text query. The matched and mismatched images are marked with green and red rectangles, respectively. (Color figure online)

in the text, indicating that SAL is capable of learning visual and textual representations with semantically fine-grained alignment across the two modalities. We also provide retrieval results on the ICFG-PEDES dataset. As illustrated in Fig. 3 (b), we compare the top-10 retrieval results between the Baseline and our proposed SAL. We can observe that SAL still achieves accurate retrieval results in the case that the Baseline fail to retrieve, which indicates the powerful ability to extract semantically aligned representations from the two modalities.

5   Conclusion

In this work, we propose a novel Cross-Modal Semantic Alignment Learning approach for text-based person search to learn discriminative aligned image-text representations. It consists of a backbone network, a Token Clustering Learning module and a Feature Alignment Learning module. The Token Clustering Learning module that clusters diverse token features extracted from the backbone into compact part prototypes, each of which represents identity-related discriminative semantics. The Feature Alignment Learning module introduces the optimal transport algorithm to align the semantic of such image-text part prototypes, towards efficient and accurate cross-modal matching. Extensive experiments on two benchmark datasets, *i.e.*, CUHK-PEDES and ICFG-PEDES show the effectiveness of our proposed SAL.

Acknowledgements. This work was supported by National Natural Science Foundation of China (NSFC) under Grants 62106245, 62225207 and U19B2038.

References

1. Chen, D., et al.: Improving deep visual representation for person re-identification by global and local image-language association. In: Ferrari, V., Hebert, M., Sminchisescu, C., Weiss, Y. (eds.) ECCV 2018. LNCS, vol. 11220, pp. 56–73. Springer, Cham (2018). https://doi.org/10.1007/978-3-030-01270-0_4

2. Chen, X., et al.: Salience-guided cascaded suppression network for person re-identification. In: CVPR, pp. 3300–3310 (2020)

3. Chen, Y.-C., et al.: UNITER: UNiversal Image-TExt Representation learning. In: Vedaldi, A., Bischof, H., Brox, T., Frahm, J.-M. (eds.) ECCV 2020. LNCS, vol. 12375, pp. 104–120. Springer, Cham (2020). https://doi.org/10.1007/978-3-030-58577-8_7

4. Chen, Y., Zhang, G., Lu, Y., Wang, Z., Zheng, Y.: TIPCB: a simple but effective part-based convolutional baseline for text-based person search. Neurocomputing **494**, 171–181 (2022)

5. Chen, Y., Zhang, G., Zhang, H., Zheng, Y., Lin, W.: Multi-level part-aware feature disentangling for text-based person search. In: ICME, pp. 2801–2806 (2023)

6. Cuturi, M.: Sinkhorn distances: lightspeed computation of optimal transport. In: Advances in Neural Information Processing Systems, vol. 26 (2013)

7. Ding, Z., Ding, C., Shao, Z., Tao, D.: Semantically self-aligned network for text-to-image part-aware person re-identification. arXiv preprint arXiv:2107.12666 (2021)

8. Farenzena, M., Bazzani, L., Perina, A., Murino, V., Cristani, M.: Person re-identification by symmetry-driven accumulation of local features. In: 2010 IEEE Computer Society Conference on Computer Vision and Pattern Recognition, pp. 2360–2367 (2010)

9. Farooq, A., Awais, M., Kittler, J., Khalid, S.S.: AXM-Net: implicit cross-modal feature alignment for person re-identification. In: AAAI, pp. 4477–4485 (2022)

10. Gao, C., et al.: Contextual non-local alignment over full-scale representation for text-based person search. arXiv preprint arXiv:2101.03036 (2021)

11. Han, X., He, S., Zhang, L., Xiang, T.: Text-based person search with limited data. arXiv preprint arXiv:2110.10807 (2021)

12. He, T., Jin, X., Shen, X., Huang, J., Chen, Z., Hua, X.S.: Dense interaction learning for video-based person re-identification. In: ICCV, pp. 1490–1501 (2021)

13. Hermans, A., Beyer, L., Leibe, B.: In defense of the triplet loss for person re-identification. arXiv preprint arXiv:1703.07737 (2017)

14. Hou, R., Ma, B., Chang, H., Gu, X., Shan, S., Chen, X.: Interaction-and-aggregation network for person re-identification. In: CVPR, pp. 9317–9326 (2019)

15. Jampani, V., Sun, D., Liu, M.-Y., Yang, M.-H., Kautz, J.: Superpixel sampling networks. In: Ferrari, V., Hebert, M., Sminchisescu, C., Weiss, Y. (eds.) ECCV 2018. LNCS, vol. 11211, pp. 363–380. Springer, Cham (2018). https://doi.org/10.1007/978-3-030-01234-2_22

16. Jing, Y., Si, C., Wang, J., Wang, W., Wang, L., Tan, T.: Pose-guided multi-granularity attention network for text-based person search. In: AAAI, pp. 11189–11196 (2020)

17. Kingma, D.P., Ba, J.: Adam: a method for stochastic optimization. arXiv preprint arXiv:1412.6980 (2014)

18. Lee, K.-H., Chen, X., Hua, G., Hu, H., He, X.: Stacked cross attention for image-text matching. In: Ferrari, V., Hebert, M., Sminchisescu, C., Weiss, Y. (eds.) ECCV 2018. LNCS, vol. 11208, pp. 212–228. Springer, Cham (2018). https://doi.org/10.1007/978-3-030-01225-0_13

19. Li, S., Lu, A., Huang, Y., Li, C., Wang, L.: Joint token and feature alignment framework for text-based person search. IEEE Signal Process. Lett. **29**, 2238–2242 (2022)
20. Li, S., Xiao, T., Li, H., Zhou, B., Yue, D., Wang, X.: Person search with natural language description. In: CVPR, pp. 1970–1979 (2017)
21. Liu, J., Zha, Z.J., Hong, R., Wang, M., Zhang, Y.: Deep adversarial graph attention convolution network for text-based person search. In: ACM MM, pp. 665–673 (2019)
22. Niu, K., Huang, Y., Ouyang, W., Wang, L.: Improving description-based person re-identification by multi-granularity image-text alignments. IEEE Trans. Image Process. **29**, 5542–5556 (2020)
23. Radford, A., et al.: Learning transferable visual models from natural language supervision. In: ICML, pp. 8748–8763 (2021)
24. Sennrich, R., Haddow, B., Birch, A.: Neural machine translation of rare words with subword units. arXiv preprint arXiv:1508.07909 (2015)
25. Shao, Z., Zhang, X., Fang, M., Lin, Z., Wang, J., Ding, C.: Learning granularity-unified representations for text-to-image person re-identification. In: ACM MM, pp. 5566–5574 (2022)
26. Shen, Y., Li, H., Yi, S., Chen, D., Wang, X.: Person re-identification with deep similarity-guided graph neural network. In: Ferrari, V., Hebert, M., Sminchisescu, C., Weiss, Y. (eds.) ECCV 2018. LNCS, vol. 11219, pp. 508–526. Springer, Cham (2018). https://doi.org/10.1007/978-3-030-01267-0_30
27. Shu, X., et al.: See finer, see more: implicit modality alignment for text-based person retrieval. In: Karlinsky, L., Michaeli, T., Nishino, K. (eds.) Computer Vision – ECCV 2022 Workshops. ECCV 2022. LNCS, vol. 13805, pp. 624–641. Springer, Cham (2022). https://doi.org/10.1007/978-3-031-25072-9_42
28. Wang, C., Luo, Z., Lin, Y., Li, S.: Text-based person search via multi-granularity embedding learning. In: IJCAI, pp. 1068–1074 (2021)
29. Wang, Z., Fang, Z., Wang, J., Yang, Y.: *ViTAA*: visual-textual attributes alignment in person search by natural language. In: Vedaldi, A., Bischof, H., Brox, T., Frahm, J.-M. (eds.) ECCV 2020. LNCS, vol. 12357, pp. 402–420. Springer, Cham (2020). https://doi.org/10.1007/978-3-030-58610-2_24
30. Wang, Z., et al.: CAIBC: capturing all-round information beyond color for text-based person retrieval. In: ACM MM, pp. 5314–5322 (2022)
31. Wang, Z., et al.: Look before you leap: improving text-based person retrieval by learning a consistent cross-modal common manifold. In: ACM MM, pp. 1984–1992 (2022)
32. Wei, L., Zhang, S., Gao, W., Tian, Q.: Person transfer GAN to bridge domain gap for person re-identification. In: CVPR, pp. 79–88 (2018)
33. Wu, B., Cheng, R., Zhang, P., Vajda, P., Gonzalez, J.E.: Data efficient language-supervised zero-shot recognition with optimal transport distillation. arXiv preprint arXiv:2112.09445 (2021)
34. Wu, Y., Yan, Z., Han, X., Li, G., Zou, C., Cui, S.: LapsCore: language-guided person search via color reasoning. In: ICCV, pp. 1624–1633 (2021)
35. Yan, S., Dong, N., Zhang, L., Tang, J.: Clip-driven fine-grained text-image person re-identification. arXiv preprint arXiv:2210.10276 (2022)
36. Zhang, Y., Liu, D., Zha, Z.J.: Improving triplet-wise training of convolutional neural network for vehicle re-identification. In: ICME, pp. 1386–1391 (2017)
37. Zhang, Y., Lu, H.: Deep cross-modal projection learning for image-text matching. In: Ferrari, V., Hebert, M., Sminchisescu, C., Weiss, Y. (eds.) ECCV 2018. LNCS,

vol. 11205, pp. 707–723. Springer, Cham (2018). https://doi.org/10.1007/978-3-030-01246-5_42

38. Zheng, K., Liu, W., He, L., Mei, T., Luo, J., Zha, Z.J.: Group-aware label transfer for domain adaptive person re-identification. In: CVPR, pp. 5310–5319 (2021)

39. Zheng, K., Liu, W., Liu, J., Zha, Z.J., Mei, T.: Hierarchical Gumbel attention network for text-based person search. In: ACM MM, pp. 3441–3449 (2020)

40. Zheng, Z., Zheng, L., Garrett, M., Yang, Y., Xu, M., Shen, Y.D.: Dual-path convolutional image-text embeddings with instance loss. ACM Trans. Multimedia Comput. Commun. Appl. (TOMM) **16**, 1–23 (2020)

Point Cloud Classification via Learnable Memory Bank

Lisa Liu[1], William Y. Wang[2], and Pingping Cai[3]([✉]) [ID]

[1] Providence Day School, Charlotte, NC 28270, USA
[2] Dutch Fork High School, Irmo, SC 29063, USA
[3] University of South Carolina, Columbia, SC 29208, USA
pcai@email.sc.edu

Abstract. Point cloud analysis has gained significant importance across various fields, in which a pivotal and formidable aspect is the classification of point clouds. To solve this problem, most existing deep learning methods follow the two-stage framework, wherein an encoder first extracts shape features and then an MLP-based classification head categorizes objects. To improve the classification performance, prior efforts have concentrated on refining the shape feature extraction through intricate encoder designs. However, the design of classification heads has remained unexplored. In this paper, we focus on the second stage and introduce a novel classification head integrated with a Learnable Memory Bank (LMB), tailored for the classification of 3D point-cloud objects. The LMB aims to learn representative category feature vectors from training objects. Subsequently, a similarity-based feature-matching mechanism is employed to generate the predicted class logits. The proposed LMB classification head can be seamlessly integrated into existing feature extraction backbones. Empirical evaluations prove the efficacy of the proposed LMB classification head, showcasing performance on par with state-of-the-art methods.

Keywords: Point Cloud · Shape Classification · Network Architecture

1 Introduction

The point cloud represents 3D objects with a set of coordinates. Due to its efficiency, point cloud has gained significant prominence in the fields of autonomous driving [9,33], virtual reality [25], and robotics [21], where the computers are supposed to understand, recognize, and depict surrounding 3D scenes. Thus, one of the fundamental computer vision tasks, point cloud classification, is proposed, where the goal is to recognize the category of the input point clouds. However, due to the complicated structure of the point cloud, it is nontrivial to achieve this task. This is because a point cloud consists of unordered points that discretely portray object surfaces in 3D space, exhibiting a distribution-irregular and permutation-invariant nature.

© The Author(s), under exclusive license to Springer Nature Switzerland AG 2024
S. Rudinac et al. (Eds.): MMM 2024, LNCS 14554, pp. 216–229, 2024.
https://doi.org/10.1007/978-3-031-53305-1_17

Recently, deep learning-based algorithms have shown promising improvement in accurately understanding the point cloud and classifying it. Especially, one of the pioneering works PointNet [16] offers a two-stage architecture by encoding the input point clouds to shape features and then using an MLP-based classification head to predict output class labels for the entire input or segment/part labels for each individual point. Since then, a variety of methods have been proposed. Based on their encoder's network architecture, we can categorize them into three types: voxel-based [8,38], graph-based [13,22,24], and even point-based methods [14,17,18]. However, these prevailing methods only focus on adding advanced local operators and scaled-up network parameters in the feature extractor, while the design of the classification head is under-explored. Different from previous methods, PointNN [34] proposes a non-learning based classification method via constructing a feature memory bank that covers all the training samples and using similarity-based matching to classify inputs as illustrated in Fig. 1(b). Although novel, the memory bank constructed in their method requires a significant amount of memory to store all the training-set features. This limitation becomes apparent when the training set is extensive, leading to increased memory costs.

To address this issue, our intuition is that instead of remembering features from all training samples, we can limit the number of learnable component sets and extract shared features for each category, greatly reducing the memory cost. Thus, we propose a novel solution called the Learnable Memory Bank (LMB) for point cloud classification tasks, as illustrated in Fig. 1. This approach introduces trainable parameters that enable the construction of a memory bank, which can be updated and optimized during the training process. One significant advantage of this design is that it allows for the customization of the size of the memory bank based on specific categories. This significantly reduces the memory costs associated with storing training set features, effectively managing memory consumption while maintaining the necessary information for accurate classification. Then, another similarity-based feature matching technique is designed to predict the output class logits.

To verify the effectiveness of the proposed LMB classification head, we integrate it into multiple encoder backbones and test it on two widely used point cloud classification datasets: ScanObjectNN [23] and ModelNet40 [27]. The experiment results prove that by using the proposed classification head, we can achieve promising classification accuracy and even outperform previous state-of-the-art baselines.

Our main contributions can be summarized below:

- We explore a novel design for classification heads that can be plugged into different deep-learning backbones to augment the performance of existing methods.
- Unlike the previous PointNN which incurs a large memory cost from storing all the training objects' feature vectors, we propose the Learnable Memory Bank that uses only a small number of feature vectors per object class.

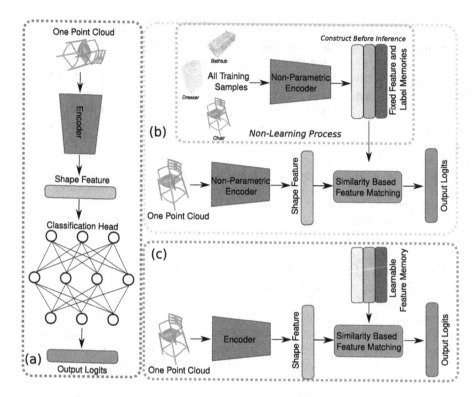

Fig. 1. Difference between the proposed Learnable Memory Bank (c) and other methods. (a) An MLP-based classification Head. (b) Non-Parametric Memory Bank Construction and Classification

2   Related Work

To classify an input point cloud, **traditional algorithms** rely on manually crafted shape features [10,15,35,36]. For example, Zhang et al. [36] devised an approach to extract multilevel Point-Cluster-Based discriminative features. This approach involves segmenting the point cloud into hierarchical clusters of varying sizes using a threshold based on a natural exponential function, extracting and encoding shape features from these multilevel point clusters via joint latent Dirichlet allocation and sparse coding. The resulting features, spanning different levels, efficiently encapsulate information regarding objects of varying dimensions, leading to a substantial enhancement in classification precision. In the work of Lin et al., [10], an eigen feature derived from weighted covariance matrices was established. This innovative technique facilitates the extraction of reliable geometric features capable of describing the local geometric attributes of a point cloud. Additionally, it offers insights into the local geometry's nature, distinguishing whether it's linear, planar, or spherical in nature. However, these traditional algorithms often struggle to perform well when the input becomes more complex.

To address this issue, **deep-learning methods** have been proposed due to their strong ability to extract features automatically [6,17,22,28,37]. One of the pioneering works, PointNet [16], offers a unified network architecture by directly taking point clouds as inputs and generating output such as class labels for the entire input or segment/part labels for each individual point with MLPs. However, local geometric clues are overlooked in the PointNet. To improve the performance, PointNet++ [17] extends from PointNet and introduces advanced techniques like Farthest Point Sampling (FPS) and local grouping operators to extract local geometries. Since then, sophisticated algorithmic have been developed to extract better local geometries. For example, RSCNN [12] extends regular grid CNN to irregular configuration for point cloud analysis, where the convolutional weight is learned from low-level geometric between a sampled point and its neighbors. As a result, it can capture better local representation with explicit reasoning about the spatial layout of points, which leads to much shape awareness and robustness. Point-GNN [22] encodes the point cloud efficiently in a fixed radius near-neighbors graph and uses a graph neural network to predict the category and shape of the object that each vertex in the graph belongs to. These approaches have demonstrated performance improvements. Instead of operating in extrinsic space or treating each point independently, DGCNN [24] introduces EdgeConv which acts on graphs dynamically computed in each layer of the network and has the ability to capture semantic characteristics over potentially long distances in the original embedding. Since discrete point cloud objects lack sufficient descriptions of shape geometries, CurvNet [28] proposes random walking for aggregating curves in point clouds, where Sequences of connected points (curves) are firstly grouped by taking guided walks in the point clouds, and then subsequently aggregated back to the point features. Point Transformer [37] investigates the application of self-attention networks to 3D point cloud processing and introduces self-attention layers to extract local geometric for point cloud tasks. Different from relying on exploring sophisticated local geometric extractors using convolution, graph, or attention mechanisms, PointMLP [14] proposes a lightweight geometric affine module and a pure residual MLP network that integrates no sophisticated local geometrical extractors but still performs very competitively. However, residual networks scale up the number of parameters and have led to increased network complexity and demands for greater computational resources.

Instead of relying solely on increasing the number of learnable parameters in a network, Zhang et al. [34], therefore, propose a novel approach called Point-NN, which employs a Non-parametric Network for point cloud analysis. Point-NN is constructed using non-learnable components, including a non-parametric encoder for 3D feature extraction and a point-memory bank for task-specific recognition. The point-memory bank serves as a cache for storing the training-set features extracted from the non-parametric encoder. During inference, the memory bank utilizes cosine feature similarity matching to output task-specific predictions for a given test point cloud. However, one drawback of this point-memory bank is that it requires a significant amount of memory to store all

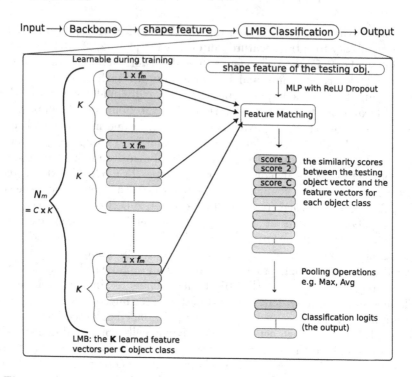

Fig. 2. Detailed algorithm pipeline of the proposed LMB Classification Head.

the training-set features, particularly when dealing with large-scale training sets. Rather than storing all the training-set features, the proposed Learnable Memory Bank only stores class-wise features and thus can significantly reduce memory costs.

3   Proposed Method

Given an input point cloud, denoted as $P \in \mathrm{R}^{N \times 3}$, containing N points, we aim to predict its category label. To accomplish this task, we first employ an encoder to extract the global shape feature, denoted as $F \in \mathrm{R}^{1 \times f_g}$, where f_g represents the dimension of the extracted shape feature vector. Then, we feed the F into the Learnable Memory Bank (LMB) Classification Head to predict the class logits.

Figure 2 illustrates the overall pipeline of the proposed LMB Classification Head. It first uses MLPs with ReLU and Dropout layers to change the dimension of the input global shape features. Then, the novel learnable memory bank denoted as $M \in \mathrm{R}^{N_m \times f_m}$ is constructed to learn the discriminative category shape features from the training samples, where N_m corresponds to the total number of learnable vectors in the memory bank and f_m represents the dimension of each learnable vector. Finally, it utilizes a similarity-based feature-matching method to calculate the similarity scores between the input shape vector and the learned shape vectors in the LMB to predict the classification labels. Notably, it offers support for various feature extractor backbones, rendering it highly adaptable across different datasets.

3.1 Learnable Memory Bank

The LMB is an essential component of the proposed classification head, consisting of a set of learnable vectors that can be automatically updated during training through the gradient backpropagation process. Inside the LMB, each vector is designed to capture a discriminative shape feature associated with a specific category. To ensure that the memory bank contains a sufficient number of vectors, we set its size to $N_m = C \times K$, where C represents the number of target categories in the training/testing sets, and K represents the minimum number of discriminative shape vectors per category. Notably, $K \geq 1$ guarantees that each category has at least one representative feature.

Moreover, to enhance the discriminative capability of the LMB, we apply additional orthogonal constraints to it using the following formula:

$$L_{const} = ||MM^T - I||_2, \tag{1}$$

where $I \in R^{N_m \times N_m}$ denotes the Identity Matrix.

The orthogonal constraints impose a condition that minimizes the deviation of the matrix product MM^T from the Identity Matrix I. This constraint encourages the learnable vectors in the memory bank to be as far away from each other as possible in the cosine similarity space, which can lead to better discrimination between different categories and improve the overall performance.

3.2 Similarity-Based Matching

After extracting the global shape feature using an encoder, we use MLPs with ReLU and Dropout to map it into the same embedding space as the learnable feature in the LMB. The mapped global feature $F' \in R^{1 \times f_m}$ is then fed into a Similarity-based Matching block, where the classification process is carried out by identifying the most similar feature in the memory bank and assigning the corresponding label to it. Figure 2 illustrates the detailed process of Similarity-based Matching.

The first step is to measure the similarity between two feature vectors. Following previous work [34], we utilize the cosine similarity. Specifically, the cosine similarity between the input feature F and the memory bank M is calculated as follows:

$$S_{cos} = \frac{F'M^T}{||F'|| \cdot ||M||} \in R^{1 \times N_m}, \tag{2}$$

where S_{cos} represents the semantic correlation of the test point cloud with the N_m samples in the memory bank.

Then, since each category may contain multiple representative feature vectors, we need to aggregate them for better performance. To do this, we reshape the predicted scores based on the number of target categories in the training/testing sets and obtain $S'_{cos} \in \mathrm{R}^{C \times K}$. Then, the aggregated similarity of each category is computed as follows:

$$Score = Avg_k(S'_{cos}) \in \mathrm{R}^{C \times 1}, \tag{3}$$

where Avg_k denotes the averaging operation across the K vectors for each category.

Finally, the output logits are obtained by applying a softmax function to $Score$ to produce the probabilities for each category. This process results in the final classification prediction for the input point cloud.

3.3   Loss Function

The LMB can be easily integrated into the different feature extraction backbones and trained through a gradient descent algorithm. Since the output of the similarity-based matching block is the logits representing the predicted probabilities for each category, we can use the cross-entropy loss, a common classification loss function, to train the network. Especially, the cross-entropy loss is formulated as follows:

$$L_{class} = -\sum_{c=1}^{|C|} y_c \log(p_c), \tag{4}$$

where L_{class} is the cross-entropy loss; $|C|$ is the total number of classes (target categories); y_c is a binary indicator that takes the value 1 if the ground truth label corresponds to class c and 0 otherwise; p_c is the predicted probability (logit) for class c, which is obtained after applying the softmax function to the output of the similarity-based matching block.

Finally, the overall loss function of the proposed method, including the classification loss (L_{class}) and the orthogonal constraint loss (L_{const}), is given by:

$$L = L_{class} + \lambda L_{const}, \tag{5}$$

where λ is the hyperparameter to balance the two components. By minimizing the overall loss function, the model simultaneously learns to make accurate category predictions while ensuring that the learnable feature vectors in the memory bank are kept orthogonal or as distinguishable from each other as possible, which can lead to improved discrimination and generalization ability.

4   Experiments

To substantiate the efficacy of the proposed approach, a series of experiments were conducted. Specifically, we rigorously tested the LMB classification head across diverse backbone architectures, including PointNet++ [17] and PointPN [34]. Moreover, the experiment covers a range of datasets, notably ModenNet40 [27] and ScanObjectNN [23], serving to show the robustness of the proposed method across varying contexts. In addition, we also systematically explore different hyper-parameter selections for the LMB classification head.

4.1 Implementation and Evaluation Metrics

We implement our networks on top of open-sourced PointNet++ [17] and PointPN [34] baselines and integrate the LMB classification head to them using Pytorch. All the networks are trained from scratch for a fair comparison with previous methods. We set hyper-parameters $f_m = 256, K = 4$, and $\lambda = 1$ for all experiments. For the PointNet++ backbone, we trained our models for 200 epochs on two Nvidia RTX A4000 GPUs with a batch size of 32. The optimizer we used is Adam with an initial learning rate of 0.001 and a weight decay of 0.0001. We also use the cosine annealing scheduler to adjust the learning rate. The training can be finished in approximately 1 h on both datasets. For the PointPN backbone, the setting is similar to the PointNet++ backbone, except that we trained our models for 600 epochs with the learning rate set to 0.01. However since the PointPN backbone is complicated, it requires more training time and the training can be finished in approximately 8 h.

Following the conventions of previous works [17,34] in the community, we set the resolution of input to 1,024 points for all data and report the metrics of class-average accuracy (mAcc) and overall accuracy (OA) on these testing sets without the voting strategy [34].

4.2 Experiment Results

ModelNet40: We first present the results on the ModelNet40 dataset [27]. It is a widely used 3D mesh dataset and includes CAD models (mostly man-made) of 40 different categories. We follow the official train/test splitting strategy with 9,843 training and 2,468 testing models.

Table 1 shows the classification accuracy of different methods as well as the model size and running time. We see that, by using the proposed LMB classification head, we can achieve promising results that are compatible with their original baselines. Notably, with PointNet++ as the backbone, we achieve an overall accuracy of 91.4%, whereas the original PointNet++ only achieves an accuracy of 90.7%. Furthermore, when employing PointPN as the backbone, the performance improves even further, with an overall accuracy of 93.6% which is close to the original PointPN baseline of 93.8%. The suggested LMB classification head also displays remarkable efficiency, causing only a marginal reduction in speed and a small increase in the number of parameters. Specifically, it achieves a testing speed of 864 samples per second, close to the original PointNet++ speed of 902 samples per second.

ScanObjectNN: Then, considering that ModelNet40 comprises only synthetic point clouds, we also conduct experiments on the ScanObjectNN benchmark [23]. Objects obtained from real-world 3D scans are significantly different from CAD models due to the presence of background noise, non-uniform density, and occlusions. ScanObjectNN dataset contains approximately 15,000 objects categorized into 15 classes, with 2,902 unique object instances, and is officially split into three subsets. Following previous work [14], we specifically focus on the hardest split (PB T50 RS).

Table 1. Point cloud classification accuracy of different methods on the synthetic ModelNet40 dataset. We also report the model parameters and speed of some open-sourced methods by samples/second tested on one RTX A4000 GPU.

Method	mAcc (%)	OA (%)	Param	Test speed
PointNN [34]	–	81.8	0.0 M	40
PointNet [16]	86	89.2		
PointCNN [7]	88.1	92.5		
PointConv [26]	–	92.5	18.6 M	9
DGCNN [24]	90.2	92.9	1.8 M	362
RS-CNN [12]	–	92.9	1.3 M	
DensePoint [11]	–	93.2		
PointASNL [32]	–	92.9		
GBNet [20]	91	93.8	8.4 M	112
GDANet [30]	–	93.8	0.9 M	14
PAConv [29]	–	93.9		
PCT [4]	–	93.2		
Point Trans. [37]	90.6	93.7		
CurveNet [28]	–	94.2	2.0 M	13
PointMLP [14]	91.3	94.1	12.6 M	101
PointNet++ [17]	–	90.7	1.4 M	902
PointPN [34]	–	93.8	0.8 M	216
Ours (PointNet++)	88.9	**91.4**	1.4 M	864
Ours (PointPN)	90.5	93.6	0.8 M	205

Table 2 shows the classification accuracy of different methods. Derived from this table, we can find similar patterns to those found in the ModelNet40 dataset. Notably, leveraging PointNet++ as the backbone, our method yields an overall accuracy of 78.1%, surpassing the performance of the original PointNet++, which is 77.9%. Alternatively, utilizing the PointPN backbone leads to an impressive overall accuracy of 87.4%. These results demonstrate the effectiveness of the proposed method in achieving compatible performance compared to previous state-of-the-art counterparts.

4.3 Ablation Study

Then, to understand the effectiveness of each component in our method, we conducted several ablation studies. In particular, we determine the optimal shape for the learnable memory bank, the best hyper-parameter λ, and the best aggregation method. For standardization, these ablation studies are conducted on the PointNet++ backbone and the ModelNet40 dataset.

Orthogonal Constraint and Hyper-parameter λ. We begin by assessing the efficacy of the orthogonal constraint, figuring out its potential to enhance performance. In Table 3, we present the experiment results. It's important to note that when λ is set to 0, the orthogonal constraint loss is omitted and we

Table 2. Point cloud classification accuracy of different methods on the real-world ScanObjectNN dataset.

Method	mAcc (%)	OA (%)
3DmFV [1]	58.1	63
PointNN [34]	–	64.9
PointNet [16]	63.4	68.2
SpiderCNN [31]	69.8	73.7
DGCNN [24]	73.6	78.1
PointCNN [7]	75.1	78.5
BGA-DGCNN [23]	75.7	79.7
BGA-PN++ [23]	77.5	80.2
DRNet [19]	78	80.3
GBNet [20]	77.8	80.5
SimpleView [3]	–	80.5
PRANet [2]	79.1	82.1
MVTN [5]	–	82.8
PointMLP [14]	83.9	85.4
PointNet++ [17]	75.4	77.9
PointPN [34]	–	87.1
Ours (PointNet++)	75.3	**78.1**
Ours (PointPN)	85.6	**87.4**

Table 3. Ablation study on different λ.

λ	0	0.1	1	100	10^4	10^6
Accuracy	90.24	90.7	**91.45**	90.84	90.32	90.07

only achieve an overall accuracy of 90.24%. Remarkably, as we increment λ, we observe a corresponding rise in overall accuracy, peaking at 91.4% with λ set to 1. Subsequent increases in λ lead to a decline in performance. This observation aligns with our intuition, suggesting that an excessively large λ might overly prioritize seeking orthogonal solutions at the expense of classification accuracy. We thus conclude that the inclusion of the orthogonal constraint stands as a pivotal design choice, capable of enhancing the memory bank's ability to discriminate between shape feature vectors.

Shape of the LMB. Given the significance of the Learnable Memory Bank in our approach, we set up a sequence of experiments to ascertain the most suitable shape for the memory bank. Especially, the memory bank M is a 2D matrix that has the shape of $N_m \times f_m$, we study both the effect of different N_m and f_m.

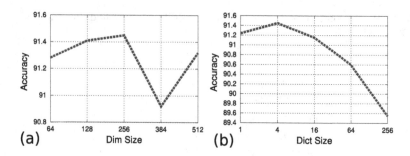

Fig. 3. Ablation study on (a) the dimension of each feature vector f_m in the memory bank, and (b) the number of vectors N_m in the memory bank.

Table 4. Effect of different aggregation methods.

Aggregation	Avg	Max	Avg + Max
Accuracy	91.4	91.1	90.7

Figure 3 displays the collective performance on the PointNet++ backbone across different shapes for the memory bank. From Fig. 3a, we see that when increasing the dimension f_m, the accuracy improves at the beginning and reaches the maximum at 256. Then, the accuracy decreases. Similar trends can be found in Fig. 3b. where the optimal performance is achieved when the N_m is set to 4. These observations serve as valuable directives in the determination of an apt shape for the memory bank.

Aggregation Methods for LMB. As stipulated in Eq. 3, the aggregation of cosine similarity scores is achieved through average pooling for each respective category. An alternative approach involves employing diverse pooling methods, such as max-pooling, to consolidate these similarity scores. This variety of pooling methodologies prompts an inquiry into their consequential effects. Consequently, we undertake an ablation study aimed at evaluating the performance under distinct pooling methods. The outcomes are presented in Table 4. Notably, with the utilization of average pooling, an overall accuracy of 91.4% is attainable on the ModelNet40 dataset.

Point-Memory Bank vs. Learnable Memory Bank. Given that the learnable memory bank is intentionally crafted to address the limitations of Point-Memory Banks, one might naturally inquire about the implications of choosing a Point-Memory Bank over a Learnable Memory Bank. To explore this, we conduct an experiment where we replace PointNN's Point-Memory Bank with the newly proposed Learnable Memory Bank and retrain the model on the ModelNet40 dataset. The comparative results are summarized in Table 5.

The results are straightforward. When utilizing the Point-Memory Bank, PointNN achieves an overall accuracy of merely 81.8%, whereas adopting the

Table 5. Performance and Memory Cost of different Memory Banks on ModelNet40 dataset with the PointNN backbone. The term "memory cost" specifically refers to the utilization of memory banks.

Aggregation	Accuracy	Memory Cost
Point-Memory Bank	81.8	59.06 MB
Learnable Memory Bank	**87.7**	0.98 MB

Learnable Memory Bank leads to a notable enhancement, pushing the accuracy to 87.7%. Furthermore, this performance boost comes with a substantial reduction in memory consumption. The memory cost diminishes from 59.06 MB to a mere 0.98 MB, a reduction of approximately 60 times compared to the Point-Memory Bank configuration.

Although the memory usage during training is comparable between LMB and Point-Memory Bank, LMB's testing stage requires drastically reduced memory due to its compact feature set representation, which makes the LMB approach more practical for real-world scenarios involving thousands of objects in the classification process.

5 Conclusion

In this paper, we investigated the design of a novel classification head for point cloud analysis. Especially, we presented the Learnable Memory Bank for learning a few representative feature vectors per object category, thus largely reducing memory costs of the original Point-Memory Bank during inference and allowing it to scale better to larger data sets. The proposed classification head can be integrated into a variety of backbones. Multiple experiments across two benchmarks show that the proposed method achieves a compatible performance with existing methods.

References

1. Ben-Shabat, Y., Lindenbaum, M., Fischer, A.: 3DmFV: three-dimensional point cloud classification in real-time using convolutional neural networks. IEEE Robot. Autom. Lett. 3(4), 3145–3152 (2018)
2. Cheng, S., Chen, X., He, X., Liu, Z., Bai, X.: PRA-Net: point relation-aware network for 3D point cloud analysis. IEEE Trans. Image Process. 30, 4436–4448 (2021)
3. Goyal, A., Law, H., Liu, B., Newell, A., Deng, J.: Revisiting point cloud shape classification with a simple and effective baseline. In: International Conference on Machine Learning, pp. 3809–3820 (2021)
4. Guo, M.H., Cai, J.X., Liu, Z.N., Mu, T.J., Martin, R.R., Hu, S.M.: PCT: point cloud transformer. Comput. Visual Media 7, 187–199 (2021)
5. Hamdi, A., Giancola, S., Ghanem, B.: MVTN: multi-view transformation network for 3D shape recognition. In: IEEE/CVF Conference on Computer Vision and Pattern Recognition (CVPR) (2021)

6. Li, R., Li, X., Heng, P.A., Fu, C.W.: PointAugment: an auto-augmentation framework for point cloud classification. In: IEEE/CVF Conference on Computer Vision and Pattern Recognition (CVPR) (2020)

7. Li, Y., Bu, R., Sun, M., Wu, W., Di, X., Chen, B.: PointCNN: convolution on X-transformed points. In: Advances in Neural Information Processing Systems (2018)

8. Li, Y., Pirk, S., Su, H., Qi, C.R., Guibas, L.J.: FPNN: field probing neural networks for 3D data. In: International Conference on Neural Information Processing Systems (2016)

9. Li, Y., et al.: Deep learning for lidar point clouds in autonomous driving: a review. IEEE Trans. Neural Networks Learn. Syst. 32(8), 3412–3432 (2021)

10. Lin, C.H., Chen, J.Y., Su, P.L., Chen, C.H.: Eigen-feature analysis of weighted covariance matrices for lidar point cloud classification. ISPRS J. Photogramm. Remote. Sens. 94, 70–79 (2014)

11. Liu, Y., Fan, B., Meng, G., Lu, J., Xiang, S., Pan, C.: DensePoint: learning densely contextual representation for efficient point cloud processing. In: IEEE International Conference on Computer Vision (ICCV) (2019)

12. Liu, Y., Fan, B., Xiang, S., Pan, C.: Relation-shape convolutional neural network for point cloud analysis. In: IEEE/CVF Conference on Computer Vision and Pattern Recognition (CVPR), pp. 8895–8904 (2019)

13. Lu, Q., Chen, C., Xie, W., Luo, Y.: PointngCNN: deep convolutional networks on 3D point clouds with neighborhood graph filters. Comput. Graph. 86, 42–51 (2020)

14. Ma, X., Qin, C., You, H., Ran, H., Fu, Y.: Rethinking network design and local geometry in point cloud: a simple residual MLP framework. In: International Conference on Learning Representations (ICLR) (2022)

15. Najafi, M., Taghavi Namin, S., Salzmann, M., Petersson, L.: Non-associative higher-order Markov networks for point cloud classification. In: Fleet, D., Pajdla, T., Schiele, B., Tuytelaars, T. (eds.) ECCV 2014. LNCS, vol. 8693, pp. 500–515. Springer, Cham (2014). https://doi.org/10.1007/978-3-319-10602-1_33

16. Qi, C.R., Su, H., Mo, K., Guibas, L.J.: PointNet: deep learning on point sets for 3D classification and segmentation. In: IEEE/CVF Conference on Computer Vision and Pattern Recognition (CVPR) (2017)

17. Qi, C.R., Yi, L., Su, H., Guibas, L.J.: PointNet++: deep hierarchical feature learning on point sets in a metric space. In: Advances in Neural Information Processing Systems (2017)

18. Qian, G., et al.: PointNext: revisiting PointNet++ with improved training and scaling strategies. In: Advances in Neural Information Processing Systems (2022)

19. Qiu, S., Anwar, S., Barnes, N.: Dense-resolution network for point cloud classification and segmentation. In: Proceedings of the IEEE/CVF Winter Conference on Applications of Computer Vision (2021)

20. Qiu, S., Anwar, S., Barnes, N.: Geometric back-projection network for point cloud classification. IEEE Trans. Multimedia 24, 1943–1955 (2022)

21. Rusu, R.B., Marton, Z.C., Blodow, N., Dolha, M., Beetz, M.: Towards 3D point cloud based object maps for household environments. Robot. Auton. Syst. 56(11), 927–941 (2008)

22. Shi, W., Rajkumar, R.: Point-GNN: graph neural network for 3D object detection in a point cloud. In: IEEE/CVF Conference on Computer Vision and Pattern Recognition (CVPR), pp. 1711–1719 (2020)

23. Uy, M.A., Pham, Q.H., Hua, B.S., Nguyen, T., Yeung, S.K.: Revisiting point cloud classification: a new benchmark dataset and classification model on real-world data.

In: IEEE/CVF Conference on Computer Vision and Pattern Recognition (CVPR) (2019)

24. Wang, Y., Sun, Y., Liu, Z., Sarma, S.E., Bronstein, M.M., Solomon, J.M.: Dynamic graph CNN for learning on point clouds. ACM Trans. Graph. (TOG) **38**, 1–12 (2019)

25. Wirth, F., Quehl, J., Ota, J., Stiller, C.: PointAtMe: efficient 3D point cloud labeling in virtual reality. In: IEEE Intelligent Vehicles Symposium (IV), pp. 1693–1698. IEEE (2019)

26. Wu, W., Qi, Z., Fuxin, L.: PointConv: deep convolutional networks on 3D point clouds. In: IEEE/CVF Conference on Computer Vision and Pattern Recognition (CVPR) (2019)

27. Wu, Z., et al.: 3D ShapeNets: a deep representation for volumetric shapes. In: IEEE/CVF Conference on Computer Vision and Pattern Recognition (CVPR) (2015)

28. Xiang, T., Zhang, C., Song, Y., Yu, J., Cai, W.: Walk in the Cloud: learning curves for point clouds shape analysis. In: IEEE International Conference on Computer Vision (ICCV) (2021)

29. Xu, M., Ding, R., Zhao, H., Qi, X.: PAConv: position adaptive convolution with dynamic kernel assembling on point clouds. In: IEEE/CVF Conference on Computer Vision and Pattern Recognition (CVPR) (2021)

30. Xu, M., Zhang, J., Zhou, Z., Xu, M., Qi, X., Qiao, Y.: Learning geometry-disentangled representation for complementary understanding of 3D object point cloud. In: AAAI Conference on Artificial Intelligence (2021)

31. Xu, Y., Fan, T., Xu, M., Zeng, L., Qiao, Yu.: SpiderCNN: deep learning on point sets with parameterized convolutional filters. In: Ferrari, V., Hebert, M., Sminchisescu, C., Weiss, Y. (eds.) ECCV 2018. LNCS, vol. 11212, pp. 90–105. Springer, Cham (2018). https://doi.org/10.1007/978-3-030-01237-3_6

32. Yan, X., Zheng, C., Li, Z., Wang, S., Cui, S.: PointASNL: robust point clouds processing using nonlocal neural networks with adaptive sampling. In: IEEE/CVF Conference on Computer Vision and Pattern Recognition (CVPR) (2020)

33. Zeng, Y., et al.: RT3D: real-time 3-D vehicle detection in LiDAR point cloud for autonomous driving. IEEE Robot. Autom. Lett. **3**(4), 3434–3440 (2018)

34. Zhang, R., Wang, L., Wang, Y., Gao, P., Li, H., Shi, J.: Parameter is not all you need: starting from non-parametric networks for 3D point cloud analysis. In: IEEE/CVF Conference on Computer Vision and Pattern Recognition (CVPR) (2023)

35. Zhang, Z., Zhang, L., Tan, Y., Zhang, L., Liu, F., Zhong, R.: Joint discriminative dictionary and classifier learning for ALS point cloud classification. IEEE Trans. Geosci. Remote Sens. **56**(1), 524–538 (2018)

36. Zhang, Z., et al.: A multilevel point-cluster-based discriminative feature for ALS point cloud classification. IEEE Trans. Geosci. Remote Sens. **54**(6), 3309–3321 (2016)

37. Zhao, H., Jiang, L., Jia, J., Torr, P.H., Koltun, V.: Point transformer. In: IEEE/CVF Conference on Computer Vision and Pattern Recognition (CVPR) (2021)

38. Zhou, Y., Tuzel, O.: VoxelNet: end-to-end learning for point cloud based 3D object detection. In: IEEE/CVF Conference on Computer Vision and Pattern Recognition (CVPR), pp. 4490–4499 (2018)

Adversarially Regularized Low-Light Image Enhancement

William Y. Wang[1], Lisa Liu[2], and Pingping Cai[3]($\boxtimes$) (iD)

[1] Dutch Fork High School, Irmo, SC 29063, USA
[2] Providence Day School, Charlotte, NC 28270, USA
[3] University of South Carolina, Columbia, SC 29208, USA
pcai@email.sc.edu

Abstract. The task of low-light image enhancement aims to generate clear images from their poorly visible counterparts taken under low-light conditions. While contemporary approaches leverage deep learning algorithms to enhance low-light images, the effectiveness of many of them heavily hinges on the availability of large amounts of normal images and their low-light counterparts to facilitate the training process. Regrettably, it is very challenging to acquire a sufficient number of such paired training images with good diversity in real-world settings. To address this issue, we present a novel approach that employs an adversarial attack process to maximize the utility of the available training data, thereby improving the network's performance while requiring far fewer images. Our key insight involves intentionally degrading the input image to create a deliberately worse version, effectively serving as an adversarial sample to the network. Moreover, we propose a novel low-light image enhancement network with specific multi-path convolution blocks, which preserve both global and localized features, resulting in better reconstruction quality. The experimental results validate that the proposed approach achieves promising low-light image enhancement quality by surpassing the performance of many previous state-of-the-art methods.

Keywords: Image Enhancement · Adversarial Attack · Low-light Images · Deep Learning

1 Introduction

Under limited ambient lighting, the images captured by camera sensors generally exhibit low intensity with fewer quantized levels, leading to poor visibility to the human eye. These low-light images seriously affect the performance of numerous downstream applications, such as object detection [2,22,23], classification [6,10,24], and video understanding [1,17,29]. This leads to the important task of low-light image enhancement, which aims to produce clear images from their poor visibility counterparts. However, low-light image enhancement is a highly challenging problem due to the lack of global object information in the darkness, as well as missing obscured details after compressing the image-intensity levels. As a result, the low-light image enhancement problem is actually ill-posed and

© The Author(s), under exclusive license to Springer Nature Switzerland AG 2024
S. Rudinac et al. (Eds.): MMM 2024, LNCS 14554, pp. 230–243, 2024.
https://doi.org/10.1007/978-3-031-53305-1_18

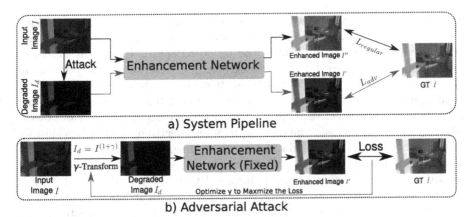

Fig. 1. An illustration of the proposed a) system pipeline and b) adversarial attack pipeline. An input low-light image is first transformed into a degraded one via adversarial attacking techniques. Then both the original input image and the degraded image are fed into the proposed network to generate the enhanced images, supervised by the adversarial loss and the regular reconstruction loss over the ground-truth image.

requires carefully designed regularization to infer the missing information. In general, it is expected that the enhanced images encompass both global semantic information and spatial details that look close to the real normal images.

In the past decades, many advanced methods have been developed for low-light image enhancement, ranging from traditional image processing approaches [5,7,9] to more recent deep learning-based approaches [15,18–20,31,36]. Generally, the prevailing trend involves leveraging deep learning and neural networks to enhance the color, tone, and contrast of low-light images, leading to the development of different network architectures for better reconstruction performance [15,31,36]. However, the success of these deep learning-based methods usually relies on the availability of abundant training images in the form of paired low-light images and their normal clear counterparts. The collection of such paired training images is not easy and can be very costly. An alternative way is to use synthetic data. This involves generating synthetic low-light images from real clear images and utilizing them as pair training data. Although synthetic data can provide a large quantity of training samples, models trained solely on synthetic data may struggle to generalize well to real-world low-light scenes because the synthetic data may not honestly reflect the spatial and semantic information loss present in real low-light images.

In this paper, we propose a novel approach that incorporates an adversarial attack process to make the most use out of a limited amount of available training images, real low-light and clear images in pairs, and enhance the robustness of the trained network. Our key insight is to transform the input low-light image into a deliberately worse version via gamma transformation, which serves as an attack to the network itself. The gamma transformation is a commonly employed and easily implemented image processing technique for adjusting image brightness. This process can be viewed as a specially designed regularization technique that generates more challenging degraded images. By feeding these worse images into the network and optimizing it, the proposed approach not only enhances the model's performance but also strengthens its resilience against adversarial attacks and various low-light degradations that may not be covered in the initial training data. At the same time, this means that each training image fed into the network is far more effective at training the network compared to previous methods, requiring fewer training images for the same effectiveness. Based on this, we further proposed a new low-light image enhancement network, with specifically designed multi-path convolution blocks, to preserve both global information and local details in enhancing low-light images. By combining adversarial attacks and multi-scale feature processing, the proposed low-light image enhancement network can be robustly optimized using fewer training images collected in the real world.

To verify the effectiveness of the proposed method, we train and evaluate it with multiple datasets: LoL-v1 [28], LOL-v2 real [33], and LOL-v2 synthetic [33]. The experiment results show that the proposed method achieves promising enhancement quality and outperforms many previous state-of-the-art methods. To sum up, the main contributions of this work are:

1. We propose a novel pipeline with an adversarial attack built upon gamma transformation to improve the performance for low-light image enhancement and that better utilizes the training images compared to previous state-of-the-art techniques.
2. We design a new low-light image enhancement network that incorporates specific multi-path convolution blocks and an adversarial regularizer.
3. We conduct extensive experiments on several datasets showing that the proposed method produces improved low-light image enhancement performance.

2   Related Work

2.1   Traditional Methods

Conventional image enhancement techniques, relying on histogram equalization and gamma correction, serve as fundamental tools for broadening the dynamic range and boosting image contrast. However, these approaches frequently yield unwanted artifacts in challenging real-world images. More advanced methods build upon the Retinex theory, such as MSRCR [21], RRM [16], and LIME [9]. The general pipeline of these methods is to estimate image illumination and construct an illumination map to adjust the exposure of each pixel. However, the estimated illumination map lacks local uniformity and tends to introduce color and brightness distortion. Ying et al. [35] effectively address this challenge by leveraging the response characteristics of cameras, where a camera response model is deployed to adjust the exposure of each pixel. Nonetheless, when dealing with real-world images, these traditional methods often lead to color distortions. As a result, more sophisticated deep-learning algorithms have been developed.

2.2   Deep Learning Methods

Numerous deep learning-based networks and methods for enhancing low-light images have emerged recently that surpass the capabilities of the previously mentioned techniques. For instance, RUAS [18] leverages the Retinex theory within the framework of deep learning. It deduces the inherent structure of low-light images and employs a search mechanism to identify the most effective enhancement program. This approach not only preserves image clarity but also reduces time and computational expenses. Given the necessity to retain both global and local spatial information in low-light image enhancement, several Multi-scale reconstruction techniques have been introduced. For example, UFormer [27] employs' a multi-scale restoration modulator to compute bias and weights. It focuses on window-based segments of the image rather than the entire image at once, resulting in improved computational efficiency and effectiveness. MIRNet [36], on the other hand, introduces a multi-scale residual block featuring parallel multi-resolution convolution streams. This architecture learns an enhanced array of features that integrate contextual details from various scales. This simultaneous preservation of high-resolution spatial nuances enhances the efficacy of low-light image enhancement. Xu et al. [31] notice the problem of dissimilarity in Signal-to-Noise Ratio (SNR) between extremely low and moderately low-light regions. To address this, they propose an approach that segregates the low-light image into sections with high and low SNR. This novel solution integrates SNR-aware transformers and convolutional models, allowing adaptive pixel enhancement through spatially variant operations. Therefore it can better enhance low-light images with varying noise and signal characteristics.

However, as mentioned earlier, these works focus on network architecture design for performance improvement and require large amounts of training images. In contrast, we propose an innovative pipeline that employs an adversarial attack process to maximize the utility of limited training images.

3   Proposed Method

To maximize the utility of training images, the proposed method involves transforming the input image into a deliberately degraded version, serving as an adversarial regularization for the network. To achieve this goal, we employ the gamma transformation T_γ as our attack model - a non-linear operation commonly used for adjusting luminance values in images. Then, the process of gradient-based attacking is applied to find a perturbation of gamma value in the predefined range that increases the model's loss, causing the network to generate inferior results. Subsequently, we apply this gamma transformation to the input image, calculate the loss for the degraded image as a regularizer, and combine it with the original image reconstruction loss to optimize the network. Moreover, since the local and global details are important to reconstruct normal clear images, we draw inspiration from previous research and incorporate multi-scale feature processing blocks into our network to capture both high and low-level features.

3.1   Adversarial Attack

Given an input low-light image I, the objective of an adversarial attack is to identify the most detrimental perturbation of γ, which transforms the input image to a degraded one as $I_d = I^{(1+\gamma)}$ for more robust network training. Considering the possible level of image degradation in practice, we consider the selection of γ in a reasonable range, e.g., [−0.1, 0.1], as used in our experiments. Figure 1b shows the proposed adversarial attacking process for finding the optimal γ in this range.

Specifically, we employ gradient-based attacks [13], which involve three key steps. In the first step, we freeze the image-enhancement network and forward propagate the input image to get the reconstruction loss with respect to the ground truth. Then, we backpropagate the loss and calculate the gradient of γ, which indicates how the loss changes concerning variations in γ. Finally, we update the value of γ in the **gradient ascent** direction, which means we incrementally adjust γ to increase the loss value and thereby push the network towards generating worse outcomes. By iteratively running these steps, we can identify the specific γ value that causes the most significant degradation in performance for the given input image I.

Adversarial Regularizer: After obtaining the degraded image I_d, we feed this degraded image into the enhancement network and obtain the output image I'. Ideally, the enhancement network should be robust to the degradation and the output image I' should be close to the ground truth. To achieve this, we calculate the Charbonnier loss [14] and the perceptual loss as adversarial regularization, which is defined as:

$$L_{adv} = \sqrt{||I' - \hat{I}||_2 + \epsilon^2} + \alpha||\Phi(I') - \Phi(\hat{I})||_1 \qquad (1)$$

where $\hat{I}$ is the ground-truth clear image, ϵ is the Charbonnier penalty to determine how closely the Charbonnier loss resembles the L1 loss, which is set to 10^{-3} in all experiments, $\Phi()$ is the operation of extracting features from the VGG network [24], α is the hyper-parameter to balance the weights between Charbonnier and perceptual losses.

The adversarial loss quantifies the discrepancy between the predicted output of the network for the degraded image and the ground-truth normal image. By evaluating the adversarial loss, we can assess the impact of the worst-case perturbation on the network's performance and use it as part of the regularization to optimize the network for improved robustness against various perturbations in real-world scenarios.

3.2   Multi-Path Convolution Block

In addition to the adversarial attack, the design of the low-light enhancement network is another important component. Our network design draws inspiration from previous methods that effectively incorporate both high and low-level information [36]. Since enhancing low-light images involves a position-sensitive process demanding pixel-to-pixel alignment between input and output images, the selective removal of undesired degraded content while preserving crucial spatial information, such as edges and textures, becomes imperative.

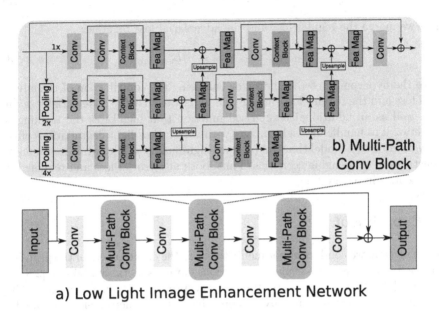

Fig. 2. The overall network architecture of a) the low-light image enhancement network, and b) the multi-path convolution block.

Toward this goal, we develop a straightforward yet efficient multi-path convolution block, as shown in Fig. 2b. It first uses multi-scale pooling to downsample the input feature map into different resolutions. Then, convolutions with residual connections and Context Block [36] are stacked together to encode multiscale context feature maps. Additionally, our approach progressively shares and merges feature maps across coarse-to-fine resolution levels while preserving the original high-resolution features within the network's hierarchy. With the primary high-resolution branch, these multi-resolution parallel branches complementarily enrich the feature representations for better reconstruction quality. As shown in Fig. 2a, our network is composed of three sets of multi-path convolution blocks. The integration of these blocks augments the model's capacity to handle multi-scale features and significantly enhances its performance in processing low-light images, even in the presence of adversarial perturbations.

3.3   Loss Function:

To train the network and ensure its convergence, we first build the regular reconstruction loss function to measure the similarity between the enhanced images

I'' generated from the original low-light inputs I and the ground-truth images. Especially, the regular loss also consists of Charbonnier loss and perceptual loss:

$$L_{regular} = \sqrt{||I'' - \hat{I}||_2 + \epsilon^2} + \alpha||\Phi(I'') - \Phi(\hat{I})||_1 \qquad (2)$$

Then we merge the regular loss and the adversarial loss together:

$$L_{all} = \beta L_{adv} + (1 - \beta)L_{regular} \qquad (3)$$

where the hyperparameter β allows for balancing the contribution of the adversarial loss and the regular loss during training. Note that the adversarial loss can be viewed as an additional regularized function that facilitates the training of the network to handle low-light images with various low-light degradations effectively, leading to a model that can produce high-quality results while maintaining robustness in challenging scenarios. Thus, both adversarial loss and regular loss are essential in our system.

4 Experiment

In this section, we conduct experiments to verify the effectiveness of the proposed approach. Specifically, we evaluate on various datasets, notably LOL-V1 [28] and LOL-V2 [33], to show the robustness of the proposed method across varying contexts. We also systematically explore different parameter selections for the adversarial regularizer.

4.1 Datasets and Implementation Details

We utilize two widely-used datasets focusing on low-light image enhancement tasks: LOLv1 [28] and LOLv2 [33]. Specifically, the **LOLv1** dataset comprises 485 pairs of images captured under low and normal lighting conditions, designated for training, along with an additional 15 pairs intended for testing. Each pair within this dataset consists of a low-light input image and an associated well-exposed reference image. Expanding further, the **LOL-v2** dataset is categorized into two subsets: **LOLv2-real** and **LOLv2-synthetic**. The LOLv2-synthetic subset is a widely used benchmark that contains 900 training image pairs and 100 testing pairs. The low-light images are generated synthetically via Adobe Lightroom. Noticing the potential disparities between synthetic and real data, the LOL-v2-real subset is gathered by manipulating exposure time and ISO settings while keeping other camera parameters unchanged. It contains a total of 689 image pairs for training and 100 pairs for testing.

Our algorithm is implemented using the PyTorch framework. We train and test the proposed algorithm on a server equipped with two Nvidia Tesla V100 GPUs. We train our model from scratch, leveraging Adam as an optimizer with a learning rate of 0.01, accompanied by a momentum value set at 0.9. We also employ the MultiStepLR technique as a scheduler and gradually decrease the learning rate by half at different steps. For all datasets, the hyper-parameter α and β are set to 0.1 and 0.5, respectively. The range of γ is set to $[-0.1, 0.1]$ for the proposed adversarial attack by considering the possible low-light degradations in practice.

4.2 Comparison with Previous Methods

Here we undertake a comprehensive performance comparison between the proposed method and several previous state-of-the-art (SOTA) methods designed for enhancing low-light images. To ensure a comprehensive evaluation, we employ both the **Peak Signal-to-Noise Ratio** (PSNR) and **Structural Similarity Index** (SSIM) as metrics. Notably, a higher SSIM score indicates the presence of enhanced high-frequency details and well-defined structures in the output.

Table 1. Experimental results on LOL-v1 dataset. The bold number indicates the best result, and the underlined number indicates the second-best result.

Methods	Dong [5]	LIME [9]	MF [7]	SRIE [8]	BIMEF [34]	DRD [28]
PSNR	16.72	16.76	18.79	11.86	13.86	16.77
SSIM	0.58	0.56	0.64	0.5	0.58	0.56
Methods	RRM [16]	SID [3]	DeepUPE [25]	KIND [38]	DeepLPF [20]	FIDE [30]
PSNR	13.88	14.35	14.38	20.87	15.28	18.27
SSIM	0.66	0.436	0.446	0.8	0.473	0.665
Methods	LPNet [15]	MIR-Net [36]	RF [12]	3DLUT [37]	A3DLUT [26]	Band [32]
PSNR	21.46	<u>24.14</u>	15.23	14.35	14.77	20.13
SSIM	0.802	0.83	0.452	0.445	0.458	0.83
Methods	EG [11]	RUAS [18]	Sparse [33]	IPT [4]	SNR-Aware [31]	Ours
PSNR	17.48	18.23	17.2	16.27	**24.61**	23.01
SSIM	0.65	0.72	0.64	0.504	**0.842**	<u>0.840</u>

We first conduct the experiment on the LOL-v1 dataset. Table 1 shows the experiment results against the comparison methods. Our method achieves an average PSNR of 23.01 and an average SSIM of 0.840, surpassing most previous methods such as RUAS, which achieved a PSNR of 18.23 and an SSIM of 0.72, as well as MIR-Net, which achieved an SSIM of 0.830.

However, LOL-v1 contains only a limited number of testing samples, (*i.e.*, 15 test samples) and the result on this dataset may not reveal the actual performance of a method adequately; we further conduct experiments on the LOL-v2 dataset to explore the performance of the proposed method. Tables 2 and 3 show the results on the LOL-v2-real and LOL-v2-synthetic datasets, respectively. Our method consistently demonstrates competitive performance when compared with other existing methods. Remarkably, in the context of the LOL-v2 real subset, our method stands out by achieving an average PSNR of 21.95 and an average

SSIM of 0.847. This performance surpasses that of prior state-of-the-art algorithms such as MIR-Net, which achieved a PSNR of 20.02 and an SSIM of 0.82, as well as SNR-Net, which achieved a PSNR of 21.48. Similarly, in the context of the LOL-v2 synthetic subset, our method still outperforms previous methods by achieving a PSNR of 25.36 and an SSIM of 0.931.

Table 2. Experimental results on the LOL-v2 real dataset. Note the results of the comparison methods are sourced either directly from the respective research papers or obtained by running the respective publicly available code.

Methods	Dong [5]	LIME [9]	MF [7]	SRIE [8]	BIMEF [34]	DRD [28]
PSNR	17.26	15.24	18.73	17.34	17.85	15.47
SSIM	0.527	0.47	0.559	0.686	0.653	0.567
Methods	RRM [16]	SID [3]	DeepUPE [25]	KIND [38]	DeepLPF [20]	FIDE [30]
PSNR	17.34	13.24	13.27	14.74	14.1	16.85
SSIM	0.686	0.442	0.452	0.641	0.48	0.678
Methods	LPNet [15]	MIR-Net [36]	RF [12]	3DLUT [37]	A3DLUT [26]	Band [32]
PSNR	17.8	20.02	14.05	17.59	18.19	20.29
SSIM	0.792	0.82	0.458	0.721	0.745	0.831
Methods	EG [11]	RUAS [18]	Sparse [33]	IPT [4]	SNR-Aware [31]	Ours
PSNR	18.23	18.37	20.06	19.8	<u>21.48</u>	**21.95**
SSIM	0.617	0.723	0.816	0.813	**0.849**	<u>0.847</u>

Table 3. Low-Light image enhancement results on the LOL-v2 synthetic dataset.

Methods	Dong [5]	LIME [9]	MF [7]	SRIE [8]	BIMEF [34]	DRD [28]
PSNR	16.9	16.88	17.5	14.5	17.2	17.13
SSIM	0.749	0.776	0.751	0.616	0.713	0.798
Methods	RRM [16]	SID [3]	DeepUPE [25]	KIND [38]	DeepLPF [20]	FIDE [30]
PSNR	17.15	15.04	15.08	13.29	16.02	15.2
SSIM	0.727	0.61	0.623	0.578	0.587	0.612
Methods	LPNet [15]	MIR-Net [36]	RF [12]	3DLUT [37]	A3DLUT [26]	Band [32]
PSNR	19.51	21.94	15.97	18.04	18.92	23.22
SSIM	0.846	0.876	0.632	0.8	0.838	0.927
Methods	EG [11]	RUAS [18]	Sparse [33]	IPT [4]	SNR-Aware [31]	Ours
PSNR	16.57	16.55	22.05	18.3	<u>24.14</u>	**25.36**
SSIM	0.734	0.652	0.905	0.811	<u>0.928</u>	**0.931**

We also provide illustrative visual samples in Fig. 3 and Fig. 4 to facilitate an intuitive comparison between our method and the comparison methods on the LOL-v2 dataset. Notably, in Fig. 3, we see that our results exhibit a notable enhancement in visual quality, consistent color rendition, and improved brightness while previous methods, such as RUAS and MIRNet, generate images with poor color consistency and low brightness. It is worth highlighting that our method maintains a heightened level of realism in the images it generates. Moreover, in Fig. 4, even in regions replete with simple textures, our outputs still demonstrate a remarkable reduction in the presence of visual artifacts and resemble the ground truths.

4.3   Ablation Study

To understand the effectiveness of each component used in our method, we consider multiple ablation settings by exploring different network components and hyperparameter settings for our framework individually. All experiments are conducted on the LOL-v2 synthetic dataset.

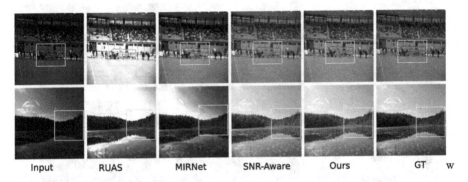

Input   RUAS   MIRNet   SNR-Aware   Ours   GT   w

Fig. 3. Samples of the enhanced images on the LOL-V2 synthetic dataset. Please zoom in for details.

Table 4. Ablation study on the adversarial attack process on the LOL-V2 synthetic dataset.

	Without Attack	Adversarial Attack	Random Sampling
PSNR	24.14	**25.36**	24.20
SSIM	0.912	**0.931**	0.921

Effectiveness of Adversarial Attack. Since the adversarial attack is the major contribution of this paper, we initiate our analysis by assessing the efficacy of the proposed adversarial regularization. In particular, we remove the adversarial loss to examine the performance change. Then, we also include a comparison between different attack methods, – the adopted gradient-based attacks and an alternative approach involving random sampling. In pursuit of a fair comparison, the random sampling randomly selects a gamma value within the range of –0.1 to 0.1 to degrade the image.

The result of this comparative analysis is shown in Table 4. It is evident that the absence of regularization leads to a noticeable performance drop, where

Fig. 4. Samples of enhanced images on the LOL-v2 real dataset.

PSNR decreases to 24.14. This verifies the positive impact of the proposed adversarial regularization. We also note that employing random sampling, which can be regarded as an alternative data augmentation technique, results in an improved PSNR of 24.20 over the baseline without an adversarial attack. However, the most impressive enhancement performance is achieved when using the proposed adversarial attack. It achieves a PSNR of 25.36, which significantly outperforms random sampling, further proving the effectiveness of the proposed method (Table 4).

One more possible concern regarding the proposed adversarial attack process is that the attacked γ value may adhere to one of the predetermined boundary values, such as –0.1 or 0.1, in order to degrade the images. If this happens, the proposed attack γ optimization process is not needed. To address this concern, we illustrate the distribution of γ after the adversarial attack over the test images of the LOL-V2 synthetic dataset. In Fig. 5, we can see that the optimal γ identified for different input training images are varied in the given range of $[-0.1, 0.1]$, with the goal of contributing most to the robust network training.

Hyperparameter Settings. Recall that we introduce two hyperparameters α and β in the loss function, where α aims to balance the weights between Charbonnier and perceptual losses and β aims to balance the regular reconstruction loss and adversarial loss. We investigate the different values of these hyperparameters. The results of this experiment are presented in Table 5.

In particular, we first fix the parameter β and explore different α values and find that the best performance is achieved when $\alpha = 0.1$. Note that $\alpha = 0$ means removing the perceptual loss. Then, we also fix the parameter α and try different β values and find that the best performance is achieved when $\beta = 0.5$, which fits our intuition that both the reconstruction of the attacked and original image are essential.

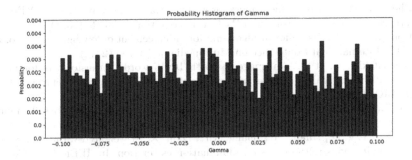

Fig. 5. Distribution of γ after attack on the LOL-V2 synthetic dataset.

Table 5. Performance with different hyperparameter settings on the LOL-V2 synthetic dataset.

α	0	0.01	0.1	1	10	100
PSNR	23.79	24.46	**25.36**	24.32	24	24.08
SSIM	0.905	0.919	**0.931**	0.913	0.905	0.906
β	0	0.1	0.3	0.5	0.7	0.9
PSNR	24.14	24.61	24.5	**25.36**	24.66	24.39
SSIM	0.912	0.922	0.920	**0.931**	0.925	0.918

5 Conclusion

In this paper, we introduced a new method of adversarial regularized low-light image enhancement. By intentionally transforming the input image to a further degraded one and treating it as an adversarial sample, we can better utilize the limited number of training samples and improve the performance and robustness of the image enhancement network. To further ensure the preservation of both global and local details for low-light images, we designed a new low-light image enhancement network with specifically designed multi-path convolution blocks. Experimental results verified that the proposed method achieves promising low-light image enhancement performance that surpasses the performance of many existing state-of-the-art counterparts.

References

1. Bertasius, G., Wang, H., Torresani, L.: Is space-time attention all you need for video understanding? In: International Conference on Machine Learning (ICML) (2021)
2. Carion, N., Massa, F., Synnaeve, G., Usunier, N., Kirillov, A., Zagoruyko, S.: End-to-end object detection with transformers. In: European Conference on Computer Vision (ECCV) (2020)
3. Chen, C., Chen, Q., Xu, J., Koltun, V.: Learning to see in the dark. In: IEEE/CVF Conference on Computer Vision and Pattern Recognition (CVPR), pp. 3291–3300 (2018)
4. Chen, H., et al.: Pre-trained image processing transformer. In: IEEE/CVF Conference on Computer Vision and Pattern Recognition (CVPR) (2021)
5. Dong, X., et al.: Fast efficient algorithm for enhancement of low lighting video. In: IEEE International Conference on Multimedia and Expo (2011)
6. Dosovitskiy, A., et al.: An image is worth 16x16 words: Transformers for image recognition at scale. In: International Conference on Learning Representations (ICLR) (2021)
7. Fu, X., Zeng, D., Huang, Y., Liao, Y., Ding, X., Paisley, J.: A fusion-based enhancing method for weakly illuminated images. Sig. Process. **129**, 82–96 (2016)
8. Fu, X., Zeng, D., Huang, Y., Zhang, X.P., Ding, X.: A weighted variational model for simultaneous reflectance and illumination estimation. In: IEEE/CVF Conference on Computer Vision and Pattern Recognition (CVPR) (2016)
9. Guo, X., Li, Y., Ling, H.: Lime: Low-light image enhancement via illumination map estimation. IEEE Trans. Image Process. **26**(2), 982–993 (2017)
10. He, K., Zhang, X., Ren, S., Sun, J.: Deep residual learning for image recognition. In: IEEE/CVF Conference on Computer Vision and Pattern Recognition (CVPR), June 2016
11. Jiang, Y., et al.: EnlightenGAN: deep light enhancement without paired supervision. IEEE Trans. Image Process. **30**, 2340–2349 (2021)
12. Kosugi, S., Yamasaki, T.: Unpaired image enhancement featuring reinforcement-learning-controlled image editing software. In: AAAI Conference on Artificial Intelligence (2020)
13. Kurakin, A., Goodfellow, I.J., Bengio, S.: Adversarial machine learning at scale. In: International Conference on Learning Representations (ICLR) (2017)
14. Lai, W.S., Huang, J.B., Ahuja, N., Yang, M.H.: Fast and accurate image super-resolution with deep laplacian pyramid networks. IEEE Trans. Pattern Anal. Mach. Intell. **41**(11), 2599–2613 (2018)
15. Li, J., Li, J., Fang, F., Li, F., Zhang, G.: Luminance-aware pyramid network for low-light image enhancement. IEEE Trans. Multimedia **23**, 3153–3165 (2021)
16. Li, M., Liu, J., Yang, W., Sun, X., Guo, Z.: Structure-revealing low-light image enhancement via robust retinex model. IEEE Trans. Image Process. **27**(6), 2828–2841 (2018)
17. Lin, J., Gan, C., Han, S.: Tsm: Temporal shift module for efficient video understanding. In: IEEE International Conference on Computer Vision (ICCV) (2019)
18. Liu, R., Ma, L., Zhang, J., Fan, X., Luo, Z.: Retinex-inspired unrolling with cooperative prior architecture search for low-light image enhancement. In: IEEE/CVF Conference on Computer Vision and Pattern Recognition (CVPR) (2021)
19. Ma, L., Ma, T., Liu, R., Fan, X., Luo, Z.: Toward fast, flexible, and robust low-light image enhancement. In: IEEE/CVF Conference on Computer Vision and Pattern Recognition (CVPR) (2022)

20. Moran, S., Marza, P., McDonagh, S., Parisot, S., Slabaugh, G.: DeepLPF: deep local parametric filters for image enhancement. In: IEEE/CVF Conference on Computer Vision and Pattern Recognition (CVPR), June 2020
21. Rahman, Z.u., Jobson, D.J., Woodell, G.A.: Retinex processing for automatic image enhancement. J. Electron. Imaging **13**(1), 100–110 (2004)
22. Redmon, J., Divvala, S., Girshick, R., Farhadi, A.: You only look once: unified, real-time object detection. In: IEEE/CVF Conference on Computer Vision and Pattern Recognition (CVPR) (2016)
23. Ren, S., He, K., Girshick, R., Sun, J.: Faster r-CNN: towards real-time object detection with region proposal networks. IEEE Trans. Pattern Anal. Mach. Intell. **39**(6), 1137–1149 (2017)
24. Simonyan, K., Zisserman, A.: Very deep convolutional networks for large-scale image recognition. arXiv preprint arXiv:1409.1556 (2014)
25. Wang, R., Zhang, Q., Fu, C.W., Shen, X., Zheng, W.S., Jia, J.: Underexposed photo enhancement using deep illumination estimation. In: IEEE/CVF Conference on Computer Vision and Pattern Recognition (CVPR), pp. 6849–6857 (2019)
26. Wang, T., Li, Y., Peng, J., Ma, Y., Wang, X., Song, F., Yan, Y.: Real-time image enhancer via learnable spatial-aware 3d lookup tables. In: IEEE/CVF Conference on Computer Vision and Pattern Recognition (CVPR) (2021)
27. Wang, Z., Cun, X., Bao, J., Zhou, W., Liu, J., Li, H.: Uformer: A general u-shaped transformer for image restoration. In: IEEE/CVF Conference on Computer Vision and Pattern Recognition (CVPR), June 2022
28. Wei, C., Wang, W., Yang, W., Liu, J.: Deep retinex decomposition for low-light enhancement. In: British Machine Vision Conference (2018)
29. Wu, C.Y., Feichtenhofer, C., Fan, H., He, K., Krahenbuhl, P., Girshick, R.: Long-term feature banks for detailed video understanding. In: IEEE/CVF Conference on Computer Vision and Pattern Recognition (CVPR) (2019)
30. Xu, K., Yang, X., Yin, B., Lau, R.W.: Learning to restore low-light images via decomposition-and-enhancement. In: IEEE/CVF Conference on Computer Vision and Pattern Recognition (CVPR), pp. 2278–2287 (2020)
31. Xu, X., Wang, R., Fu, C.W., Jia, J.: SNR-aware low-light image enhancement. In: IEEE/CVF Conference on Computer Vision and Pattern Recognition (CVPR) (2022)
32. Yang, W., Wang, S., Fang, Y., Wang, Y., Liu, J.: Band representation-based semi-supervised low-light image enhancement: bridging the gap between signal fidelity and perceptual quality. IEEE Trans. Image Process. **30**, 3461–3473 (2021)
33. Yang, W., Wang, W., Huang, H., Wang, S., Liu, J.: Sparse gradient regularized deep retinex network for robust low-light image enhancement. IEEE Trans. Image Process. **30**, 2072–2086 (2021)
34. Ying, Z., Li, G., Gao, W.: A bio-inspired multi-exposure fusion framework for low-light image enhancement. arXiv preprint arXiv:1711.00591 (2017)
35. Ying, Z., Li, G., Ren, Y., Wang, R., Wang, W.: A new low-light image enhancement algorithm using camera response model. In: Proceedings of the IEEE International Conference on Computer Vision Workshops (2017)
36. Zamir, S.W., Arora, A., Khan, S.H., Munawar, H., Khan, F.S., Yang, M.H., Shao, L.: Learning enriched features for fast image restoration and enhancement. IEEE Trans. Pattern Anal. Mach. Intell. **45**(2), 1934–1948 (2022)
37. Zeng, H., Cai, J., Li, L., Cao, Z., Zhang, L.: Learning image-adaptive 3d lookup tables for high performance photo enhancement in real-time. IEEE Trans. Pattern Anal. Mach. Intell. **44**(4), 2058–2073 (2020)
38. Zhang, Y., Zhang, J., Guo, X.: Kindling the darkness: a practical low-light image enhancer. In: ACM International Conference on Multimedia, pp. 1632–1640 (2019)

Advancing Incremental Few-Shot Semantic Segmentation via Semantic-Guided Relation Alignment and Adaptation

Yuan Zhou[1], Xin Chen[2], Yanrong Guo[1], Jun Yu[3], Richang Hong[1], and Qi Tian[2](✉)

[1] Hefei University of Technology, Hefei, China
2018110971@mail.hfut.edu.cn, yrguo@hfut.edu.cn
[2] Huawei Inc, Shenzhen, China
tian.qi1@huawei.com
[3] Hangzhou Dianzi University, Hangzhou, China
yujun@hdu.edu.cn

Abstract. Incremental few-shot semantic segmentation aims to extend a semantic segmentation model to novel classes according to only a few labeled data, while preserving its segmentation capability on learned base classes. However, semantic aliasing between base and novel classes severely limits the quality of segmentation results. To alleviate this issue, we propose a *semantic-guided relation alignment and adaptation* method. Specifically, we first conduct *semantic relation alignment* in the base step, so as to align base class representations to their semantic information. Thus, base class embeddings are constrained to have relatively low semantic correlations to classes that are different from them. Afterwards, based on semantically aligned base classes, we further conduct *semantic-guided adaptation* during incremental learning, which aims to ensure affinities between visual and semantic embeddings of encountered novel classes, thereby making feature representations be consistent with their semantic information. In this way, the semantic-aliasing issue can be suppressed. We evaluate our model on PASCAL VOC 2012 and COCO datasets. The experimental results demonstrate the effectiveness of the proposed method.

Keywords: Incremental few-shot semantic segmentation · Semantic alignment · Semantic-guided adaptation

1 Introduction

Semantic segmentation has achieved impressive performance by using deep neural networks. However, conventional semantic segmentation models generally have a fixed output space. Therefore, when encountering new classes, they need to be re-trained from scratch to update their segmentation capability. Moreover, these models require large-scale pixel-level labeled data, which are expensive

© The Author(s), under exclusive license to Springer Nature Switzerland AG 2024
S. Rudinac et al. (Eds.): MMM 2024, LNCS 14554, pp. 244–257, 2024.
https://doi.org/10.1007/978-3-031-53305-1_19

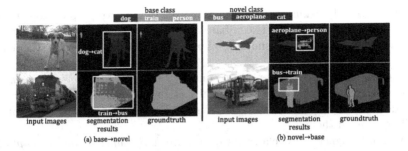

Fig. 1. Typical examples for semantic-aliasing issues in IFSS, which are obtained from the PASCAL VOC 2012 dataset under the 1-shot setting. In the figure, "A→B" indicates that regions belonging to A are incorrectly segmented as B.

and laborious to obtain. These issues limit their applicability in open-ended real-world scenarios. In this context, Cermelli et al. [4] proposed a *Incremental Few-shot Semantic Segmentation (IFSS)* task, which aims to effectively extend a semantic segmentation model to new classes using a few labeled data, while maintaining its segmentation capability on learned old ones. In this way, the extendibility and flexibility of the model can be improved, which is critical for many real-world applications, such as autonomous driving and human-machine interaction.

However, semantic-aliasing issues between base and novel classes severely limit the performance of IFSS models. As exemplified in Fig. 1, the semantic confusion between the base class "dog" and the encountered novel class "cat" misleads IFSS models to draw incorrect segmentation results. Recently, semantic information has been successfully introduced in the few-shot classification task [10,26–28]. Inspired by these works, we propose a *Semantic-guided Relation Alignment and Adaptation (SRAA)* approach, which alleviates semantic-aliasing issues in IFSS by fully considering the guidance of semantic information. Specifically, as shown in Fig. 2, on one hand, we conduct *Semantic Relation Alignment (SRA)* in the base step, so as to semantically align base classes in representation space. Thereby, base class embeddings are constrained to have relatively low semantic correlations to classes that are different from them. Meanwhile, cross entropy is utilized to measure discrepancy between segmentation results and groundtruth labels. As a result, the model is trained to segment base classes, while being aware of their semantic information. Based on aligned base classes, we conduct *Semantic-Guided Adaptation (SGA)* to adapt the model to novel classes, which aims to ensure affinities between visual and semantic embeddings of novel classes, thereby making feature representations be consistent with their semantic information. By considering the semantic information of both base and novel classes, semantic-aliasing issues can be alleviated. We evaluate our method on PASCAL VOC 2012 and COCO datasets, which validates the effectiveness of the proposed method. The contributions of this paper can be summarized below:

- Aiming to make segmentation results more accurate, we propose to suppress semantic-aliasing issues between base and novel classes in IFSS by fully considering the guidance of semantic information, which is generally ignored by previous methods.
- To realize this goal, we propose to conduct *Semantic Relation Alignment* for base step learning, so as to semantically align base class representations and guide them to have relatively low semantic correlations to classes that are different from them.
- Based on aligned base classes, we propose a *Semantic-Guided Adaptation* strategy for incremental learning stages, which can guide the embeddings of novel classes to be consistent with their semantic information. In this way, the semantic-aliasing issue between base and novel classes can be alleviated.

2   Related Work

Few-shot Learning. Few-shot learning aims to transfer models to unseen classes according to only a few training data, so as to reduce expenses cost on data collection and annotation. Currently, few-shot learning methods are mainly based on metric learning, which aims at learning an effective metric classifier from low-shot support samples. For example, Santoro et al. [20] proposed memory-augmented neural networks that leverage memory to make metric classification more accurate. Li et al. [10] and Zhang et al. [28] proposed to additionally consider semantic attributes encoded by GloVe [17] or word2vec [15], thereby making visual embeddings more representative. Besides, Xu et al. [26] and Yang et al. [27] proposed to further consider the semantic guidance from CLIP [18], as they found that semantic information from CLIP is more effective in learning representative visual embeddings.

Incremental Learning. Incremental learning aims to transfer models to new classes while effectively maintaining learned old knowledge. Knowledge distillation has shown advantages in overcoming catastrophic forgetting problems [11,19]. To achieve a better balance between knowledge maintenance and class adaptation, a cross-distillation and a balanced finetuneing strategy were utilized in [2], while [9] proposed an adaptive feature consolidation policy to restrict the representation drift of old class embeddings. Furthermore, Wang et al. [24] advanced incremental learning models by using a gradient-boosting policy, which guides models to effectively learn their residuals to target ones. Liu et al. [13] proposed to enhance data replay by designing a reinforced memory management mechanism, thereby dynamically adjusting stored memory information in each incremental step and helping to overcome a sub-optimal memory allocation problem.

Incremental Few-shot Semantic Segmentation. Incremental Few-shot Semantic Segmentation (IFSS), which was proposed by Cermelli et al. [4], aims at enduing semantic segmentation models with the capability of few-shot incremental learning, thereby making them more suitable to be deployed in open-ended real-world scenarios. In order to tackle this task, Cermelli et al. [4] proposed a

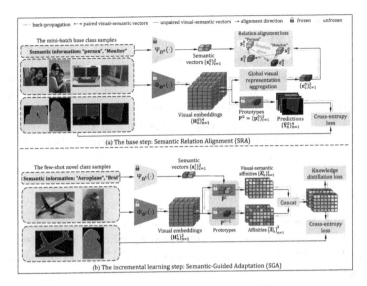

Fig. 2. An overview of our proposed method. In the above figure, "$\Psi_{\Omega^t}(\cdot)$" and "$\Phi_{\Theta^t}(\cdot)$" represent a semantic encoder and a visual encoder respectively.

prototype-based knowledge distillation, which alleviates the catastrophic forgetting problem by constraining the invariance of old class predictions. Shi et al. [21] proposed to build hyper-class representations, so as to suppress representation drift during incremental learning stages as much as possible. Also, Shi et al. adopted a different evaluation protocol than the one employed in [4]. However, despite the progress achieved by these methods, the guidance of semantic information is ignored in them, which has been proven to play an important role in low-data tasks [26,27]. Different from these approaches, in this paper, we study on how to exploit prior semantic information to furtehr boost the performance of IFSS models.

3   Methodology

As described in Fig. 2, our model consists of two phases, i.e., Semantic Relation Alignment (SRA) and Semantic-Guided Adaptation (SGA), which concentrate on guiding models to be aware of semantic information of both base and novel classes. In the following subsections, we first introduce the preliminaries about the proposed method. Then, we give the details about these two different phases.

3.1   Preliminaries

The semantic space of IFSS models is generally expanded over time. We define C^t as the classes appearing in the step t. Thereby, after learning in this step, the semantic space of IFSS models is expanded to $S^t = S^{t-1} \bigcup C^t$, where $S^{t-1} = $

$\bigcup_{i=0}^{t-1} \mathbf{C}^i$ denotes the semantic space learned at the step $t-1$. Moreover, in each step, a dataset $\mathbf{D}^t = \{\mathbf{X}_n^t, \mathbf{Y}_n^t\}_{n=1}^{N_t}$ is provided to update learnable parameters. $\mathbf{X}_n^t$ represents the n-th support image and $\mathbf{Y}_n^t$ denotes the label map of $\mathbf{X}_n^t$. Specially, in the base step ($t = 0$), a base dataset $\mathbf{D}^0$ is provided to initialize models, which contains relatively more labeled samples. After the base step, $\mathbf{D}^t$ is only in the few-shot setting, which satisfies the condition $N_t << N_0$ for $\forall t >= 1$. In this paper, we term classes appearing in the base step as "base classes", while classes encountered in incremental learning stages are termed "novel classes".

3.2 Semantic Relation Alignment

During the SRA phase, we aim to align base classes and guide models to generate semantic-consistent visual representations. Specifically, we first extract the visual embeddings $\{\mathbf{H}_n^0\}_{n=1}^{N_b}$ of the input images $\{\mathbf{X}_n^0\}_{n=1}^{N_b} \subset \mathbf{D}^0$ using the visual encoder $\Phi_{\mathbf{\Theta}^0}(\cdot)$,

$$\{\mathbf{H}_n^0\}_{n=1}^{N_b} = \Phi_{\mathbf{\Theta}^0}(\{\mathbf{X}_n^0\}_{n=1}^{N_b}) \tag{1}$$

where $\mathbf{\Theta}^0$ indicates the parameters of the visual encoder in the base step, and N_b is the number of images in a mini-batch. Meanwhile, the semantic encoder embeds the semantic vectors $\{s_c^0\}_{c=1}^{|\mathbf{C}_b|}$ of the classes $\mathbf{C}_b$ involved in the inputs,

$$\{s_c^0\}_{c=1}^{|\mathbf{C}_b|} = \Psi_{\mathbf{\Omega}^0}(\mathbf{C}_b) \tag{2}$$

where $\mathbf{\Omega}^0$ indicates the parameters of the semantic encoder, while s_c^0 denotes the semantic vector about the class c. Afterwards, the global visual representations of each class are aggregated,

$$\varepsilon_c^0 = \frac{\sum_{n=1}^{N_b} \sum_{i=1}^{H} \sum_{j=1}^{W} (\mathbf{H}_{n,[i,j]}^0 * \mathbf{\Delta}_{n,[i,j]}^{0,c})}{\sum_{n=1}^{N_b} \sum_{i=1}^{H} \sum_{j=1}^{W} \mathbf{\Delta}_{n,[i,j]}^{0,c}} \tag{3}$$

in which $\mathbf{H}_n^0 \in \mathbb{R}^{H \times W \times D}$ denotes the feature map encoded by the visual encoder, and $\mathbf{\Delta}_n^{0,c} \in \mathbb{R}^{H \times W}$ represents a binary mask about the class c. H, W, and D represent the height, width, and dimension of feature maps. In particular, if the pixel at the position $[i,j]$ of $\mathbf{X}_n^0$ belongs to the class c, $\mathbf{\Delta}_{n,[i,j]}^{0,c} = 1$; otherwise, $\mathbf{\Delta}_{n,[i,j]}^{0,c} = 0$.

In order to align base class representations with their semantic information, the relation alignment loss $\mathcal{L}_{align}$ is employed, which jointly considers the correlations between visual and semantic embeddings that are paired and unpaired,

$$\mathcal{L}_{align} = \underbrace{\sum_{c_1=1}^{|\mathbf{C}_b|} \sum_{c_2=1,c_2\neq c_1}^{|\mathbf{C}_b|} \frac{\varepsilon_{c_1}^0 * s_{c_2}^0}{|\mathbf{C}_b| \times (|\mathbf{C}_b| - 1)}}_{\text{Unpaired}} - \underbrace{\sum_{c=1}^{|\mathbf{C}_b|} \frac{\varepsilon_c^0 * s_c^0}{|\mathbf{C}_b|}}_{\text{Paired}}. \tag{4}$$

Paired visual-semantic embeddings indicate that the visual vector ε_c^0 and the semantic vector s_c^0 belong to the same class, and thus ε_c^0 should be aligned to match s_c^0 in embedding space. Otherwise, if $\varepsilon_{c_1}^0$ and $s_{c_2}^0$ are unpaired ($c_1 \neq c_2$), the correlations between them should be suppressed, so as to ensure inter-class discrimination. We minimize $\mathcal{L}_{align}$ w.r.t. the parameters of the visual encoder, therefore guiding the model to generate semantic-consistent visual representations.

Furthermore, the segmentation results $\{\bar{\mathbf{Y}}_n^0 \in \mathbb{R}^{H \times W \times |\mathbf{C}^0|}\}_{n=1}^{N_b}$ with respect to the base classes $\mathbf{C}^0$ are drawn by using the learnable base class prototypes $\mathbf{P}^0 = \{\mathbf{p}_c^0\}_{c=1}^{|\mathbf{C}^0|}$,

$$\bar{\mathbf{Y}}_{n,[i,j,c]}^0 = P(c|\mathbf{X}_{n,[i,j]}^0, \mathbf{P}^0, \mathbf{\Theta}^0) = \frac{\exp(\mathrm{Sim}(\mathbf{H}_{n,[i,j]}^0, \mathbf{p}_c^0))}{\sum_{c' \in \mathbf{C}^0} \exp(\mathrm{Sim}(\mathbf{H}_{n,[i,j]}^0, \mathbf{p}_{c'}^0))}. \quad (5)$$

In the equation, $P(c|\mathbf{X}_{n,[i,j]}^0, \mathbf{P}^0, \mathbf{\Theta}^0)$ represents the probability that the pixel $\mathbf{X}_{n,[i,j]}^0$ is inferred as the class c according to $\mathbf{P}^0$ and $\mathbf{\Theta}_v^0$. $\mathrm{Sim}(\cdot, \cdot)$ is a similarity metric function that measures consine similarity between embeddings. Cross entropy $\mathcal{L}_{ce}$ is used to measure discrepancies between segmentation results and label maps,

$$\mathcal{L}_{ce} = \frac{1}{N_b} \sum_{n=1}^{N_b} \mathrm{CE}(\bar{\mathbf{Y}}_n^0, \mathbf{Y}_n^0). \quad (6)$$

By minimizing $\mathcal{L}_{align}$ and $\mathcal{L}_{ce}$, the visual encoder is guided to be aware of semantic information, and the base class prototypes are optimized to segment out base categories.

3.3 Semantic-Guided Adaptation

To alleviate semantic aliasing between base and novel classes, we hope the model can also be aware of novel class semantic information. To this end, we propose to boost affinities between visual and semantic embeddings of novel classes. Taking the step t as an example, we first extract the visual embeddings of the images from $\mathbf{D}^t$

$$\{\mathbf{H}_n^t\}_{n=1}^{N_t} = \Phi_{\mathbf{\Theta}^t}(\{\mathbf{X}_n^t\}_{n=1}^{N_t}), \quad (7)$$

where $\mathbf{\Theta}^t$ indicates the parameters of the visual encoder in the step t. Moreover, the semantic encoder encodes the semantic embeddings of the encountered new classes $\mathbf{C}^t$

$$\{s_c^t\}_{c=1}^{|\mathbf{C}^t|} = \Psi_{\mathbf{\Omega}^t}(\mathbf{C}^t), \quad (8)$$

which are utilized to imprint the weights of the semantic prototypes $\tilde{\mathbf{P}}^t = \{\tilde{\mathbf{p}}_c^t\}_{c=1}^{|\mathbf{C}^t|}$.

Then, the affinities $\{\bar{\mathbf{A}}_n^t \in \mathbb{R}^{H \times W \times |\mathbf{C}^t|}\}_{n=1}^{N_t}$ between visual and semantic embeddings of novel classes are calculated by using Eq. 9,

$$\bar{\mathbf{A}}_{n,[i,j,c]}^t = \frac{\mathbf{H}_{n,[i,j]}^t * \tilde{\mathbf{p}}_c^t}{|\mathbf{H}_{n,[i,j]}^t| * |\tilde{\mathbf{p}}_c^t|}, \quad \text{s.t.,} 0 < c <= |\mathbf{C}^t|. \quad (9)$$

$\bar{\mathbf{A}}_n^t$ denotes the dense visual-semantic affinities about the sample $\mathbf{X}_n^t$, and $\bar{\mathbf{A}}_{n,[i,j,c]}^t$ is the affinity between the visual features at the position $[i,j]$ and the semantic embeddings of the class c. Visual-semantic affinities reflect the relation between the visual embeddings and the semantic information of the encountered novel classes.

Also, the prototypes $\mathbf{P}^{t-1} = \left\{\mathbf{p}_i^{t-1}\right\}_{i=1}^{|\bigcup_{j=0}^{t-1}\mathbf{C}^j|}$ learned in previous steps are utilized to calculate the affinities of the current feature maps to the old classes $\left\{\tilde{\mathbf{A}}_n^t \in \mathbb{R}^{H\times W\times|\bigcup_{j=0}^{t-1}\mathbf{C}^j|}\right\}_{n=1}^{N_t}$,

$$\tilde{\mathbf{A}}_{n,[i,j,c]}^t = \frac{\mathbf{H}_{n,[i,j]}^t * \mathbf{p}_c^{t-1}}{|\mathbf{H}_{n,[i,j]}^t| * |\mathbf{p}_c^{t-1}|}, \text{s.t.}, 0 < c <= |\bigcup_{j=0}^{t-1}\mathbf{C}^j|. \tag{10}$$

For each sample, the affinity maps $\tilde{\mathbf{A}}_n^t$ and $\bar{\mathbf{A}}_n^t$ are concatenated together

$$\mathbf{A}_n^t = \tilde{\mathbf{A}}_n^t \oplus \bar{\mathbf{A}}_n^t, \tag{11}$$

thereby producing $\left\{\mathbf{A}_n^t \in \mathbb{R}^{H\times W\times|\bigcup_{j=0}^t\mathbf{C}^j|}\right\}_{n=1}^{N_t}$. We use cross entropy to constrain the correctness of these affinities

$$\mathcal{L}_{ce}' = \frac{1}{N_t}\sum_{n=1}^{N_t}\text{CE}(\mathbf{A}_n^t, \mathbf{Y}_n^t), \tag{12}$$

thus guiding the visual embeddings of the encountered new classes to have high correlations to their semantic information, while suppressing their affinities to the old classes. Moreover, knowledge distillation [8] is utilized to suppress overfitting,

$$\mathcal{L}_{kd} = \frac{1}{N_t}\sum_{n=1}^{N_t}\text{KD}(\mathbf{A}_n^t, \mathbf{A}_n^{t-1}). \tag{13}$$

By minimizing $\mathcal{L}_{ce}'$ and $\mathcal{L}_{kd}$, the model is guided to be aware of novel class semantic information. After the step t, the prototypes are updated: $\mathbf{P}^t \leftarrow \hat{\mathbf{P}}^{t-1}\bigcup\hat{\tilde{\mathbf{P}}}^t$, where $\hat{\mathbf{P}}^{t-1}$ and $\hat{\tilde{\mathbf{P}}}^t$ indicate $\mathbf{P}^{t-1}$ and $\tilde{\mathbf{P}}^t$ after being optimized in the current step. The updated prototypes and visual encoder are used to draw segmentation results for all encountered classes, which is as same as the procedure in Eq. 5.

4   Experiments

4.1   Implementation Details

We evaluate our model on two public datasets, which are PASCAL VOC 2012 [5] and COCO [1,12]. For the details about these two datasets, please refer to [4]. On PASCAL VOC 2012, we set the number of epochs as 30 on the base step and 400 during incremental learning. Moreover, in each phase, the initial learning rate is set as 0.01. On the COCO dataset, we train the model on the

Table 1. Experimental results on the PASCAL VOC 2012 dataset. In the table, "FT" represents directly finetuning the model on novel classes, while "HM" denotes the harmonic mean of the mIoU on base and novel classes. "SFS" and "MFS" indicate the single few-shot step setting and the multiple few-shot step setting respectively.

Method	SFS						MFS					
	1-shot			5-shot			1-shot			5-shot		
	mIoU (%)			mIoU (%)			mIoU (%)			mIoU (%)		
	base	novel	HM	base	novel	HM	base	novel	HM	base	novel	HM
FT	58.3	9.7	16.7	55.8	29.6	38.7	47.2	3.9	7.2	58.7	7.7	13.6
WI [16]	62.7	15.5	24.8	63.3	21.7	32.3	66.6	16.1	25.9	66.6	21.9	33.0
DWI [6]	64.3	15.4	24.8	64.9	23.5	34.5	67.2	16.3	26.2	67.6	25.4	36.9
RT [23]	59.1	12.1	20.1	60.4	27.5	37.8	49.2	5.8	10.4	45.1	10.0	16.4
AMP [22]	57.5	16.7	25.8	51.9	18.9	27.7	58.6	14.5	23.2	57.1	17.2	26.4
SPN [25]	59.8	16.3	25.6	58.4	33.4	42.5	49.8	8.1	13.9	61.6	16.3	25.8
LWF [11]	61.5	10.7	18.2	59.7	30.9	40.8	42.1	3.3	6.2	59.8	7.5	13.4
ILT [14]	64.3	13.6	22.5	61.4	32.0	42.1	43.7	3.3	6.1	59.0	7.9	13.9
MIB [3]	61.0	5.2	9.7	65.0	28.1	39.3	43.9	2.6	4.9	60.9	5.8	10.5
PIFS [4]	60.9	18.6	28.4	60.0	33.4	42.8	64.1	16.9	26.7	64.5	27.5	38.6
Our SRAA	65.2	19.1	29.5	63.8	36.7	46.6	66.4	18.8	29.3	64.3	28.7	39.7

base set for 50 epochs with the initial learning rate 0.02. In addition, during incremental learning, the epochs are set as 400, and the initial learning rate is set as 0.01. As same as the previous work [4], we evaluate our method in both the Single Few-shot Step (SFS) setting and the Multiple Few-shot Step (MFS) setting based on the cross validation protocol. In particular, SFS indicates that novel classes are provided at once in one step, while MFS represents novel classes are progressively given in multiple steps. Following [19], data replay is utilized to further relieve catastrophic forgetting issues. We build our semantic encoder using CLIP [18] due to its powerful capability in encoding semantic information [26,27], and utilize resnet101 [7] to build the visual encoder. To guarantee the stability, we freeze the parameters of the semantic encoder during training, i.e., $\Omega^t = \Omega^0$ for $\forall t >= 1$.

4.2 Main Experimental Results

In this subsection, we report the performance of our model on PASCAL VOC 2012 and COCO datasets. According to the results in Table 1 and Table 2, we have the following observations. On the PASCAL VOC 2012 dataset, our method can achieve higher mIoU on both base and novel classes than that of FT, RT, AMP, SPN, and PIFS under the single few-shot step setting. Despite the performance of MIB, ILT, LWF, DWI, and WI on base classes is better than that of ours in some cases, our method has obviously higher segmentation accuracy for novel classes. For example, under the 1-shot setting, the novel class mIoU of

Table 2. Experimental results on the COCO dataset.

Method	SFS						MFS					
	1-shot			5-shot			1-shot			5-shot		
	mIoU (%)			mIoU (%)			mIoU (%)			mIoU (%)		
	base	novel	HM	base	novel	HM	base	novel	HM	base	novel	HM
FT	41.2	4.1	7.5	41.6	12.3	19.0	38.5	4.8	8.5	39.5	11.5	17.8
WI [16]	43.8	6.9	11.9	43.6	8.7	14.6	46.3	8.3	14.1	46.3	10.3	16.9
DWI [6]	44.5	7.5	12.8	44.9	12.1	19.1	46.2	9.2	15.3	46.6	14.5	22.1
RT [23]	46.2	5.8	10.2	46.9	13.7	21.2	38.4	5.2	9.2	44.1	16.0	23.5
AMP [22]	37.5	7.4	12.4	34.6	11.0	16.7	36.6	7.9	13.0	33.2	11.0	16.5
SPN [25]	43.5	6.7	11.7	43.7	15.6	22.9	40.3	8.7	14.3	41.4	18.2	25.3
LWF [11]	43.9	3.8	7.0	44.6	12.9	20.1	41.0	4.1	7.5	42.3	12.6	19.4
ILT [14]	46.2	4.4	8.0	47.0	11.0	17.8	43.7	6.2	10.9	45.3	15.3	22.9
MIB [3]	43.8	3.5	6.5	44.7	11.9	18.8	40.4	3.1	5.8	43.8	11.5	18.2
PIFS [4]	40.8	8.2	13.7	42.8	15.7	23.0	40.4	10.4	16.5	41.1	18.3	25.3
Our SRAA	41.2	9.3	15.2	42.6	17.1	24.4	40.7	11.3	17.7	41.0	19.7	26.6

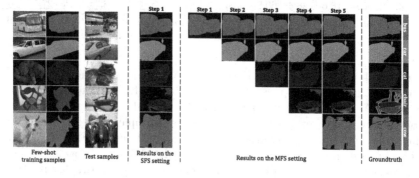

Fig. 3. Visualization for step-by-step segmentation results of our method in both the Single Few-shot Step (SFS) setting and the Multiple Few-shot Step (MFS) setting according to only a labeled sample per category.

our method is 13.9%, 5.5%, 8.4%, 3.7%, and 3.6% higher than that of MIB, ILT, LWF, DWI, and WI. In the meantime, our method also shows its superiority on the PASCAL VOC 2012 dataset under the multiple few-shot step setting, e.g., the novel class mIoU of PIFS is lower than that of our method on both the 1-shot and the 5-shot task. Similar results can also be found on the experiments of the COCO dataset. For example, in the single few-shot step setting, the performance of our method is better than that of PIFS and AMP on both base and novel classes. Although the base class mIoU of our method is lower than that of some counterparts, it can show consistently better performance for novel classes, e.g., on the 1-shot task, the novel class mIoU of our model is 5.8%, 4.9%, 5.5%, 2.6%, 3.5%, 1.8%, and 2.4% higher than that of MIB, ILT, LWF,

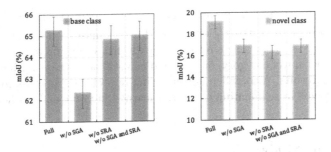

Fig. 4. Ablation study for our proposed method, which is conducted on the 1-shot task of the PASCAL VOC 2012 dataset under the single few-shot step setting.

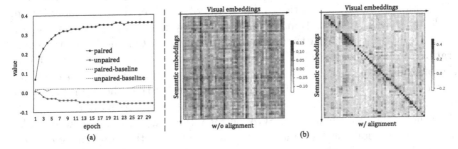

Fig. 5. Analysis for the influence of our SRA. (a): The values of the paired and the unpaired term in $\mathcal{L}_{align}$ during training. To better reflect value change, the baseline curves are also drawn in the figure, which reflect the values of these two terms when $\mathcal{L}_{align}$ does not participate in model optimization. (b): The mean affinities between visual and semantic embeddings with or without being aligned by our SRA, which is conducted on the COCO dataset. The values in the diagonal are the affinities between visual and semantic embeddings that belong to the same class, while the others are the affinities between them that are unpaired.

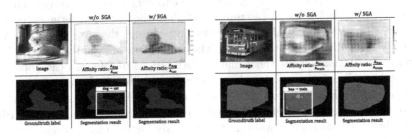

Fig. 6. Analysis for the influence of SGA. In the figure, $\mathbf{A}_c$ represents the affinity map of the image to the class c.

SPN, RT, DWI, and WI. Under the multiple few-shot step setting, the novel class mIoU of our method is higher than that of all the compared methods in the table, e.g., the novel class mIoU of our method is 0.9% and 1.4% higher than that of PIFS on the 1-shot and the 5-shot task respectively. In Fig. 3, we also give our step-by-step segmentation results for encountered novel classes under the two different settings. These results indicate that according to only one training instance per novel category, our method can still achieve promising semantic segmentation results. Also, when encountering new classes, it can still maintain high effectiveness in segmenting categories learned in previous few-shot learning steps.

4.3   Ablation Study

Ablation study is provided for our proposed method in this subsection. We first study the influence of SGA and SRA on accuracy in Fig. 4. In the figure, "w/o SGA" indicates that semantic guidance is not considered during adaptation, while "w/o SRA" denotes that SRA is not employed and base class representations are not aligned with their semantic information. On one hand, the cooperation of our SRA and SGA (i.e., "Full") can achieve higher mIoU than that of the baseline model (i.e., "w/o SGA and SRA") on both base and novel classes under the two different settings, which demonstrates that the appropriate use of prior semantic information can make segmentation results more accurate. On the other hand, the removal of SGA or SRA (i.e., "w/o SGA" or "w/o SRA") leads to an obvious performance drop, which validates the importance of these two components. In particular, the above results also indicate that semantic information should be considered in both the base and the incremental learning stage. Otherwise, inconsistency across training phases will reduce segmentation accuracy.

In Fig. 5 (a), we also visualize the values of the paired and the unpaired term in the relation alignment loss $\mathcal{L}_{align}$ during the training process. The results indicate that $\mathcal{L}_{align}$ can be optimized stably. On one hand, with the epoch increases, the paired term is maximized progressively, thereby constraining that visual and semantic embeddings belonging to the same category have relatively high correlations. On the other hand, the minimization of the unpaired term suppresses the similarity between visual and semantic embeddings that are unpaired. As a result, visual embeddings belonging to the same class are guided to have high semantic correlations, while the semantic correlations of different categories are suppressed. In Fig. 5 (b), we also visualize the mean affinities between visual and semantic embeddings that are aligned or not aligned by our proposed method. The results in Fig. 5 (b) validate the effectiveness of SRA again, e.g., our SRA can calibrate visual representations to obviously better match their semantic information.

Furthermore, we also analyze the influence of our SGA in Fig. 6. According to the results in the figure, we have the following observations. On one hand, without using SGA, the instance about "dog" is incorrectly segmented as the class "cat" due to semantic aliasing, e.g., the affinity ratio map $\frac{A_{dog}}{A_{cat}}$ has low values in target regions. On the other hand, the use of SGA can boost the affinities of the image to the target class obviously, while suppressing its affinities to "cat" as much as possible. In this way, the segmentation results can become more accurate. Similar results can be found in the other example as well. For example, when not using SGA, the affinity ratio map $\frac{A_{bus}}{A_{train}}$ exhibits low values for a part of the regions belonging to "bus". Moreover, for background areas, the ratio map $\frac{A_{bus}}{A_{train}}$ has incorrect high values. In contrast, by further considering semantic guidance, our method can rectify these incorrect relationships, and makes segmentation results more accurate. Finally, in Fig. 7, we quantitatively analyze the influence of our SRAA method on final segmentation results. The experimental results consistently validate its superiority as well. For example, when not using our method, semantic aliasing between base and novel classes severely limits the quality of final segmentation results. However, in contrast, our proposed SRAA approach can obviously make final segmentation results more accurate.

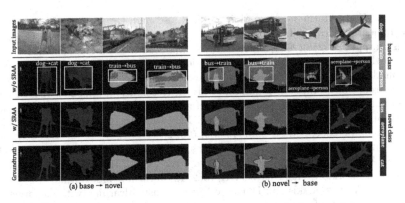

Fig. 7. Visualized analysis for the influence of our method on alleviating the semantic-aliasing issue in both the "base→novel" and the "novel→base" scenario, which is conducted on the PASCAL VOC 2012 dataset under the 1-shot setting. In particular, "A→B" indicates that regions belonging to A are incorrectly segmented as B.

5   Conclusion

In this paper, we propose to alleviate semantic-aliasing issues in IFSS, so as to make segmentation results more accurate. To realize this goal, we propose a method called *Semantic-guided Relation Alignment and Adaptation*. We propose to conduct *Semantic Relation Alignment* in the base step, aiming to semantically align base class representations and guide the model to generate semantic-consistent feature representations. Also, *Semantic-Guided Adaptation* is adopted to incrementally adapt the model to novel classes. It ensures the visual-semantic affinities of encountered novel categories, making their feature representations be consistent with corresponding semantic information. By considering the semantic information of both base and novel classes, the semantic-aliasing problem can be relieved. Currently, it is still very challenging to achieve accurate segmentation results for IFSS when objects have complex boundaries. In the future, we plan to tackle this issue by fully considering the fine-grained information of local features.

Acknowledgment. This work was supported by the National Key Research and Development Program (Grant No. 2019YFA0706200), the National Nature Science Foundation of China under Grant No. 62072152, 62172137, 72188101.

References

1. Caesar, H., Uijlings, J., Ferrari, V.: Coco-stuff: thing and stuff classes in context. In: Proceedings of the Conference on Computer Vision and Pattern Recognition, pp. 1209–1218 (2018)
2. Castro, F.M., Marín-Jiménez, M.J., Guil, N., Schmid, C., Alahari, K.: End-to-end incremental learning. In: Proceedings of the European Conference on Computer Vision, pp. 233–248 (2018)
3. Cermelli, F., Mancini, M., Bulo, S.R., Ricci, E., Caputo, B.: Modeling the background for incremental learning in semantic segmentation. In: Proceedings of the Conference on Computer Vision and Pattern Recognition, pp. 9233–9242 (2020)
4. Cermelli, F., Mancini, M., Xian, Y., Akata, Z., Caputo, B.: Prototype-based incremental few-shot semantic segmentation. In: Proceedings of the British Machine Vision Conference, BMVC 2021, p. 484. (2021)
5. Everingham, M., Van Gool, L., Williams, C.K.I., Winn, J., Zisserman, A.: The PASCAL visual object classes challenge 2012 (VOC2012) results. http://www.pascal-network.org/challenges/VOC/voc2012/workshop/index.html
6. Gidaris, S., Komodakis, N.: Dynamic few-shot visual learning without forgetting. In: Proceedings of the Conference on Computer Vision and Pattern Recognition, pp. 4367–4375 (2018)
7. He, K., Zhang, X., Ren, S., Sun, J.: Deep residual learning for image recognition. In: Proceedings of the IEEE Conference on Computer Vision and Pattern Recognition, pp. 770–778 (2016)
8. Hinton, G., Vinyals, O., Dean, J.: Distilling the knowledge in a neural network. arXiv preprint arXiv:1503.02531 (2015)
9. Kang, M., Park, J., Han, B.: Class-incremental learning by knowledge distillation with adaptive feature consolidation. In: Proceedings of the Conference on Computer Vision and Pattern Recognition, pp. 16071–16080 (2022)
10. Li, A., Huang, W., Lan, X., Feng, J., Li, Z., Wang, L.: Boosting few-shot learning with adaptive margin loss. In: Proceedings of the Conference on Computer Vision and Pattern Recognition, pp. 12576–12584 (2020)
11. Li, Z., Hoiem, D.: Learning without forgetting. IEEE Trans. Pattern Anal. Mach. Intell. **40**(12), 2935–2947 (2017)
12. Lin, T.-Y., et al.: Microsoft COCO: common objects in context. In: Fleet, D., Pajdla, T., Schiele, B., Tuytelaars, T. (eds.) ECCV 2014. LNCS, vol. 8693, pp. 740–755. Springer, Cham (2014). https://doi.org/10.1007/978-3-319-10602-1_48
13. Liu, Y., Schiele, B., Sun, Q.: RMM: reinforced memory management for class-incremental learning. Adv. Neural. Inf. Process. Syst. **34**, 3478–3490 (2021)
14. Michieli, U., Zanuttigh, P.: Incremental learning techniques for semantic segmentation. In: Proceedings of the International Conference on Computer Vision Workshops (2019)

15. Mikolov, T., Chen, K., Corrado, G., Dean, J.: Efficient estimation of word representations in vector space. arXiv preprint arXiv:1301.3781 (2013)
16. Nichol, A., Achiam, J., Schulman, J.: On first-order meta-learning algorithms. arXiv preprint arXiv:1803.02999 (2018)
17. Pennington, J., Socher, R., Manning, C.D.: Glove: global vectors for word representation. In: Proceedings of the Conference on Empirical Methods in Natural Language Processing, pp. 1532–1543 (2014)
18. Radford, A., et al.: Learning transferable visual models from natural language supervision. In: Proceedings of the International Conference on Machine Learning, pp. 8748–8763. PMLR (2021)
19. Rebuffi, S.A., Kolesnikov, A., Sperl, G., Lampert, C.H.: ICARL: incremental classifier and representation learning. In: Proceedings of the Conference on Computer Vision and Pattern Recognition, pp. 2001–2010 (2017)
20. Santoro, A., Bartunov, S., Botvinick, M., Wierstra, D., Lillicrap, T.: Meta-learning with memory-augmented neural networks. In: Proceedings of the International Conference on Machine Learning, pp. 1842–1850. PMLR (2016)
21. Shi, G., Wu, Y., Liu, J., Wan, S., Wang, W., Lu, T.: Incremental few-shot semantic segmentation via embedding adaptive-update and hyper-class representation. In: Proceedings of the ACM International Conference on Multimedia, pp. 5547–5556 (2022)
22. Siam, M., Oreshkin, B., Jagersand, M.: Adaptive masked proxies for few-shot segmentation. arXiv preprint arXiv:1902.11123 (2019)
23. Tian, Y., Wang, Y., Krishnan, D., Tenenbaum, J.B., Isola, P.: Rethinking few-shot image classification: a good embedding is all you need? In: Vedaldi, A., Bischof, H., Brox, T., Frahm, J.-M. (eds.) ECCV 2020. LNCS, vol. 12359, pp. 266–282. Springer, Cham (2020). https://doi.org/10.1007/978-3-030-58568-6_16
24. Wang, FY., Zhou, DW., Ye, HJ., Zhan, DC.: FOSTER: feature boosting and compression for class-incremental learning. In: Avidan, S., Brostow, G., Cissé, M., Farinella, G.M., Hassner, T. (eds.) Computer Vision - ECCV 2022. ECCV 2022. LNCS, vol. 13685, pp. 398–414. Springer, Cham (2022). https://doi.org/10.1007/978-3-031-19806-9_23
25. Xian, Y., Choudhury, S., He, Y., Schiele, B., Akata, Z.: Semantic projection network for zero-and few-label semantic segmentation. In: Proceedings of the Conference on Computer Vision and Pattern Recognition, pp. 8256–8265 (2019)
26. Xu, J., Le, H.: Generating representative samples for few-shot classification. In: Proceedings of the Conference on Computer Vision and Pattern Recognition, pp. 9003–9013 (2022)
27. Yang, F., Wang, R., Chen, X.: Semantic guided latent parts embedding for few-shot learning. In: Proceedings of the Winter Conference on Applications of Computer Vision, pp. 5447–5457 (2023)
28. Zhang, B., Li, X., Ye, Y., Huang, Z., Zhang, L.: Prototype completion with primitive knowledge for few-shot learning. In: Proceedings of the Conference on Computer Vision and Pattern Recognition, pp. 3754–3762 (2021)

PMGCN:Preserving Measuring Mapping Prototype Graph Calibration Network for Few-Shot Learning

Zhengye Shen, Guangtong Lu, Qian Qiao, and Fanzhang Li[✉]

School of Computer Science and Technology, Soochow University, Suzhou 215006, China
{zyshen111,gtlu,qqiao}@stu.suda.edu.cn, lfzh@suda.edu.cn

Abstract. The aim of few-shot classification is to learn the discriminative features from a limited sample of labeled data. Due to the few number of labeled data, the model can not learn the discriminative knowledge, and can usually become overfitted and achieve a bad generalization effect. In previous work, the training data is not mentioned after the model is trained. However, considering the few-shot learning paradigm, we try to use the training data in more places when the data set itself is not sufficient. In this paper, we propose a new method **P**reservation measuring **M**apping Prototype **G**raph **C**alibration **N**etwork(**PMGCN**) based on the prototypical networks to fully utilize the training dataset to obtain a more accurate prototype. In order to get more discriminative feature, we first use a Convolutional Block Attention Module(CBAM) to focus on the key parts of the data and ignore the noisy parts, which is a simple yet effective attention module. Then, in order to fully utilize the limited labeled data, we propose to use a graph convolutional network(GCN) combined with preserving measuring mapping(PMM) in ergodic theory to select some similar base class prototypes to calibrate the new prototypes to obtain better generalization and make the model more interpretable. We validate our model on mimiimagenet, tieredimagenet and cub. The performance of the proposed method is improved by 2.24% and 2.86%, 1.35% and 2.19%, 1.42% and 0.82% separately.

Keywords: few-shot learning · ergodic theory · graph conventional network · attention mechanism

1 Introduction

At present, with the support of a large number of data samples, the image classification method based on deep neural networks has achieved great success. However, deep neural networks have some limitations. On the one hand, deep learning requires large training data, but in many sample scarcity areas, deep neural networks are often unable to obtain enough samples for training, so that it can lead to poor performance in classification tasks. Deep learning, on the

© The Author(s), under exclusive license to Springer Nature Switzerland AG 2024
S. Rudinac et al. (Eds.): MMM 2024, LNCS 14554, pp. 258–272, 2024.
https://doi.org/10.1007/978-3-031-53305-1_20

other hand, due to the complexity of its network, is often unable to judge the correctness of the results. For this reason, deep learning may perform worse on specific tasks, especially in tasks with limited samples.

In order to solve those problems, a large number of researchers have made different attempts before this. Thru et al. [1,2] use meta-learning to allow the model to quickly adapt to novel tasks with only few training samples. Herranz et al. [3,4] try to synthesize data or features by learning a generative model to alleviate the data insufficiency problem. Gidaris et al. [5] propose to utilize unlabeled data and predict pseudo labels to improve the performance of few-shot learning.

To sum up, most of the previous work has focused on how to design more effective models. The further use of data samples, especially the support set, is ignored by most researchers. In fact, prototypes formed by few samples are not very accurate in distinguishing all data samples belonging to the category. Therefore, in this paper, we propose to use a graph convolutional network (GCN) [6–9] to utilize the valid information in the support set. In addition, in order to make more efficient use of the sample, we also add an attention mechanism called the Convolutional Block Attention Module (CBAM) [10]. CBAM allows the model to more accurately determine the discriminant features in the image. At the same time, it allows the model to ignore undistinguished features in the image.

Furthermore, we find that although using GCN to utilize effective information in the support set image data can achieve good results compared to the original PN, it is not as effective as other few-shot learning methods [11–13]. Therefore, we improve the original GCN by applying the preserving measuring mapping (PMM) [14,15] in ergodic theory to enhance its ability to utilize effective information. We name this method PMGCN.

In this paper, we propose a method to obtain the discriminant features of samples mainly by using attention mechanisms, graph neural networks, and preservation measuring mapping in ergodic theory, and modify the current class prototypes through the previous class prototypes. After a feature extractor and the CBAM, through PMGCN, we will establish edges based on the cosine similarity between the current class prototypes and the previous class prototypes, and take the sample features as vertexes, so that the class prototype is calibrated by the PMGCN. Finally, we only need to use metric methods to categorize the query set, and the work in this paper can be summarized as follows:

- We propose a method to use the CBAM to obtain image discriminant features, use the PMGCN to calibrate the prototype, and use the metric method to identify the sample labels.
- We analyze the role of each module of the method and discuss hyperparameters influence.
- We use practical experiments to prove that in the field of few-shot learning, our PMGCN has a obvious improvement over the original prototypical networks, and has competitiveness compared with the existing few-shot classification methods.

2 Related Works

2.1 Meta-Learning

Meta-learning offers a paradigm in which the model can obtain knowledge from a large number of related but unique tasks and then use this experience to effectively solve similar but previously unseen tasks [16]. This two stage paradigm has been acknowledged to have clearly progress, especially in the situation where the data is not adequate. Thus the meta-learning is commonly employed for few-shot classification to learn from a few samples. It can be summarized classified into two categories: optimization-based [1,4,17,18], and metric-based [11–13,19, 20].

Optimization-based approaches involve training the meta-learner either by training a generic optimization algorithm as an alternative to the SGD optimizer, which updates the model during training, or by training a good initialization which allows rapid adaptation to new problems in models update with a small amount of training data. Representative works in this area comprise LSTM [21] and MAML [17].

Metric-based approaches typically train a feature extractor for mapping the features of an image into a vector feature space, preferably the data points which are belongs to the same category should close to each other and data points which are belongs to different categories should be away. In a word, metric-based approaches first map the image data into feature vectors and then use the resulting feature vectors to select appropriate classifiers to classify the query image. For example, Vinyals et al. [11] proposed matching networks, which employ an attention mechanism to derive a weighted k-NN classifier. Snell et al. [20] used the mean vector of support images belonging to the same class as a prototype and introduced a nearest centroid classifier, the prototypical networks. Our main idea is based on the prototypical networks.

2.2 Attentional Mechanism

Soft and hard attention [22] mechanisms have been widely used in vision-related tasks. Soft attention mechanism by selectively ignoring some information to reweight the rest of the information for aggregation calculation and all information is reweighted in an adaptive manner before being aggregated. This separates important information and avoids the interference by unimportant information, thus improving accuracy while hard attention mechanism means selecting information at a certain position in the input sequence, such as selecting a random message or selecting the message with the highest probability. Existing attention models are mainly based on soft attention. Jie er al. [23] propose SENet, a network which empowers the features of important channels and erodes the features of unimportant channels by weighting each piece of features with a soft attention mechanism. Furthermore, there is an attention module which combines spatial attention and channel attention which is called convolutional block attention module (CBAM) [24].

2.3   Graph Convolutional Network

Graph convolutional network (GCN) [7–9,25], is a type of convolutional neural network that directly processes information about graphs and their structures. GCN constructs a graph by taking the target data according to certain rules and then use this graph as an input. Convolutional neural networks extract the features of a sample from the sample data itself. GCN extracts the features of a sample by traversing the information on the graph structure, and each node's features will be augmented by its own information as well as that of other nodes by traversing the information on the graph. Unlike other previous methods [24,26,27], that directly use graph convolutional neural networks to extract features and then directly classify them, in this paper we aim to calibrate the prototype by using class prototypes that have been previously obtained to give the prototype a better generalization.

2.4   Ergodic Theory

Ergodic theory [14,15] is the branch of mathematics that studies the asymptotic properties of guaranteed transformations. It originated from the study of the "ergodic hypothesis" that provided the basis for statistical mechanics, and has close ties with mathematical branches such as probability theory, information theory, functional analysis, and number theory.

Among ergodic theory, the preserving measuring mapping(PMM) is a major application. It is a mapping between measure spaces that keeps the magnitude of measures between spaces constant. PMM has a wide range of applications in the fields of measure theory and dynamical systems. We study this concept usually in a probability space, which is defined as follows:

Theorem 1. *Let* X *be a set, consider a set class* $\mathcal{B}$ *consisting of some subset of* X. *Call* $\mathcal{B}$ σ *algebra if it satisfies the following three conditions:*

(i) $X \in \mathcal{B}$;
(ii) $B \in \mathcal{B} \Rightarrow X \setminus B \in \mathcal{B}$;
(iii) $B_n \in \mathcal{B}$ *(*$n \in \mathbb{N}$*)* $\Rightarrow \bigcup\limits_{n=1}^{\infty} B_n \in \mathcal{B}$.

In other words, σ *algebra is a class of sets that contain the complete set* X, *and are closed to the remainder operations, the countable union operations, and the countable intersection operations. We call* $(X, \mathcal{B})$ *the measurable space and the elements in* $\mathcal{B}$ *the measurable set.*

Theorem 2. *The probability measure on* $(X, \mathcal{B})$ *refers to a function* $m : \mathcal{B} \to [0,1]$ *that satisfies the following three conditions:*

(i) $m(\emptyset) = 0$;
(ii) $m(X) = 1$;

(iii) If $\{B_n\}_1^\infty \subset \mathcal{B}$ is a sequence of disjointed subsets, then $m(\bigcup\limits_{n=1}^{\infty} B_n) = \sum\limits_{n=1}^{\infty} m(B_n)$.

We call $(X, \mathcal{B}, m)$ the probability measure space, referred to as the probability space.

Through the definition of probability space, we treat the set of sample features as a probability space in the process of prototype calibration, and carry out preserving measuring mapping in the probability space to ensure the invariability and stability of the features (Fig. 1).

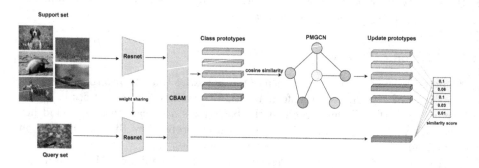

Fig. 1. Process of our method. The first module is CBAM after a feature extractor. The Next module is PMGCN which purpose is to modify the prototypes. The last module is a metric-based discriminator to determine the category of the samples.

3   Methods

3.1   Problem Definition

In few-shot classification, we are given a small support set of N labeled examples $S = \{(x_1, y_1), ...(x_n, y_n)\}$ where each $x_i \in \mathbb{R}^D$ is the D-dimensional feature vector of an example and $y_i = \{1, ...K\}$ is the corresponding label. S_k denotes the set of examples labeled with class k.

We define the N-way K-shot problem using the episodic formulation of Snell et al. [20]. Each task instance T_i is a classification problem sampled from a task distribution $p(T_i)$. The tasks are divided into a support set $S = (x_n^i, y_n^i)_{i=0}^{N \times K}$ and a query set $Q = (x_n^i, y_n^i)_{i=N \times K}^{N \times K + N \times Q}$.

3.2   Convolutional Block Attention Module

In our method, we first use ResNet10 as a feature extractor to extract the feature of the image. In order to make better use of the images and for our model to learn to focus on the important parts of the images, we consider the attention mechanism [28] for the features obtained by the feature extractor. In few-shot learning, the data is limited, meanwhile the transformer has many parameters, which inevitably causes overfitting during the training, so we choose another lightweight attention mechanism module Convolutional Block Attention Module(CBAM) [24] as show in Fig. 2, it is a combination of channel attention mechanism and spatial attention mechanism. In this paper, the feature map obtained after the feature extractor is utilized to sequentially generate attentional feature map information in both channel and spatial dimensions. Then, these two feature map information are multiplied with the feature map to produce the final feature map.

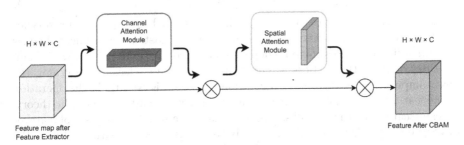

Fig. 2. The Illustration of the CBAM. The module has two sequential submodules, channel and space. On each convolutional block of the deep network, the intermediate feature maps are adaptively refined by our module.

3.3   Preservation Measuring Mapping Graph Calibration Module

After getting the class prototype, we want to make the feature vector described by the prototypes more precisely. Therefore, we propose a method for prototype calibration by using PMGCN through base class prototypes from the support set. Thus we need to construct a graph $G(\mathbf{V}, \mathbf{E})$ where the $\mathbf{V} \in \mathbb{R}^{N \times +N}$ denotes the feature vectors of the data meanwhile the $\mathbf{E}$ indicate whether there are consecutive edges between different vertices. First, we need to construct the adjacency matrix $\mathbf{A}$ of the graph G. We calculate the cosine similarity between different vertices to decide whether to construct a connected edge between the different vertices:

$$\mathbf{A}[i,j] = \begin{cases} 1 & cos(V_i, V_j) >= k \\ 0 & otherwise \end{cases} \qquad (1)$$

where $\mathbf{A}[i,j] = 1$ denotes that there is a connected edge between vertex V_i and vertex V_j, otherwise it means that there is no connected edge between the two vertices. The k refers the threshold value of cosine similarity between two vertices, when the cosine similarity between different vertices is greater than k, the two vertices are considered to have connected edges. Next we need to compute the degree diagonal matrix of the graph G which is defined as follow:

$$\mathbf{D}[i,i] = \sum_j \mathbf{A}[i,j] \tag{2}$$

Finally, in order to solve this problem of computational distortion after multiple rounds of aggregation where there is only one neighbor vertex and the vertex has many neighborhood vertices, according to the [25] we also use the degree matrix to normalize the adjacency matrix in terms of rows and columns. Thus the resulting adjacency is as follows:

$$\mathbf{A} = D^{-1/2}\mathbf{A}D^{-1/2} \tag{3}$$

Eventually, the vertices of the graph denote all base category prototypes and prototypes in the current task. Meanwhile, the existence of connected edges between different classes of circles depends on the cosine similarity between them.

After defining the vertices, edges, and adjacency matrices of the graph, the next computation in the graph convolutional network is essentially the operation of the feature properties and the adjacency matrix. According to ergodic theory, we can treat the features of all samples in a task as a set, where the set of features of the same sample is a subset of the above set, and the similarity between the features forms a measure, so that we can form a probability space. In each layer of the GCN, we make a PMM of the feature vectors obtained in the probability space to ensure the similarity of the features, and then input the obtained results into the next layer of the GCN for operation. In the process of PMM, we need to ensure the characteristics of the invariant measure, which are defined as follows:

$$V \, preserve \, measuring \, m \iff \int f \circ T dm = \int f dm = m(T^{-1}B) \tag{4}$$

where V represents the feature vector of the data, f represents any feature in the probability space composed of sample features, and T represents measurable mapping. Besides, we can also guarantee the correctness of this process by proof. By mapping of preserving measurability on the original task sample composition, we can further improve the accuracy of the prototypes.

Next, we perform operations on the feature attributes and adjacency matrices on the graph convolutional network. Specifically, the adjacency matrix under each layer of convolution and the feature attributes of the layer, are multiplied based on layer weights, and then go through a layer set of activation functions such as Relu. The result is used as the new feature attributes to continue the cyclic operation. The convolution kernel of GCN is weight sharing in each layer, that is, the weight table of any vertex on the graph under the same order is the same for the weighted summation of the surrounding neighborhood vertices. Thus, the purpose of our class prototype calibration is achieved after we perform feature propagation using a graph convolutional neural network. As shown in the following equation:

$$V_{calibration} = \sigma(\widetilde{D}^{-1/2}\widetilde{A}\widetilde{D}^{-1/2}VW) \tag{5}$$

The adjacency matrix mentioned above only takes into account the information about the neighbors of the current vertex and ignores the information about the vertex itself. To solve this problem, the value of the diagonal of the adjacency matrix is set to 1, which means that each vertex will point to itself. This is also the meaning of $\widetilde{A}$ meanwhile $\widetilde{D}$ is the corresponding degree matrix. The σ denotes the nonlinear activation function, in our work we select Relu as the function. Finally the W denotes the learnable parameter matrix.

3.4   Loss Definition

In this paper, we generate a class distribution for the query image data based on a softmax of distances to the calibration prototype in the feature vector space, the formula is as follows:

$$p_\phi(y = k \mid x_q) = \frac{\exp(-d(f_\phi(x_q), k))}{\sum_{k'} \exp(-d(f_\phi(x_q), k'))} \tag{6}$$

where the $f_\phi(x)$ denote the feature vector of the query image data x after Resnet, CBAM and PMGCN. The k denotes the category of the query image data x_q. Thus the loss formula can be defined as follows:

$$J(\phi) = -\log(p_\phi(y = k' \mid x)) \tag{7}$$

The probability of correctly classifying the query data is maximized by minimizing the loss function. Meanwhile the paramater of our CBAM and PMGCN are optimized.

4   Experiments

4.1   Datasets

We evaluate our method on 3 classic few-shot classification datasets: miniImageNet [11], tieredImageNet [25] and CUB [29]. All three datasets share a common feature in that all of their categories are divided into two broad classes which are called base classes and novel classes. We perform a run of 10,000 random draws to obtain an accuracy score and indicate confidence scores 95% when relevant.

miniImageNet: The miniImageNet dataset [11] is extracted from the ImageNet dataset. MiniImageNet contains 60,000 colour images in 100 categories, with 600 samples per category and each image is 84 × 84. In general, the training and test sets of this dataset are divided into categories of 80 : 20.

tieredImageNet: The tieredImageNet dataset [25] is a few-shot classification task dataset. Like miniImagenet, it is a subset of ILSVRC-12. However, tiered-ImageNet represents a larger subset of ILSVRC-12. Similar to Omniglot, which groups characters into letters, tieredImageNet divides categories into broader classes corresponding to nodes at higher levels in the ImageNet hierarchy.

CUB: The CUB dataset [29] has a total of 11,788 bird images of size $84 \times 84 \times 3$ containing 200 bird subclasses, with 5,994 images in the training dataset and 5,794 images in the test set. The 200 classes is divided into 100 base classes to train and 50 novel classes to test.

4.2 Implementation Details

The first feature extractor in our experiments is ResNet10 which contains a total of 10 convolutional layers, the kernel size of them is 3×3 pixels where the kernel size of CBAM module is 7×7 pixels. Considering that the number of all categories in each of our dataset samples is not very large, our graph convolution module is set up with only two layers, and the latent hidden dimension is 512 as same as the output of the feature extractor and CBAM module. So our new class prototype can be well calibrated after a graph convolution. The nonlinear activation function ReLU is consistently used in both feature extractor, CBAM module and PMGCN module. We train our module for a total of 400 epochs meanwhile use Adam optimizer [33] with a learning rate of 0.0001 with early stopping based on the accuracy of the validation set. The value of k is used to determine the similarity between any two class prototypes. We set 0.7 for miniInageNet dataset and 0.9 for the CUB dataset.

Table 1. 1-shot and 5-shot accuracy of state-of-the art methods in the literature, compared with our proposed solution on miniImageNet.

Method	Backbone	miniImageNet	
		1-shot	5-shot
MAML [17]	ResNet10	54.79 ± 0.87	65.62 ± 0.93
ProtoNet [20]	ResNet10	61.73 ± 0.38	78.27 ± 0.77
TADAM [30]	ResNet10	$\mathbf{64.78 \pm 0.74}$	79.67 ± 0.43
Matching Networks [11]	ResNet10	62.98 ± 0.78	74.60 ± 0.99
Relation Network [13]	ResNet10	51.24 ± 0.76	72.20 ± 0.58
Baseline++ [31]	ResNet10	51.78 ± 0.74	75.86 ± 0.59
Meta SGD [32]	ResNet10	52.41 ± 0.66	73.20 ± 0.78
CMLA [22]	ResNet10	64.20 ± 0.64	80.24 ± 0.52
FL-GCN [26]	ResNet10	62.39 ± 0.45	79.86 ± 0.52
PMGCN(Ours)	ResNet10	63.97 ± 0.77	$\mathbf{82.13 \pm 0.32}$

Table 2. 1-shot and 5-shot accuracy of state-of-the art methods in the literature, compared with our proposed solution on tieredImageNet.

Method	Backbone	tieredImageNet	
		1-shot	5-shot
MAML [17]	ResNet10	56.67 ± 0.97	70.62 ± 0.37
ProtoNet [20]	ResNet10	65.43 ± 0.38	81.27 ± 0.88
TADAM [30]	ResNet10	65.78 ± 0.74	80.76 ± 0.34
Relation Network [13]	ResNet10	55.43 ± 0.66	80.31 ± 0.64
Matching Networks [11]	ResNet10	64.78 ± 0.75	81.60 ± 0.79
Baseline++ [31]	ResNet10	65.89 ± 0.46	82.66 ± 0.51
LEO [34]	ResNet10	62.41 ± 0.66	81.20 ± 0.78
FL-GCN [26]	ResNet10	60.39 ± 0.45	79.86 ± 0.50
PMGCN(Ours)	ResNet10	**66.78 ± 0.67**	**83.46 ± 0.72**

4.3 Experimental Results

We compare the performance of our work with state-of-the-art solutions on the 5-way 1-shot task and 5-way 5-shot task. The results with 95% confidence intervals is shown in Table 1, Table 2 and Table 3. As mentioned before we set the k value to 0.7 for miniimageNet and tieredimageNet while 0.9 for CUB dataset because they gave the best results experimentally and we will analyze the effect of different k values on the experimental results later in the ablation experiment. Based on the experimental results, it can be seen that our proposed method achieves state-of-the-art performance for both 1-shot and 5-shot classification results on any of the datasets. The result is better than using GCN only, which further demonstrates the effectiveness of the preserving measuring mapping. In particular, our method works best on the CUB dataset because the miniImageNet dataset contains a very large number of categories and therefore the similarity between categories is low while CUB dataset is a fine-grained bird dataset and

Table 3. 1-shot and 5-shot accuracy of state-of-the art methods in the literature, compared with our proposed solution on CUB.

Method	Backbone	CUB	
		1-shot	5-shot
MAML [17]	ResNet10	70.67 ± 0.21	81.64 ± 0.65
ProtoNet [20]	ResNet10	73.97 ± 0.88	88.46 ± 0.72
Matching Networks [11]	ResNet10	74.39 ± 0.92	87.67 ± 0.53
Relation Network [13]	ResNet10	73.34 ± 0.94	87.84 ± 0.68
Baseline++ [31]	ResNet10	73.53 ± 0.83	88.14 ± 0.61
PMGCN(Ours)	ResNet10	**75.39 ± 0.92**	**89.28 ± 0.65**

there is a high similarity between different categories of birds. This also reflects
the effectiveness of our proposed class prototype calibration.

4.4   Importance of Similarity Judgment

In this paper, we consider that if the value of cosine similarity between any two
class prototypes is greater than or equal to 0.7, there exists similarity between
these two class prototypes. Then there will be an edge between the vertices
V representing these two classes of prototypes respectively in our graph G.
To demonstrate the validity of our selected values, we used different k values
for 5-way 1-shot and 5-way 5-shot experiments on the miniImagenet dataset,
tieredimageNet dataset and CUB dataset, and the result is shown in Fig. 3.

From the data in the figure, the model classification accuracy is highest for
the miniimageNet and tieredimageNet when the value of k is 0.7 while 0.9 is the
best choice for the CUB dataset. When the value of k is small, it means that the
judgment of the similarity of any two class prototypes is not strict enough, then
it will inevitably bring a lot of redundant feature data to the class prototypes
and thus reduce the classification ability of the model. On the contrary, when
the value of k is large, the judgment of similarity becomes strict, which helps
to improve the classification ability of the model. Specially, we observe that the
optimal k-value of the CUB dataset is higher than that of miniimageNet. The
reason for this gap is that the CUB dataset itself is a fine-grained bird dataset
with a large degree of similarity among different categories, so if the value of k
is not strict enough, it will also cause redundant features and thus interfere with
the model judgment. It should be noted that when the value of k is 1, our model
degenerates to the prototypical network [20] and the accuracy is still lower than
when the value of k is 0.7 or 0.9, which also illustrates the effectiveness of our
model.

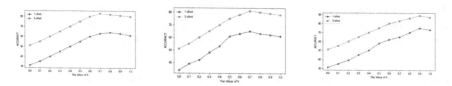

Fig. 3. The effect of different k values on the accuracy of the miniimageNet, tieredimageNet and CUB

4.5   Ablation Study

In this paper, we propose to use the attention module to obtain better feature
vectors for images and the base class prototypes are continuously updated during
the training phase. Next, we will conduct experiments on the miniimgaeNet
dataset for these two components to verify their validity.

The essence of the attention mechanism is to locate the information of interest and suppress useless information. In order to visualize the data after the CBAM module, we use t-SNE to downscale the high-dimensional feature vectors of the image data.

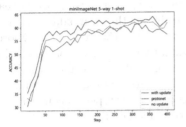 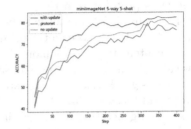

Fig. 4. The accuracy of 5-way 1-shot and 5-way 5-shot on miniimageNet with update mechanism and not and the accuray on the protonet.

Besides, we propose to update the base class prototype in the dictionary during the model training process to achieve better prototype calibration results, thus we conduct an ablation experiment to validate its validity. We perform 5-way 1-shot and 5-way 5-shot experiments on the miniimageNet dataset using the class prototype update mechanism and without the update mechanism respectively, and in order to illustrate the advantages of our model without using the update mechanism we also conducted an experiment using protoNet. The result is shown in Fig. 4. It can be observed that our model achieves better results than the prototype network with or without the update mechanism while the results with the update mechanism are better than without it.

Table 4. Ablation experiment on miniimageNet with 5-way 1-shot and 5-way 5-shot

UPDATE	CBAM	1-shot	5-shot
✓		60.99 ± 0.23	78.64 ± 0.97
	✓	62.33 ± 0.21	80.12 ± 0.65
✓	✓	63.97 ± 0.77	81.67 ± 0.32

Finally, we perform ablation experiments on the miniimageNet dataset to verify the effectiveness of our modules, UPDATE for the use of the PMGCN in our model and CBAM for the use of the CBAM module. The result is shown in the Table 4. The first row represents just the class prototype update mechanism without the attention mechanism, and we can notice a large decrease in accuracy on both 1-shot and 5-shot. The second row represents the use of the attention mechanism without the class prototype update mechanism, where the accuracy also decreases but to a slightly better degree compared to the first set of experiments.

5   Conclusion

In this paper, we propose a few-shot classification model using base class prototypes to calibrate the novel class prototypes which is able to obtain class prototypes with better representational power. Meanwhile our model will remember
the base class prototypes in the training phase, which is an important feature
that distinguishes it from other few-shot learning methods. In addition, in order
for our model to focus more on the primary content of the images and ignore the
secondary content to obtain a more discriminative feature vector for the image
data, we also use an attention mechanism module called CBAM, our model eventually achieves good results in some standard datasets in few-shot learning. Our
next work will focus on directly using base class prototypes to construct graph
G, rather than generating it through hyperparameters k.

Acknowledgement. This work is supported by National Key R&D Program of China
(2018YVFA0701700, 2018YFA0701701), NSFC (62176172, 61672364).

References

1. Thrun, S., Pratt, L.: Learning to Learn. Springer Science & Business Media, New
   York (2012). https://doi.org/10.1007/978-1-4615-5529-2
2. Hospedales, T., Antoniou, A., Micaelli, P., Storkey, A.: Meta-learning in neural
   networks: a survey. IEEE Trans. Pattern Anal. Mach. Intell. **44**(9), 5149–5169
   (2021)
3. Wang, Y., Wu, C., Herranz, L., Van de Weijer, J., Gonzalez-Garcia, A., Raducanu,
   B.: Transferring GANs: generating images from limited data. In: Proceedings of
   the European Conference on Computer Vision (ECCV), pp. 218–234 (2018)
4. Zhao, Y., Ding, H., Huang, H., Cheung, N.M.: A closer look at few-shot image
   generation. In: Proceedings of the IEEE/CVF Conference on Computer Vision
   and Pattern Recognition, pp. 9140–9150 (2022)
5. Gidaris, S., Singh, P., Komodakis, N.: Unsupervised representation learning by
   predicting image rotations. arXiv preprint arXiv:1803.07728 (2018)
6. Kipf, T.N., Welling, M.: Semi-supervised classification with graph convolutional
   networks. arXiv preprint arXiv:1609.02907 (2016)
7. He, X., Deng, K., Wang, X., Li, Y., Zhang, Y., Wang, M.: LightGCN: simplifying
   and powering graph convolution network for recommendation. In: Proceedings of
   the 43rd International ACM SIGIR Conference on Research and Development in
   Information Retrieval, pp. 639–648 (2020)
8. Wang, X., He, X., Wang, M., Feng, F., Chua, T.S.: Neural graph collaborative
   filtering. In: Proceedings of the 42nd International ACM SIGIR Conference on
   Research and Development in Information Retrieval, pp. 165–174 (2019)
9. Liu, F., Cheng, Z., Zhu, L., Liu, C., Nie, L.: An attribute-aware attentive GCN
   model for attribute missing in recommendation. IEEE Trans. Knowl. Data Eng.
   34(9), 4077–4088 (2020)
10. Woo, S., Park, J., Lee, J.Y., Kweon, I.S.: CBAM: convolutional block attention
    module. In: Proceedings of the European Conference on Computer Vision (ECCV),
    pp. 3–19 (2018)

11. Vinyals, O., Blundell, C., Lillicrap, T., Wierstra, D., et al.: Matching networks for one shot learning. Adv. Neural Inf. Process. Syst. **29** (2016)
12. Koch, G., Zemel, R., Salakhutdinov, R., et al.: Siamese neural networks for one-shot image recognition. In: ICML Deep Learning Workshop. Lille (2015)
13. Sung, F., Yang, Y., Zhang, L., Xiang, T., Torr, P.H., Hospedales, T.M.: Learning to compare: relation network for few-shot learning. In: Proceedings of the IEEE Conference on Computer Vision and Pattern Recognition, pp. 1199–1208 (2018)
14. Sun, W.: Ergodic Theory. Peking University Press, Beijing (2018)
15. Walters, P.: An Introduction to Ergodic Theory, vol. 79. Springer Science & Business Media, New York (2000)
16. Xu, H., Zhang, C., Wang, J., Ouyang, D., Zheng, Y., Shao, J.: Exploring parameter space with structured noise for meta-reinforcement learning. In: Proceedings of the Twenty-Ninth International Conference on International Joint Conferences on Artificial Intelligence, pp. 3153–3159 (2021)
17. Finn, C., Abbeel, P., Levine, S.: Model-agnostic meta-learning for fast adaptation of deep networks. In: International Conference on Machine Learning, pp. 1126–1135. PMLR (2017)
18. Ravi, S., Larochelle, H.: Optimization as a model for few-shot learning. In: International Conference on Learning Representations (2016)
19. Li, G., Zheng, C., Su, B.: Transductive distribution calibration for few-shot learning. Neurocomputing **500**, 604–615 (2022)
20. Snell, J., Swersky, K., Zemel, R.: Prototypical networks for few-shot learning. Adv. Neural Inf. Process. Syst. **30** (2017)
21. Santoro, A., Bartunov, S., Botvinick, M., Wierstra, D., Lillicrap, T.: Meta-learning with memory-augmented neural networks. In: International Conference on Machine Learning, pp. 1842–1850. PMLR (2016)
22. Xu, K., et al.: Show, attend and tell: neural image caption generation with visual attention. In: International Conference on Machine Learning, pp. 2048–2057. PMLR (2015)
23. Hu, J., Shen, L., Sun, G.: Squeeze-and-excitation networks. In: Proceedings of the IEEE Conference on Computer Vision and Pattern Recognition, pp. 7132–7141 (2018)
24. Zhang, H., Zou, J., Zhang, L.: EMS-GCN: an end-to-end mixhop superpixel-based graph convolutional network for hyperspectral image classification. IEEE Trans. Geosci. Remote Sens. **60**, 1–16 (2022)
25. Ren, M., et al.: Meta-learning for semi-supervised few-shot classification. arXiv preprint arXiv:1803.00676 (2018)
26. Liu, F., Qian, X., Jiao, L., Zhang, X., Li, L., Cui, Y.: Contrastive learning-based dual dynamic GCN for SAR image scene classification. IEEE Transactions on Neural Networks and Learning Systems (2022)
27. Ding, Y., Zhao, X., Zhang, Z., Cai, W., Yang, N.: Multiscale graph sample and aggregate network with context-aware learning for hyperspectral image classification. IEEE J. Sel. Top. Appl. Earth Obs. Remote Sens. **14**, 4561–4572 (2021)
28. Vaswani, A., et al.: Attention is all you need. Adv. Neural Inf. Process. Syst. **30** (2017)
29. Wah, C., Branson, S., Welinder, P., Perona, P., Belongie, S.: The caltech-ucsd birds-200-2011 dataset (2011)
30. Oreshkin, B., Rodríguez López, P., Lacoste, A.: TADAM: task dependent adaptive metric for improved few-shot learning. Adv. Neural Inf. Process. Syst. **31** (2018)
31. Chen, W.Y., Liu, Y.C., Kira, Z., Wang, Y.C.F., Huang, J.B.: A closer look at few-shot classification. arXiv preprint arXiv:1904.04232 (2019)

32. Li, Z., Zhou, F., Chen, F., Li, H.: Meta-SGD: learning to learn quickly for few-shot learning. arXiv preprint arXiv:1707.09835 (2017)

33. Kingma, D.P., Ba, J.: Adam: a method for stochastic optimization. arXiv preprint arXiv:1412.6980 (2014)

34. Rusu, A.A., et al.: Meta-learning with latent embedding optimization. arXiv preprint arXiv:1807.05960 (2018)

ARE-CAM: An Interpretable Approach to Quantitatively Evaluating the Adversarial Robustness of Deep Models Based on CAM

Zituo Li, Jianbin Sun[✉], Yuqi Qin, Lunhao Ju, and Kewei Yang

College of Systems Engineering, National University of Defense Technology, Changsha 410073, China
sunjianbin@nudt.edu.cn

Abstract. Evaluating the adversarial robustness of deep models is critical for training more robust models. However, few methods are both interpretable and quantifiable. Interpretable evaluation methods cannot quantify adversarial robustness, leading to unobjective evaluation results. On the other hand, quantifiable evaluation methods are often unexplainable, making it difficult for evaluators to trust and trace the results. To address this issue, an adversarial robustness evaluation approach based on class activation mapping (ARE-CAM) is proposed. This approach utilizes CAM to generate heatmaps and visualize the areas of concern for the model. By comparing the difference between the original example and the adversarial example from the perspective of visual and statistical characteristics, the changes in the model after being attacked are observed, which enhances the interpretability of the evaluation. Additionally, four metrics are proposed to quantify adversarial robustness: the average coverage coincidence rate (ACCR), average high activation coincidence rate (AHCR), average heat area difference (AHAD) and average heat difference (AHD). Comprehensive experiments are conducted based on 14 deep models and different datasets to verify ARE-CAM's efficiency. To the best of our knowledge, ARE-CAM is the first quantifiable and interpretable approach for evaluating adversarial robustness.

Keywords: Adversarial robustness evaluation · Alass activation mapping · Adversarial attack and defense · Deep neural network

1 Introduction

In the field of image classification, deep neural networks are vulnerable to adversarial perturbations, which poses a security risk in real-world applications [8,13]. Therefore, it is essential to improve the ability of a deep neural network to resist adversarial perturbations, that is, to enhance adversarial robustness. Improving adversarial robustness scientifically requires reliable evaluation methods. However, the current evaluation methods for adversarial robustness are difficult to

© The Author(s), under exclusive license to Springer Nature Switzerland AG 2024
S. Rudinac et al. (Eds.): MMM 2024, LNCS 14554, pp. 273–285, 2024.
https://doi.org/10.1007/978-3-031-53305-1_21

quantify and explain. Interpretable evaluation methods lack the ability to accurately quantify adversarial robustness, while quantifiable evaluation methods cannot explain the internal representation and decision results of the model. Therefore, a new evaluation method is needed to scientifically improve adversarial robustness.

Adversarial robustness evaluation methods are categorized into two groups, i.e., ranking evaluation methods and metric evaluation methods [6,10]. The ranking evaluation methods compare the performance of the model under attack and defense algorithms and provide robustness evaluation results in the form of rankings. Platforms such as Ares [5], Cleverhans [14], AdverTorch [4] and Deep-Robust [9] evaluate the adversarial robustness of the model by ranking attack results and defense results. However, these ranking evaluation methods lack a consistent metric to quantify adversarial robustness since the evaluation results are derived by comparison.

In metric evaluation methods, many studies have focused on evaluating the entire image classification process from the perspectives of model output and neuron response. For metrics based on model output, most papers further optimize accuracy (ACC) and attack success rate (ASR) and describe the behavior of the model under adversarial conditions by prediction confidence [11,12]. In addition, ACC and the ASR are plotted into curves to visualize the evaluation results [5,7]. For metrics based on neuron response, researchers have attempted to extract information about neurons and loss function inside the model [3,18]. These metrics evaluate the adversarial robustness from the model itself, with the most authoritative being the CLEVER score and its second-order form [16,17]. The metrics use the lower bound of the minimum adversarial disturbance distance to estimate the adversarial robustness of the model. However, existing metric evaluation methods have limitations in understanding and interpreting evaluation results because the model is a black box. For example, although the position of the curve can be observed to compare the robustness of different models [7], it is difficult to trace the reasons for the poor robustness of the model based on this result. Therefore, it is challenging to effectively improve the ability of the model to resist attacks.

To improve the interpretability of deep learning models, class activation mapping (CAM) [19] has been proposed and widely adopted in various fields of weakly supervised learning [1]. CAM generates a class activation graph related to the target class by weighting and adjusting the last layer of the feature graph. It is easy to understand the process of model classification by examining the highlighted areas of this graph. However, CAM cannot be applied to all models, which limits its universality. To address this issue, gradient-weighted class activation mapping (Grad-CAM) [15] has been proposed. Grad-CAM generates heatmaps that provide visual explanations of deep model predictions and is more universal than CAM. A variation of Grad-CAM, Grad-CAM++, which provides even better visual explanations of deep model predictions, has also been proposed [2]. In conclusion, heatmaps generated by CAM and its variants provide a way to interpret the feature analysis and decision results of deep learning models.

These heatmaps can help researchers and practitioners understand the internal representation and decision-making process of deep learning models.

Four metrics are proposed based on the visual and statistical characteristics of heatmaps. These metrics include the average coverage coincidence rate (ACCR), average high activation coincidence rate (AHCR), average heat area difference (AHAD) and average heat difference (AHD). Furthermore, an interpretable and quantifiable method named ARE-CAM has been proposed to evaluate adversarial robustness. ARE-CAM uses CAM to generate heatmaps that visualize weighted combinations of the resulting feature maps. ARE-CAM then evaluates adversarial robustness by comparing heatmaps before and after adversarial attacks. This method can quantify the robustness of the model, mine the tacit knowledge of the model, and enhance the interpretability of the evaluation. Overall, this method provides a way to quantify the robustness and interpret the decision-making process of deep learning models. This method can help researchers and practitioners better understand and improve the performance of deep learning models in various applications.

The main contributions of this paper are summarized as follows:

1. Difficulty in quantifying and interpreting adversarial robustness evaluation is identified. Quantifiable methods lack interpretability, while on the other hand, interpretable methods cannot be quantified.
2. CAM technology is introduced to enhance the interpretability of the adversarial robustness evaluation process.
3. Four metrics, AHCR, ACCR, AHAD and AHD, which enable the evaluation of adversarial robustness to be quantified, are provided.
4. ARE-CAM, a new approach that is both calculable and interpretable for evaluating adversarial robustness, is proposed. Its practicability and effectiveness are verified by experiments on 14 state-of-the-art neural network models.

The structure of the paper is illustrated as follows: Sect. 2 defines and provides details of our proposed approach, ARE-CAM: An interpretable approach to calculably evaluating the adversarial robustness based on CAM; Sect. 3 verifies the availability and effectiveness of ARE-CAM by experiments; and Sect. 4 summarizes and concludes the paper.

2   ARE-CAM: Adversarial Robustness Evaluation Based on CAM

2.1   Preliminaries

The ARE-CAM method allows for the use of any CAM technology. For the original example x_i and adversarial example x_i^{adv}, Grad-CAM++ is utilized to produce two heatmaps in this paper due to its strong positioning ability.

There are N heatmaps of size $m \times n$. For the heatmap of x_i, the heat level of each pixel (i, j) is k_{ij}, while for the heatmap of x_i^{adv}, it is k_{ij}^{adv}. Each pixel in the heatmap is assigned a corresponding HSV value, which is then divided into eight heat levels based on the HSV color table presented in Table 1. These eight heat levels correspond to eight distinct colors, with red indicating higher heat levels. This approach enables the quantification and visualization of the significance of information in an image using the heatmaps displayed in Fig. 1. In this study, the area with a heat level greater than 5 is identified as the high activation area, which facilitates the differentiation of objects in the image from the background. Note that this parameter can be adjusted based on the specific requirements of the study.

Table 1. HSV color table

	Red	Orange	Yellow	Green	Cyan	Blue	Purple	Other
H_{min}	0/156	11	26	35	78	100	125	0
H_{max}	10/180	25	34	77	99	124	155	180
S_{min}	43	43	43	43	43	43	43	0
S_{max}	255	255	255	255	255	255	255	255/43/30
V_{min}	46	46	46	46	46	46	46	0/46/221
V_{max}	255	255	255	255	255	255	255	46/220/255
Heat levels	7	6	5	4	3	2	1	0

(a) Quantified heatmap before attack   (b) Quantified heatmap after attack

Fig. 1. Quantization based on heatmap

2.2 Metrics of ARE-CAM

There are significant differences between the two heatmaps of the original and adversarial examples. To quantify these differences, four metrics are proposed to evaluate adversarial robustness by integrating the visual and statistical characteristics. These metrics include the average coverage coincidence rate (ACCR), average high activation coincidence rate (AHCR), average heat area difference (AHAD) and average heat difference (AHD).

The ACCR is defined as the percentage of overlapping areas with the same heat level in two heatmaps. It describes the model's ability to locate objects and backgrounds in images before and after being attacked. A high ACCR means that the model can still accurately distinguish and locate the object and background in the image after being attacked, indicating that the model has strong adversarial robustness.

$$ACCR = \begin{cases} \frac{1}{N} \times \frac{\sum_{i=1}^{m} \sum_{j=1}^{n} n_{ij}^{adv}}{m \times n}, & k_{ij} = k_{ij}^{adv}, n_{ij}^{adv} = 1 \\ 0, & n_{ij}^{adv} = 0 \end{cases}, \qquad (1)$$

where if two heatmaps have the same heat level at (i,j), n_{ij}^{adv} is equal to 1. Otherwise, n_{ij}^{adv} is 0.

High activation areas are often the focus of attention in image classification tasks, as they correspond to the most important regions for the classification task. Therefore, the AHCR is proposed as a metric that represents the percentage of areas overlapped with high activation areas in the two heatmaps.

$$AHCR = \begin{cases} \frac{1}{N} \times \frac{\sum_{i=1}^{m} \sum_{j=1}^{n} N_{ij}^{adv}}{S_{high}}, & k_{ij} = k_{ij}^{adv} \geq K, N_{ij}^{adv} = 1 \\ 0, & N_{ij}^{adv} = 0 \end{cases}, \qquad (2)$$

where K represents the heat levels of high activation areas. S_{high} is the high activation area in the original example. If two heatmaps have the same heat level at (i,j) and the heat level is greater than K, N_{ij}^{adv} is equal to 1. Otherwise, N_{ij}^{adv} is 0. Because the heat level of the high activation area is artificially given, different values will yield unique evaluation results.

In addition to the ACCR and AHCR, there are two such metrics AHAD and AHD that calculate adversarial robustness from the perspective of statistical characteristics, which provide a more accurate evaluation result. AHAD and AHD can overcome the limitations of the ACCR and AHCR, which may lose some information during the calculation due to human factors.

The AHAD is defined as the area difference between regions of each heat level in two heatmaps. The AHAD describes changes in the ability to accurately perceive the importance of information after a model is attacked. When the change decreases, the model's ability to identify whether the information is important becomes stronger, and the adversarial robustness increases.

$$AHAD = \frac{\sum_{q=0}^{7} \sum_{i=1}^{m} \sum_{j=1}^{n} \left[Num_q(k_{ij} = q) - Num_q^{adv}(k_{ij}^{adv} = q) \right]}{N}, \qquad (3)$$

where $Num_q(\cdot)$ denotes the number of pixel points with heat level q in the heatmap of the original example and $Num_q^{adv}(\cdot)$ denotes the number of pixel points with heat level q in the heatmap of the adversarial example.

The AHD is defined as the difference in heat levels at the same location in two heatmaps. Compared with the AHAD, the AHD incorporates location information and computes each coordinate point, which provides keener observations of the changes in model robustness. As the difference decreases, the sensitivity of the model to the adversarial examples decreases, and the robustness of the model increases.

$$AHD = \frac{1}{N} \times \frac{\sum_{i=1}^{m} \sum_{j=1}^{n} \left| k_{ij} - k_{ij}^{adv} \right|}{m \times n}, \tag{4}$$

In theory, the accuracy of the four metrics, from high to low, should be the AHD, ACCR, AHCR, and AHAD. Furthermore, according to the definition, the computational efficiency of the four metrics ranges from high to low is AHCR, AHAD, ACCR and AHD. Therefore, it is important to consider the trade-off between accuracy and computational cost when choosing metrics to evaluate adversarial robustness.

In summary, the four metrics of ARE-CAM can quantify the adversarial robustness from both a visual perspective and statistical perspective. The evaluator can choose metrics to evaluate the adversarial robustness according to the demand.

2.3   Workflow of ARE-CAM

The workflow of ARE-CAM is presented in Fig. 2. This workflow enables the evaluation of the adversarial robustness of a model by comparing the differences in the heatmaps of the original and adversarial examples. By using this workflow, the metrics provide a quantitative and interpretable evaluation of the model's ability to maintain its prediction accuracy even when faced with adversarial attacks.

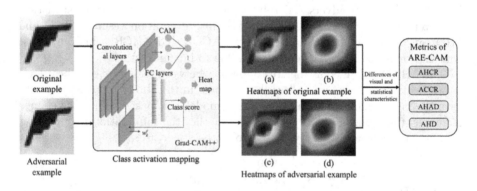

Fig. 2. ARE-CAM workflow for evaluating adversarial robustness

The first step is to generate adversarial examples using attack algorithms. Second, Grad-CAM++ is employed to generate heatmaps of the original and adversarial examples. These heatmaps illustrate the changes in the neurons within the model and the shift in the model's focus area before and after the attack. This step extracts implicit information from the black box model and enhances the interpretability of the evaluation process. Note that other CAM technologies are available in addition to Grad-CAM++. Here, Grad-CAM++

with strong recognition and positioning ability is selected for verification. The ARE-CAM value is calculated based on the generated heatmaps. By utilizing this feature, four metrics based on heatmaps are calculated to quantify the changes in the attacked model, and subsequently, adversarial robustness is evaluated.

3 Experiments

3.1 Settings

Datasets and Models. We conduct experiments on the benchmark standard datasets MNIST[1] and CIFAR-10[2]. For each dataset, 400 images are randomly sampled to evaluate the robustness metrics. We extensively calculate the four metrics on the 14 neural network models, which are pretrained models provided in PyTorch, including ResNet18 (Res-18), ResNet34 (Res-34), ResNet50 (Res-50), ResNet101 (Res-101), SqueezeNet1_0 (SN-V1.0), SqueezeNet1_1 (SN-V1.1), DenseNet121 (DN-121), DenseNet169 (DN-169), DenseNet201 (DN-201), MobileNetV2 (MN-V2), VGG11, VGG13, VGG16, and VGG19. They are deep neural network models with different structures and layers that are frequently utilized in image classification tasks.

Attack and Defense Algorithms. Both white-box attacks and black-box attacks are adopted to test the model, including the Fast Gradient Sign Method (FGSM), the Projected Gradient Descent (PGD) method, DeepFool and One-pixel attack. PGD adversarial training (PAT) is applied as a defense method and achieves state-of-the-art accuracy against a wide range of attacks on models. All of the attack and defense algorithms used the open-source DeepRobust package.

3.2 Results and Analysis

In this section, the ACCR, AHCR, AHAD and AHD are separately calculated on both natural models and robust models to conduct further analysis on ARE-CAM.

Analysis of Metrics Before and After Defense. A robust model can maintain a stable classification capability after being attacked, which means that the two heatmaps will not significantly change before and after the attack. Based on the statistical characteristics of different regions in the heatmap, ARE-CAM can describe the differences between the original examples and the adversarial examples. Consequently, the changes in four metrics before and after PAT are calculated to verify the effectiveness of the metrics. Table 3 shows the differences between two heatmaps generated based on the natural model and robust model.

---

[1] http://yann.lecun.com/exdb/mnist/.
[2] http://www.cs.toronto.edu/~kriz/cifar.html.

The changes in four metrics before and after PAT are calculated to verify the effectiveness of the metrics. Figure 3 and Fig. 4 compare four metrics of ARE-CAM on natural and robust models for MNIST and CIFAR-10 under different attacks.

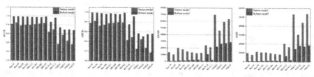

(a) Comparison of ARE-CAM's four metrics between natural models and robust models under the Deepfool attack

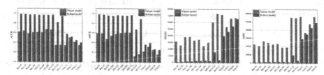

(b) Comparison of ARE-CAM's four metrics between natural models and robust models under the FGSM attack

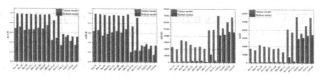

(c) Comparison of ARE-CAM's four metrics between natural models and robust models under the PGD attack

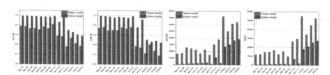

(d) Comparison of ARE-CAM's four metrics between natural models and robust models under the One-pixel attack

Fig. 3. Comparison of ARE-CAM's four metrics between natural models and robust models on MNIST

As shown in Fig. 3 and Fig. 4, after PAT, ACCR and AHCR increased by different degrees, while AHAD and AHD decreased by different amounts. These results indicate that the proposed approach is effective in evaluating the adversarial robustness of the model. The AHCR and ACCR metrics increase, while the AHAD and AHD metrics decrease with the enhancement of adversarial robustness.

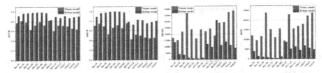

(a) Comparison of ARE-CAM's four metrics between natural
models and robust models under the DeepFool attack

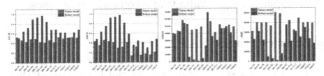

(b) Comparison of ARE-CAM's four metrics between natural
models and robust models under the FGSM attack

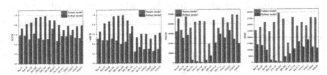

(c) Comparison of ARE-CAM's four metrics between natural
models and robust models under the PGD attack

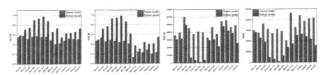

(d) Comparison of ARE-CAM's four metrics between natural
models and robust models under the One-pixel attack

Fig. 4. Comparison of ARE-CAM's four metrics between natural models and robust
models on CIFAR-10

In addition, the relationship among model structure, number of layers and
adversarial robustness has been investigated. A comparison of models with a
similar number of layers in Fig. 3 and Fig. 4 reveals that four metrics of VGG11,
VGG13, VGG16 and VGG19 are lower on average than Res-18, SN-V1.0 and
SN-V1.1. Based on the structure of these models, it can be concluded that the
pooling layer can improve the adversarial robustness to some extent. VGG models only have a convolution layer and fully connected layer and lack the pooling
layer that other models have. Therefore, the AHCR and ACCR of the VGG
models are lower, while the AHAD and AHD are higher. On the other hand,
models with a larger number of layers are generally more robust. A comparison
of DN-121, DN-169, DN-201, SN-V1.0 and SN-V1.1 reveals that the AHCR and
ACCR of the DN model are higher than those of SN, while the AHAD and

AHD are lower than SN. However, adding too many layers to the model may not improve its adversarial robustness. For example, the AHCR and ACCR of DN-201 are not higher than those of DN-169, and the AHAD and AHD are not lower than those of DN-169. These findings can help guide the evaluation and improvement of the model's adversarial robustness.

In summary, the CAM-based metrics proposed by ARE-CAM can effectively measure and explain the adversarial robustness of the model. It is a calculable and interpretable approach for evaluating adversarial robustness.

Analysis of Metrics Under PAT with Different Epochs. The effectiveness of the metrics is verified by observing the changes under PAT every 20 epochs from 20 epochs to 80 epochs. Due to the large size of the experiment, only two models, DN-169 and Res-34, with different levels of complexity are selected and tested on MNIST or CIFAR-10. The results are not affected by whether the attack algorithm is a white-box algorithm or black-box algorithm. Therefore, the FGSM and DeepFool, which have high attack efficiency, are chosen for the experiment. Figure 5(a) shows the changes in the four metrics of DN-169 under the FGSM attack on MNIST with increasing epochs. Figure 5(b) shows the changes in the four metrics of Res-34 under the DeepFool attack on CIFAR-10 with increasing epochs.

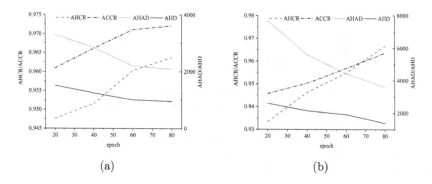

(a)   (b)

Fig. 5. Changes in the four metrics under defense training with different epochs. Figure (a) shows the changes in four metrics of DN-169 under the FGSM attack on MNIST with increasing epochs. Figure (b) shows the changes in the four metrics of Res-34 under the DeepFool attack on CIFAR-10 with increasing epochs.

Figure 5 shows that ACCR and AHCR monotonically increase with increasing epochs, while AHAD and AHD monotonically decrease. In Fig. 5(a), after 60 epochs of PAT, the AHCR increases by 1.6%, the ACCR increases by 1.1%, the AHAD decreases by 1208.975, and the AHD decreases by 564.25. In Fig. 5(b), the AHCR increases by 3.3%, the ACCR increases by 1.8%, the AHAD decreases by 4086.500, and the AHD decreases by 1214.460. The increase in training epochs is helpful to improve the adversarial robustness of the model. Therefore, a positive correlation is observed between adversarial robustness and the ACCR and AHCR. Moreover, there is a negative correlation between adversarial robustness and AHAD and AHD. In summary, these results show that the proposed metrics reflect the change in adversarial robustness.

3.3   Comparison with Baseline Metrics

The CLEVER score, ROBY, neuron sensitivity(NS) and ASR are the most widely adopted evaluation metrics of adversarial robustness. If there is a strong correlation between these metrics and ARE-CAM, then ARE-CAM is considered a feasible evaluation approach. To further validate the practicability of ARE-CAM, the Pearson correlation coefficient is used to measure the strength of the correlation between ARE-CAM and three metrics.

First, the baseline metrics of fourteen robust models are first calculated. Second, the ARE-CAM method is used to calculate the four metrics of these models. Third, the robustness of the model is ranked according to the index size. Finally, the Pearson correlation coefficients of the baseline metrics and the proposed metrics are computed based on the rank and presented in Table 2. For the CIFAR-10 and MNIST datasets, six metrics are the average values under four attack algorithms.

Table 2. Pearson correlation coefficients of the baseline metrics and the proposed metrics

Datasets	Pearson correlation coefficient	ACCR	AHCR	AHAD	AHD
MNIST	ASR	−0.883	−0.902	0.877	0.868
	NS	−0.844	−0.925	0.899	0.907
	ROBY	−0.801	−0.854	0.858	0.860
	CLEVER score	−0.883	−0.865	0.885	0.886
CIFAR-10	ASR	−0.926	−0.979	0.780	0.627
	NS	−0.936	−0.952	0.810	0.845
	ROBY	−0.820	−0.891	0.679	0.801
	CLEVER score	−0.812	−0.963	0.680	0.723

Table 2 presents the results of the Pearson correlation coefficients between the proposed metrics and the baseline metrics for both the MNIST and CIFAR-10 datasets. For both datasets, the Pearson correlation coefficients of the AHCR and ACCR with the baseline metrics are in the range of -0.80 to -0.98. Additionally, for the MNIST dataset, both AHAD and AHD are above 0.85, while for CIFAR-10, their values range from 0.62 to 0.85. Further analysis shows that the AHCR, ACCR AHAD and AHD have good consistency with the baseline metrics. In summary, ARE-CAM can efficiently evaluate adversarial robustness by calculating the difference in heatmaps before and after attack.

Notably, the AHCR and ACCR have high Pearson correlation coefficients with the baseline metrics in CIFAR-10. However, the correlation coefficients among the AHAD, AHD, and the baseline metrics in CIFAR-10 are unsatisfactory. One possible reason is that the images in the CIFAR-10 dataset include noise as well as real-world objects, rendering them susceptible to noise interference during the evaluation process. This interference leads to the metrics based on statistical characteristics being unable to better evaluate adversarial robustness. Therefore, the AHAD and AHD are more suitable for evaluation in a digital environment, while the AHCR and ACCR are more suitable for evaluation in a physical environment.

Furthermore, the correlation coefficients from high to low are the AHCR, ACCR, AHAD, and AHD. From this point of view, the AHCR is more suitable for evaluation since it measures the changes in the core areas of the image before and after the model is attacked. The AHCR is closely related to the principle that attack algorithms will shift the focus of the model. However, in certain scenarios, such as image multiobject classification tasks, other metrics may be more suitable than the AHCR.

4   Conclusion

A new approach to evaluate adversarial robustness, ARE-CAM, has been proposed. This approach is both calculable and interpretable and utilizes four quantitative metrics: AHCR, ACCR, AHAD, and AHD. ARE-CAM evaluates adversarial robustness from the perspective of visual and statistical characteristics by using the difference between the heatmaps of the original example and those of the adversarial example. ARE-CAM enhances the interpretability of the adversarial robustness evaluation process. Extensive experiments have demonstrated that ARE-CAM matches baseline metrics such as ASR and CLEVER scores on deep models. Thus, ARE-CAM can serve as an efficient approach for evaluating the adversarial robustness of models. We hope that our study provides a pathway to quantifiable, interpretable evaluation and lays the foundation for building robust models.

Future studies should explore more experiments to validate the universality and reliability of ARE-CAM and reveal a quantitative relationship between them and adversarial robustness. In addition, improving the ARE-CAM to make the evaluation more interpretable is also worthy of investigation.

Acknowledgment. This work was supported in part by the National Natural Science Foundation of China under Grant No. 71901212 and No. 72071206, in part by Key Projects of the National Natural Science Foundation of China under Grant No. 72231011.

References

1. Bae, W., Noh, J., Kim, G.: Rethinking class activation mapping for weakly supervised object localization. vol. 12360 LNCS, pp. 618–634. Glasgow, United kingdom (2020)
2. Chattopadhay, A., Sarkar, A., Howlader, P., Balasubramanian, V.N.: Gradcam++: generalized gradient-based visual explanations for deep convolutional networks. vol. 2018-January, pp. 839–847. Lake Tahoe, NV, United states (2018)
3. Chen, S.H., Shen, H.J., Wang, R., Wang, X.Z.: Relationship between prediction uncertainty and adversarial robustness. J. Softw. **33**(2), 524–538 (2022)
4. Ding, G.W., Wang, L., Jin, X.: Advertorch v0. 1: an adversarial robustness toolbox based on pytorch. arXiv preprint arXiv:1902.07623 (2019)
5. Dong, Y., Fu, et al.: Benchmarking adversarial robustness on image classification, pp. 318–328. Virtual, Online, United states (2020)
6. Guo, J., et al.: A comprehensive evaluation framework for deep model robustness: an evaluation framework for model robustness. Pattern Recogn. **137**, 109308 (2023)
7. Šircelj, J., Skoaj, D.: Accuracy-perturbation curves for evaluation of adversarial attack and defence methods, pp. 6290–6297. Virtual, Milan, Italy (2020)
8. Ju, L., Cui, R., Sun, J., Li, Z.: A robust approach to adversarial attack on tabular data for classification algorithm testing, pp. 371–376. Guiyang, China (2022)
9. Li, Y., Jin, W., Xu, H., Tang, J.: Deeprobust: a pytorch library for adversarial attacks and defenses. arXiv preprint arXiv:2005.06149 (2020)
10. Li, Z., Sun, J., Yang, K., Xiong, D.: A review of adversarial robustness evaluation for image classification. J. Comput. Res. Develop. **59**(10), 2164–2189 (2022)
11. Ling, X., et al.: Deepsec: A uniform platform for security analysis of deep learning model, vol. 2019-May, pp. 673–690. San Francisco, CA, United states (2019)
12. Luo, B., Liu, Y., Wei, L., Xu, Q.: Towards imperceptible and robust adversarial example attacks against neural networks, pp. 1652–1659. New Orleans, LA, United states (2018)
13. Moosavi-Dezfooli, S.M., Fawzi, A., Frossard, P.: Deepfool: a simple and accurate method to fool deep neural networks, vol. 2016-December, pp. 2574–2582. Las Vegas, NV, United states (2016)
14. Papernot, N., et al.: Technical report on the cleverhans v2. 1.0 adversarial examples library. arXiv preprint arXiv:1610.00768 (2016)
15. Selvaraju, R.R., Cogswell, M., Das, A., Vedantam, R., Parikh, D., Batra, D.: Gradcam: visual explanations from deep networks via gradient-based localization. Int. J. Comput. Vision **128**(2), 336–359 (2020)
16. Weng, T.W., Zhang, H., Chen, P.Y., Lozano, A., Hsieh, C.J., Daniel, L.: On extensions of clever: a neural network robustness evaluation algorithm, pp. 1159–1163. Anaheim, CA, United states (2018)
17. Weng, T.W., et al.: Evaluating the robustness of neural networks: an extreme value theory approach. Vancouver, BC, Canada (2018)
18. Zhang, C., et al.: Interpreting and improving adversarial robustness of deep neural networks with neuron sensitivity. IEEE Trans. Image Process. **30**, 1291–1304 (2021)
19. Zhou, B., Khosla, A., Lapedriza, A., Oliva, A., Torralba, A.: Learning deep features for discriminative localization. vol. 2016-December, pp. 2921–2929. Las Vegas, NV, United states (2016)

SSK-Yolo: Global Feature-Driven Small Object Detection Network for Images

Bei Liu$^{(\boxtimes)}$, Jian Zhang, Tianwen Yuan, Peng Huang, Chengwei Feng, and Minghe Li

Beijing University of Technology, Beijing, China
liubei@emails.bjut.edu.cn

Abstract. Timely and effective locust detection to prevent locust plagues is crucial for safeguarding agricultural production and ecological balance. However, under natural conditions, the "colour mixing mechanism" of locusts and the small scale of locusts in high-resolution images make it difficult to detect locusts. In this study, we propose a multi-scale prediction network SSK-Yolo based on YoloV5 to effectively solve the above two problems. Firstly, in the data preprocessing stage, in order to better adapt to the relatively small-scale targets, we use the k-means algorithm to cluster the a priori frames to obtain anchor frames with appropriate scale sizes. Secondly, in the backbone, we still use the traditional convolution to extract the shallow graphical features, and we use swin-transformer to extract the deep semantic features, so as to improve the accuracy of feature extraction and fusion for small targets in high-resolution images. In addition, in the data post-processing stage, we replace the NMS algorithm with the soft-nms algorithm by setting a Gaussian function for the neighbouring detection frames based on the overlapping part instead of suppressing all of them. A series of experimental results on the publicly available East Asian locust dataset demonstrate that SSK-Yolo outperforms YoloV5 with a 5% improvement in precision, 1.64% in recall, 12% in mAP, and 2.66% in F1-score. SSK-Yolo provides an efficient and viable solution for locust detection in the field of pest and disease control.

Keywords: Swin-Transformer · YoloV5 · soft-nms · k-means · Locust detection

1 Introduction

Throughout human civilization, locusts have been one of the most destructive and threatening pests. Early detection of locust populations is of paramount importance as it can effectively prevent the development of locust swarms into full-fledged locust plagues. Locust detection is a crucial means for studying the mechanisms of locust outbreaks and establishing predictive models. Establishment of a locust detection network that regularly detects areas where

© The Author(s), under exclusive license to Springer Nature Switzerland AG 2024
S. Rudinac et al. (Eds.): MMM 2024, LNCS 14554, pp. 286–299, 2024.
https://doi.org/10.1007/978-3-031-53305-1_22

locusts are likely to be present and issues early warnings when locust populations exceed thresholds, so that follow-up measures can be taken to reduce the probability of locust infestation. Traditional Pest and Disease Recognition algorithms [1,3,9,19,21,23,24] typically involve manual feature engineering and common classifiers (such as Random Forest, region selection, support vector machines, and naive Bayes classifiers) for image classification. However, these methods require more manual engineering and specialized techniques, making them less scalable to other tasks and datasets. They are only effective in scenarios with small datasets or specific task requirements. Therefore, in the field of pest and disease recognition, several object detection algorithms based on convolutional neural networks (CNNs) have been proposed [8,17,28]. These algorithms significantly outperform traditional methods in terms of accuracy, stability, and practical application. They utilize CNNs for feature extraction and bounding box regression, combining feature pyramids from different scales to detect objects of varying sizes. However, due to the complex outdoor conditions, the object detection model described above trained on locust images captured in laboratory settings are not suitable for detecting locusts in natural environments. This primarily involves two issues: Firstly, Locusts in high-resolution images are relatively small in size, and the aforementioned algorithmic models do not address the problem of small object detection. Secondly, In natural environments, there can be instances of occlusion among locusts, which the aforementioned algorithmic models do not consider. These two categories of issues pose significant challenges for locust detection tasks in natural conditions. To address the aforementioned issues, this paper introduces a detection network mechanism driven by both geometric and semantic features, referred to as SSK-Yolo.

The main contributions of this paper are as follows: Firstly, we incorporate the Swim-Transformer into the C3 module of the backbone network. Through sliding window operations and self-attention computations within the window, the model allows for hierarchical adaptation to images of different scales. This enhances the accuracy of feature extraction and fusion for small targets in high-resolution images. This advancement represents a qualitative leap in the model's capability for feature extraction of small targets. Secondly, in the preprocessing stage of the data, we employ the k-means method for clustering operations on the dataset to obtain the optimal prior anchor box sizes. In the post-processing of object detection, we replace the NMS algorithm with Gaussian-penalized soft-nms. soft-nms enhances the model's success rate in detecting occluded targets. Thirdly, we propose a comprehensive multi-scale prediction network, SSK-Yolo, which demonstrates a certain level of advancement in addressing both small target and occlusion issues in the field of locust detection.

Comparative experiments and ablation studies demonstrate that the proposed SSK-Yolo network exhibits a certain level of advancement, enabling the detection of locust images in real-world scenarios. This contribution provides a reference for the subsequent field of pest detection in agricultural contexts.

2 Related Work

2.1 Machine Learning

Machine Learning algorithms were first used to solve problems in the field of pest and disease identification. Al Bashish et al. [2] used the color co-occurrence method to extract texture features for five plant diseases and then classified them in a neural network. Rothe et al. [23] proposed an operational method for extracting leaf symptom features from digital images using the Otsu segmentation method and then using SVM to classify three cotton leaf diseases. Ahmad et al. [1] extracts features from seven chilli pest and disease images using six conventional and six deep learning methods. The extracted features were fed into three machine learning classifiers, namely SVM, RF, and ANN for the identification task. Traditional image recognition algorithms are based on small sample datasets. As the number of samples increases and the sample types become more complex, the application results of these algorithmic models become unstable.

2.2 Deep Learning

With the advent of convolutional neural networks, the shortcomings of traditional image recognition have been addressed, and there has been a great deal of research into the use of convolutional neural networks to solve pest detection problems. Compared to traditional image recognition algorithms, deep learning algorithms can yield more advanced feature information, better application results, and greater stability when faced with multiple types of samples. Chen et al. [4] proposed a pest detection system that used Deep-learning-based object detection models to identify pest images, and finally verified that yolov4 was the most effective in identifying scale pests. Karar et al. [11] implementation of a mobile application for the automatic classification of pests, proving that Faster-Rcnn outperforms SSD in identifying five categories of pests: Aphids, Cicadellidae, Flax Budworm, Flea Beetles, and Red Spider. Li et al. [13] tested three deep convolutional neural network models on the IP102 dataset and Baidu AI Insect Detection dataset. The experimental results show that YoloV5 outperforms Faster-Rcnn and Mask-Rcnn. Liu et al. [16] constructed the DFF-ResNet to improve the raw residual block capacity, and eventually DFF-ResNet surpassed the performance of ResNet on the IP102 dataset. Dong et al. [6] proposed the MCPD-net, which includes a multiscale feature pyramid network (MFPN) and a novel adaptive feature region proposal network (AFRPN). The results showed that the MCPD-net model outperformed the other models in the MPD2021 dataset. Khalifa et al. [12] used AlexNet, GoogleNet, and SqueezNet to classify and recognize the IP102 dataset and also used image enhancement techniques to overcome the overfitting problem during training. Wang et al. [27] proposed a S-RPN network. The network structure solves the category imbalance problem by using a spare sampling method and proposes an adaptive RoI selection method (ARS). Experiments show that S-RPN outperforms other algorithms in terms of recognition accuracy on AgriPest21 datasets. Liu et al. [14]

optimized the feature layer of the YOLOV3 network structure to achieve multi-scale feature detection, which accurately and quickly detects the location and class of tomato pests and diseases. Experimental results show that the improved YOLOV3 outperforms YOLOV3-Darknet53, SSD, and Faster-Rcnn.

Since most of the datasets used in the above pest detection studies were taken under controlled conditions, the trained models do not work well for detecting small objects and for detecting occluded pests in large scenes, which does not apply to the detection of Locusta migratoria manliness in real scenes.

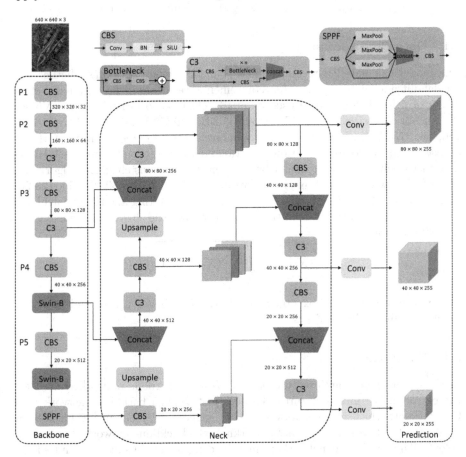

Fig. 1. Structure of the SSK-Yolo network.

3   SSK-Yolo

YoloV5 is the most widely used of the Yolo series, first proposed by Jocher in 2021, with a high level of recognition accuracy and fast end-to-end detection [10]. The relative scale of the grasshoppers is very small in comparison to the whole high-resolution image in the dataset, so the ability of the model to detect small objects is essential. In this paper, YoloV5 is improved to compensate for the deficiency of YoloV5 in detecting small objects. The improved YoloV5 model is shown in Fig. 1. Firstly, the size of the prior frame is obtained by clustering the anchor frames by the k-means algorithm. Secondly, the improved backbone network is used to extract features from the image, and then spatial pyramid pooling is used to achieve the fusion of graphical and semantic features and multi-scale target detection. Finally, the detection accuracy of the model for occluded locusts is improved by utilizing the soft-nms algorithm.

3.1   Data Pre-processing

. The model calculated a recall of more than 0.98 when checking whether the set default anchor boxes met the requirements, resulting in no automatic clustering of priori boxes, so we used manually clustered priori boxes. YoloV5 set the anchor box sizes to (10,13, 16,30, 33,23), (30,61, 62,45, 59,119), (116,90, 156,198, 373,326), which are clearly too large for locust detection in the GHCID dataset [5]. In the locust detection scenario under natural conditions, the GHCID dataset was clustered with anchor boxes using the k-means clustering algorithm, and the following priori box sizes were obtained:(12,20, 18,39, 23,11), (23,20, 32,61, 38,33), (45,18, 59,81, 86,37). The k-means clustering algorithm calculates the Euclidean distance between the data object and the cluster centre using the following formula:

$$d(X, C_i) = \sqrt{\sum_{j=1}^{m} (X_j - C_{ij})^2} \tag{1}$$

where X is the anchor box object, C_i is the i_{th} cluster centre and m is the number of anchor box objects.

3.2   Backbone

In the overall model architecture of YoloV5, the backbone network is used for the extraction of image features. The original YoloV5 backbone network uses CSPdarknet, which is a network structure generated by borrowing ideas from the CSPnet network on the basis of darknet, the backbone network that follows yolov3. This backbone network is mainly composed of Conv, C3 and SPPF modules.

The size of the Receptive Field is particularly important in extracting image features during object detection. Conventional convolution has a small receptive field when extracting shallow features, which results in the extraction of graphical features such as texture colour. When extracting deep features, although the receptive field is large and the model is theoretically able to acquire semantic features, but we can't rule out the possibility that the model is easy to take shortcuts, even when the receptive field becomes large, it still chooses to learn the features that are easier to get, such as texture and colour, which will lead

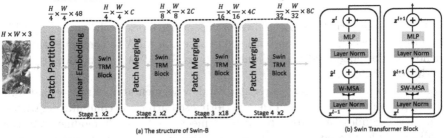

Fig. 2. (a) The structure of a Swin-Transformer (Swin-B); (b) Swin-Transformer Blocks.

to the formation of a large shape bias [7]. However, when detecting locusts in the GHCID dataset, since locusts are small targets, then how well the model extracts semantic features determines the accuracy of the final locust location and category obtained. Not only does the Swin-Transformer have a large receptive field in the form of a sliding window to compute global self attention, it is also capable of constructing multi-scale feature maps like traditional convolution.

Based on the above principles and analysis, we choose to abandon the use of traditional convolution when extracting deep features from the backbone network and instead use the Swin-Transformer, i.e. we use the Swin-Transformer to process the higher order information and the underlying features are still processed using traditional convolution. The Swin-Transformer structure is shown in Fig. 2(a).

It uses the idea of hierarchical design to construct four levels of feature mapping. Each stage reduces the resolution of the input feature map, which is similar to the CNN operation of increasing the receptive field layer by layer. Images are first sliced into patches, and each patch is used as a token that is projected to an arbitrary dimension through a linear embedding layer. After that, each stage first performs a patch merging operation to merge adjacent patches before performing self attention computation to reduce the number of tokens and increase the feature dimensions. In this way, at each stage, the resolution of the image decreases and the number of channels increases, resulting in the creation of hierarchical features. It should be emphasised here that there will always be an even number of Swin-Transformer blocks. The Swin Transformer block is composed of multi-head self attention module based on shift windows [18]. The relevant operations of the multi-head self attention mechanism are as follows.

$$MultiHead(Q, K, V) = Concat(head_1, head_2, \ldots, head_h)W^O \qquad (2)$$

$$head_i = Attention(QW_i^Q, KW_i^K, VW_i^V) \qquad (3)$$

$$Attention(Q, K, V) = softmax(\frac{QK^T}{\sqrt{d_k}})V \qquad (4)$$

Q,K,V denote query and key of dimension d_k and value of dimension d_v respectively. W_i^Q, W_i^K, W_i^V are learning parameters and h denotes the number of attention layers.

The internal structure of the Swin Transformer block is shown in Fig. 2(b). Each block consists of two sub-blocks, one based on MSA to compute attention for non-overlapping windows and one based on SW-MSA to compute attention for shifted windows.

MSA-based block employs a regular window segmentation strategy that uniformly segments the image and then performs self attention computation within a localised window and uses residual connections for each module, but the localised window-based self attention module lacks information transfer between windows, so in order to maintain efficient computation of self attention within non-overlapping windows while keeping the windows connected, Swin Transformer employs a shifted window partitioning approach [18]. To complete the information transfer, The SW-MSA based block calculates the self attention between windows. A window configuration offset from the window configuration of the previous layer is used by moving the window by pixels from the window of the regular partition [18].

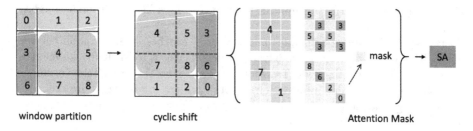

window partition   cyclic shift   Attention Mask

Fig. 3. Batch calculation of self attention in shifted window partitioning.

The core structure of the Swim Transformer is the W-MSA and SW-MSA. A key design of the Swin Transformer is the window-movement segmentation between successive layers of self attention. The batch calculation method of cyclic-shifting is used by cyclic-shifting to the top-left as shown in Fig. 3. A batch window may consist of non-adjacent sub-windows in the feature map, using a masking mechanism to limit the calculation of self-attention to within each sub-window [18]. This also solves the problem of moving window partitions causing more sub-windows. This paper improves the YoloV5 backbone network using the Swin-B base model.

3.3   Data Post-processing

NMS is the search for local maxima from the many boxes predicted. The model focuses on doing IOU operations with other boxes using the maximum scoring prediction box in the form of constant iteration and then filtering out those prediction boxes whose IOUs exceed a set threshold. However, Nms algorithm excludes all prediction boxes with the IOU greater than a specified threshold. If the two locusts cover each other with a lot of overlapping parts, it will result in a missed detection phenomenon. The improved YoloV5 model uses the soft-nms algorithm to solve the post-processing problems. The main idea of the soft-nms algorithm is to reduce the confidence level of the prediction, instead of removing all prediction boxes larger than the IOU threshold. Below is a comparison of the two algorithms:

$$S_i = \begin{cases} S_i & iou(M, b_i) < N_t \\ 0 & iou(M, b_i) \geq N_t \end{cases} \quad (5)$$

$$S_i = \begin{cases} S_i & iou(M, b_i) < N_t \\ S_i e^{-\frac{iou(M, b_i)^2}{\sigma}} & iou(M, b_i) \geq N_t \end{cases} \quad (6)$$

The first formula is the rescoring function in the NMS pruning step, where S_i is the prediction box score, M is the highest scoring prediction box, b_i is the one remaining prediction box, N_t is the set threshold. When the calculated IOU is greater than the threshold, the score of that prediction frame is directly set to 0. The second formula is the rescoring function in the soft-nms pruning step, where we use Gaussian weighting to update the pruning rule. Which decays instead of setting the box score to 0 if the calculated IOU is above a threshold.

The advantages of the algorithm are that it improves the traditional NMS algorithm without adding additional parameters, the algorithm complexity is the same as the traditional NMS algorithm, which is efficient to use and no additional training of the model is required and it can be integrated directly into the model validation stage.

4   Experiments

4.1   Datasets

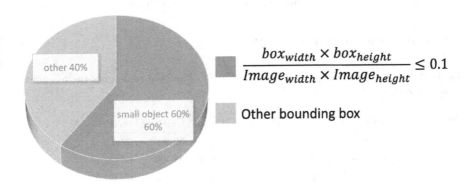

Fig. 4. Relative scale distribution of the bounding box.

In this paper, the public dataset of the GHCID was used and the sample type was mainly adult Asian grasshoppers (Orthoptera: Locustidae), all images are

3968 × 2976 resolution. The image acquisition process took into account the effect of different lighting conditions on the detection effect, so it was done at different times to collect the locust images. We analysed the distribution of the relative scales of the grasshoppers to further demonstrate the detection of small objects(see Fig. 4). To evaluate the performance of each experimental model, we randomly selected 2682 images from this dataset as a training set, 536 images as a validation set, and the remaining images as a test set.

4.2 Experimental Configuration

Table 1. Parameter settings of different models.

Model	Image Size	Learning Rate	Epoch/Iteration	Batch Size
YoloV5	640 × 640	0.01	300epoch	16
YoloV5-MobileNet	640 × 640	0.01	300epoch	16
YoloV5-RegNet	640 × 640	0.01	300epoch	16
YoloV5-EfficientNet	640 × 640	0.01	300epoch	16
Faster-Rcnn	600 × 600	0.01	300epoch	4
SSK-Yolo	640 × 640	0.01	300epoch	16
SSD	640 × 640	0.0001	13500iter	32

YoloV5 is employed as a baseline for comparison against other experimental methods to validate the effectiveness and accuracy of SSK-Yolo. We primarily utilize network models with higher accuracy for comparison, namely SSD [15], Faster-Rcnn [22], and YoloV5 (wherein CSPDarknet in the YoloV5 model is replaced with MobileNet [25], RegNet [20], and EfficientNet [26] as backbone networks respectively). The model parameters are set in the Table 1.

4.3 Comparation Study

Table 2. Quantitative comparison on GHCID(%). The bold represents the maximum value per column.

Model	Precision	Recall	F1-score	mAP@.5
YoloV5	72	41.86	52.94	69.5
YoloV5-MobileNet	64	39.02	48.48	64
YoloV5-RegNet	70	41.18	51.85	69.4
YoloV5-EfficientNet	75	42.86	54.54	72.9
Faster-Rcnn	74	42.53	54.01	71.3
SSD	71	41.52	52.40	70
SSK-Yolo	**77**	**43.50**	**55.60**	**81.5**

Quantitative Results. Precision, recall, F1-score and mAP@.5 were used to evaluate model performance for locust detection. Table 2 reports the quantitative results comparing the methods on the GHCID dataset. We can see from Table 2 that SSK-Yolo, because it extracts semantic features using Swin-Transformer and Gaussian-weighted soft-nms to cull the prediction frames, obtains the best Precision (77%), Recall (43.50%), F1-score (55.60%), mAP@.5 (81.5%). YoloV5-EfficientNet ranks second on all four metrics and YoloV5-MobileNet is the worst. On the GHCID dataset, SSK-Yolo improves on YoloV5-EfficientNet by 2%, 0.64%, 1.06% and 8.6% on precision, recall, F1-score and mAP@.5 respectively.

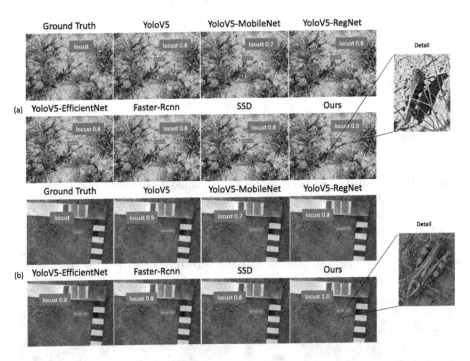

Fig. 5. Comparison of visualisations on the GHCID dataset. (a) the detection performance of the different methods under high illumination and the detailed view of the detected object (b)the detection performance of the different methods under low illumination and the detailed view of the detected object.

Qualitative Results. After the quantitative analysis, to further demonstrate the ability of SSK-Yolo to extract small objects in complex scenarios, we performed a visual comparative analysis of the detection results. Figure 5 shows some examples of qualitative results obtained by different methods. From the detection results, it can be seen that our proposed SSK-Yolo method detects the best results in both high and low light conditions. The confidence of the prediction boxes of the other methods is lower than that of SSK-Yolo, which reaches 1.0 in low light conditions, indicating that the ability of SSK-Yolo to extract features is higher than all other methods.

4.4   Ablation Study

Quantitative Results. In addition to the comparison experiments, we also designed ablation experiments of the relevant algorithmic models to verify the effectiveness of the three algorithmic improvement points. The Table 3 shows how the three improvement points affect the model's performance in detecting Locusta migratoria manilensis in terms of the model's mAP and F1-score. The quantitative results confirm the improved detection performance of the model by introducing Swin-Transformer to extract semantic features, k-means clustering of prior boxes and soft-nms to filter redundant prediction boxes.

Table 3. Ablation results on the dataset of GHCID (%). The bold represents the maximum value per column.

Model	Swin-Transformer	k-means	soft-nms	F1-score	mAP@.5
YoloV5				52.94	69.5
S-YoLo	✓			55.28	79.6
SK-YoLo	✓	✓		55.28	80.6
SSK-Yolo	✓	✓	✓	**55.60**	**81.5**

Qualitative Results. Figure 6 shows the visualisation of the ablation experiment. Before the introduction of swin-Transformer, the model incorrectly identified vegetation as locusts. Before the introduction of soft-nms, for occluded locusts, the model recognised redundant prediction boxes. This qualitative result further demonstrates the effectiveness of the C3STR module for extracting semantic features of small objects and the soft-nms algorithm for solving the occlusion problem.

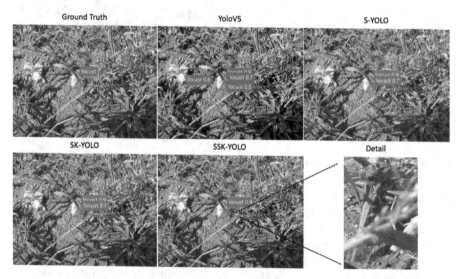

Fig. 6. Visualisation of ablation experiments. S-YoLo is the introduction of the swin-transformer into YoloV5. SK-YoLo is the introduction of k-means into S-YoloV5.

5   Conclusion

This paper introduces a globally-featured-driven small object detection network named SSK-Yolo to address the urgent demand for automated locust detection in agricultural informatization, particularly for high-resolution images. Firstly, we employ the k-means clustering algorithm to determine the sizes of prior boxes. Next, at the feature extraction level of the network, we utilize the swin-Transformer to extract global semantic features, thereby enhancing the capability to detect small objects in high-resolution scenarios. We fuse features from different scales of conceptual layers and visual layers to form multi-scale feature predictions. Finally, we apply soft-nms with a Gaussian penalty term to suppress detection boxes, effectively resolving the issue of occluded targets. Comparative and ablation studies on the GHCID dataset demonstrate the superiority of SSK-Yolo over other detection models.

References

1. Ahmad Loti, N.N., Mohd Noor, M.R., Chang, S.W.: Integrated analysis of machine learning and deep learning in chili pest and disease identification. J. Sci. Food Agric. **101**(9), 3582–3594 (2021)
2. Al Bashish, D., Braik, M., Bani-Ahmad, S.: A framework for detection and classification of plant leaf and stem diseases. In: 2010 International Conference on Signal and Image Processing, pp. 113–118. IEEE (2010)
3. Aurangzeb, K., Akmal, F., Attique Khan, M., Sharif, M., Javed, M.Y.: Advanced machine learning algorithm based system for crops leaf diseases recognition. In: 2020 6th Conference on Data Science and Machine Learning Applications (CDMA), pp. 146–151 (2020). https://doi.org/10.1109/CDMA47397.2020.00031
4. Chen, C., Liang, Y., Zhou, L., Tang, X., Dai, M.: An automatic inspection system for pest detection in granaries using yolov4. Comput. Electron. Agric. **201**, 107302 (2022)
5. Chudzik, P., Mitchell, A., Alkaseem, M., Wu, Y., Fang, S., Hudaib, T., Pearson, S., Al-Diri, B.: Mobile real-time grasshopper detection and data aggregation framework. Sci. Rep. **10**(1), 1150 (2020)
6. Dong, S., et al.: Automatic crop pest detection oriented multiscale feature fusion approach. Insects **13**(6), 554 (2022)
7. Geirhos, R., Rubisch, P., Michaelis, C., Bethge, M., Wichmann, F.A., Brendel, W.: ImageNet-trained CNNs are biased towards texture; increasing shape bias improves accuracy and robustness. arXiv preprint arXiv:1811.12231 (2018)
8. Gong, H., et al.: Based on FCN and DenseNet framework for the research of rice pest identification methods. Agronomy **13**(2), 410 (2023)
9. Islam, M.A., Islam, M.S., Hossen, M.S., Emon, M.U., Keya, M.S., Habib, A.: Machine learning based image classification of papaya disease recognition. In: 2020 4th International Conference on Electronics, Communication and Aerospace Technology (ICECA), pp. 1353–1360 (2020). https://doi.org/10.1109/ICECA49313.2020.9297570
10. Jocher, G., et al.: ultralytics/yolov5: v6. 0-YOLOv5n 'nano' models, roboflow integration, tensorflow export, OpenCV DNN support. Zenodo (2021)

11. Karar, M.E., Alsunaydi, F., Albusaymi, S., Alotaibi, S.: A new mobile application of agricultural pests recognition using deep learning in cloud computing system. Alex. Eng. J. **60**(5), 4423–4432 (2021)

12. Khalifa, N.E.M., Loey, M., Taha, M.H.N.: Insect pests recognition based on deep transfer learning models. J. Theor. Appl. Inf. Technol. **98**(1), 60–68 (2020)

13. Li, W., Zhu, T., Li, X., Dong, J., Liu, J.: Recommending advanced deep learning models for efficient insect pest detection. Agriculture **12**(7), 1065 (2022)

14. Liu, J., Wang, X.: Tomato diseases and pests detection based on improved yolo V3 convolutional neural network. Front. Plant Sci. **11**, 898 (2020)

15. Liu, W., et al.: SSD: single shot multibox detector. In: Leibe, B., Matas, J., Sebe, N., Welling, M. (eds.) Computer Vision - ECCV 2016, pp. 21–37. Springer International Publishing, Cham (2016)

16. Liu, W., Wu, G., Ren, F., Kang, X.: DFF-ResNet: an insect pest recognition model based on residual networks. Big Data Min. Analytics **3**(4), 300–310 (2020)

17. Liu, Y., et al.: Forest pest identification based on a new dataset and convolutional neural network model with enhancement strategy. Comput. Electron. Agric. **192**, 106625 (2022)

18. Liu, Z., et al.: Swin transformer: hierarchical vision transformer using shifted windows. In: Proceedings of the IEEE/CVF International Conference on Computer Vision, pp. 10012–10022 (2021)

19. Madhavan, M.V., Thanh, D.N.H., Khamparia, A., Pande, S., Malik, R., Gupta, D.: Recognition and classification of pomegranate leaves diseases by image processing and machine learning techniques. Comput. Mater. Continua **66**(3), 2939–2955 (2021)

20. Radosavovic, I., Kosaraju, R.P., Girshick, R., He, K., Dollar, P.: Designing network design spaces. In: Proceedings of the IEEE/CVF Conference on Computer Vision and Pattern Recognition (CVPR) (2020)

21. Ramesh, S., et al.: Plant disease detection using machine learning. In: 2018 International Conference on Design Innovations for 3Cs Compute Communicate Control (ICDI3C), pp. 41–45 (2018). https://doi.org/10.1109/ICDI3C.2018.00017

22. Ren, S., He, K., Girshick, R., Sun, J.: Faster R-CNN: towards real-time object detection with region proposal networks. In: Advances in Neural Information Processing Systems, vol. 28 (2015)

23. Rothe, P., Kshirsagar, R.: Automated extraction of digital images features of three kinds of cotton leaf diseases. In: 2014 International Conference on Electronics, Communication and Computational Engineering (ICECCE), pp. 67–71 (2014)

24. Shuhan, L., Ye, S.J.: Using an image segmentation and support vector machine method for identifying two locust species and instars. J. Integr. Agric. **19**(5), 1301–1313 (2020)

25. Sinha, D., El-Sharkawy, M.: Thin MobileNet: an enhanced mobilenet architecture. In: 2019 IEEE 10th Annual Ubiquitous Computing, Electronics & Mobile Communication Conference (UEMCON), pp. 0280–0285 (2019). https://doi.org/10.1109/UEMCON47517.2019.8993089

26. Tan, M., Le, Q.: EfficientNet: Rethinking model scaling for convolutional neural networks. In: Chaudhuri, K., Salakhutdinov, R. (eds.) Proceedings of the 36th International Conference on Machine Learning. Proceedings of Machine Learning Research, vol. 97, pp. 6105–6114. PMLR (2019). https://proceedings.mlr.press/v97/tan19a.html

27. Wang, R., Jiao, L., Xie, C., Chen, P., Du, J., Li, R.: S-RPN: sampling-balanced region proposal network for small crop pest detection. Comput. Electron. Agric. **187**, 106290 (2021)

28. Xiao, Z., Yin, K., Geng, L., Wu, J., Zhang, F., Liu, Y.: Pest identification via hyperspectral image and deep learning. SIViP **16**(4), 873–880 (2022)

MetaVSR: A Novel Approach to Video Super-Resolution for Arbitrary Magnification

Zixuan Hong[1,2], Weipeng Cao[2(✉)] (ID), Zhiwu Xu[1], Zhenru Chen[2], Xi Tao[2], Zhong Ming[1,2], Chuqing Cao[3], and Liang Zheng[3]

[1] College of Computer Science and Software Engineering, Shenzhen University, Shenzhen 518060, China
[2] Guangdong Laboratory of Artificial Intelligence and Digital Economy (Shenzhen), Shenzhen 518107, China
caoweipeng@gml.ac.cn
[3] Anhui Province Key Laboratory of Machine Vision Inspection, Yangtze River Delta HIT Robot Technology Research Institute, Wuhu 241000, China

Abstract. Video super-resolution is a pivotal task that involves the recovery of high-resolution video frames from their low-resolution counterparts, possessing a multitude of applications in real-world scenarios. Within the domain of prevailing video super-resolution models, a majority of these models are tailored to specific magnification factors, thereby lacking a cohesive architecture capable of accommodating arbitrary magnifications. In response to this lacuna, this study introduces "MetaVSR", a novel video super-resolution model devised to handle arbitrary magnifications. This model is structured around three distinct modules: inter-frame alignment, feature extraction, and upsampling. In the inter-frame alignment module, a bidirectional propagation technique is employed to attain the alignment of adjacent frames. The feature extraction module amalgamates superficial and profound video features to enhance the model's representational prowess. The upsampling module serves to establish a mapping correlation between the desired target resolution and the input provided in lower resolution. An array of empirical findings attests to the efficacy of the proposed MetaVSR model in addressing this challenge.

Keywords: Video super-resolution · Deep learning · Neural network

1 Introduction

The domain of computer vision extensively employs super-resolution technology to achieve the transformation of low-resolution images into high-resolution counterparts. This technology finds widespread application across various domains, including but not limited to medical imaging [20], surveillance [2], and security [1]. As it stands, it has solidified its position as a cornerstone within the realm of computer vision methodologies.

© The Author(s), under exclusive license to Springer Nature Switzerland AG 2024
S. Rudinac et al. (Eds.): MMM 2024, LNCS 14554, pp. 300–313, 2024.
https://doi.org/10.1007/978-3-031-53305-1_23

Conventional up-sampling methods predominantly rely on interpolation techniques to enhance image resolution. Examples include nearest-neighbor, bilateral, and bicubic interpolations. However, these traditional methodologies often engender the predicament of image blurring, thereby constraining their overall efficacy within practical applications. In light of these limitations, the emergence of deep learning has ushered in a revolutionary shift in super-resolution techniques.

Deep learning methodologies have accomplished more than simply mitigating the blurring issue that plagues conventional techniques. These methods have introduced a realm of visually arresting effects. Consequently, deep learning-based super-resolution models exhibit the proficiency not only to rectify the challenge of image blurring endemic to traditional techniques but also to engender breathtaking visual enhancements. It follows that these models underpinned by deep learning can effectively and accurately restore high-resolution images, thus effecting a substantial and noteworthy enhancement in the visual quality of images.

The application of neural networks in the field of imaging provides efficient solutions to computer vision problems. SRCNN [6] and VSRnet [12] perform well in Single-Image Super-Resolution (SISR) and Video Super-Resolution (VSR) tasks, laying the foundation for subsequent research. The difference between video and image lies in the fact that video contains more spatiotemporal information, and simply super-resolution video frames may lead to artifacts. Therefore, designing an effective frame alignment method has a significant impact on the final video super-resolution results. In recent years, many VSR methods [3–5,8,10,23,27–29] have been proposed. They focus on video super-resolution at specific magnification and require retraining the network when other magnification super-resolution is required. This process is time-consuming and limits the practical application in different scenarios. In addition, existing VSR methods usually adopt a simple convolutional layer stacking approach for feature extraction, which fails to fully utilize shallow and deep feature information, thus affecting performance.

Therefore, in an attempt to overcome the above mentioned limitations of existing models, this paper proposes MetaVSR, a model that can realize video super-resolution at arbitrary magnification. It mainly consists of inter-frame alignment, feature extraction, and upsampling modules. In the inter-frame alignment module, we use a bidirectional propagation method to achieve frame alignment of neighboring frames in a pyramid structure, which provides a better frame alignment effect to improve the model's performance. In the feature extraction module, we utilize a dense residual block to fuse the shallow and deep feature information to obtain a more detailed representation of video frames for better feature learning capability. The upsampling module generates the final super-resolution results by establishing the mapping relationship between the target and low resolutions and calculating the corresponding weights. We evaluate the proposed model through extensive experiments and test it on widely used benchmark datasets. The experimental results show that our method is remarkably effective and produces excellent visual results.

In summation, this study presents a collection of substantive contributions, encapsulated as follows:

- We introduce a novel model named MetaVSR, adept in effectuating video super-resolution for arbitrary magnification scenarios. Comprising inter-frame alignment, feature extraction, and upsampling modules, this model constitutes a pioneering stride toward enhancing video quality.
- To refine the alignment process, we proffer a novel approach-bidirectional frame alignment employing a pyramid structure. By amplifying information interchange among neighboring frames, this innovative methodology bolsters alignment precision, thereby serving as a cornerstone for optimizing video super-segmentation outcomes.
- We advocate the deployment of dense residual blocks to amalgamate shallow and profound feature information, thereby engendering a more intricate feature representation. The ultimate super-resolution outputs materialize through the establishment of a mapping nexus between target and low resolutions. Extensive experimental results on benchmarks demonstrate the effectiveness of the proposed MetaVSR model.

2   Related Works

2.1   Single Image Super Resolution

The application of neural networks to image super-resolution helps to learn image features to overcome the image blurring problem caused by the traditional up-sampling method. SRCNN [6] represents a turning point within the filed of single image super-resolution, laying the groundwork for subsequent research endeavors. Existing research in the field of single image super-resolution can be categorized primarily into three groups: methods based on residual networks [15,16,31] which leverage their robust feature learning capabilities for performance enhancement; methods employing attention mechanisms [25,30], which elevate the significance of attended regions; and other approaches [7,9,22,26]. These techniques have made remarkable progress in the domain of single-image super-resolution, continually advancing image quality through the exploration of diverse network architectures and technical modalities.

Residual-Based Methods. The approaches based on residual networks integrate image features through skip connections to enhance the network's capacity to learn image characteristics. For instance, in [16], the authors addressed image super-resolution by combining residual learning and sub-sampling layers [24]. However, this method falls short of effectively merging shallow and deep-level image information. As a result, in [31], the authors introduced Residual Dense Blocks (RDBs), which concatenate shallow and deep-level image information along the channel dimension to enhance image learning. Nonetheless, these methods are constrained to feature extraction on a single scale, inevitably leading to the loss of certain local information. Hence, in [15], the authors proposed a method to adaptively extract image features across different scales to achieve superior visual effects.

Attention-Based Methods. Incorporating attention mechanisms into the single-image super-resolution task enables networks to selectively focus on crucial information within features. In [30], the authors introduced the RCAN network, incorporating designed Residual Channel Attention Blocks (RCABs) to introduce channel-wise attention mechanisms, allowing neural networks to better exchange information across different channels, thereby enhancing network performance. In [25], the authors presented a novel architecture that combines RCAN with LSTM to enhance network performance. However, due to attention mechanisms potentially disregarding inter-dimensional correlations, the approach proposed by [19] effectively addressed this issue. This method merges layer-wise attention modules with channel-spatial attention modules and employs a residual block approach to tackle this concern.

Others Methods. In recent years, with the surge in popularity of stable diffusion models, numerous scholars have incorporated such models into the realm of single-image super-resolution tasks [7,22]. These approaches commonly utilize the DDPM and U-Net architectures to accomplish image super-resolution tasks. Nevertheless, all of the aforementioned methods are confined to specific magnification factors. Therefore, in an attempt to achieve arbitrary magnification in single-image super-resolution, meta-learning techniques are used to implement arbitrary magnification super-resolution [9,26].

2.2   Video Super Resolution

With the rapid development of image super-resolution technology, the field of video super-resolution is also indirectly driven. VSRnet [12], based on the concept of SRCNN [6], introduced deep learning to video super-resolution tasks for the first time, which laid the foundation of the field and spawned a series of evolved video super-resolution work. However, videos have more information in the spatio-temporal dimension. Therefore, fully utilizing this spatio-temporal information to achieve better frame alignment results will have a profound impact on the final video super-resolution. Currently, video super-resolution work can be categorized into two main groups: Motion Estimation and Motion Compensation (MEMC) methods [3–5,8,10,23,29] and Deformable Convolution [27,28].

MEMC-Based Methods. The motion information between video frames is estimated using optical flow and applied with corresponding distortions. FRVSR [23] employs a recurrent framework, utilizing motion estimates from previous frames for the current frame. This allows the network to handle information propagation over extensive temporal ranges, resulting in continuous video super-resolution outcomes. TOFlow [29] explores task-oriented motion and utilizes self-supervised and task-specific approaches for motion representation learning. TecoGAN proposed by [5] incorporated spatiotemporal discriminators and a Ping-Pong loss function to achieve coherent video super-resolution results. BasicVSR [3] and BasicVSR++ [4] designed a bidirectional propagation mechanism, enabling each frame to acquire alignment information from a wide range of frames. This approach facilitates the super-resolution of target frames, yielding significantly enhanced visual effects. STARnet [8] leveraged the temporal-spatial correlations to provide more accurate motion compensation information, leading to more precise alignment results. In [10], the authors designed recurrent detail structure blocks and hidden state adaptation blocks and then combined them through unidirectional propagation to finally enhance the quality of video super-resolution.

Deformable-Convolution-Based Methods. Due to MEMC's shortcomings in dealing with large-scale motion and occluded regions, deformable convolution has been introduced. By calculating offset values, the network can more accurately capture local information, thereby enhancing the performance of traditional convolutions. EDVR [28] combined pyramid structures with deformable convolutions to achieve video frame alignment across different scales, resulting in more accurate alignment outcomes and improved performance. TDAN [27] utilized deformable convolutions and offset values at the feature level of video frames within the same scale for alignment.

3   Proposed Method

3.1   Overview

Given (2N+1) video frames $I_t^{LR}, t \in [-N, N]$, as an example, we notate the middle frame as the target frame I_0^{LR} and the rest of the frames as neighboring frames. Our goal is to utilize the proposed video super-resolution model to reconstruct the high-resolution target frame I_0^{HR} by using the information of the neighboring frames, which is given to our target frame by bidirectional propagation strategy. The proposed high-resolution network architecture MetaVSR can be seen shown in Fig. 1, which is mainly composed of inter-frame alignment, feature extraction, and upscale modules. In the inter-frame alignment module, we use a bidirectional propagation strategy to achieve frame alignment of neighboring frames to provide better super-resolution results. The feature extraction module utilizes dense residual blocks to form a continuous memory mechanism to form multilevel features and fuses shallow and deep information to obtain better video frame feature learning capability. The upsampling module generates the final super-resolution results by establishing the mapping relationship between the target resolution and the lower resolution and calculating the corresponding positional weights.

3.2   Inter-frames Alignment

In the inter-frame alignment module, we introduce bidirectional propagation for information transfer between neighboring and target frames, improving alignment results. For optical flow computation, we use the pre-trained SpyNet [21] module denoted as S, combining convolutional neural networks with pyramid structures to efficiently compute accurate flow information across scales. Such an optical flow computation is not only more accurate, but also capable of capturing motion information at different scales, thus providing more comprehensive information to support the frame alignment process.

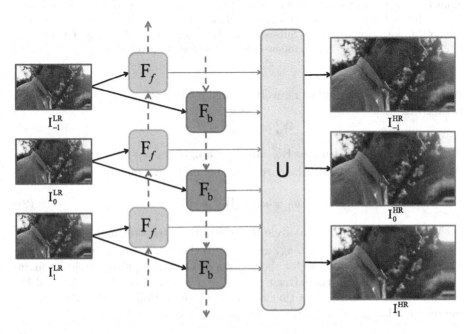

Fig. 1. This figure shows the network architecture of MetaVSR. It is a bidirectional propagation architecture, where F_f, F_b, and U are the forward propagation, backpropagation, and upscale modules, respectively. The architecture of forward propagation and backpropagation is the same, which mainly consists of inter-frame alignment module and feature extraction module, and they differ only in the input information of the modules.

In the propagation process, it is divided into two directions, forward propagation and backward propagation, which are denoted as F_f and F_b, respectively. Taking the forward propagation process of the target frame I_0^{LR} as an example, in the forward propagation process, we will process the feature information of the previous t frame $t \in [-N, -1]$ through layer-by-layer propagation to generate the feature h_{-1}^{LR} of the previous frame I_{-1}^{LR}, which is then inputted into the forward propagation module. At the same time, the target frames I_0^{LR} and I_{-1}^{LR}

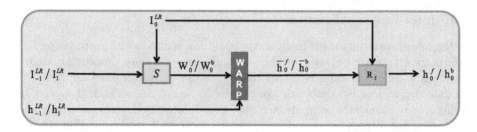

Fig. 2. This figure shows the structure of the bidirectional propagation F_f and F_b, which are identical. The features h_{-1} and h_1 from the previous frame propagation are input into the forward and backward propagation, respectively, and the corresponding low-resolution frames are input into module S for optical flow computation to generate the optical flow estimate $W_0^{f,b}$. Subsequently, $W_0^{f,b}$ and $h_{0\pm1}^{LR}$ are utilized for warping to generate $\bar{h}_0^{f,b}$ results. This is followed by R_f for feature extraction.

are inputted into the SpyNet network to predict the optical flow estimation W_0^f between them.

$$W_0^f = S(I_0^{LR}, I_{-1}^{LR}) \tag{1}$$

The optical flow estimates W_0^f and h_{-1}^{LR} are subsequently utilized to obtain the feature $\bar{h}_0^f$ aligned with the current frame after a warping operation.

$$\bar{h}_0^f = warp(W_0^f, h_{-1}^{LR}) \tag{2}$$

Finally, it is input to the module R_f for feature extraction, and the generated features of the current frame, h_0^f, are input to the forward propagation process of the next target frame. As for the process of backward propagation, its main input is the features of the latter t frames after propagation, $t \in [1, N]$, and its overall process is the same as that of forward propagation. For the flowchart of the propagation process, it can be seen as shown in Fig. 2.

$$h_o^f = R_f(I_0^{LR}, \bar{h}_0^f) \tag{3}$$

3.3   Feature Extraction Module

In the feature extraction module R_f, we fully borrow the design idea of Residual Dense Blocks (RDBs) from [31]. The uniqueness of this block is that it can efficiently fuse shallow and deep information to effectively capture multi-level features of an image. By interrelating multiple residual blocks in a densely connected manner, we achieve dense extraction of local features and also establish a continuous information transfer mechanism. This allows the current RDB block to be organically linked with the previous blocks, which in turn stabilizes the entire network training process. However, since the feature extraction module inputs the current frame I_0^{LR} together with the feature $\bar{h}_0^f$ aligned to it, we need to stitch them in the channel dimension beforehand, followed by fusion through

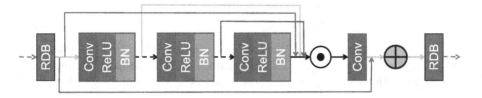

Fig. 3. This figure shows the structure of the feature extraction module. With the help of RDBs block, the features at all levels are fused to utilize the shallow and deep information to form a continuous memory extreme for the network's richer features for learning.

a convolutional layer to maintain feature consistency and integration. In addition, the introduction of RDB increases the depth of the network, which may bring about the problem that the model is difficult to converge. For this reason, we introduce Batch Normalization after each convolutional block to enhance the training speed and stability of the model. For more details on the feature extraction module, see Fig. 3.

3.4   Upscale Module

After bidirectional propagation and feature extraction, we obtain $\bar{h}_0^f$ and $\bar{h}_0^b$, respectively. In order to further fuse these features, we concat them according to the channel dimensions and then fuse them by convolutional layers to obtain the feature map F^{LR}. To realize arbitrary magnification super-resolution, inspired by MetaSR [9], we design a mapping relationship between target resolution and low resolution. By calculating the corresponding weights, we are able to obtain the final target super-resolution results. Its structure can be seen as shown in Fig. 4.

In the target resolution-low resolution mapping relationship, for each pixel (i, j) in the target resolution, the coordinates $(i^{'}, j^{'}) = (floor(\frac{i}{r}), floor(\frac{j}{r}))$ in the low-resolution image are generated by inverse mapping according to the selected scaling factor r and constructed into a position matrix by using its offset.

$$
PositionMatrix = \begin{pmatrix} R(0) & R(0) \\ R(0) & R(1) \\ R(0) & R(2) \\ \dots & \dots \\ R(i) & R(j-1) \\ R(i) & R(j) \\ R(i) & R(j+1) \\ \dots & \dots \end{pmatrix} \tag{4}
$$

where $R(i) = \frac{i}{r} - floor(\frac{i}{r}), R(j) = \frac{i}{r} - floor(\frac{i}{r})$, and the size of position matrix is $HW \times 2$. Then, Position matrix is input to the weight prediction network ϕ, and the weight $W(i, j)$ corresponding to each position of the target resolution image can be obtained.

$$W(i,j) = \phi(PositionMatrix, \theta) \tag{5}$$

where θ is the weight value of the weight prediction network. After obtaining the corresponding weights, the feature results can be obtained by multiplying the weights with the corresponding positions on the low-resolution image.

$$I_0^{HR}(i,j) = F^{LR}(i',j') W(i,j) \tag{6}$$

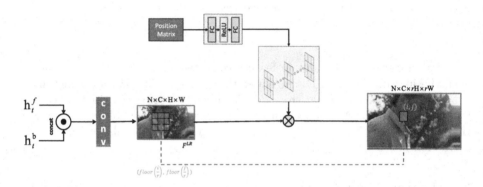

Fig. 4. Inspired by MetaSR's meta-upsampling module [9]. This figure illustrates the structure of the upsampling module. This involves establishing a mapping between target and low resolutions, and calculating positional weights to derive feature values for each pixel in the target resolution.

3.5 Loss Function

In the training process, we mainly use two loss functions, Charbonnier Loss [14] and VGG Loss [11], denoted as L_{ch} and L_{vgg}, respectively, so the overall loss function is shown in Eq. 7:

$$L(I_0^{HR}, I_0^{GT}) = L_{ch}(I_0^{HR}, I_0^{GT}) + \lambda L_{vgg}(I_0^{HR}, I_0^{GT}) \tag{7}$$

where $\mathcal{L}_c(x,y) = \sqrt{\|x - y\|^2 + \epsilon^2}$, ϵ and λ are constants. In our all experiments, we set ϵ and λ to 1e$-$6 and 1e$-$4, respectively.

4   Experiments

4.1   Implementation Details

Training Datasets. During the training phase, we employed the Vimeo90K [29] dataset with a resolution of 256×448 as our training set. To assess the performance of our model, we conducted testing using the Vimeo90K, REDS4 [18], and Vid4 [17] datasets. We quantified the performance using metrics such as PSNR and SSIM.

Training Details. We generate a set of scaling factors ranging from 1 to 4 with an increment of 0.1 and set the batch size to 8, while keeping the dimensions of the low-resolution frames fixed at 50 × 50. During each iteration, a target scaling factor is randomly selected from the set of scaling factors, and the target cropping size is obtained by multiplying the selected scaling factor with the pre-set low-resolution size, and then the video frame is randomly cropping according to the target cropping size, and the cropping video frame is downsampled by bicubic. For optimization, we use the Adam optimizer [13], setting the parameters β_1 and β_2 to 0.9 and 0.99, respectively. The initial learning rate is set at 1e−4, and we employ PyTorch's CosineAnnealing function to adjust the learning rate every 300 epochs, with a minimum rate of 1e−6. In our experiments, the number of blocks in our RDBs is 16, with 8 layers in each block.

4.2 Comparisons with State-of-the-Art Methods

We compare our model with SOTA models, e.g., EDVR [28], STARnet [8], RSDN [10], and calculate the corresponding PSNR and SSIM evaluation metrics. All methods are tested and compared on three datasets Vimeo90K [29], REDS4 [18], and Vid4 [17]. And all of them use the Bicubic downsampling method to generate low-resolution images and perform experiments with 4x super-resolution. The quantitative metrics of the experimental results can be seen as shown in Table 1. Note that in the experimental results, the values in red and blue represent the best and the second-best results, respectively. PSNR and SSIM metrics are represented by PSNR/SSIM. The above experimental results demonstrate that compared to EDVR, STARnet, and RSDN, our method shows a small improvement in both PSNR and SSIM on the Vimeo90K dataset. On the REDS4 and Vid4 datasets, the quantization results are comparable to their results. Thus, in general, our method is comparable to the results of the compared methods in terms of metrics computation for Vimeo90K and REDS4 compared to the three chosen methods, with a small improvement especially on the Vimeo90K dataset. In addition, our method is effective in providing a better description of the contours of individual objects in the scene compared to other methods. And to better illustrate the effectiveness of the methods, Fig. 5 shows the corresponding visual comparison results, and it can be seen that our method is able to obtain better visual results in general.

Table 1. The table shows the quantitative metrics results obtained by performing 4x super-resolution video on the Vimeo90K, REDS4, and Vid4 datasets.

		bicubic	EDVR	RSDN	STARnet	Our
Vimeo90K	RGB	29.97/0.867	33.32/0.928	33.80/0.923	33.99/0.923	34.90/0.936
	Y-channel	31.32/0.887	35.10/0.942	35.81/0.946	35.80/0.945	36.16/0.948
REDS4	RGB	24.95/0.718	27.24/0.811	27.93/0.812	27.58/0.804	27.70/0.809
	Y-channel	26.28/0.743	29.58/0.829	29.27/0.821	29.95/0.830	29.59/0.819
Vid4	RGB	20.67/0.604	25.19/0.795	25.52/0.804	25.57/0.805	24.66/0.788
	Y-channel	22.05/0.628	26.62/0.802	27.01/0.821	26.87/0.813	25.80/0.787

In addition, our method is able to realize the video super-resolution task at arbitrary magnification, rather than a super-resolution network for a specific magnification, and thus it is possible to realize video super-resolution at arbitrary magnification using only our trained network. Therefore, in order to better illustrate the effectiveness of our method for arbitrary magnification super-resolution, we performed the video super-resolution task at 2x and 3x respectively, and the experimental results can be seen as shown in Figs. 6 and 7.

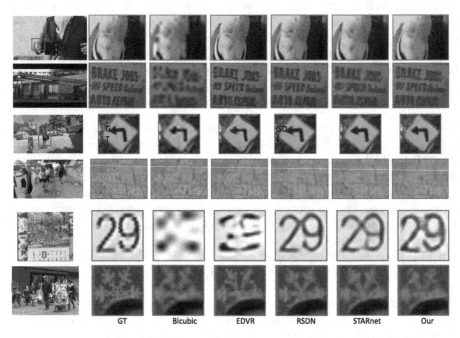

Fig. 5. This figure shows the visualization of our method with 4x super-resolution on the dataset. From top to bottom are Vimeo90K, REDS4 and Vid4, two representative video frames were selected for each dataset. In the first, second and fourth rows, our method shows better restoration for text and texture. In the fifth line, there is a better description of the object contours.

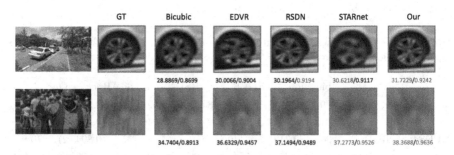

Fig. 6. This figure shows the results of performing a 2x super-resolution experiment and calculating the corresponding PSNR and SSIM metrics. Our method is able to obtain better metrics results and has a better description of the contours of the objects, e.g., the restoration of the tire contours and details in the first row.

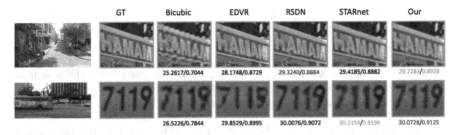

Fig. 7. This figure shows the results of performing a 3x super-resolution experiment and calculating the corresponding PSNR and SSIM metrics. Our method yields relatively smooth and textured results for text reduction.

5 Conclusions

To rectify the issue of necessitating network retraining for specific super-resolution magnifications-resulting in substantial temporal expenditures and confining the method's applicability within practical contexts—this paper introduces an innovative arbitrary magnification video super-resolution model named MetaVSR. The MetaVSR model is designed to establish a robust video super-resolution framework through the realization of inter-frame alignment via bidirectional propagation, the fusion of deep and shallow feature information, and computations grounded in the mapping relationship between the target and low-resolution frames. Extensive experimental validation substantiates our approach, revealing commendable quantitative metrics and visually appealing outcomes across widely accessible benchmark datasets. Additionally, our technique boasts expeditious inference speeds, versatility across various prevalent scenarios, and the attainment of favorable visual results.

Nevertheless, it is worth noting that the visual efficacy of the method still requires enhancement, particularly when confronted with intricate scenes featuring distant subjects. The recognized limitations of this model pave the way for meaningful avenues of research in the forthcoming endeavors. Indeed, we hold a conviction that the MetaVSR model is poised to exhibit even more remarkable performance in the realm of video reconstruction, offering promising prospects for the future.

Acknowledgement. This work was supported by the National Natural Science Foundation of China (62106150, 61836005, 62372304) and the Open Research Fund of Anhui Province Key Laboratory of Machine Vision Inspection (KLMVI-2023-HIT-01).

References

1. Cao, W., Wu, Y., Chakraborty, C., Li, D., Zhao, L., Ghosh, S.K.: Sustainable and transferable traffic sign recognition for intelligent transportation systems. IEEE Trans. Intell. Transp. Syst. **24**, 15784–15794 (2022)

2. Cao, W., Zhou, C., Wu, Y., Ming, Z., Xu, Z., Zhang, J.: Research progress of zero-shot learning beyond computer vision. In: Qiu, M. (ed.) ICA3PP 2020. LNCS, vol. 12453, pp. 538–551. Springer, Cham (2020). https://doi.org/10.1007/978-3-030-60239-0_36

3. Chan, K.C., Wang, X., Yu, K., Dong, C., Loy, C.C.: BasicVSR: the search for essential components in video super-resolution and beyond. In: Proceedings of the IEEE/CVF Conference on Computer Vision and Pattern Recognition, pp. 4947–4956 (2021)

4. Chan, K.C., Zhou, S., Xu, X., Loy, C.C.: BasicVSR++: improving video super-resolution with enhanced propagation and alignment. In: Proceedings of the IEEE/CVF Conference on Computer Vision and Pattern Recognition, pp. 5972–5981 (2022)

5. Chu, M., Xie, Y., Leal-Taixé, L., Thuerey, N.: Temporally coherent GANs for video super-resolution (tecogan). arXiv preprint arXiv:1811.09393 (2018)

6. Dong, C., Loy, C.C., He, K., Tang, X.: Image super-resolution using deep convolutional networks. IEEE Trans. Pattern Anal. Mach. Intell. 38(2), 295–307 (2015)

7. Gao, S., et al.: Implicit diffusion models for continuous super-resolution. In: Proceedings of the IEEE/CVF Conference on Computer Vision and Pattern Recognition, pp. 10021–10030 (2023)

8. Haris, M., Shakhnarovich, G., Ukita, N.: Space-time-aware multi-resolution video enhancement. In: Proceedings of the IEEE/CVF Conference on Computer Vision and Pattern Recognition, pp. 2859–2868 (2020)

9. Hu, X., Mu, H., Zhang, X., Wang, Z., Tan, T., Sun, J.: Meta-SR: a magnification-arbitrary network for super-resolution. In: Proceedings of the IEEE/CVF Conference on Computer Vision and Pattern Recognition, pp. 1575–1584 (2019)

10. Isobe, T., Jia, X., Gu, S., Li, S., Wang, S., Tian, Q.: Video super-resolution with recurrent structure-detail network. In: Vedaldi, A., Bischof, H., Brox, T., Frahm, J.-M. (eds.) ECCV 2020. LNCS, vol. 12357, pp. 645–660. Springer, Cham (2020). https://doi.org/10.1007/978-3-030-58610-2_38

11. Johnson, J., Alahi, A., Fei-Fei, L.: Perceptual Losses for Real-Time Style Transfer and Super-Resolution. In: Leibe, B., Matas, J., Sebe, N., Welling, M. (eds.) ECCV 2016. LNCS, vol. 9906, pp. 694–711. Springer, Cham (2016). https://doi.org/10.1007/978-3-319-46475-6_43

12. Kappeler, A., Yoo, S., Dai, Q., Katsaggelos, A.K.: Video super-resolution with convolutional neural networks. IEEE Trans. Comput. Imaging 2(2), 109–122 (2016)

13. Kingma, D.P., Ba, J.: Adam: a method for stochastic optimization. arXiv preprint arXiv:1412.6980 (2014)

14. Lai, W.S., Huang, J.B., Ahuja, N., Yang, M.H.: Deep laplacian pyramid networks for fast and accurate super-resolution. In: Proceedings of the IEEE Conference on Computer Vision and Pattern Recognition, pp. 624–632 (2017)

15. Li, J., Fang, F., Mei, K., Zhang, G.: Multi-scale residual network for image super-resolution. In: Proceedings of the European Conference on Computer Vision (ECCV), pp. 517–532 (2018)

16. Lim, B., Son, S., Kim, H., Nah, S., Mu Lee, K.: Enhanced deep residual networks for single image super-resolution. In: Proceedings of the IEEE Conference on Computer Vision and Pattern Recognition Workshops, pp. 136–144 (2017)

17. Liu, C., Sun, D.: On Bayesian adaptive video super resolution. IEEE Trans. Pattern Anal. Mach. Intell. 36(2), 346–360 (2013)

18. Nah, S., et al.: Ntire 2019 challenge on video deblurring and super-resolution: Dataset and study. In: Proceedings of the IEEE/CVF Conference on Computer Vision and Pattern Recognition Workshops, pp. 0–0 (2019)

19. Niu, B., et al.: Single image super-resolution via a holistic attention network. In: Vedaldi, A., Bischof, H., Brox, T., Frahm, J.-M. (eds.) ECCV 2020. LNCS, vol. 12357, pp. 191–207. Springer, Cham (2020). https://doi.org/10.1007/978-3-030-58610-2_12

20. Patwary, M.J., Cao, W., Wang, X.Z., Haque, M.A.: Fuzziness based semi-supervised multimodal learning for patient's activity recognition using RGBDT videos. Appl. Soft Comput. **120**, 108655 (2022)

21. Ranjan, A., Black, M.J.: Optical flow estimation using a spatial pyramid network. In: Proceedings of the IEEE Conference on Computer Vision and Pattern Recognition, pp. 4161–4170 (2017)

22. Saharia, C., Ho, J., Chan, W., Salimans, T., Fleet, D.J., Norouzi, M.: Image super-resolution via iterative refinement. IEEE Trans. Pattern Anal. Mach. Intell. **45**(4), 4713–4726 (2022)

23. Sajjadi, M.S., Vemulapalli, R., Brown, M.: Frame-recurrent video super-resolution. In: Proceedings of the IEEE Conference on Computer Vision and Pattern Recognition, pp. 6626–6634 (2018)

24. Shi, W., et al.: Real-time single image and video super-resolution using an efficient sub-pixel convolutional neural network. In: Proceedings of the IEEE Conference on Computer Vision and Pattern Recognition, pp. 1874–1883 (2016)

25. Shoeiby, M., Armin, A., Aliakbarian, S., Anwar, S., Petersson, L.: Mosaic super-resolution via sequential feature pyramid networks. In: Proceedings of the IEEE/CVF Conference on Computer Vision and Pattern Recognition Workshops, pp. 84–85 (2020)

26. Soh, J.W., Cho, S., Cho, N.I.: Meta-transfer learning for zero-shot super-resolution. In: Proceedings of the IEEE/CVF Conference on Computer Vision and Pattern Recognition, pp. 3516–3525 (2020)

27. Tian, Y., Zhang, Y., Fu, Y., Xu, C.: TDAN: temporally-deformable alignment network for video super-resolution. In: Proceedings of the IEEE/CVF Conference on Computer Vision and Pattern Recognition, pp. 3360–3369 (2020)

28. Wang, X., Chan, K.C., Yu, K., Dong, C., Change Loy, C.: EDVR: video restoration with enhanced deformable convolutional networks. In: Proceedings of the IEEE/CVF Conference on Computer Vision and Pattern Recognition Workshops, pp. 0–0 (2019)

29. Xue, T., Chen, B., Wu, J., Wei, D., Freeman, W.T.: Video enhancement with task-oriented flow. Int. J. Comput. Vision **127**, 1106–1125 (2019)

30. Zhang, Y., Li, K., Li, K., Wang, L., Zhong, B., Fu, Y.: Image super-resolution using very deep residual channel attention networks. In: Proceedings of the European Conference on Computer Vision (ECCV), pp. 286–301 (2018)

31. Zhang, Y., Tian, Y., Kong, Y., Zhong, B., Fu, Y.: Residual dense network for image super-resolution. In: Proceedings of the IEEE Conference on Computer Vision and Pattern Recognition, pp. 2472–2481 (2018)

From Skulls to Faces: A Deep Generative Framework for Realistic 3D Craniofacial Reconstruction

Yehong Pan[1], Jian Wang[1], Guihong Liu[2], Qiushuo Wu[2], Yazi Zheng[2], Xin Lan[1], Weibo Liang[2], Jiancheng Lv[1(✉)], and Yuan Li[2(✉)]

[1] College of Computer Science, Sichuan University,
Chengdu, People's Republic of China
lvjiancheng@scu.edu.cn

[2] West China School of Basic Medical Sciences and Forensic Medicine,
Sichuan University, Chengdu, People's Republic of China
liyuan530060@163.com

Abstract. The shape of the human face is largely determined by the underlying skull morphology. Craniofacial reconstruction (CfR), or the process of reconstructing the face from the skull, is a challenging task with applications in forensic science, criminal investigation, and archaeology. Traditional craniofacial reconstruction methods suffer from subjective interpretation and simple low-dimensional learning approaches, resulting in low reconstruction accuracy and realism. In this paper, we present a deep learning-based framework for CfR based on conditional generative adversarial networks. Unlike conventional methods that adopt 3D representations directly, we employ 2D depth maps to represent faces and skulls as the model's input and output. It can provide enough face geometric information and may mitigate the potential risk of dimensionality issues. Our framework is capable of modeling both local and global details of facial appearance through a novel discriminator structure that leverages multi-receptive field features in one output, thus generating realistic and individualized faces from skulls. Furthermore, to explore the impact of conditional information such as age and gender on facial appearance, we develop a conditional CfR paradigm that incorporates an improved residual block structure with conditional information modulation and a conditional information reconstruction loss function. Extensive experiments and comparisons are conducted to demonstrated the effectiveness and superiority of our method for CfR.

Keywords: Craniofacial Reconstruction · Generative Adversarial Networks · Depth Map

This work is supported by the National Natural Science Foundation of China (No. 82202079) and Natural Science Foundation of Sichuan Province (No. 2022NSFSC1403).

© The Author(s), under exclusive license to Springer Nature Switzerland AG 2024
S. Rudinac et al. (Eds.): MMM 2024, LNCS 14554, pp. 314–326, 2024.
https://doi.org/10.1007/978-3-031-53305-1_24

1   Introduction

Craniofacial reconstruction (CfR) is a techniqual method that aims to reproduce the original facial appearance of an unknown skull based on the interrelationship between the soft tissue and the facial morphology and the skull morphology [1]. Highly realistic three-dimensional (3D) reconstructed faces with a wealth of facial shape data are the direct basis for individual recognition [2–4], providing valuable leads for criminal investigations. Furthermore, CfR also holds significant importance in fields such as forensic and archaeology.

CfR assumes that there is a predictable relationship between the skull and facial morphology. However, this relationship is not simple or linear, but rather complex and high-dimensionally non-linear, involving many hidden factors. Therefore, CfR methods struggle to produce accurate and realistic facial images that can be used for identification. Traditional manual CfR [5] uses clay or rubber to simulate facial muscles and fat tissues on a skull replica. It highly relies on the expertise and intuition of the practitioners, and are often time-consuming, intricate, and non-reproducible, resulting in suboptimal efficiency and accuracy in practical applications. Modern CfR methods use computer-assisted techniques that employ imaging technologies such as CT and MRI to obtain 3D data of skulls and faces. These techiques [6–8] use algorithms based on template deformation or statistical models to generate facial images from skull data. Due to their low capacity, however, these algorithms may produce faces that are too average or simplistic, lacking the individuality and realism of natural faces.

In this paper, we propose a novel CfR framework that uses generative adversarial networks (GANs) [9], capable of learning the intricate high-dimensonal mapping from a large dataset, enabling the accurate reconstruction from skull to face. Our framework uses depth maps which capture the 3D information of skulls and faces by representing the distance of each pixel from the camera [10] as the input and output of the model. Compared to other 2D or 3D representations, depth maps contain abundant geometric information while possessing lower dimensions. This data form effectively avoid overfitting and curse of dimensionality, and is suitable for convolutional operations [11,12]. We note that the skull to face mapping is ill-posed, as a single skull may correspond to multiple faces with different ages and genders. Therefore, a conditional CfR paradigm is developed to reduce the ambiguity of the mapping between the skull and face. In this paradigm, we use an improved residual block to incorporate conditional information into the model, which is achieved by modulating the weights of convolutional layers in residual blocks. To ensure that the conditional information is not ignored, we also introduce a conditional reconstruction loss. To enhance the accuracy and realism of facial reconstruction, we introduce two crucial components in the framework: a CLIP-based [13] perceptual module and a multi-receptive field discriminator (MRFD). The CLIP-based perceptual module leverages a pre-trained large-scale model to assess the similarity between the reconstructed and real faces on high-level semantics. On the other hand, MRFD improves the quality of the reconstructed face by capturing diverse receptive scales of features in the input image, thereby emphasizing both local and global details. Finally, we evaluate our framework by comparing it with other methods, showing its superiority in terms of accuracy, realism, and diversity. Additionally, ablation experiments are conducted to demonstrate the effectiveness of each component of our framework.

Our contributions are summarized as follows:

- We propose a novel GAN-based CfR framework that can generate precise and realistic faces from skulls using depth maps as input and output.
- A conditional CfR paradigm is developed that can incorporate conditional information such as age and gender into the model and reduce the ambiguity of the skull-face mapping.
- To improve the similarity and realism of the reconstructed faces, we introduce a CLIP-based perceptual loss and a multi-receptive field discriminator into the framework.
- Comparative and ablation experiments are conducted to validate the superiority of our framework and the necessity of its individual components.

2   Related Work

2.1   Traditional Craniofacial Reconstruction

Traditional manual craniofacial reconstruction involves molding the skull and placing soft tissue depth markers on the selected skull anatomical landmarks. Clay or rubber is then used to simulate facial muscles, fat, and soft tissue to build up the face. This approach is commonly known as the Manchester method [5], which is widely applied in the field.

Access to digital skull and face resources has enabled computer-based three-dimensional image processing and visualization tasks [14]. As a result, craniofacial reconstruction has entered the phase of computer-assisted craniofacial reconstruction [15]. Generally, the traditional computer-assisted craniofacial reconstruction methods can be devided into three categories:

(1) Sparse or dense soft tissue thickness reconstruction [14,15] utilizes reference skull templates and average thickness at landmarks for deformation. This results in "averaged" reconstructed faces.
(2) Registration-based craniofacial reconstruction [16,17] uses deformation between an unknown skull and reference template to copy facial tissue distribution, yielding a "template-like" face.
(3) Statistical model-based craniofacial reconstruction [6–8] relies on intrinsic skull-face relationships learned from shape parameters and regression models. Limited dimensionality and simplistic learning challenge modeling high-dimensional skull-face correspondence, constraining its effectiveness.

2.2 Deep Image Generation

CNN-based methods have made progress in image generation and transformation, but challenges persist in achieving realism. Traditional CNNs [18] yield smooth images, lacking fine details compared to real images, particularly in tasks like generating images from sketches.

Generative Adversarial Networks (GANs) [9] have greatly overcome the limitations of traditional CNN methods in image generation tasks. By introducing the adversarial training mechanism between the generator an discriminator, GANs is enabled to generate more realistic and detailed images. Building upon GANs, pix2pix [19] incorporates conditional guidance, enabling accurate transformation of input images into target images. On the other hand, CycleGAN [20] excels in cross-domain image translation by using cycle-consistency loss to maintain image transformation consistency.

In previous studies, several deep learning-based methods have been applied to the task of craniofacial reconstruction. For instance, CR-GAN [21] achieved promising 2D craniofacial reconstruction with deep generative adversarial networks, applied to identity recognition. MM-CGAN [12] incorporates Body Mass Index (BMI) as conditional information for skull-to-face depth map mapping. These methods demonstrate neural networks' potential but face challenges due to data limitations and model capacity, requiring accuracy improvement.

3 Proposed Method

3.1 Overview

In this work, we build a skull-to-facial depth map mapping based on GANs. Our proposed model primarily consists of a reconstruction network (G), composed of convolutions and residual blocks [22], capable of reconstructing skull-to-face transformations specific to different ages and genders. Besides, a CLIP-based perceptual module and a multi-receptive field discriminator (D) is employed to supervise and guide the reconstruction network in generating more satisfactory results. Given the depth map of a skull, denoted as x, along with its corresponding identity information c as input, the objective of the framework is to reconstruct the facial depth map:

$$\hat{y} = G(x, c) \tag{1}$$

The architecture of the proposed framework is illustrated in Fig. 1.

3.2 Details of Model Architecture

To be more specific, the reconstruction network within the framework incorporates a conditional code to achieve gender and age-specific facial reconstruction. This process is combined with an improved residual block structure to fuse it with image features. The multi-receptive field discriminator employs dilated convolutions to capture features with varying receptive fields in the reconstructed faces. Additionally, the CLIP image encoder [13] further enhances the efficacy of facial reconstructions generated by G.

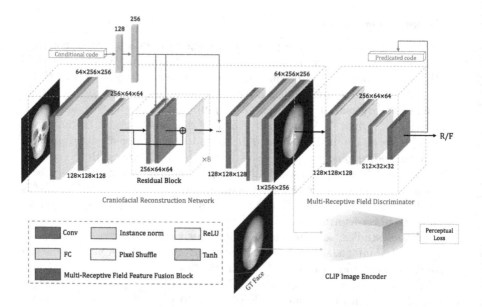

Fig. 1. Overview of our proposed architecture. The reconstruction network achieves identity-specific reconstruction from skull to facial images. Meanwhile, the discriminator is capable of distinguishing the authenticity of the reconstructed images. CLIP Image Encoder further enhances the quality of the reconstruction.

Conditional Craniofacial Reconstruction Paradigm. Just as the appearance of the same individual changes with different ages, similarly, for the model, a single skull should correspond to multiple faces with varying ages and genders. To enhance the effectiveness and accuracy of the reconstructed faces, we integrate these identity-specific information into the reconstruction process. Initially, we transform these information into conditional encodings with fully connected layers. Subsequently, these encodings are merged with the extracted skull features from the reconstruction network. In order to make the network better adapt to different conditional information, we employ an enhanced residual block structure named M-D residual block. Instead of directly concatenating features, the conditional information is incorporated into the model by modulating the weights of the convolutional layers. This approach enables the network to dynamically adjust weights based on varying conditions, thereby enhancing the model's adaptability and expressive capacity. The M-D residual block and the reconstructed facial images for different ages are depicted in Fig. 2.

To ensure effective utilization of identity information and reduce the ambiguity of the mapping between the skull and face, the conditional code loss is introduced into the training of both the reconstruction network and the discriminator, as shown in Eq. 2.

$$L_c = \mathbf{E}_{c,\hat{c}}[\|c - \hat{c}\|_1] \tag{2}$$

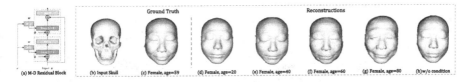

Fig. 2. (a) represents the M-D residual block. (b) is the input skull. (c) denotes the ground truth with an age of 59 and a female gender. (d) to (g) depict the reconstruction results for different ages. (h) is the reconstruction without condition information.

c is the corresponding identity information (such as gender, age, etc.) encodings, $\hat{c}$ denotes the conditional encodings predicted by the discriminator. By introducing this loss, we improve the individuality and diversity of the reconstructions.

CLIP-Based Perceptual Module. Perceptual loss utilizes a pre-trained convolutional neural networks to calculate the feature distance between the generated image and the target image, thereby promoting the better retention of the content and style [23]. As a large-scale pre-trained model, CLIP has been demonstrated to possess exceptional image feature extraction and advanced semantic understanding capabilities [24,25]. To ensure consistency between the reconstructed facial images and the ground truth at a higher semantic and content level, thereby enhancing the realism and naturalness of the reconstructions, we introduce perceptual loss in the model training as shown in Eq. 3, where F denotes the pre-trained feature extractor, and in this paper, the CLIP Image Encoder is utilized.

$$L_{pec} = \mathbf{E}_{c,x,y}[\|F(y) - F(G(x,c))\|_1] \tag{3}$$

y is the real facial image, x denotes the input skull image, and G represents the reconstruction network.

Multi-Receptive Field Discriminator. The task of the discriminator is to discern the authenticity of the reconstructed images. To achieve accurate discrimination, we construct a multi-receptive field discriminator based on ASPP (Atrous Spatial Pyramid Pooling) [26]. After the image is input into the discriminator, it undergoes feature extraction through three convolutional blocks. Subsequently, the features are passed through a multi-receptive field feature fusion block composed of dilated convolutions. Finally, the discriminator evaluates the features from multiple receptive fields to achieve precise assessment of the reconstructed facial images and provide informative feedback to the reconstruction network. The structure of the multi-receptive field feature fusion block is illustrated in Fig. 3.

With such a structure, our discriminator is able to simultaneously capture local and global features at various receptive fields, thus enhancing the realism and accuracy of the reconstructions.

320 Y. Pan et al.

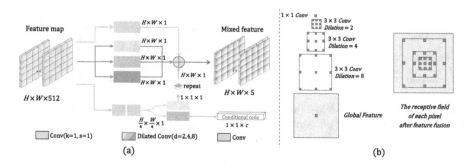

Fig. 3. (a)Detail structure of multi-receptive field feature fusion block. (b)The receptive field of each pixel after passing through this block.

Loss Function. To ensure high-quality reconstruction of facial images and the proper training of the model, we also introduce the following losses:

Pixel Loss. Due to our paired skull and facial depth map data, we adopted the pixel loss as Eq. 4. It ensures that our reconstructed facial images closely match the groud truth at the pixel level.

$$L_{pix} = \mathbf{E}_{c,x,y}[\|y - G(x,c)\|_1] \tag{4}$$

Adversarial Loss. Like most GAN-based frameworks, we also incorporate adversarial loss. For the reconstruction network, the adversarial loss is as shown in Eq. 5, where D denotes the discriminator.

$$L_{adv} = \mathbf{E}_{c,x}[-\log(D(G(x,c)))] \tag{5}$$

For the discriminator, the adversarial loss is calculated as shown in Eq. 6.

$$L_{adv} = \mathbf{E}_y[-\log(D(y)] + \mathbf{E}_{c,x}[-\log(1 - D(G(x,c))] \tag{6}$$

Inspired by CycleGAN, we simultaneously train the model for facial-to-skull transformations. By introducing the cycle consistency loss as shown in Eq. 7, we ensure the consistency of image transformations between the skull and facial domains, resulting in more natural and accurate reconstruction outcomes.

$$L_{cyc} = \mathbf{E}_{y,c}[\|y - \hat{G}(G(y,c))\|_1] \tag{7}$$

$\hat{G}$ is the generator for the transformation from facial to skull. L_{cyc} is finally added to the reconstruction network's adversarial loss.

By optimizing the above losses, our model can simultaneously consider the pixel-level similarity, realism, and high-level semantic and content consistency of the images. The final loss function is formulated as shown in Eq. 8.

$$L_G = \lambda_{pix}L_{pix} + \lambda_{adv}L_{adv} + \lambda_c L_c + \lambda_{pec}L_{pec} \tag{8}$$

λ denotes the weights of each loss.

4   Experiments and Analysis

In this section, we present the detailed implementation of the model and the corresponding evaluation metrics for assessing the reconstructed face. Based on these metrics, comparative experiments and ablation experiments are conducted to demonstrate the effectiveness and superiority of our approach.

4.1   Datasets

The source data used in our study consists of thin slices of head CT scans stored in DICOM format, which are obtained from patients scanned at West China Hospital of Sichuan University. In order to convert the data into the depth map, it is necessary to reconstruct the thin slices CT scans into meshes. In this work, we employe the method proposed by [21] utilizing MIMICS software to acquire paired skull and facial meshes. Afterwards, the coordinate system alignment is performed on the paired meshes to ensure consistency. The meshes are then scaled to fit within a cubic space ranging form -1 to 1. At this stage, the meshes with uniform posture and size are obtained.

Once the aligned meshes are obtained, we utilize the pyrender library to import the meshes and create corresponding scenes. By setting the camera parameters, including camera coordinates, normal directions, field of view(FOV) and lighting properties, we capture the well-unified pairs of skull and facial depth maps. Since the pixel values in the depth maps are concentrated in the higher range, and to ensure that the enhancement is reversible, we apply a linear transformation to the regions with lower pixel values and a gamma transformation to the higher regions. This approach helps to enhance the details in the depth maps and improve their overall quality for better model training.

A total of 780 pairs of cranio-facail data are collected, of which 150 pairs are randomly selected as the test dataset. To ensure effective model training, data augmentation is necessary. For every pair of meshes, we perform rotations of ± 15 degrees around the X, Y, and Z axes, and then apply the aforementioned approach to obtain the corresponding depth maps. This data augmentation not only increase the number of data samples, resulting in a total of 4410 image pairs for model training, but also allow the model to learn the mapping between skull and facial depth maps at different angles, thus enhancing the model's robustness.

4.2   Implementation Details

Our framework is inspired by pix2pix and CycleGAN and is implemented by PyTorch. Instance Normalization is utilized in the model, which has been proven to enhance the performance of deep neural networks in image generation tasks [27]. In addition to the ReLU activation applied after each convolutional layer, we also utilize the tanh activation in the final layer of the network to ensure that the output values are within the range of $[0, 255]$. Our model and all comparative methods are implemented using a NVIDIA GeForce RTX 3090 of 24GB VRAM, with CUDA version 11.3 and PyTorch version 1.11. During training, we utilize the Adam optimization algorithm with a batch size of 1. The initial learning rate is set to 10^{-4}, and after reaching 50 epochs, we decrease the learning rate by 50% every 10 epochs. We set the weights for different losses as follows: $\lambda_{pix} =1$, $\lambda_{adv}=0.02$, $\lambda_{c}=0.01$, $\lambda_{pec}=0.01$. Variations in hyperparameters do not significantly compromise the model's performance.

4.3   Evaluation Metrics

To assess the quality of the model's reconstructions, we calculate the Reconstruction Error(MAE) between the ground truth and the reconstructions from the test dataset. Additionally, we compute the Fréchet Inception Distance (FID) [28] to gauge the quality of the reconstructed images. For each model, we independently trained it three times to obtain the average and range of their FID and Reconstruction Error. Both FID and reconstruction error are better when lower.

Furthermore, we visualize the reconstruction errors in the form of heatmaps, providing a more intuitive representation of the model's reconstruction performance. The color in the heatmap shifts towards blue indicating smaller errors, while red signifies larger errors.

4.4   Experimental Comparison and Analysis

Comparative Experiment. To validate the superiority of the proposed framework, we compare it against other deep learning methods in the field, including MM_CGAN [12], Pix2Pix [19], and CycleGAN [20]. Corresponding statistical data is presented in Table 1 and Fig. 4.

Table 1. Statics of Reconstuction errors and FID for 150 samples in comparative study.

Model Composition	FID ↓	Reconstruction Error ↓		
		Min	Max	Mean
Pix2Pix	42.06±0.94	**0.017±0.004**	0.172±0.007	0.084±0.006
CycleGAN	41.74±0.46	0.034±0.001	0.237±0.002	0.073±0.001
MM_CGAN	37.01±0.55	0.035±0.003	0.157±0.004	0.073±0.002
Our model	**29.97 ±0.48**	0.019±0.002	**0.137±0.001**	**0.058±0.002**

While Pix2Pix outperforms our approach in terms of minimum reconstruction error, the comparison of FID and mean reconstruction error values in Table 1 indicates that the quality and realism of our framework's reconstructions is superior to other deep models. This is also corroborated by Fig. 4, where our framework achieves lower reconstruction errors for the test data, effectively keeping them within a reduced range.

To provide a more intuitive presentation of our model's reconstructions compared to other models, we randomly select several test outcomes and present

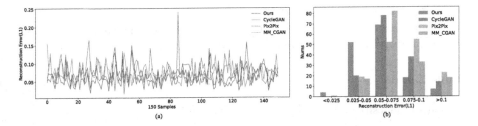

Fig. 4. (a)Reconstruction errors of 150 reconstructed faces on the test dataset in the comparative experiments. (b)Histogram distribution of reconstruction errors for 150 reconstructed faces.

the reconstruction error using heatmaps, as illustrated in Fig. 5. In this figure, from left to right, we have the ground truth, reconstructed faces corresponding to each model, and the heatmap illustrating the differences between the reconstructions and the ground truth. It's evident that our model exhibits smaller renconstruction errors and significantly higher reconstruction quality compared to others.

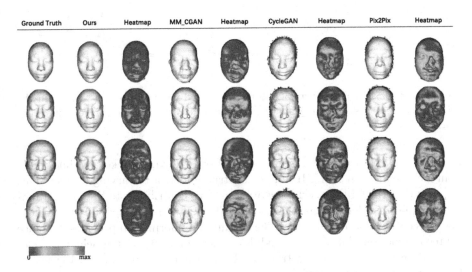

Fig. 5. Comparison of Heatmaps Showing Differences in Reconstructed Results of Different Models

Ablation Experiments. Comparative experiments demonstrate the superior performance of the proposed framework over other models. Additionally, for the purpose of validating the rationality of the model design, as well as the influence and necessity of each module on overall performance, we conduct a comparison between the full model and the models with the perception module, multi-receptive field discriminator block, and conditional encoding information removed separately. The experimental results are shown in Table 2 and Fig. 6.

Table 2. Statics of Reconstuction errors and FID for 150 samples in ablation study.

Model Composition	FID ↓	Reconstruction Error ↓		
		Min	Max	Mean
w/o perceptual	37.31±0.95	0.033±0.002	0.155±0.003	0.072±0.002
w/o condition	37.56±0.82	0.031±0.001	0.152±0.002	0.071±0.001
w/o multi-D	35.67±0.55	0.029±0.001	0.147±0.004	0.072±0.001
full model	**29.97±0.48**	**0.019±0.002**	**0.137±0.001**	**0.058±0.002**

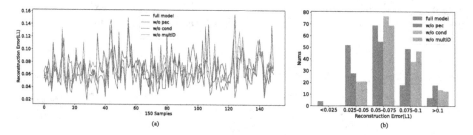

Fig. 6. (a)Reconstruction errors of 150 reconstructed faces on the test dataset in the ablation experiments. (b)Histogram distribution of reconstruction errors for 150 reconstructed faces.

From Table 2, it can be observed that after removing any key module, there is an increasement in both FID and reconstruction error to varying degrees. This suggests that the CLIP-based perceptual module contributes to the understanding of high-dimensional semantics, the MRFD enhances the capability to capture multi-receptive field features, and the conditional information provides guidance for reconstruction. All of these modules play significant roles in achieving realism and accurate facial reconstruction. Furthermore, as depicted in Fig. 6, our framework, under the combined influence of various modules, effectively minimizes the fluctuation of the reconstruction error and maintains it within a lower range.

We also provide an intuitive heatmap visualization for the ablation experiments, shown as Fig. 7. From this figure, we can observe that when the conditional information is absent, CfR exhibits ambiguity, resulting in reconstructions that closely resemble the average age of the samples rather than being consistent with ground truth. When the multi-receptive field discriminator and CLIP-based perceptual module are missing, the model generates higher reconstruction errors and produces less realistic facial images.

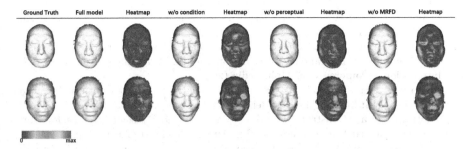

Fig. 7. Comparison of heatmaps showing the reconstructions when different key modules are missing.

5   Conclusion and Discussion

In this paper, we construct a novel GAN-based depthmap craniofacial reconstruction framework. The key modules designed by us are capable of effectively utilizing skull features, combining identity information to remove the ambiguity of reconstruction, and capturing global and local features with understanding high-dimensional semantics, so as to achieve realistic and accurate face reconstruction. For future work, there is potential for improvement in reconstructing facial features such as nose and eyes. Additionally, more direct three-dimensional representations, such as mesh-based facial reconstruction techniques, remains an area for further exploration.

References

1. Claes, P., et al.: Computerized craniofacial reconstruction: conceptual framework and review. Forensic Sci. Int. **201**(1–3), 138–145 (2010)
2. Masi, I., et al.: Deep face recognition: a survey. In: 2018 31st SIBGRAPI Conference on Graphics, Patterns and Images (SIBGRAPI). IEEE (2018)
3. A-masiri, P., Kerdvibulvech, C.: Anime face recognition to create awareness. Int. J. Inf. Technol. **15**, 3507–3512 (2023)
4. Hörmann, S.: Robust Face Recognition Under Adverse Conditions: Technische Universität München (2023)
5. Wilkinson, C.: Forensic Facial Reconstruction: Cambridge University Press (2004)
6. Vandermeulen, D., et al.: Computerized craniofacial reconstruction using CT-derived implicit surface representations. Forensic Sci. Int. **159**, S164–S174 (2006)
7. Deng, Q., et al.: A novel skull registration based on global and local deformations for craniofacial reconstruction. Forensic Sci. Int. **208**(1–3), 95–102 (2011)
8. Berar, M., et al.: Craniofacial reconstruction as a prediction problem using a Latent Root Regression model. Forensic Sci. Int. **210**(1–3), 228–236 (2011)
9. Creswell, A., et al.: Generative adversarial networks: an overview. IEEE Signal Process. Mag. **35**(1), 53–65 (2018)
10. Eigen, D., Puhrsch, C., Fergus, R.: Depth map prediction from a single image using a multi-scale deep network. In: Advances in Neural Information Processing Systems, vol. 27 (2014)

11. Zhang, C., et al.: A study on overfitting in deep reinforcement learning. arXiv preprint arXiv:1804.06893 (2018)
12. Zhang, N., et al.: An end-to-end conditional generative adversarial network based on depth map for 3D craniofacial reconstruction. In: Proceedings of the 30th ACM International Conference on Multimedia (2022)
13. Radford, A., et al.: Learning transferable visual models from natural language supervision. In: International Conference on Machine Learning. PMLR (2021)
14. Tilotta, F., et al.: Construction and analysis of a head CT-scan database for craniofacial reconstruction. Forensic Sci. Int. 191(1–3), 112-e1 (2009)
15. De Greef, S., Guy, W.: Three-dimensional cranio-facial reconstruction in forensic identification. J. Forensic Sci. 50(1), JFS2004117 (2005)
16. Prieels, F., Hirsch, S., Hering, P.: Holographic topometry for a dense visualization of soft tissue for facial reconstruction. Forensic Sci. Med. Pathol. 5, 11–16 (2009)
17. Claes, P., et al.: Craniofacial reconstruction using a combined statistical model of face shape and soft tissue depths: methodology and validation. Forensic Sci. Int. 159, S147–S158 (2006)
18. Krizhevsky, A., Sutskever, I., Hinton, G.E.: ImageNet classification with deep convolutional neural networks. In: Advances in Neural Information Processing Systems, vol. 25 (2012)
19. Isola, P., et al.: Image-to-image translation with conditional adversarial networks. In: Proceedings of the IEEE Conference on Computer Vision and Pattern Recognition (2017)
20. Zhu, J.-Y., et al.: Unpaired image-to-image translation using cycle-consistent adversarial networks. In: Proceedings of the IEEE International Conference on Computer Vision (2017)
21. Li, Y., et al.: CR-GAN: automatic craniofacial reconstruction for personal identification. Pattern Recogn. 124, 108400 (2022)
22. He, K., et al.: Deep residual learning for image recognition. In: Proceedings of the IEEE Conference on Computer Vision and Pattern Recognition (2016)
23. Johnson, J., Alahi, A., Fei-Fei, L.: Perceptual losses for real-time style transfer and super-resolution. In: Leibe, B., Matas, J., Sebe, N., Welling, M. (eds.) ECCV 2016. LNCS, vol. 9906, pp. 694–711. Springer, Cham (2016). https://doi.org/10.1007/978-3-319-46475-6_43
24. Yao, L., et al.: Filip: fine-grained interactive language-image pre-training. arXiv preprint arXiv:2111.07783 (2021)
25. Shen, S., et al.: How much can clip benefit vision-and-language tasks? arXiv preprint arXiv:2107.06383 (2021)
26. Chen, L.-C., et al.: DeepLab: semantic image segmentation with deep convolutional nets, Atrous convolution, and fully connected CRFs. IEEE Trans. Pattern Anal. Mach. Intell. 40(4), 834–848 (2017)
27. Ulyanov, D., Vedaldi, A., Lempitsky, V.: Instance normalization: the missing ingredient for fast stylization. arXiv preprint arXiv:1607.08022 (2016)
28. Heusel, M., et al.: GANs trained by a two time-scale update rule converge to a local nash equilibrium. In: Advances in Neural Information Processing Systems, vol. 30 (2017)

Structure-Aware Adaptive Hybrid Interaction Modeling for Image-Text Matching

Wei Liu[1,2], Jiahuan Wang[1], Chao Wang[3,4(✉)], Yan Peng[2,3,4],
and Shaorong Xie[1]

[1] School of Computer Engineering and Science, Shanghai University, Shanghai
200444, China
{liuw,jiahwang,Srxie}@shu.edu.cn
[2] Shanghai Artificial Intelligence Laboratory, Shanghai 201114, China
[3] School of Future Technology, Shanghai University, Shanghai 200444, China
{cwang,pengyan}@shu.edu.cn
[4] Institute of Artificial Intelligence, Shanghai University, Shanghai 200444, China

Abstract. Image-text matching is a rapidly evolving field in multi-modal learning, aiming to measure the similarity between image and text. Despite significant progress in image-text matching in recent years, most existing methods for image-text interaction rely on static ways, often overlooking the substantial variations in scene complexity among different samples. Actually, multimodal interaction strategies should be flexibly adjusted according to the scene complexity of different inputs. For instance, excessive multimodal interactions may introduce noise when dealing with simple samples. In this paper, we propose a novel **S**tructure-aware **A**daptive **H**ybrid **I**nteraction **M**odeling (SAHIM) network, which can adaptively adjust the image-text interaction strategies based on varying inputs. Moreover, we design the Multimodal Graph Inference (MGI) module to explore potential structural connections between global and local features, as well as the Entity Attention Enhancement (EAE) module to filter out irrelevant local segments. Finally, we align the image and text features with the bidirectional triplet loss function. To validate the proposed SAHIM model, we design and conduct comprehensive experiments on Flickr30K and MSCOCO. Experimental results show that SAHIM outperforms state-of-the-art methods on both datasets, demonstrating the superiority of our model.

Keywords: Cross-Modal Retrieval · Graph Convolutional Network · Image-Text Matching · Multimodal Interaction Modeling

1 Introduction

Image-text matching is an increasingly important research field in multimodal learning, which aims to retrieve data between different modalities by exploring

© The Author(s), under exclusive license to Springer Nature Switzerland AG 2024
S. Rudinac et al. (Eds.): MMM 2024, LNCS 14554, pp. 327–341, 2024.
https://doi.org/10.1007/978-3-031-53305-1_25

their semantic association. In recent years, image-text matching has encountered significant challenges due to the exponential expansion of multimedia data, along with evident semantic and heterogeneous gaps that exist between different modalities [24]. Therefore, how to bridge the gap of heterogeneous data is the key step to overcoming these challenges. In recent years, although some progress has been made in the field of image-text matching, many overlooked issues still need to be addressed. In the following, we enumerate three major issues that currently exist in this field.

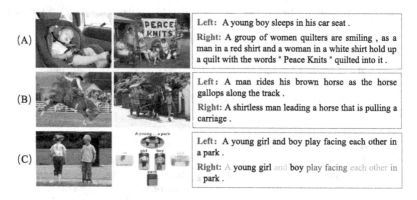

Fig. 1. Illustration of the importance of adaptive interaction modeling (A), structural information (B), and attention filtering (C) in visual semantic matching.

(1) *The complex interaction patterns between images and texts lack research.* As shown in Fig. 1 (a), it demonstrates that since the scene complexity varies for different images, the interaction mode should be adaptively adjusted according to the input. In detail, the left image contains fewer objects, and its corresponding sentence is also brief, so its scene is relatively simple. In this instance, too many modal interactions may introduce noise and affect the matching accuracy. By contrast, the right image contains a large number of objects, and the corresponding sentence is also lengthy, in which case it is essential to perform a complex process of modal interaction.

(2) *The importance of structural information for image-text matching is ignored.* As shown in Fig. 1 (b), the relation triple in the left image is *man-ride-horse*, while the right image is *man-lead-horse*. Even though the two images have consistent key objects (man and horse), they have significant differences due to their different structural information. This confirms that the structural information is crucial for the correct matching of image and text.

(3) *The interference caused by meaningless alignments is not considered.* As shown in Fig. 1 (c), the current works fail to take into account the interference caused by alignments constructed with meaningless words such as "the", "is", and "of" [5,14]. Actually, the alignments of these irrelevant words undermine the effectiveness of entity fragments, thus impeding the accuracy

of image-text matching. Therefore, it is important to design a model that can set different attention weights for words with different meanings.

To address these issues, we propose a novel **Structrue-aware Adaptive Hybrid Interaction Modeling (SAHIM)** model for image-text matching. The SAHIM model can adaptively adjust the image-text interaction strategy based on different inputs. Our contributions are summarized as follows:

- We propose the SAHIM model that can adaptively adjust the multimodal modeling strategy according to the scene complexity of different inputs, which achieves more flexible multimodal interaction patterns.
- We design the MGI module to consider structural information between global and local features, along with the EAE module to effectively filter out irrelevant fragments that may introduce interference.
- Extensive comparative experiments demonstrate that the SAHIM model outperforms state-of-the-art image-text retrieval models on two benchmark datasets. Moreover, sufficient ablation experiments further verify the effectiveness of each module in SAHIM.

2   Related Work

2.1   Image-Text Matching

Most existing works learn the correspondence between image and text based on object co-occurrence, which can be categorized into two groups: global alignment and local alignment. The main objective of global alignment methods [5,14] is to map the global features of image and text into a shared space for measuring the similarity score. The local alignment methods [2,9,21] focus more on fine-grained features, such as salient regions in the image and words in the text. However, both global and local alignment methods heavily rely on static ways for multimodal alignment, ignoring the significant data diversity among different samples. To solve the limitations of these methods, we propose an image-text matching model that can adaptively adjust the multimodal alignment approaches based on different inputs. In this way, the model can automatically choose to focus on global alignment or local alignment according to the scene complexity of inputs.

2.2   Multimodal Interaction Modeling

Multimodal interaction modeling plays a pivotal role in accurately establishing semantic connections across diverse modalities, which can be broadly categorized into three main groups. First, intra-modal relationship modeling [10,26] refers to the process of capturing relationships within a specific modality, aiming to discover the similarities and correlations among elements within the same modality. In contrast to this, cross-modal interaction modeling [6,12] is dedicated to exploring the relationships between different modalities. To simultaneously enhance the intra-modal and cross-modal relationship modeling, some approaches based on hybrid interaction modeling [16,19] have been proposed. In this paper, we adopt the hybrid interaction modeling approach, which can not only capture the internal correlation of vision or text but also consider the interaction between local information of different modalities.

3   Methodology

In this section, we present the overall framework of our proposed SAHIM, as shown in Fig. 2.

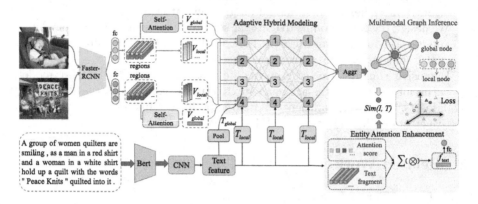

Fig. 2. The overall architecture of the SAHIM framework.

3.1   Feature Representation

Visual Representation. Given an image I, we represent its features as a combination of M salient regions, which are obtained by extracting region-level visual features from the Faster R-CNN [17] model pre-trained on Visual Genomes [8]. Then, a fully connected layer is applied to convert it into the d-dimensional feature space, which is regarded as the visual local feature representation $V_{local} = \{v_1, \ldots, v_i, \ldots, v_M\}$, where $v_i \in \mathbb{R}^d$. Afterward, we aggregate the local regions through the self-attention mechanism [18] to obtain the visual global feature representation V_{global}.

Textual Representation. Given a sentence S, we first extract textual features via the pre-trained BERT [4] encoder and represent it as $S = \{s_1, \ldots, s_j, \ldots, s_N\}$, where s_j denotes the representation of the j-th word, N is the number of words in sentence S. We then utilize convolution kernels of different scales to capture the phrase-level features of the sentence and concatenate the results as text features. Moreover, a linear layer is applied to map the features into the d-dimensional space as sentence local features, denoted as $T_{local} = \{t_1, \ldots, t_N\}$. Finally, the global sentence features T_{global} is obtained by pooling operation.

3.2   Adaptive Hybrid Modeling

To adaptively adjust the multimodal interaction patterns based on the scene complexity of different inputs, we design four different units. As shown in Fig. 2, these four units are the Basic Correction Unit (BCU), the Intra-modal Interaction Unit (IIU), the Cross-modal Interaction Unit (CIU), and the Global Enhancement Unit (GEU). They correspond to the four units "1", "2", "3", and "4" in the Adaptive Hybrid Modeling module, respectively.

Basic Correction Unit. Complex modal interaction strategies for inputs with simple scenarios are unnecessary [1]. Accordingly, we design a basic correction unit to preserve the original feature information of inputs. To alleviate the problem of gradient disappearance, we design the unit as $Output = \text{ReLU}(X)$, where X indicates the input and $\text{ReLU}(\cdot)$ stands for the Rectified Linear Unit.

Intra-modal Interaction Unit. To improve the representation capability of a single modality, we propose an intra-modal interaction unit. Specifically, we adopt the multi-head self-attention mechanism to enhance the modality representation, which is formulated as:

$$\text{MultiHead}(X) = \text{Concat}\left(head_1, \ldots, head_i, \ldots, head_h\right) W + X, \quad (1)$$

where $\text{Concat}(\cdot)$ stands for concatenating the outputs of h heads, h is the number of attention heads, W is the learnable parameter matrix, and $head_i$ represents the output of the i-th head. We then employ a Multilayer Perceptron (MLP) to combine the features extracted by multiple attention heads, which is defined as $Output = \text{MLP}(\text{MultiHead}(X))$.

Cross-Modal Interaction Unit. To better utilize the association information between image and text modalities, we propose a cross-modal interaction unit to facilitate fine-grained interaction between different modalities. Similar to the method in [9], we first compute the similarity between each image region and each word, which is formulated as:

$$s_{ij} = \frac{v_i^\top t_j}{\|v_i\| \, \|t_j\|}, i \in [1, M], j \in [1, N]. \quad (2)$$

There are M detection regions in a picture, N words in a sentence, and s_{ij} denotes the similarity between the i-th visual region and the j-th textual word. Next, we set the negative values in the similarity matrix to zero and then normalize the similarity scores, which can be defined as $\bar{s}_{ij} = [s_{ij}]_+ / \sqrt{\sum_{i=1}^{M} [s_{ij}]_+^2}$, where $[x]_+$ is equivalent to $max[\cdot, 0]$. We then apply a weighted combination of word representations to concentrate on words for each image region as follows:

$$\begin{cases} \alpha_{ij} = \dfrac{\exp\left(\lambda \bar{s}_{ij}\right)}{\sum_{j=1}^{N} \exp\left(\lambda \bar{s}_{ij}\right)}, \\ r_i = \sum_{j=1}^{N} \alpha_{ij} t_j, \end{cases} \quad (3)$$

where λ is the temperature parameter, α_{ij} is the attention weight of the image region v_i with the sentence word t_j after normalization, and r_i represents the correlation between the image region v_i and the entire sentence obtained by weighted aggregation of word t_j.

To leverage the complementarity between different modalities, we adopt a conditional modulation strategy to guide the modal features. The correlation r_i is utilized to precisely guide the original visual features v_i. In detail, the image region feature r_i is first mapped to generate the scaling vector μ and offset vector η, which is formulated as follows:

$$
\begin{cases}
\mu_i = \text{Tanh}\left(\text{FC}_\mu\left(r_i\right)\right), \\
\eta_i = \text{FC}_\eta\left(r_i\right).
\end{cases}
\tag{4}
$$

where $\text{Tanh}(\cdot)$ is the hyperbolic tangent activation function, and $\text{FC}(\cdot)$ is the fully connected layer. After passing through an MLP layer, the image region v_i undergoes feature transformation to produce the enhanced image region feature $\tilde{v}_i$, which can be expressed by $\tilde{v}_i = \text{MLP}\left(v_i \odot \mu_i + \eta_i\right) + v_i$, and $Output = \tilde{V} = [\tilde{v}_1, \tilde{v}_2, \ldots, \tilde{v}_M]$. In this way, we obtain the image features $\tilde{V}$ by the cross-modal interaction unit. Symmetrically, employing analogous procedures allows us to acquire enhanced text features.

Global Enhancement Unit. To enhance the quality of the feature representation, we utilize global information to complement local information. Here, we use sentence global features $\bar{t}$ to enhance image region features v_i as an example, and the symmetric methods can be applied to obtain enhanced local text features. This way is formally defined as follows:

$$
\begin{cases}
e_i = \text{Norm}(\text{FC}\left(v_i\right) \odot \bar{t}), \\
\tilde{v}_i = (1 + e_i) \odot v_i,
\end{cases}
\tag{5}
$$

where e_i denotes the enhancement rate of sentence global features $\bar{t}$ to image region features v_i, $\odot$ stands for Hadamard product, also known as element-wise product, and $\text{Norm}(\cdot)$ represents the normalization of features. Finally, we obtain the enhanced image features $\tilde{V} = [\tilde{v}_1, \tilde{v}_2, \ldots, \tilde{v}_M]$ by the global enhancement unit.

Adaptive Path Selection. To adaptively select the multimodal interaction modeling mode for inputs with different scene complexity, we adopt an adaptive path selection mechanism to select the optimal path for different inputs. Specifically, we combine four different interaction units into L layers in a fully connected manner and then train the network to estimate the probability of each path. The path probability of each unit through the individual units of the next layer can be calculated as:

$$
p^{(l)} = \delta\left(\text{FC}\left(\text{ReLU}\left(\text{FC}\left(\frac{1}{n}\sum_{i=1}^{n} x_i\right)\right)\right)\right),
\tag{6}
$$

where x_i denotes a local segment of the image or text, $\delta(\cdot)$ stands for the activation function, and $p^{(l)}$ represents the path probability output by the l-th layer. In addition, the output feature of the i-th unit in the l-th layer can be computed using the following equation. Subsequently, the final output is obtained by weighing the interaction units in the last layer, which can be formulated as:

$$X_b^{l+1} = \begin{cases} X, & l = 0 \\ \sum\limits_{a=1}^{U} p_{ab}^l X_a^l, & l > 0 \end{cases},$$ (7)

where X represents the input feature of a unit, U denotes the number of units in each layer of the adaptive hybrid modeling module, p_{ab}^l stands for the path probability from the a-th unit in the l-th layer to the b-th unit in the $(l+1)$-th layer, and X_a^l represents the features of the a-th unit in the l-th layer.

In this manner, the feature X_b^{l+1} of the b-th unit in the $(l+1)$-th layer is derived by employing weighted aggregation of features from the l-th layer. In particular, for the last layer (the L-th layer), there is only one output after aggregation, which represents the output feature for adaptive hybrid modeling.

3.3 Multimodal Graph Inference

We construct a heterogeneous graph to explore the potential structural information among different hierarchical features. In this work, the heterogeneous graph includes global and local nodes, which represent two distinct levels of feature information. Specifically, we propagate node information in Z steps to obtain higher-order association information. We first calculate the edge values of the graph represented by an adjacency matrix, which is defined as:

$$A = \text{Softmax}\left(f_W(X) f_W(X)^\top\right) + E,$$ (8)

where X denotes the global or local features extracted by the adaptive hybrid modeling module, the global feature is obtained by averaging the local features, f_W represents a nonlinear projection with the learnable matrix W, and E is the identity matrix to preserve information about the node itself. Each element a_{ij} of the adjacency matrix A represents the correlation between the i-th node and the j-th node.

We then apply graph convolution to aggregate messages and propagate them across nodes, thereby capturing underlying structural information among global and local features, which is defined as $X^{(z)} = \sigma\left(A X^{(z-1)} W^{(z)}\right)$, where z denotes the z-th step of the multimodal graph inference and $\sigma(\cdot)$ represents the activation function. We iteratively perform Z steps of inference, and the global node in the last step serves as the global feature representation.

3.4 Entity Attention Enhancement

In multimodal representation learning, certain meaningless segments may impede the effectiveness of multimodal understanding. Therefore, we develop an EAE module to filter out irrelevant local segments. We first calculate the attention weights ρ between each local segment and the overall feature as follows:

$$\rho_i = \frac{\delta\left(\mathrm{BN}\left(W_f x_i\right)\right)}{\sum_{x_j \in \mathcal{N}} \delta\left(\mathrm{BN}\left(W_f x_j\right)\right)}, \tag{9}$$

where x_i is a visual or textual local fragment, $\delta(\cdot)$ is the sigmoid function, $\mathrm{BN}(\cdot)$ denotes the batch normalization, W_f stands for the linear transformation, and $\mathcal{N}$ indicates all the local fragments. We then utilize the attention weight ρ to aggregate the local segments, which is defined as $\tilde{X} = \sum_{x_i \in \mathcal{N}} \rho_i x_i$. In this way, we obtain the enhanced features $\tilde{X}$ after filtering out irrelevant segments using the EAE module.

3.5   Objective Functions

Image-Text Retrieval Loss Function. We adopt the bidirectional triplet loss as the image-text retrieval loss function. Given a matched image-text pair (v, t), along with the corresponding hardest negative image v^- and the hardest negative text t^- within a minibatch, our goal is to optimize the hard negative samples that result in the highest loss. The optimization objective can be formulated as:

$$\mathcal{L}_r = \left[\gamma - \mathcal{S}_r(v, t) + \mathcal{S}_r\left(v, t^-\right)\right]_+ + \left[\gamma - \mathcal{S}_r(v, t) + \mathcal{S}_r\left(v^-, t\right)\right]_+, \tag{10}$$

where γ is the margin parameter, $\mathcal{S}_r(v, t)$ represents the cosine similarity between the image v and text t, the $[\cdot]_+$ function is equivalent to $max[\cdot, 0]$, and $\mathcal{L}_r$ is the loss function for image-text retrieval.

Path Regularization Loss Function. We introduce a path regularization loss to select similar paths for inputs with similar semantics. To be specific, we aggregate the paths generated by all units to form a path selection vector, denoted as V^{path}. Meanwhile, we employ BERT to extract textual word features and derive the semantic embedding vector V^{sem} via pooling operation. Therefore, during the training phase, the loss function for path regularization is defined as follows:

$$\begin{cases} V_{emb}^{path} = \mathrm{FC}(V^{path}), \ V^{path} \in \mathbb{R}^{b \times \left(U^2 \times (L-1) + U\right)} \\ V_{emb}^{sem} = \mathrm{FC}(V^{sem}), \ V_{sem} \in \mathbb{R}^{b \times 768} \end{cases}, \tag{11}$$

and

$$\mathcal{L}_p = \mathrm{Mean}([cos(V_{emb}^{path}) - cos(V_{emb}^{sem})]^2), \tag{12}$$

where b denotes the mini-batch size, L corresponds to the number of layers within the adaptive hybrid modeling module, U represents the number of units per layer in this module, and $\mathcal{L}_p$ is the loss function of path regularization.

Training Objective. Finally, based on the image-text retrieval loss function $\mathcal{L}_r$ and the path regularization loss function $\mathcal{L}_p$ obtained above, the total objective function is expressed as $\mathcal{L} = \mathcal{L}_r + \lambda_p \mathcal{L}_p$, where λ_p is the balance parameter.

4 Experiments

4.1 Datasets and Implementation Details

Datasets. Flickr30K [15] consists of 31,783 images collected from the Flickr website. Each image contains five textual descriptions. Following the setting in [21], we allocate 1,000 images for validation, 1,000 images for testing, and the rest for training. MSCOCO [13] contains 123,287 images, with each image corresponding to five manually annotated sentences. As for the dataset partitioning, we apply 113,287 images for training, and both the validation set and the test set contain 5,000 instances. The evaluation results of MSCOCO are computed on 5-fold test images.

Evaluation Metrics. To quantify the retrieval performance, we utilize Recall@K ($K = 1, 5, 10$), a commonly employed evaluation metric in information retrieval. It is defined as the proportion of correct matchings among the top-K retrieved results. The higher Recall@K indicates better retrieval performance.

Implementation Details. All experiments are implemented in PyTorch with one NVIDIA Tesla A100 GPU. The Adam [7] optimizer is employed to train the SAHIM network with a mini-batch size of 64, with 40 epochs on Flickr30K and MSCOCO. The initial learning rate is set to 0.0005 with decaying 10% every 28 epochs on Flickr30K and decaying 10% every 15 epochs on MSCOCO. In addition, the dimension of visual features is 2,048 and the number of visual regions is 36. Furthermore, the number of steps Z for multimodal graph inference, the number of layers L for adaptive hybrid modeling, and the weight λ_p of the path regularization loss are set to 1, 3, and 0.5, respectively.

4.2 Experimental Results

We compare the SAHIM model with several state-of-the-art models on two benchmark datasets. Similar to the compared models, we improve the performance by integrating two models with different directions (SAHIM-i2t and SAHIM-t2i). The comparison results are presented in Table 1.

Results on Flickr30K. As shown in Table 1, SAHIM model outperforms the state-of-the-art methods in all evaluation metrics on Flickr30K. Specifically, the SAHIM model achieves remarkable performance of 82.8% and 65.0% on R@1 for sentence retrieval and image retrieval, respectively. Compared with the latest method RAAN, our method attains 5.7% and 9.0% relative R@1 gains in two directions. It is worth noting that our method achieves highly competitive retrieval performance even with a single model, thereby fully demonstrating the superiority of the SAHIM model.

Table 1. Comparison of bidirectional retrieval results (R@K(%)) on Flickr30K and MSCOCO. Where * represents the result of model integration.

Methods	Flickr30K dataset						MSCOCO (1K) dataset					
	Sentence Retrieval			Image Retrieval			Sentence Retrieval			Image Retrieval		
	R@1	R@5	R@10	R@1	R@5	R@10	R@1	R@5	R@10	R@1	R@5	R@10
SCAN [9]	67.4	90.3	95.8	48.6	77.7	85.2	72.7	94.8	98.4	58.8	88.4	94.8
VSRN [10]	71.3	90.6	96.0	54.7	81.8	88.2	76.2	94.8	98.2	62.8	89.7	95.1
CVSE [20]	73.5	92.1	95.8	52.9	80.4	87.8	74.8	95.1	98.3	59.9	89.4	95.2
IMRAM [2]	74.1	93.0	96.6	53.9	79.4	87.2	76.7	95.6	98.5	61.7	89.1	95.0
RRTC [23]	72.7	93.8	96.8	54.2	79.4	86.1	76.2	96.3	**98.9**	61.6	89.3	94.6
WCGL [22]	74.8	93.3	96.8	54.8	80.6	87.5	75.4	95.5	98.6	60.8	89.3	95.3
SHAN [6]	74.6	93.5	96.9	55.3	81.3	88.4	76.8	96.3	98.7	62.6	89.6	95.8
BiKA [28]	75.2	91.6	97.4	54.8	82.5	88.6	77.6	**96.5**	98.6	62.8	90.3	95.8
CGMN [3]	77.9	93.8	96.8	59.9	85.1	90.6	76.8	95.4	98.3	63.8	90.7	95.7
VSRN++ [11]	79.2	94.6	97.5	60.6	85.6	91.4	77.9	96.0	98.5	64.1	91.0	96.1
GLFN [27]	75.1	93.8	97.2	54.5	82.8	89.9	78.4	96.0	98.5	62.6	89.6	95.4
MIMN [25]	75.8	93.9	97.2	57.1	82.3	88.7	**78.7**	95.8	98.8	62.8	90.0	95.2
RAAN [21]	77.1	93.6	97.3	56.0	82.4	89.1	76.8	96.4	98.3	61.8	89.5	95.8
SAHIM-i2t	79.6	94.5	97.7	61.7	86.1	91.1	78.3	96.1	98.5	64.0	91.0	96.2
SAHIM-t2i	78.0	94.4	97.8	59.7	85.3	90.8	77.8	96.3	98.6	62.1	90.4	96.0
SAHIM*	**82.8**	**95.6**	**98.6**	**65.0**	**88.4**	**93.0**	77.8	96.4	98.6	**64.2**	**91.2**	**96.6**

Results on MSCOCO. For the larger dataset MSCOCO, the SAHIM model outperforms the state-of-the-art models in most metrics. Here, some metrics do not achieve the optimum, potentially due to the vast MSCOCO dataset with more than 120,000 images. Abundant data leads to overlaps and confusion among samples, like multiple images in similar scenes. Even for humans, discerning correct matches becomes challenging, resulting in a level of matching randomness. Compared to the classical fine-grained image-text matching method SCAN, the relative Recall@1 in two directions is increased by 5.1% and 5.4% on SAHIM. The improvements verify the effectiveness of our structure-aware adaptive hybrid interaction modeling model for image-text matching.

4.3 Ablation Studies

Impact of Different Modules. To explore the impact of each module on visual semantic matching, we conduct ablation studies on Flickr30K using the SAHIM-i2t model, as shown in Table 2. The results reveal that removing the BCU unit leads to performance degradation, emphasizing the significance of retaining visual and semantic original features. Similarly, performance drops upon the removal of the IIU unit, underscoring the necessity for intricate information interaction within a single modality. Eliminating the CIU unit causes a substantial performance decline, indicating the pivotal role of cross-modal interaction in image-text matching. Removing the GEU unit results in a reduction in accuracy, highlighting the effectiveness of global information in guiding local details. Additionally, the removal of the EAE module leads to a decrease in model performance, emphasizing the importance of filtering out irrelevant local fragments.

Table 2. Ablation studies on Flickr30K of SAHIM-i2t model to investigate the effect of different modules. The best result is shown in bold.

Models	Image-to-Text			Text-to-Image			rsum
	R@1	R@5	R@10	R@1	R@5	R@10	
SAHIM-w/o BCU	77.9	94.4	**97.7**	59.5	84.8	91.0	505.3
SAHIM-w/o IIU	76.8	95.1	97.6	60.3	85.0	90.6	505.4
SAHIM-w/o CIU	69.0	89.3	94.2	53.5	81.1	87.7	474.8
SAHIM-w/o GEU	76.2	**94.8**	97.0	60.2	85.2	90.4	503.8
SAHIM-w/o EAE	76.5	93.8	96.9	60.2	85.7	**91.1**	504.2
SAHIM-Full	**79.6**	94.5	**97.7**	**61.7**	**86.1**	**91.1**	**510.7**

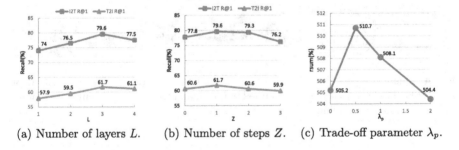

(a) Number of layers L.   (b) Number of steps Z.   (c) Trade-off parameter λ_p.

Fig. 3. Influence of different parameters on Flickr30K of SAHIM-i2t model.

Impact of Different Parameters. In the following, we analyze the important parameters in SAHIM through specific experiments.

The impact of the number of layers L in the AHM module. We set the number of layers from 1 to 4 for experiments, as shown in Fig. 3(a). The Recall@1 value of cross-modal retrieval rises with more layers up to a point but declines after 4 layers. This suggests that increasing layers (within a limit) enhances multimodal interaction in hybrid modeling. Excessive depth, however, harms multimodal learning and degrades image-text matching performance.

The impact of the number of steps Z in the MGI module. In Fig. 3(b), we vary the number of steps from 0 to 3. The best performance is achieved when the number of inference steps is set to 1. This indicates that the multimodal graph inference module captures structural information between global and local nodes. However, excessive propagation steps may cause nodes to converge, leading to a compromise in their distinctiveness.

The impact of different path regularization trade-off values λ_p. Figure 3(c) shows that when λ_p is set to 0 (no path regularization), the model performance suffers a significant decline, which proves the effectiveness of the path regularization. Moreover, the best performance is achieved when λ_p is set to 0.5, indicating excessive path regularization weakens multimodal data characteristics.

4.4   Case Studies

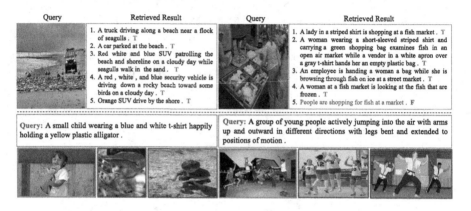

Fig. 4. Qualitative results of image-to-text retrieval and text-to-image retrieval on Flickr30K. The correct results are marked in green, and the incorrect results are marked in red. (Color figure online)

To qualitatively validate the effectiveness of SAHIM, we visualize the image-to-text and text-to-image matching results on Flickr30K, as shown in Fig. 4. For each image query, we display the top-5 retrieved sentences ranked by similarity scores predicted by the SAHIM model. For each sentence query, we show the top-3 retrieved images. The results indicate that regardless of whether the scenes are simple or complex, the SAHIM model can always retrieve matching answers. Note that the only incorrectly matched text in image-to-text retrieval is closely related to the subject of the query image, despite not being the correct answer.

5   Conclusion

In this paper, we propose a novel structure-aware adaptive hybrid interaction modeling network for image-text matching. Specifically, we incorporate four units into the adaptive hybrid modeling module to achieve flexible interaction modeling patterns between image and text. Subsequently, the multimodal graph inference module captures the structural relationship between global and local features, while the entity attention enhancement module filters out irrelevant local segments. Extensive experiments on two benchmark datasets demonstrate that the SAHIM model can always achieve excellent retrieval performance regardless of the complexity of data scenes. In the future, we intend to extend the SAHIM model to encompass additional modalities such as video and audio.

Acknowledgements. This work was supported by the Major Program of the National Natural Science Foundation of China (No. 61991410), the Natural Science Foundation of Shanghai (No. 23ZR1422800), and the Program of the Pujiang National Laboratory (No. P22KN00391).

References

1. Cao, M., Li, S., Li, J., Nie, L., Zhang, M.: Image-text retrieval: a survey on recent research and development. arXiv preprint arXiv:2203.14713 (2022)
2. Chen, H., Ding, G., Liu, X., Lin, Z., Liu, J., Han, J.: IMRAM: iterative matching with recurrent attention memory for cross-modal image-text retrieval. In: Proceedings of the IEEE/CVF Conference on Computer Vision and Pattern Recognition, pp. 12655–12663 (2020)
3. Cheng, Y., Zhu, X., Qian, J., Wen, F., Liu, P.: Cross-modal graph matching network for image-text retrieval. ACM Trans. Multimedia Comput. Commun. Appl. **18**(4), 1–23 (2022)
4. Devlin, J., Chang, M.W., Lee, K., Toutanova, K.: BERT: pre-training of deep bidirectional transformers for language understanding. arXiv preprint arXiv:1810.04805 (2018)
5. Huang, Z., Zeng, Z., Huang, Y., Liu, B., Fu, D., Fu, J.: Seeing out of the box: end-to-end pre-training for vision-language representation learning. In: Proceedings of the IEEE/CVF Conference on Computer Vision and Pattern Recognition, pp. 12976–12985 (2021)
6. Ji, Z., Chen, K., Wang, H.: Step-wise hierarchical alignment network for image-text matching. In: IJCAI International Joint Conference on Artificial Intelligence, pp. 765–771 (2021)
7. Kingma, D.P., Ba, J.: Adam: a method for stochastic optimization. arXiv preprint arXiv:1412.6980 (2014)
8. Krishna, R.: Visual genome: connecting language and vision using crowdsourced dense image annotations. Int. J. Comput. Vision **123**, 32–73 (2017)
9. Lee, K.H., Chen, X., Hua, G., Hu, H., He, X.: Stacked cross attention for image-text matching. In: Proceedings of the European Conference on Computer Vision (ECCV), pp. 201–216 (2018)
10. Li, K., Zhang, Y., Li, K., Li, Y., Fu, Y.: Visual semantic reasoning for image-text matching. In: Proceedings of the IEEE/CVF International Conference on Computer Vision, pp. 4654–4662 (2019)
11. Li, K., Zhang, Y., Li, K., Li, Y., Fu, Y.: Image-text embedding learning via visual and textual semantic reasoning. IEEE Trans. Pattern Anal. Mach. Intell. **45**(1), 641–656 (2022)

12. Li, W.H., Yang, S., Wang, Y., Song, D., Li, X.Y.: Multi-level similarity learning for image-text retrieval. Inf. Process. Manage. **58**(1), 102432 (2021)
13. Lin, T.-Y., et al.: Microsoft COCO: common objects in context. In: Fleet, D., Pajdla, T., Schiele, B., Tuytelaars, T. (eds.) ECCV 2014. LNCS, vol. 8693, pp. 740–755. Springer, Cham (2014). https://doi.org/10.1007/978-3-319-10602-1_48
14. Miyawaki, S., Hasegawa, T., Nishida, K., Kato, T., Suzuki, J.: Scene-text aware image and text retrieval with dual-encoder. In: Proceedings of the 60th Annual Meeting of the Association for Computational Linguistics: Student Research Workshop, pp. 422–433 (2022)
15. Plummer, B.A., Wang, L., Cervantes, C.M., Caicedo, J.C., Hockenmaier, J., Lazebnik, S.: Flickr30k entities: collecting region-to-phrase correspondences for richer image-to-sentence models. In: Proceedings of the IEEE International Conference on Computer Vision, pp. 2641–2649 (2015)
16. Qu, L., Liu, M., Wu, J., Gao, Z., Nie, L.: Dynamic modality interaction modeling for image-text retrieval. In: Proceedings of the 44th International ACM SIGIR Conference on Research and Development in Information Retrieval, pp. 1104–1113 (2021)
17. Ren, S., He, K., Girshick, R., Sun, J.: Faster R-CNN: towards real-time object detection with region proposal networks. In: Advances in Neural Information Processing Systems, vol. 28 (2015)
18. Vaswani, A., et al.: Attention is all you need. In: Advances in Neural Information Processing Systems, vol. 30 (2017)
19. Wang, G., Xu, X., Shen, F., Lu, H., Ji, Y., Shen, H.T.: Cross-modal dynamic networks for video moment retrieval with text query. IEEE Trans. Multimedia **24**, 1221–1232 (2022)
20. Wang, H., Zhang, Y., Ji, Z., Pang, Y., Ma, L.: Consensus-aware visual-semantic embedding for image-text matching. In: Vedaldi, A., Bischof, H., Brox, T., Frahm, J.-M. (eds.) ECCV 2020. LNCS, vol. 12369, pp. 18–34. Springer, Cham (2020). https://doi.org/10.1007/978-3-030-58586-0_2
21. Wang, Y., et al.: Rare-aware attention network for image-text matching. Inf. Process. Manage. **60**(3), 103280 (2023)
22. Wang, Y., et al.: Wasserstein coupled graph learning for cross-modal retrieval. In: 2021 IEEE/CVF International Conference on Computer Vision (ICCV), pp. 1793–1802. IEEE (2021)
23. Wu, J., Wu, C., Lu, J., Wang, L., Cui, X.: Region reinforcement network with topic constraint for image-text matching. IEEE Trans. Circuits Syst. Video Technol. **32**(1), 388–397 (2021)
24. You, S., et al.: What image do you need? A two-stage framework for image selection in e-commerce. In: Companion Proceedings of the ACM Web Conference 2023, pp. 452–456 (2023)
25. Yu, R., Jin, F., Qiao, Z., Yuan, Y., Wang, G.: Multi-scale image-text matching network for scene and spatio-temporal images. Future Gener. Comput. Syst. **142**, 292–300 (2023)
26. Zhang, J., He, X., Qing, L., Liu, L., Luo, X.: Cross-modal multi-relationship aware reasoning for image-text matching. Multimedia Tools Appl. **81**, 12005–12027 (2022)

27. Zhao, G., Zhang, C., Shang, H., Wang, Y., Zhu, L., Qian, X.: Generative label fused network for image-text matching. Knowl.-Based Syst. **263**, 110280 (2023)
28. Zhu, J., Li, Z., Zeng, Y., Wei, J., Ma, H.: Image-text matching with fine-grained relational dependency and bidirectional attention-based generative networks. In: Proceedings of the 30th ACM International Conference on Multimedia, pp. 395–403 (2022)

Using Saliency and Cropping to Improve Video Memorability

Vaibhav Mudgal[1], Qingyang Wang[1], Lorin Sweeney[2],
and Alan F. Smeaton[2]($\boxtimes$) (iD)

[1] School of Computing, Dublin City University, Dublin, Ireland
[2] Insight Centre for Data Analytics, Dublin City University, Dublin, Ireland
alan.smeaton@dcu.ie

Abstract. Video memorability is a measure of how likely a particular video is to be remembered by a viewer when that viewer has no emotional connection with the video content. It is an important characteristic as videos that are more memorable are more likely to be shared, viewed, and discussed. This paper presents results of a series of experiments where we improved the memorability of a video by selectively cropping frames based on image saliency. We present results of a basic fixed cropping as well as the results from dynamic cropping where both the size of the crop and the position of the crop within the frame, move as the video is played and saliency is tracked. Our results indicate that especially for videos of low initial memorability, the memorability score can be improved.

1 Introduction

The saturation of contemporary society with digital content has rendered the ability to capture and sustain human attention increasingly elusive. In this replete landscape, the concept of "video memorability" surfaces as a valuable construct. At its core, video memorability is commonly defined as encapsulating the propensity of a viewer to recognise a video upon subsequent viewing [6,22,24], a phenomenon that transcends the boundaries of emotional biases or personal connections. This assertion may appear counter-intuitive, given the prevailing inclination to associate memory with emotional resonance or subjective biases. However, the cognitive processes underlying memorability reveal it as an emotionally impersonal cognitive metric, intrinsically woven into the fabric of the content itself [5], and hence, impervious to the viewer's unique emotional landscape or individual predilections. This conceptualisation of video memorability as an emotionally impersonal entity underscores the notion that certain videos innately possess characteristics that enhance their likelihood of being remembered, irrespective of the viewer's cognitive milieu. This intriguing facet of human cognition necessitates a more profound investigation, as it not only elucidates the complexities of our cognitive machinery, but also bears significant implications for a multitude of domains, including content creation, digital marketing, and education, among others. An exploration of the literature reveals a

© The Author(s), under exclusive license to Springer Nature Switzerland AG 2024
S. Rudinac et al. (Eds.): MMM 2024, LNCS 14554, pp. 342–355, 2024.
https://doi.org/10.1007/978-3-031-53305-1_26

paucity of research specifically dedicated to video memorability manipulation, despite a sizeable body of work on its corollary, image memorability. This disparity is, in part, attributable to the inherent complexities associated with videos, which, unlike static images, encompass dynamic spatial-temporal information. This additional layer of complexity engenders a host of challenges that have yet to be elegantly surmounted. Ultimately, a deeper appreciation of video memorability holds the promise of not only advancing our understanding of the human mind but also revolutionising the way we create, consume, and interact with digital content in an increasingly digital world.

This paper presents the findings from a series of experiments designed to augment memorability in short, 3-second videos, using a technique of selective frame cropping guided by visual saliency. Defined as the distinctiveness of certain elements or areas within an image or video frame that capture human visual attention, visual saliency [7] serves as a cornerstone for our exploration into diverse cropping strategies. These range from fundamental fixed cropping to the more nuanced dynamic cropping, which not only adjusts the dimensions of the cropping frame but also its position, aligning with the shifts and movements of saliency throughout the video.

2   Background

The characteristics of images that make them more or less memorable than others were first explored from a computational viewpoint more than a decade ago in [11]. That work opened up a new domain for researchers to explore the field of image memorability and why some still images are more memorable than others. The work in [11] posited that the memorability of an image is an intrinsic property and it aimed at finding visual attributes that make an image more, or less, memorable than others. It was found that images with people in them tend to be more memorable than those without and that image memorability further depends upon more precise attributes such as the peoples' ages, hair colour and clothing.

Driven primarily by the MediaEval Media Memorability tasks [20,22], computational memorability has since evolved to encompass more complex visual stimuli, such as videos. In 2018, a video memorability annotation procedure was established, and the first ever large video memorability dataset—10,000 short soundless videos with both long-term and short-term memorability scores—was created [6]. The integration of deep visual features with semantically rich attributes, such as captions, emotions, and actions, has been identified as a particularly efficacious strategy for predicting video memorability [4,17,19,23]. This confluence of modalities not only amplifies prediction precision but also furnishes a comprehensive perspective on the myriad factors that collectively shape video memorability. Furthermore, dimensionality reduction has been shown to enhance prediction outcomes, and certain semantic categories of objects or locales have been found to be intrinsically more memorable than others [6].

Building upon this foundation, recent research has further expanded the multifaceted approach by incorporating auditory features into the prediction model.

A study by [21] illuminated the contextual role of audio in either augmenting or diminishing memorability prediction in a multimodal video context, thereby accentuating the necessity of a holistic approach that integrates diverse modalities, including visual, auditory, and semantic features, for more robust and accurate memorability predictions.

Moreover, the practical applications of video memorability prediction have been explored in real-world scenarios. A study by [3] conducted a comprehensive analysis of the memorability of video clips from the Crime Scene Investigation (CSI) TV series. Utilising a fine-tuned Vision Transformer architecture, the study predicted memorability scores based on multiple video aspects, dissecting the relationship between the characters in the TV series and the memorability of the scenes in which they appear. This analysis also probed the correlations between the significance of a scene and its visual memorability, revealing intriguing insights into the nexus between narrative importance and visual memorability.

Despite these advancements, the manipulation of video memorability remains a relatively uncharted territory. Existing literature predominantly focuses on predicting memorability scores rather than actively modifying them. This gap in the research underscores the need for novel approaches that not only predict but also enhance video memorability. The current study aims to address this gap by exploring the potential of saliency-based cropping as a technique for manipulating video memorability. By selectively cropping frames based on visual saliency, we hypothesise that it is possible to highlight the most memorable regions of a video, thereby enhancing its overall memorability.

2.1 The MediaEval Benchmarking Task on Predicting Video Memorability

Much of the work on computational prediction of the memorability of short form videos has been brought together as a task within the annual MediaEval benchmark. Each year the task organisers share a collection of videos with participants who are asked to compute and submit runs which predict the memorability of each video in the collection. Once runs are submitted, the organisers compare submitted runs against human annotated, ground-truth memorability scores, and announce performance evaluation metrics for each run. The task has run for several years [9,20,22] and has led to significant improvements in the performance of automatic memorability prediction of short form videos.

Computation of memorability scores in this work are from an updated and adapted version of a video memorability prediction model presented in 2021 [2] and compared against manual groundtruth memorability scores which are human evaluations. This uses a Bayesian Ridge Regressor (BRR) to model memorability prediction as regression task. The BRR model was trained on CLIP (Contrastive Language-Image Pre-training) features [18] extracted from the Memento10k training dataset. Given an input image, the model outputs a memorability scores ranging from 0 to 1, where higher scores indicate higher memorability. A single memorability score is generated for a video by averaging the memorability scores for selected video frames.

2.2 Image Saliency

The concept of image saliency pertains to the extent to which an object or a specific region within an image or video frame differentiates itself from the surrounding elements in a visual scene [25]. Essentially, it quantifies the capacity of a particular segment of a visual scene to capture the attention of a typical human observer. This concept is of paramount importance in the domains of computer vision and image processing, as it facilitates the strategic allocation of computational resources to regions deemed visually significant. Models of saliency endeavour to ascertain the regions of an image that are most likely to captivate human attention, and are thus instrumental in an array of applications, ranging from image compression and object recognition to visual search [1].

The first DeepGaze saliency model was introduced in 2015. This work developed a novel approach to improving fixation prediction models by reusing existing neural networks trained on object recognition [1]. DeepGaze II used different feature spaces but the same readout architecture to predict human fixations in an image. This highlighted the importance of understanding the contributions of low and high-level features to saliency prediction [13]. DeepGaze IIE [14], the current version, showed that by combining multiple backbones in a principled manner, a good confidence calibration on objects in unseen image datasets can be achieved, resulting in a significant leap in benchmark performance with a 15% improvement over DeepGaze II reaching an AUC of 88.3% on the MIT/Tuebingen Saliency Benchmark [14].

Several studies have found a small correlations between visual salience and image memorability in specific contexts [10,11,16]. This relationship is strongest when images feature a single or limited number of objects presented in an uncluttered, close-up context. However, the introduction of multiple objects and fixation points considerably diminishes the association between the two [10], thereby signifying the distinctiveness of memorability and salience. Building upon this understanding, we leverage saliency to isolate specific sections of video frames. The underlying premise is that by magnifying the most salient part of a video frame—achieved by cropping the surrounding areas—the resultant cropped video, now more concentrated on the salient elements, could potentially enhance its memorability.

2.3 The Memento10k Dataset and Memorability Scores

Since memorability experiments are usually done on a dataset we use the Memento10k dataset [17], the most substantial video memorability dataset available for public use. The dataset as a whole comprises over 10,000 short form videos and nearly 1 million human annotations, providing a rich information source for analysing video media memorability. The videos cover diverse everyday events captured in a casual, homemade manner, enhancing the real-world

applicability of the findings of those who use it. To ensure a robust evaluation process, we used a subset of 1,500 videos from the dataset for testing and
evaluation, the same 1,500 videos as used in the evaluations in the MediaEval
benchmark.

Each video in Memento10k has a duration of approximately 3 s or 90 frames
per video. We selected every 10th frame of each video for analysis to reduce
computational complexity thus each video has 9 representative frames. We used
the model in [2] to compute memorability scores for each of the 1,500 test videos.
The results of this are shown in Fig. 1 which is a comparison between the ground
truth of manually determined memorability scores provided with the dataset and
memorability scores generated by the model in [2]. The manually determined
memorability scores for the test videos ranges from 0.38 to 0.99, where a lower
score implies lower memorability of a video and vice-versa. Figure 1 shows the
performance of the model for the 1,500 videos to be quite accurate though it
is more conservative than the ground truth. For our work we use the generated
scores as a baseline for improving memorability by cropping.

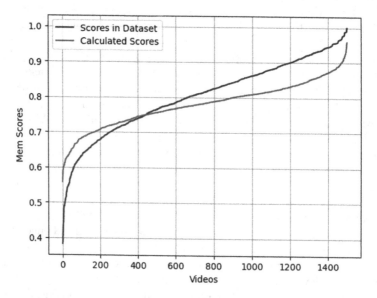

Fig. 1. Memorability scores calculated from [2] vs. manually determined memorability
scores in the Memento10k dataset provided for 1,500 test videos

To illustrate, Fig. 2 shows sample frames from just 3 of the 1,500 videos,
corresponding memorability scores for individual frames, and cumulative memorability scores for each video. We can see in Fig. 2 that the first two videos, which
seem more aesthetically pleasing, surprisingly exhibit lower memorability scores
than the third video in contrast to findings in [10,11,16] about how salience and
memorability have a stronger relationship when images are simpler.

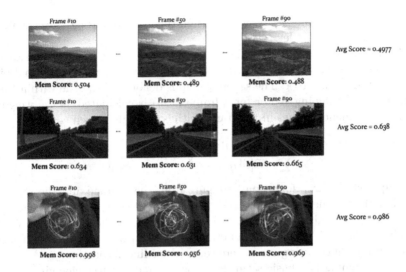

Fig. 2. Sample frames from 3 videos, memorability scores for those frames, and average memorability scores for the videos.

3 Experimental Methodology

3.1 Predicting Video Memorability

We used a vision transformer model fine-tuned on the task of predicting video memorability to predict memorability [3]. A quantitative investigation of the memorability of a popular crime-drama TV series, CSI was made through the use this model. The CLIP [18] pre-trained image encoder was used to train a Bayesian Ridge Regressor (BRR) on the Memento10k dataset. This work found that video shot memorability scores for lead cast members in the TV series heavily correlate with the characters' importance, considering on-screen time as importance.

3.2 Saliency-Based Cropping

When people observe an image without a specified task to complete, we do not focus the same level of intensity on each part of the image. Instead, attentive mechanisms direct our attention to the most important and relevant elements of an image [8]. The importance of different parts of an image can be computed yielding a heatmap of importance, a saliency map. This can assist in understanding the significance of different parts of an image. Cropping a video around the centre of the saliency heatmap of individual frames and moving that crop around the screen from frame to frame, and/or changing the size of the crop as the saliency map changes, may increase the memorability of the overall video. Our rationale for employing frame cropping is eliminating noise from video frames and making the viewer focus on the most salient parts of the frame by removing the remainder.

We used a pre-trained DeepGaze IIE model [14] to compute the saliency map of video frames, the output of which is a saliency map with individual pixel values ranging from 0 to 1. The saliency map is denoted S and the within-frame location of the cropping center for a frame (x, y) is calculated as the weighted center of all saliency values in the frame with coordinates given as:

$$x = \frac{\sum_{i,j} i * S[i][j]}{\sum_{i,j} S[i][j]} \qquad y = \frac{\sum_{i,j} j * S[i][j]}{\sum_{i,j} S[i][j]} \qquad (1)$$

We computed the area for frame cropping from the average and the standard deviation of the saliency map of every frame in a video, denoted as $a_0, a_1, ... a_n$ where n is the number of frames. We then fit frame indices and used $\sqrt{a}$ for smoothing purposes to a linear function using a linear least squares method. Videos with smaller fitting error were selected for linear zooming.

Linear zooming is a basic setup for zooming and involves changing the size of the frame cropping throughout the duration of a video. When we add a zoom to the cropping of a video we do so from the first to the last frame and in a liner fashion. We did not investigate variable crop zooming, or even zooming followed by reverse zoom within a video as this would be too disorientating for viewers because videos in Memento10k have short duration.

We investigated three approaches to combining saliency-based cropping and zooming. In the first, we fixed the crop to the centre of the frame in order to allow us to measure the impact of zooming on memorability. In the second approach we use tracking to follow the movement of the most salient part of frames while keeping the crop size fixed and the final approach allowed the crop size to change as the saliency moved and increased/decreased throughout the video. We now describe each approach.

1. Cropping at the Video Centrepoint: Here we fixed the cropping of each video frame at the centrepoint and analysed the changes in memorability scores resulting from this. We systematically cropped frames at percentage levels ranging from 10% to 90%, along a linear x- or y-axis, i.e. not by area. It is important to note that not all videos had their salient part in the middle of the frame and thus this should be considered a preliminary approach, laying the groundwork for more sophisticated video cropping.

2. Tracking Saliency with a Fixed Crop Size: In the second approach we determined the centrepoints of the saliency from the output of the DeepGaze IIE saliency model [14] for each frame. We then tracked their movement around the frame throughout each video while maintaining a fixed size for the bounding box surrounding this region. The saliency map was generated at different thresholds in order to binarise it as shown below and the results at different saliency thresholds are shown in the bottom-half of Fig. 3. The thresholds used are shown below and an illustration of this cropping is shown in the top sequence of frames in Fig. 4.

$$saliency\_map > saliency\_mean + saliency\_std \qquad (2)$$
$$saliency\_map > saliency\_mean + 2 \times saliency\_std \qquad (3)$$

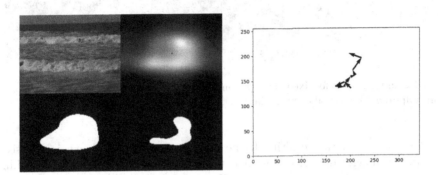

Fig. 3. Video frame (top left) and its generated saliency map with the centre point of the saliency spread marked as a point (top right). The image on the left also shows the saliency map at two different thresholds. The graph on the right shows the movement of the centrepoint of saliency for the duration of the video.

An illustrative example of the resulting frames can be observed in the top row of images corresponding to three frames from a video in Fig. 4. It can be seen in this example that the size of the bounding box shown in red remains constant throughout the video. The main limitation of fixing the bounding box is that it does not consider any changes in the spread or contraction of the most salient part of the frames as the video progresses and might allow either a non-salient part in the cropped frame or might crop out a salient part.

In practice, the thresholding condition forces the viewer to focus on areas of the frame with higher saliency values corresponding to regions of interest compared to the background noise or the less visually significant regions. It is important to note that the spread of the saliency within the frame might vary over time, either increasing and spreading or reducing and contracting, thus introducing dynamic changes in the salient region size. However, the fixed cropping size used for the bounding box in this approach did not account for temporal variations in saliency spread.

3. Tracking Saliency with a Variable Crop Size: In the third approach, tracking involved monitoring the salient part with the crop size dynamically adjusted by linearly increasing, or decreasing, or neither, based on the size of the identified salient region.

Fig. 4. Sample frames for fixed-sized (second approach) and for variable-sized cropping (third approach) with saliency tracking.

An example of the resulting frame for this approach is shown in the second row of images in Fig. 4. Here it can be seen that the size of the bounding box increases as the video progresses. To accommodate changes in the spread or decrease in saliency throughout the video, only videos with a consistent increase, or decrease, in thresholded saliency size were cropped linearly thus taking changes in saliency spread into consideration. Finally, in all cases where we used cropping we incorporated padding around the crop in order to provide some context for the most salient parts of the video frames.

4   Experimental Results

4.1   Cropping at Video Centrepoints

The range plot on the left in Fig. 5 shows the ranges for 90% of the memorability prediction scores from our initial centrepoint cropping across all 1,500 test videos. The image on the right illustrates that cropping. As we reduce the sizes of the frames by cropping from 90% down to 10% of their original sizes, the 90% ranges of memorability scores show a noticeable decrease. This is probably caused by cropping ignoring the within-frame locations of the most salient parts of the frame and most likely cropping through, rather than around, objects of interest. The result of simple cropping just from centrepoints justifies a further exploration into a more sophisticated saliency-informed approach to cropping.

4.2   Cropping with Saliency Tracking

The graphs in Fig. 6 show results when comparing predicted memorability scores after cropping with scores for the original 1,500 videos for both fixed (top graph) and variable (bottom graph) crop sizes. Videos in the graphs are sorted left to

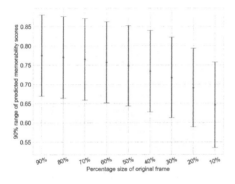

Fig. 5. Changes in predicted video memorability as a result of varying crop sizes where a crop size of 90% means discarding 10% of the frame

right by increasing values of their initial memorability scores. Blue lines show where cropping improved the score, orange lines show where it was reduced.

For each graph it can be seen that for low initial memorability scores blue lines tend to dominate, and this reverses as scores increase with orange lines dominating. For the fixed crop size with tracking, 707 videos have improved memorability prediction scores while 794 have decreased scores. When we introduce variable crop sizes with tracking we find that 718 videos have improved scores with 783 with decreased and this explains why the graphs are so similar, the variable crop size did not have much impact on the results. The distribution of these changes in scores for variable crop sizes with tracking is summarised in Table 1 where we see 83.1% of videos with an initial score above 0.7 have improvements, falling to less than half when the threshold is 0.9.

Table 1. Distribution of videos with improved scores from variable sized cropping by initial memorability scores.

Threshold memorability score	0.70	0.75	0.80	0.85	0.90	0.95
Videos with improved scores	83.1%	73.0%	58.8%	51.7%	45.7%	47.8%

One explanation for this is that for videos which are more memorable, cropping removes memorable information or at best it removes some of the context for memorable information in the video frame thus reducing the video's overall memorability. For videos which already have lower memorability scores, cropping generates improved memorability because it removes visual noise from the frames, allowing the viewer to concentrate on the most salient, and probably more memorable, aspects of the videos.

This interpretation of the results can be shown by the graph in Fig. 7 which is the distribution of the "cumulative mean" for memorability score changes as a result of both fixed and variable cropping of videos. In calculating this, the

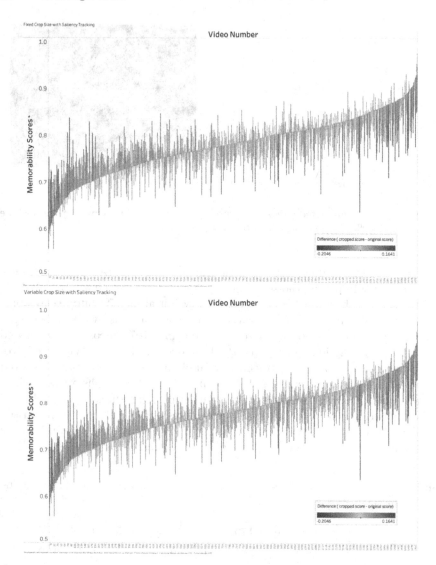

Fig. 6. Results for changes in predicted memorability from fixed crop size (top) and variable crop size (bottom), both with saliency tracking. Blue lines show where scores after cropping are improvements on the original memorability score and orange lines are the opposite. (Color figure online)

cumulative mean at step $i = \frac{\sum_{k=1}^{i} x_k}{i}$ where x_k represents the data point at step k in the sequence and i is the current step for which the cumulative mean is being calculated. We removed outliers from this using the inter-quartile method to make an unbiased plot.

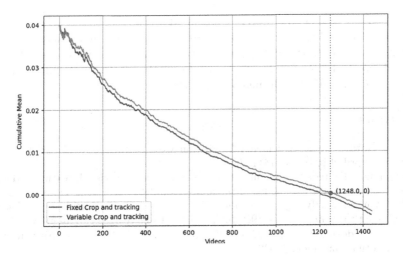

Fig. 7. Line graph showing score comparison between fixed and variable cropping – the x-axis shows 1,500 videos and the y-axis is the cumulative mean of changes in memorability score. (Color figure online)

There are two notable conclusions from Fig. 7 which are that the variable cropping method (orange line) performs slightly better than the fixed cropping method and that the cumulative mean decreases as the original memorability score increases for a video. Hence we can say that cropping improves memorability scores for most videos with an original memorability score lower than a threshold and beyond that, cropping starts to remove salient part of the frames.

5 Conclusions

The investigation outlined in this paper explores the potential of improving video memorability prediction by applying frame cropping based on visual saliency. Using a state-of-the-art saliency model, DeepGaze IIE [14], we detected and tracked saliency through short duration videos, cropped frames around the most salient parts, and re-generated videos. Our results on the Memento10k test set [17] demonstrate a notable improvement in memorability prediction for the majority of videos with initial memorability scores up to a certain threshold.

While this approach is undoubtedly effective for short videos with lower memorability scores—eliminating noise or less salient parts of a video—it has limitations. Short videos already exhibiting high memorability scores are likely to have inherently lower levels of noise and a more prominent narrative. In such cases, exploring more sophisticated visual manipulations, such as temporal segmentation (selectively retaining contextually important segments while removing less crucial/incoherent ones), employing advanced object recognition models to identify and alter key elements within the frame with video in-painting [12], or applying techniques like video super-resolution [15] to improve the quality of the videos, is necessary. Furthermore, employing dynamic scene composition adjustments to better highlight key actions or objects in the video may also prove fruitful. Ultimately, a multifaceted approach that explores various advanced visual manipulation techniques will be necessary to optimise the memorability of videos which already manifest high memorability scores.

References

1. Deep Gaze I: Boosting Saliency Prediction with Feature Maps Trained on ImageNet, April 2015. arXiv:1411.1045
2. Anonymised. Predicting media memorability: comparing visual, textual and auditory features. In: Proceedings of the MediaEval 2021 Workshop, December 2021
3. Anonymised. Analysing the memorability of a procedural crime-drama TV series, CSI. In: Proceedings of the 19th International Conference on Content-Based Multimedia Indexing, CBMI 2022, pp. 174–180, New York, NY, USA. Association for Computing Machinery (2022)
4. Azcona, D., Moreu, E., Hu, F., Ward, T., Smeaton, A.F.: Predicting media memorability using ensemble models. In: Proceedings of MediaEval 2019, Sophia Antipolis, France. CEUR Workshop Proceedings, October 2019
5. Bainbridge, W.A., Dilks, D.D., Oliva, A.: Memorability: a stimulus-driven perceptual neural signature distinctive from memory. NeuroImage **149**, 141–152 (2017)
6. Cohendet, R., Demarty, C., Duong, N.Q.K., Engilberge, M.: VideoMem: constructing, analyzing, predicting short-term and long-term video memorability. In: Proceedings of the IEEE/CVF International Conference on Computer Vision, pp. 2531–2540 (2019)
7. Cong, R., Lei, J., Huazhu, F., Cheng, M.-M., Lin, W., Huang, Q.: Review of visual saliency detection with comprehensive information. IEEE Trans. Circuits Syst. Video Technol. **29**(10), 2941–2959 (2018)
8. Cornia, M., Baraldi, L., Serra, G., Cucchiara, R.: Predicting human eye fixations via an LSTM-based saliency attentive model. IEEE Trans. Image Process. **27**(10), 5142–5154 (2018)
9. Seco de Herrera, A.G., et al.: Overview of MediaEval 2020 predicting media memorability task: what makes a video memorable? In: MediaEval Multimedia Benchmark Workshop Working Notes (2020)
10. Dubey, R., Peterson, J., Khosla, A., Yang, M.-H., Ghanem, B.: What makes an object memorable? In: Proceedings of the IEEE International Conference on Computer Vision, pp. 1089–1097 (2015)
11. Isola, P., Parikh, D., Torralba, A., Oliva, A.: Understanding the intrinsic memorability of images. In: Advances in Neural Information Processing Systems, vol. 24 (2011)
12. Kim, D., Woo, S., Lee, J.-Y., Kweon, I.S.: Deep video inpainting. In: Proceedings of the IEEE/CVF Conference on Computer Vision and Pattern Recognition (CVPR), June 2019
13. Kummerer, M., Wallis, T.S.A., Gatys, L.A., Bethge, M.: Understanding low- and high-level contributions to fixation prediction. In: 2017 IEEE International Conference on Computer Vision (ICCV), pp. 4799–4808, Venice. IEEE, October 2017
14. Linardos, A., Kümmerer, M., Press, O., Bethge, M.: DeepGaze IIE: calibrated prediction in and out-of-domain for state-of-the-art saliency modeling. In: Proceedings of the IEEE/CVF International Conference on Computer Vision, pp. 12919–12928 (2021)

15. Liu, C., Yang, H., Fu, J., Qian, X.: Learning trajectory-aware transformer for video super-resolution. In: Proceedings of the IEEE/CVF Conference on Computer Vision and Pattern Recognition, pp. 5687–5696 (2022)
16. Mancas, M., Meur, O.L.: Memorability of natural scenes: the role of attention. In: 2013 IEEE International Conference on Image Processing, pp. 196–200. IEEE (2013)
17. Newman, A., Fosco, C., Casser, V., Lee, A., McNamara, B., Oliva, A.: Multimodal memorability: modeling effects of semantics and decay on video memorability. In: Vedaldi, A., Bischof, H., Brox, T., Frahm, J.-M. (eds.) ECCV 2020. LNCS, vol. 12361, pp. 223–240. Springer, Cham (2020). https://doi.org/10.1007/978-3-030-58517-4_14
18. Radford, A., et al.: Learning transferable visual models from natural language supervision. In: International Conference on Machine Learning, pp. 8748–8763. PMLR (2021)
19. Reboud, A., Harrando, I., Laaksonen, J., Troncy, R., et al.: Predicting media memorability with audio, video, and text representations. In: Proceedings of the MediaEval 2020 Workshop, December 2020
20. Kiziltepe, R.S., et al.: Overview of the MediaEval 2021 predicting media memorability task. In: CEUR Workshop Proceedings, vol. 3181 (2021)
21. Sweeney, L., Healy, G., Smeaton, A.F.: The influence of audio on video memorability with an audio gestalt regulated video memorability system. In: 2021 International Conference on Content-Based Multimedia Indexing (CBMI), pp. 1–6. IEEE, June 2021
22. Sweeney, L., et al.: Overview of the MediaEval 2022 predicting video memorability task. In: Proceedings of MediaEval 2022 (2022)
23. Sweeney, L., Healy, G., Smeaton, A.F.: Leveraging audio gestalt to predict media memorability. In: MediaEval Multimedia Benchmark Workshop Working Notes, arXiv preprint arXiv:2012.15635 (2020)
24. Sweeney, L., Healy, G., Smeaton, A.F.: Diffusing surrogate dreams of video scenes to predict video memorability. In: Proceedings of the MediaEval 2022 Workshop, December 2022
25. Wang, M., Konrad, J., Ishwar, P., Jing, K., Rowley, H.: Image saliency: from intrinsic to extrinsic context. In: CVPR 2011, pp. 417–424. IEEE (2011)

Contextual Augmentation with Bias Adaptive for Few-Shot Video Object Segmentation

Shuaiwei Wang[1], Zhao Liu[3], Jie Lei[1(✉)], Zunlei Feng[2], Juan Xu[3], Xuan Li[3], and Ronghua Liang[1]

[1] College of Computer Science, Zhejiang University of Technology, Hangzhou, China
{swwang,jasonlei,rhliang}@zjut.edu.cn
[2] College of Computer Science, Zhejiang University, Hangzhou, China
zunleifeng@zju.edu.cn
[3] Ping An Life Insurance of China, Ltd., Shanghai, China
{liuzhao556,xujuan635,lixuan208}@pingan.com.cn

Abstract. Few-shot video object segmentation (FSVOS) is a challenging task that aims to segment new object classes across query videos with limited annotated support images. Typically, meta learner is the main approach to handle few-shot tasks. However, the current meta learner ignores contextual information and lacks the use of temporal information in videos. Moreover, the trained models are biased towards the segmentation of novel classes, favoring the seen class, which hinders the recognition of novel classes. To address these problems, we propose contextual augmentation with bias adaptive for few-shot video object segmentation, consisting of a context augmented learner (CAL) and a bias adaptive learner (BAL). The context augmented learner processes the contextual information in the video and guides the meta learner to obtain rough prediction results. Afterwards, the bias adaptive learner adapts to the bias of the novel classes. The BAL branch utilizes a base class learner to identify the base classes and compute the similarity between the query video and the support set, guiding the adaptive integration of coarse-robust results to generate accurate segmentation. Experiments conducted on the Youtube-VIS dataset demonstrate that our approach achieves state-of-the-art performance.

Keywords: few-shot video object segmentation · contextual augmentation · bias adaptive

1 Introduction

With the popularity of digital devices, video data is growing exponentially and the demand for processing video content is increasing. Video object segmentation (VOS) attracts more and more attention as a key technology for video processing. Most of the current VOS techniques are unsupervised [9,15,24] and

---

S. Wang and Z. Liu—Authors contributed equally to this work.

© The Author(s), under exclusive license to Springer Nature Switzerland AG 2024
S. Rudinac et al. (Eds.): MMM 2024, LNCS 14554, pp. 356–369, 2024.
https://doi.org/10.1007/978-3-031-53305-1_27

semi-supervised [6,10,29]. However, unsupervised VOS has no control over the specific objects to be segmented, while semi-supervised VOS requires pixel-level annotation of the first frame of each video, which limits the scalability to handle large amounts of video. In contrast, few-shot learning (FSL) [4,5,21] allows to identify new objects by a handful of examples. Therefore, we undertake the application of FSL in the field of VOS, termed as few-shot video object segmentation (FSVOS). The goal of FSVOS is to segment new object classes in query videos while using only a small number of annotated support images. The support images can be randomly selected outside the query video.

Currently, most existing few-shot segmentation approaches seek to attain generalization through a Meta-learning framework. However, for video data, most methods focus on features over the single frames, thus overlook the contextual information of the video and fail to learn the complete information of the video. To address this issue, we propose a context augmented learner to enhance contextual information in the video. This module optimizes the segmentation outcome of the meta learner for the present frame by utilizing the contextual information. Furthermore, there are serious shortcomings in sampling a series of learning tasks from the base dataset to mimic few-shot scenarios of novel classes. The network is trained on a large quantity of annotated base class datasets during meta-training, causing it to be biased towards the segmentation of base classes and diminishing its performance when segmenting novel classes [7]. According to [12], promising results have been achieved by utilizing a base learner in few-shot segmentation(FSS). Therefore, to tackle this issue, we propose the identification of confusable regions within query videos using a high-capacity segmentation model trained under the traditional paradigm, thereby improving recognition of novel objects. Specifically, we segment the base class targets in the video using a trained base learner and compute the similarity between the query video and support set. Then, the coarse results of the outputs of these two parts are adaptively integrated to generate precise predictions under the constraint of similarity.

In this paper, we present contextual augmentation with bias adaptive for few-shot video object segmentation, consisting of a context augmented learner (CAL) and a bias adaptive learner (BAL). Experimental results on Youtube-VIS benchmarks indicated that the proposed method outperforms previous state-of-the-art methods. Our methods improves 2.6% in $\mathcal{F}\_Mean$ and 0.6% in $\mathcal{J}\_Mean$. Our primary contributions can be concluded as follows:

(1) We propose a novel and effective scheme to solve Few-shot video object segmentation by contextual augmentation and adaptive bias.
(2) A contextual module CAL is proposed in the meta learner to enhance the contextual information of the video and optimize the segmentation results.
(3) We propose a BAL consisting of a base class learner and the similarity between the query video and the support set to adaptive biases in the meta learner.

2 Related Work

2.1 Video Object Segmentation

Video object Segmentation (VOS) can be divided into two categories: unsupervised video object segmentation and semi-supervised video object segmentation. Unsupervised VOS directly segments the main target objects located in the salient regions of the video without any manual intervention. The main approach is to segment the objects in the video by heuristic rules [18,24] and zero-shot learning [9,15]. Semi-supervised VOS requires providing the true mask of the first frame and propagating the labeled object information to subsequent frames to segment the video's subsequent frames. Semi-supervised video object segmentation methods mainly include propagation-based methods [10,11], detection-based methods [1,25], and use the memory network [6,14,29]. Propagation-based methods use the consistency of the object to learn object mask propagators to work, while detection-based methods find the best correspondence between the first frame and the query frame by using the object appearance on the first frame for propagation. Besides, other methods use the memory work to store the feature information of past frames and segment the current frame.

2.2 Few-Shot Semantic Segmentation

The task of few-shot semantic segmentation involves the learning of segmentation for a new class using only few examples. Shaban et al. [19] have made significant contributions to this field by proposing a two-branch network, referred to as the OSLSM, which comprises a condition branch and a query branch. The condition branch extracts features from support images that guide the query branch for segmentation prediction. Several approaches [12,22] are based on the notion of a prototype derived from metric learning to address this challenge. More recently, some relevant research has focused on building powerful blocks on a fixed network to improve performance. such as ASGNet [13], CANet [26], and PFENet [20].

2.3 Few-Shot Video Object Segmentation

Few-shot video object segmentation (FSVOS) aims to segment objects in query videos by specifying specific classes in a few marker-supported images. Currently, there is relatively little research in this area. This problem was first proposed by Chen [2] in 2021. The authors argue that the key to solve this task is to model the relationship between the query videos and the support images for propagating the object information. They treat this as a many-to-many problem and propose a novel Domain Agent Network (DAN) to break down the full-rank attention, which reduces the space and time complexity of full-rank attention and improves the performance of the network. DAN is poorly considered in the contextual information of the video. Therefore, the study of FSVOS is not enough and more in-depth research is needed.

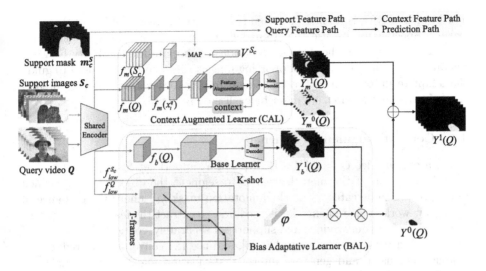

Fig. 1. The overall architecture of our proposed approach. CAL augmented the meta learner with the context module to learn video contextual information. BAL consists of a base learner and a video-image similarity φ. Adaptive integration of the coarse results of the two learners is performed under the guidance of φ. The superscript "0" and "1" represent the background and foreground. Y_m and Y_b denote the predictions of the meta learner and base learner.

3 Problem Definition

The goal of FSVOS is to segment objects with the same class in the query video by a small number of labeled support images. To achieve this, we partition the video dataset into two distinct sets: D_{train} and D_{test}, based on class labels. The D_{train} set is utilized for the training phase, while the D_{test} set is reserved for evaluation. It is worth noting that there is no overlap between the class labels in D_{train} and D_{test} datasets. Both D_{train} and D_{test} sets are further sub-divided into several episodes. Each episode comprises of a support set S_c and a query set Q for a specific class label c. The query set $Q = \{x_t^q | t \in [1, T]\}$ is comprised of T frames,where x_t^q is the RGB frame at time step t. while the support set $S_c = \{x_{c,k}^s, m_{c,k}^s | k \in [1, K]\}$ is made up of K labeled image-mask pairs under class label c. The network is trained to predict the mask $Y_c = \{m_{c,t}^q | t \in [1, T]\}$ for each frame present in the query set.

4 Proposed Method

As shown in the Fig. 1, the overall framework of our network is described, which consists of two parts: **Context Augmented Learner (CAL) and Bias Adaptative Learner (BAL)**. First, we augment the existing meta learner with a contextual module, which allows the meta learner to learn the temporal dimensional information of the video and obtain rough prediction results. Then, A new branch of BAL is further proposed to alleviate the bias problem of the meta learner. Specifically, we use the trained base learner to predict the confusion-prone regions and use the distance function to compute the similarity between the query video and the set of supporting images. The similarity is used to guide the adaptive integration of the two rough results to generate accurate predictions. In the next section, we describe each part in detail.

4.1 Context Augmented Learner

Given a query video $Q = \{x_t^q | t \in [1, T]\}$ and a support set $S_c = \{x_{c,k}^s, m_{c,k}^s | k \in [1, K]\}$, The goal of the meta learner is to segment the objects c in Q, which belong to the same category as the annotation mask m_c^S under the guidance of S_c. In our work, a shared encoder is used first to extract the features of different dimensions of the query video and support set separately, connect the features of different dimensions according to [20, 26], and use 1x1 convolution to reduce the channel dimension and generate intermediate feature maps $f_m(Q = \{x_t^q\}_{t=1}^T)$ and $f_m(S_c = \{x_{c,k}^s\}_{k=1}^K)$. Furthermore, to provide the crucial class-related cues to the query video, we calculate the prototype through the masked average pooling (MAP) [27]. For the K shot task, use the prototype of average support branch extraction to obtain the final support prototype V^{S_c}. Then, V^{S_c} is used to activate the target area in the video frame.

In addition, in order to learn video contextual information for better segmentation, We introduce the context augmentation module to enhance the contextual information of the video $f_{context}^t(Q)$, As shown in the Fig. 1. For a frame t in the video, we stitch together V^{S_c}, $f_m(x_t^q)$, $f_{context}^{t-1}(Q)$ and align for information expansion.

$$f_m'(x_t^q) = F_{augment}\left(f_m(x_t^q) \oplus f_{context}^{t-1}(Q) \oplus V^{S_c}\right), \tag{1}$$

where $\oplus$ denotes the concatenation operation along channel dimension. $F_{augment}$ denotes feature enrichment module that passes annotation information and context information into the query branch to provide specific segmentation cues. When processing the first video frame, we treat $f_m(x_t^q)$ as $f_{context}^{t-1}(Q)$, while $t = 1$. Next, $f_{context}^t(Q)$ is re-updated based on the newly generated $f_m'(x_t^q)$ and the $f_{context}^{t-1}(Q)$ of the context module, $RULE$ is an activation function.

$$f_{context}^t(Q) = RULE\left(F_{1\times1}\left(f_{context}^{t-1}(Q) \oplus f_m'(x_t^q)\right)\right). \tag{2}$$

Finally we use the decoder to generate a rough prediction Y_m of all frames of the query video. We calculate the BCE Loss L_{meta} between Y_m and the ground-truth m^q to update all parameters of the meta learner.

4.2 Bias Adaptive Learner

CAL is able to process video information but the segmented objects obtained are rough. The meta learner still has the problem of bias for new classes and the problem of sensitivity to support sets. Therefore, we introduce the Bias Adaptive Learner (BAL). As shown in the Fig. 1. Firstly, a base learner is introduced to predict the region of the base class in the video, which is then used to tune the meta learner bias through the base learner's base class goals. Secondly, we apply the Dynamic Time Warping (DTW) algorithm [17] to compute the similarity between the query video and the support images set to optimize the coarse predictions provided by the meta learner. The details are as follows.

The base learner is to predict the region of the base classes in the query video. This requires the introduction of a video supervised network. However, it is not practical to add an extra large network to the original meta learner, which would introduce too many parameters and slow down the inference. Therefore, we design a unified framework in which both learners share the same backbone network. And according to [12], a two-stage training strategy is used to train base learner and meta learner separately. Specifically, given a query video $Q = \{x_t^q | t \in [1, T]\}$, we first use the encoder network shared with meta learner and the convolution block to extract the intermediate feature maps $f_b(Q = \{x_t^q\}_{t=1}^T)$. Then, the decoder D_b is used to gradually enlarge the intermediate feature maps $f_b(Q)$ and eventually yield predictions for the base classes.

$$Y_b(Q) = softmax(D_b(f_b(Q))) \in \mathcal{R}^{T \times (1+N_b) \times H \times W}, \tag{3}$$

where N_b denotes the number of base categories. We use a supervised learning paradigm to train base learner, and use cross-entropy loss to evaluate the difference between the prediction $Y_b(Q)$ and the ground-truth m_b^q to obtain the loss of base learner L_{base}.

Merging the two results alone would be a very rough segmentation result, with the boundary parts of the objects not being well distinguished. To better adaptively integrate the coarse results predicted by the two learners, we apply the Dynamic Time Warping (DTW) algorithm to compute the similarity between the query video and the support set to constrain them. Firstly, our approach employs a shared encoder to extract low-level features f_{low}^Q, $f_{low}^{S_c}$ from the query video Q and the support set S_c. Specifically, low-level features are derived from bolck-2 in ResNet-50 [8]. Then calculate the frame-level distance matrix $D \in \mathcal{R}^{T \times K}$ as

$$D(t, k) = 1 - \frac{f_{low}^Q(x_t^q) \cdot f_{low}^{S_c}(x_k^s)}{\| f_{low}^Q(x_t^q) \| \| f_{low}^{S_c}(x_k^s) \|}, \tag{4}$$

where $D(t, k)$ is the frame-level distance value between the t-th frame of video Q and the k-th image in the support set S_c.

We define $\omega \subset \{0, 1\}^{K \times T}$ to be the set of possible binary similarity matrices, where $\forall M \in \omega$, $\omega_{tk} = 1$ denotes the t-th frame of video Q is exactly similar to the k-th image in the support set S_c. Our goal is to find the best similarity matrices $M' \in \omega$.

$$argmin \langle M, D \rangle \rightarrow M', M \in \omega, \tag{5}$$

The ideal similarity matrices M' would minimize the inner product between the similarity matrices M and the frame-level distance matrix D defined in Eq. 4.

The similarity between the query video and support set is thus given by:

$$\varphi(Q, S_c) = \langle M', D \rangle. \tag{6}$$

We propose to use the Dynamic Time Warping (DTW) algorithm to solve Eq. 5. We achieve this by solving a cumulative distance function

$$\gamma(t, k) = D(t, k) + min\{\gamma(t - 1, k - 1), \gamma(t - 1, k), \gamma(t, k - 1)\}, \tag{7}$$

where $M(1, 1) = 1$ and $M(T, T) = 1$.

After that, foreground for base learner and background for meta learner are integrated under the guidance of adjustment factor $\varphi(Q, S_c)$, get a clear background area $Y^0(Q)$:

$$Y^0(Q) = F_{1 \times 1}\left(F_\varphi\left(Y_m^0(Q)\right), Y_b^1(Q)\right), \tag{8}$$

where the superscript "0" and "1" represent the background and foreground. Y_m and Y_b denote the predictions of the meta learner and base learner. F_φ is a convolution operation, and to Optimize the coarse results generated by the meta learner.

4.3   Segmentation Fusion

Finally, the background region obtained $Y^0(Q)$ by BAL is combined with the foreground region obtained by meta learner to obtain the final prediction result $Y^1(Q)$:

$$Y^1(Q) = Y^0(Q) \oplus F_\varphi(Y_m^1(Q)). \tag{9}$$

The overall loss during the meta-training phase can be evaluated by:

$$L = L_{final} + \lambda L_{meta}, \tag{10}$$

where λ is set to 1.0 in all experiments, L_{final} denotes the BCE Loss between $Y^1(Q)$ and the ground-truth m^q and L_{meta} is the loss function of the meta learner.

5   Experiments

5.1   Dataset and Metrics

Given that FSVOS continues to be an under-explored problem with few relevant benchmarks, we chose to assess the performance of our proposed method solely on the Youtube-VIS dataset [23]. This dataset comprises a total of 2238 YouTube videos and 3774 instances, spanning 40 different object categories. To evaluate the proposed method, we partitioned the object categories equally into four folds and conducted cross-validation experiments. The performance of the proposed method was evaluated using the region similarity ($\mathcal{J}$) and contour accuracy ($\mathcal{F}$) metrics. For a fair comparison with other existing methods, we adopted the same experimental setup as in [2]. Specifically, we randomly selected K images from a single category as the support set, and utilized consecutive frames from other videos within the same category as the query set. The default setting K is 5 in our experiments. We run 5 experiments on each fold and report the average performance to ensure the reliability and accuracy of the results.

5.2 Implementation Details

The training process is divided into two phases: pre-training and meta-training. In the pre-training phase, we employ a standard supervised learning approach to train the base learner in BAL. Specifically, we randomly sample 500 frames from each of the 31 categories (including background) extracted from Youtube-VIS videos. The pre-training lasts for 100 epochs using PSPNet [28] as the base learner. We utilize the SGD optimizer with an initial learning rate of 2.5e-3 and a training batch size of 12 to update the parameters. Then, we train both the meta learner and BAL jointly while fixing the parameters of the base learner. Note that we use shared the same encoder to extract the features of the input images for generalization. We train 100 epochs with SGD,the batch size and learning rate set to 8 and 5e-2. We follow the data augmentation techniques in [20] for training. We use a variant of PFENet [20] as our meta learner, and replace FEM with ASPP [3] to reduce the complexity. The proposed model is implemented in PyTorch and runs on NVIDIA RTX 3090 GPUs.

5.3 Comparisons to Existing Methods

As shown in Table 1, which present the $\mathcal{J}$ and $\mathcal{F}$ results of different approaches on Youtube-VIS benchmarks. The top three methods are state-of-the-art image-based methods, which can be adapted to FSVOS task by processing each frame

Table 1. Comparisons to existing applicable methods. Results in bold denote the best performance, while the underlined ones indicate the second best.

	Methods	Fold-1	Fold-2	Fold-3	Fold-4	Mean
$\mathcal{F}$	PMMs [22]	34.2	56.2	49.4	51.6	47.9
	PFENet [20]	33.7	55.9	48.7	48.9	46.8
	PPNet [16]	35.9	50.7	47.2	48.4	45.6
	DAN [2]	<u>42.3</u>	<u>62.6</u>	<u>60.6</u>	<u>60.0</u>	<u>56.3</u>
	Ours	**46.0**	**64.2**	**64.9**	**60.5**	**58.9**
$\mathcal{J}$	PMMs [22]	32.9	61.1	56.8	55.9	51.7
	PFENet [20]	37.8	64.4	56.3	56.4	53.7
	PPNet [16]	**45.5**	63.8	60.4	58.9	57.1
	DAN [2]	43.2	<u>65.0</u>	62.0	**61.8**	<u>58.0</u>
	Ours	<u>43.3</u>	**67.5**	62.2	<u>61.4</u>	**58.6**

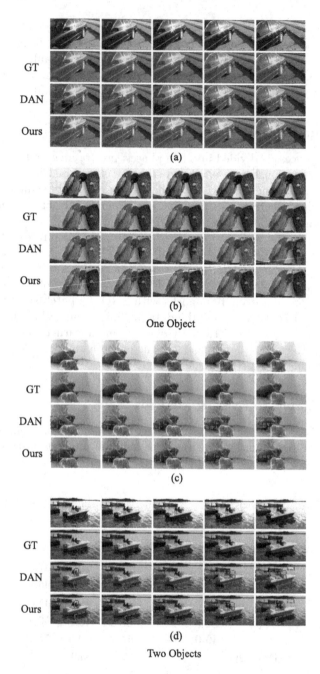

Fig. 2. Qualitative results of our method with baseline and DAN [2]. Video (a) and (b) contain only one object (sedan/parrot), and Video (c) and (d) indicate that two objects are included. (novel class: rabbit/boat; base class: people). The red dashed box indicates the spatial extent of the error in the video. Each row from top to bottom represents query video frames, query video with GT masks, DAN results, and our results, respectively. (Color figure online)

one by one. The method of DAN [2] is basically the same as our task setup. It can be found that our method usually outperforms the above methods, espercially on the metric of $\mathcal{F}$. Compared with the DAN, the proposed method improves 2.6% in $\mathcal{F}\_Mean$ and 0.6% in $\mathcal{J}\_Mean$, which fully demonstrates the effectiveness of our proposed method. Note that in Fold-4, our method is lower in the metric of $\mathcal{J}$ than the DAN, and we argue that the main reason is that the number of videos containing multiple objects in Fold-4 is much less than in other folds, causing the BAL in our method to not work well.

In order to analyze and comprehend our model more effectively, we visualize the segmentation outcomes of both our proposed method and other existing methods, as depicted in Fig. 2. The four presented videos are categorized based on the number of objects contained within: two videos feature a single object ((a) and (b)), while the remaining two videos exhibit two objects ((c) and (d)). Upon observing our results (4th row in each video), it is evident that the segmentation masks generated by our method exhibit greater compactness with respect to the target objects. Our method demonstrates improved robustness compared to the latest techniques, presumably due to its consideration of the video's context. Notably, when the video contains two objects, the false activation targets of the base classes are effectively suppressed, thereby validating the effectiveness of our proposed method.

5.4 Ablation Study

Table 2. Ablation studies of different components. None: no components are used. CAL: only the CAL module is used. BAL: only the BAL module is used.

Methods		Fold-1	Fold-2	Fold-3	Fold-4	Mean
$\mathcal{F}$	None	33.7	55.9	48.7	48.9	46.8
	CAL	44.9	61.4	61.3	59.7	56.8
	BAL	45.8	62.4	62.5	60.0	57.6
	ALL	**46.0**	**64.2**	**64.9**	**60.5**	**58.9**
$\mathcal{J}$	None	37.8	64.4	56.3	56.4	53.7
	CAL	41.2	65.3	60.5	60.7	56.9
	BAL	41.5	64.8	58.2	59.7	56.0
	ALL	**43.3**	**67.5**	**62.2**	**61.4**	**58.6**

We conduct a series of ablation studies to investigate the impact of each component on segmentation performance. As shown in Table 2, $\mathcal{J}$ and $\mathcal{F}$ for each component in each fold are reported. "None" denotes that no components are used. *i.e.*, the original PFENet is used for this task; CAL indicates that only the meta learner with the context module added is used to output the prediction results; BAL indicates that the context module in the meta learner is not used and the output results of the variant meta learner are used to input to BAL for optimization. From the experimental results, it can be seen that CAL and BAL show significant improvements in both the $\mathcal{J}$ and $\mathcal{F}$. In comparison to the original method, our proposed two modules exhibit a noteworthy improvement of 4.9% in $\mathcal{J}\_Mean$ and a significant enhancement of 12.1% in $\mathcal{F}\_Mean$.

Figure 3 depicts the segmentation results obtained by leveraging different components of our proposed method. The input query video comprises two distinct objects, a human (base class) and a dog (novel class). Without utilizing any components, the resulting segmentation erroneously includes a partial region of the person. However, after incorporating both the CAL and BAL modules, the segmentation results are significantly improved. Worth noting is that the BAL module effectively suppresses the segmentation of erroneous regions, especially in video sequences with multiple targets. And CAL module enables more accurate segmentation results. Overall, the proposed two modules BAL and CAL can make the segmentation results more complete and compact.

Fig. 3. Qualitative comparisons of our method using different components.

6   Conclusion

In this paper, we present a novel approach called contextual augmentation with bias adaptive for few-shot video object segmentation. Our method aims to enhance the meta learner by incorporating contextual information from the video. To address the bias problem of the meta learner, we introduce a new branch known as BAL (Bias Adaptive Learner). Extensive experiments conducted on the Youtube-VIS dataset validate the effectiveness of our approach, as it achieves state-of-the-art performance. We hope that our work provides valuable insights and can inspire future research in the field of few-shot video object segmentation.

Acknowledgements. This work was supported in part by Zhejiang Provincial Natural Science Foundation of China (No. LDT23F0202, No. LDT23F02021F02, No. LQ22F020013) and the National Natural Science Foundation of China (No. 62036009, No. 62106226).

References

1. Bao, L., Wu, B., Liu, W.: CNN in MRF: video object segmentation via inference in a CNN-based higher-order spatio-temporal MRF. In: Conference on Computer Vision and Pattern Recognition (CVPR) (2018)
2. Chen, H., Wu, H., Zhao, N., Ren, S., He, S.: Delving deep into many-to-many attention for few-shot video object segmentation. In: Computer Vision and Pattern Recognition (2021)
3. Chen, L.C., Papandreou, G., Kokkinos, I., Murphy, K., Yuille, A.L.: Deeplab: semantic image segmentation with deep convolutional nets, atrous convolution, and fully connected CRFs. IEEE Trans. Pattern Anal. Mach. Intell. **40**(4), 834–848 (2018). https://doi.org/10.1109/TPAMI.2017.2699184
4. Chen, S., Wang, C., Liu, W., Ye, Z., Deng, J.: Pseudo-label diversity exploitation for few-shot object detection. In: Dang-Nguyen, D.T., et al. (eds.) MMM 2023. LNCS, vol. 13834, pp. 289–300. Springer, Cham (2023). https://doi.org/10.1007/978-3-031-27818-1_24
5. Cheng, G., Li, R., Lang, C., Han, J.: Task-wise attention guided part complementary learning for few-shot image classification. SCIENCE CHINA Inf. Sci. **64**(2), 14 (2021)
6. Cheng, H.K., Schwing, A.G.: XMem: long-term video object segmentation with an Atkinson-Shiffrin memory model. In: Avidan, S., Brostow, G., Cissé, M., Farinella, G.M., Hassner, T. (eds.) ECCV 2022. LNCS, vol. 13688, pp. 640–658. Springer, Cham (2022). https://doi.org/10.1007/978-3-031-19815-1_37
7. Fan, Z., Ma, Y., Li, Z., Sun, J.: Generalized few-shot object detection without forgetting. In: 2021 IEEE/CVF Conference on Computer Vision and Pattern Recognition (CVPR), Los Alamitos, CA, USA, pp. 4525–4534. IEEE Computer Society (2021)
8. He, K., Zhang, X., Ren, S., Sun, J.: Deep residual learning for image recognition. In: Conference on Computer Vision and Pattern Recognition (CVPR) (2016)
9. Jain, S.D., Bo, X., Grauman, K.: Fusionseg: learning to combine motion and appearance for fully automatic segmentation of generic objects in videos. In: 2017 IEEE Conference on Computer Vision and Pattern Recognition (CVPR) (2017)
10. Khoreva, A., Benenson, R., Ilg, E., Brox, T., Schiele, B.: Lucid data dreaming for object tracking. Int. J. Comput. Vis. **127** (2017)
11. Khoreva, A., Perazzi, F., Benenson, R., Schiele, B., Sorkine-Hornung, A.: Learning video object segmentation from static images. In: 2017 IEEE Conference on Computer Vision and Pattern Recognition (CVPR) (2017)
12. Lang, C., Cheng, G., Tu, B., Han, J.: Learning what not to segment: a new perspective on few-shot segmentation. In: Conference on Computer Vision and Pattern Recognition (CVPR) (2022)

13. Li, G., Jampani, V., Sevilla-Lara, L., Sun, D., Kim, J., Kim, J.: Adaptive prototype learning and allocation for few-shot segmentation. In: Conference on Computer Vision and Pattern Recognition (CVPR) (2021)
14. Li, G., Gong, S., Zhong, S., Zhou, L.: Spatial and temporal guidance for semi-supervised video object segmentation. In: Tanveer, M., Agarwal, S., Ozawa, S., Ekbal, A., Jatowt, A. (eds.) ICONIP 2022. LNCS, vol. 13625, pp. 97–109. Springer, Heidelberg (2023). https://doi.org/10.1007/978-3-031-30111-7_9
15. Li, S., Seybold, B., Vorobyov, A., Fathi, A., Kuo, C.: Instance embedding transfer to unsupervised video object segmentation. In: Conference on Computer Vision and Pattern Recognition (CVPR) (2018)
16. Liu, Y., Zhang, X., Zhang, S., He, X.: Part-aware prototype network for few-shot semantic segmentation. In: Vedaldi, A., Bischof, H., Brox, T., Frahm, J.-M. (eds.) ECCV 2020. LNCS, vol. 12354, pp. 142–158. Springer, Cham (2020). https://doi.org/10.1007/978-3-030-58545-7_9
17. Müller, M.: Dynamic time warping. In: Müller, M. (ed.) Information Retrieval for Music and Motion, pp. 69–84. Springer, Heidelberg (2007). https://doi.org/10.1007/978-3-540-74048-3_4
18. Ochs, P., Malik, J., Brox, T.: Segmentation of moving objects by long term video analysis. IEEE Trans. Pattern Anal. Mach. Intell. **36**(6), 1187–1200 (2014). https://doi.org/10.1109/TPAMI.2013.242
19. Shaban, A., Bansal, S., Liu, Z., Essa, I., Boots, B.: One-shot learning for semantic segmentation. In: British Machine Vision Conference 2017 (2017)
20. Tian, Z., Zhao, H., Shu, M., Yang, Z., Jia, J.: Prior guided feature enrichment network for few-shot segmentation. IEEE Trans. Pattern Anal. Mach. Intell. **44**(2), 1050–1065 (2020)
21. Wang, H., Lian, J., Xiong, S.: Few-shot classification with transductive data clustering transformation. In: Yang, H., Pasupa, K., Leung, A.C.-S., Kwok, J.T., Chan, J.H., King, I. (eds.) ICONIP 2020. LNCS, vol. 12533, pp. 370–380. Springer, Cham (2020). https://doi.org/10.1007/978-3-030-63833-7_31
22. Yang, B., Liu, C., Li, B., Jiao, J., Ye, Q.: Prototype mixture models for few-shot semantic segmentation. In: Vedaldi, A., Bischof, H., Brox, T., Frahm, J.-M. (eds.) ECCV 2020. LNCS, vol. 12353, pp. 763–778. Springer, Cham (2020). https://doi.org/10.1007/978-3-030-58598-3_45
23. Yang, L., Fan, Y., Xu, N.: Video instance segmentation. In: International Conference on Computer Vision (2019)
24. Yong, J.L., Kim, J., Grauman, K.: Key-segments for video object segmentation. In: IEEE International Conference on Computer Vision, ICCV 2011, Barcelona, Spain, 6–13 November 2011 (2011)
25. Zeng, X., Liao, R., Gu, L., Xiong, Y., Fidler, S., Urtasun, R.: DMM-Net: differentiable mask-matching network for video object segmentation. In: 2019 IEEE/CVF International Conference on Computer Vision (ICCV) (2019)
26. Zhang, C., Lin, G., Liu, F., Yao, R., Shen, C.: CANet: class-agnostic segmentation networks with iterative refinement and attentive few-shot learning. In: Proceedings of the IEEE/CVF Conference on Computer Vision and Pattern Recognition, pp. 5217–5226 (2019)
27. Zhang, X., Wei, Y., Yang, Y., Huang, T.S.: SG-one: similarity guidance network for one-shot semantic segmentation. IEEE Trans. Cybern. **50**(9), 3855–3865 (2020)

28. Zhao, H., Shi, J., Qi, X., Wang, X., Jia, J.: Pyramid scene parsing network. IEEE Computer Society (2016)
29. Zhao, R., Zhu, K., Cao, Y., Zha, Z.-J.: AS-Net: class-aware assistance and suppression network for few-shot learning. In: Þór Jónsson, B., et al. (eds.) MMM 2022. LNCS, vol. 13142, pp. 27–39. Springer, Cham (2022). https://doi.org/10.1007/978-3-030-98355-0_3

A Lightweight Local Attention Network for Image Super-Resolution

Feng Chen, Xin Song, and Liang Zhu$^{(\boxtimes)}$

Hebei University, Baoding 071002, Hebei, China
zhu@hbu.edu.cn

Abstract. For many years, deep neural networks have been used for Single Image Super-resolution (SISR) tasks. However, more extensive networks require higher computing and storage costs, obstructing their deployment on resource-constrained devices. How to ensure the model lightweight while improving its performance is an important direction of SISR research at present. This paper proposed a lightweight local attention network (LLAN) mainly composed of lightweight residual attention groups (LRAGs). LRAG contains lightweight self-calibrated residual blocks with pixel attention (LSC-PAs) and average local attention blocks (ALABs); it utilizes the advantages of the residual connection and attention mechanism. The LSC-PA with strong expression ability can propagate and fuse features better. The ALAB can combine global features and accelerate the network. Furthermore, we discussed three parts of a general SISR network: feature extraction, feature fusion, and reconstruction. Besides, we carried out extensive experiments using five benchmark datasets. The experimental results demonstrated that our method outperforms other compared state-of-the-art techniques, and a better balance between the complexity and performance of the SISR algorithms is achieved.

Keywords: Lightweight · Single Image Super-resolution · Self-calibrated convolution · Pixel attention · Global average pooling

1 Introduction

Single Image Super-resolution (SISR) is a task that has the most straightforward input-output relationship, and it aims to produce a High-resolution (HR) image from one Low-resolution (LR) image [12]. As a typical inverse problem, SISR can be modeled as LR = DS × BL × HR + NI, where LR is the observed data, DS is the down-sampling operator, BL represents the blurring operator, HR is the unknown original image, and NI means the image noise. Super-resolution has many significant applications, such as low-level image processing, medical imaging, and remote sensing [4, 24].

In recent years, many deep learning-based methods have achieved excellent results. Since super-resolution convolutional neural network (SRCNN) was put forward [9], many CNN-based methods have been born. Subsequently, a series of investigations focus on increasing the depth of the networks, and the imaging effect increases with

© The Author(s), under exclusive license to Springer Nature Switzerland AG 2024
S. Rudinac et al. (Eds.): MMM 2024, LNCS 14554, pp. 370–384, 2024.
https://doi.org/10.1007/978-3-031-53305-1_28

network complexity due to the proposed residual network. Then, the generative adversarial network (GAN) is applied to the field of super-resolution reconstruction, and the obtained images have more detailed textures.

Nowadays, the research on SR mainly focuses on self-supervised, multi-scale, real-scene, and lightweight [6, 7]. The lightweight research of SISR is related to the deployment of SR technology, which is the leading research content of this article.

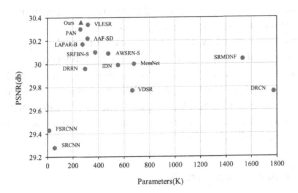

Fig. 1. The results of × 3 SR for various state-of-the-art SISR methods on the Set14 test dataset

This paper proposes a lightweight local attention network (LLAN). We propose a lighter multi-way structured convolutional block to make our model lightweight and effective enough. In addition, we design a new attention scheme, average local attention (ALA), which contains very few parameters but helps generate better reconstruction results. The contributions of our work can be summarized as follows: (1) Propose a lightweight local attention network (LLAN), which has a better balance of performance and complexity (as shown in Fig. 1, "Δ" is our method). (2) Propose a lightweight self-calibrated convolution with pixel attention (LSC-PA), which is effective in lightweight SR networks. (3) Propose an average local attention block (ALAB), which can exploit the clues of the feature information to combine features better.

The rest of the paper is organized as follows. In Sect. 2, we review related works. In Sect. 3, we describe our method. Section 4 illustrates detailed experimental results and comparisons against other state-of-the-art methods. Section 5 concludes the paper.

2 Related Works

In this section, we have reviewed some classic works, mainly focusing on lightweight networks. Some techniques for feature fusion and reconstruction were also explored.

2.1 Lightweight SISR Methods Based on CNNs

After the emergence of the residual idea, many deep convolutional neural networks have achieved good results [18]. However, a large number of convolutional layers will undoubtedly lead to an excessive number of parameters.

Kim et al. proposed a Deeply Recursive Convolutional Network [16], and the form of recursive network sharing training parameters helps reduce the amounts of parameters. Besides, some recursive networks combined with feedback connections or transformers were also proposed [11]. Knowledge distillation learns the loss of the teacher network to improve reconstruction performance [13], and neural architecture search selects a balanced solution among many component combinations [8], both of which are also conducive to the realization of lightweight super-resolution methods.

Channel splitting and efficient module design are more commonly used than the above methods. Ahn et al. used group convolution to reduce the amounts of parameters and operation time [1]. Inspired by lattice filter banks, Luo et al. proposed a lattice block conducive to lightweight model design [21]. Li et al. proposed a multi-scale feature fusion block, which uses convolution kernels of different sizes to collect multi-scale features [18]. Zhang et al. and Song et al. used channel attention [32] and local-global attention [25], respectively, no longer treating each channel feature equally. In this paper, we used a multi-way architecture, local attention, and the removal of some redundant designs to make the model lightweight.

2.2 Feature Fusion and Reconstruction

Feature fusion is considered essential since features may be redundant or lost in deep networks. To fully utilize feature information, some methods were proposed, such as residual connection, dense connection, and channel concatenation. Furthermore, the fusion operation often follows a point-wise convolution [2, 23]. In 2020, Luo et al. proposed a backward fusion module where the high-frequency features are backward sequentially fused with the low-frequency features [21]. In 2023, Gao et al. also grouped features and adopted feature secondary fusion [14].

Except for some networks that directly use the interpolation method at the beginning [26], the general network uses techniques such as deconvolution and pixel-shuffle at the final stage [22, 35]. Moreover, the network that employed interpolation methods at the end of the network can also obtain good performance. In recent works, the Pixel attention network (PAN) adopted nearest neighbor up-sampling and attention-based convolution layers [34]. Inspired by PAN, we also used nearest neighbor up-sampling in the reconstruction stage for better reconstruction.

3 Proposed Method

In this section, we describe the architecture of our proposed model lightweight local attention network (LLAN) and introduce the composition of each part in detail.

3.1 Network Architecture

As shown in Fig. 2, the network architecture of our LLAN mainly consists of two operations: feature extraction and reconstruction. The feature extraction module contains M lightweight residual attention groups ($LRAG_1, LRAG_2, ..., LRAG_M$).

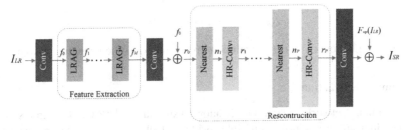

Fig. 2. The architecture of the proposed lightweight local attention network (LLAN)

In every LRAG, there are N lightweight self-calibrated residual blocks with pixel attention (LSC-PA$_1$, LSC-PA$_2$, ..., LSC-PA$_N$) and an average local attention block (ALAB), as shown in Fig. 3. At last, the reconstruction module contains P high-resolution convolutional blocks (HR-Conv$_1$, HR-Conv$_2$, ..., HR-Conv$_P$). Note that the symbol "$\oplus$" denotes a counterpoint addition operation.

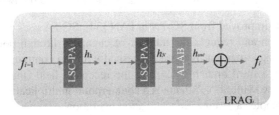

Fig. 3. The architecture of the lightweight residual attention group (LRAG)

Assuming the input LR image is denoted as I_{LR}. The I_{LR} first passes through a convolutional layer which has been used in previous research to extract shallow features. The output f_0 of a convolution layer with a 3×3 kernel is fed into LRAGs for further deep feature extraction. For each $1 \leq i \leq M$, $F_{LRAG}^i(\cdot)$ denotes the i-th LRAG function, f_i is the output of the i-th LRAG, and it can be described as Eq. (1).

$$f_i = F_{LRAG}^i(f_{i-1}) \tag{1}$$

The final fusion feature f_M is input to the up-sampler to reconstruct the target image after passing through a backbone convolutional layer with a skip connection:

$$n_1 = F_{Nearest}(F_{3 \times 3}(f_M) + f_0) \tag{2}$$

where $F_{Nearest}(\cdot)$ denotes the nearest neighbor interpolation method, and n_1 is the reconstructed information.

We utilize the HR-Conv to process the simple reconstructed features as Eq. (3). For each $1 \leq i \leq P$, $F_{HR-Conv}^i(\cdot)$ denotes the i-th HR-Conv block. Finally, the reconstructed information r_P is fed into a convolutional layer, which has also been used in previous research and can be expressed as Eq. (4). In addition, we add a global connection $F_{UP}(\cdot)$, in which a bilinear interpolation is performed to accelerate the convergence speed.

$$r_i = F_{HR-Conv}^i(n_i) \tag{3}$$

$$I_{SR} = F_{3\times3}(r_P) + F_{UP}(I_{LR}) \qquad (4)$$

where $F_{3\times3}(\cdot)$ denotes a 3×3 convolution layer, and I_{SR} is the generated image.

3.2 Feature Extraction

In the backbone network, the commonly adopted structure is RIR (residual in residual). Each residual group comprises multiple residual blocks with short skip connections. This paper utilizes the RIR structure for the feature extraction module.

The feature extraction module consists of M lightweight residual attention groups (LRAGs). Each LRAG contains N lightweight self-calibrated residual blocks with pixel attention (LSC-PAs) and an average local attention block (ALAB), as shown in Fig. 3. Let $F_{ALAB}(\cdot)$ is the average local attention block function, $F_{LSC-PA}^{N}(\cdot)$ denotes the N-th LSC-PA function. The output of the i-th LRAG can be expressed as:

$$f_i = F_{LRAG}^{i}(f_{i-1}) = f_{i-1} + F_{ALAB}\left(F_{LSC-PA}^{N}\left(F_{LSC-PA}^{N-1}\left(\cdots F_{LSC-PA}^{1}(f_{i-1})\right)\right)\right) \qquad (5)$$

The LSC-PA is improved from SC-PA in PAN [34]. SC-PA uses pixel attention for calibration operation and reduces the number of channels through point-wise convolution. However, pixel attention is more concerned with the combination of channels, so this paper adds a third branch to expand the receptive field (the lower branch in Fig. 4). Note that the symbol "⊗" denotes a counterpoint multiplication operation, and the symbol "©" denotes a channel concatenation operation.

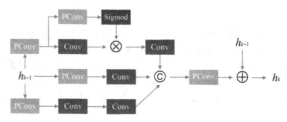

Fig. 4. LSC-PA in LRAG

The LSC-PA contains three branches, each with a 1×1 convolution layer at the beginning, as shown in Fig. 4. The upper branch contains two 3×3 convolution layers, where the first one is equipped with a pixel attention module. The middle branch has only one 3×3 convolution layer. Moreover, the lower branch contains two 3×3 convolution layers. Next, a 1×1 point-wise convolution layer integrates the concatenated features. At last, a skip connection is adopted to learn the local residual information. The output of the upper branch can be expressed as:

$$F_{upper}(h_{k-1}) = F_{3\times3}\left(F_{3\times3}(F_{1\times1}(h_{k-1})) \otimes F_{sig}(F_{1\times1}(F_{1\times1}(h_{k-1})))\right) \qquad (6)$$

where $F_{1\times1}(\cdot)$ is the point-wise convolutional function, $F_{sig}(\cdot)$ is the sigmoid function to obtain the attention maps, $F_{3\times3}(\cdot)$ denotes a 3×3 convolution layer.

The number of the feature channels fed to LSC-PA was E. Then, the number of channels on the lower branch was $E/2$, and the number on both the upper and the middle branches was $E/4$. Compared with SC-PA, the lower branch expands the receptive field, and the shunting operation reduces computational complexity and parameters. For each $1 \leq k \leq N$, h_{k-1}^{con} is the concatenated feature in the k-th LSC-PA function, and the output h_k of the k-th LSC-PA function can be expressed as:

$$h_k = F_{LSC-PA}^k(h_{k-1}) = h_{k-1} + F_{1\times1}\left(h_{k-1}^{con}\right) \tag{7}$$

$$h_{k-1}^{con} = \psi\left(F_{upper}(h_{k-1}), F_{3\times3}(F_{1\times1}(h_{k-1})), F_{3\times3}(F_{3\times3}(F_{1\times1}(h_{k-1})))\right) \tag{8}$$

where $F_{1\times1}(\cdot)$ is the point-wise convolutional function, $F_{3\times3}(\cdot)$ denotes a convolution layer with a 3×3 kernel, and $\psi(\cdot, \cdot, \cdot)$ denotes the channel concatenation operation.

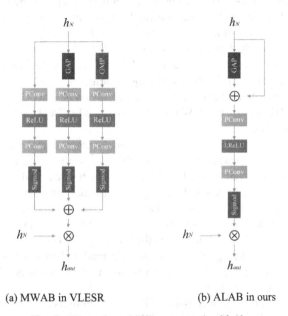

(a) MWAB in VLESR   (b) ALAB in ours

Fig. 5. Illustration of different attention blocks

The LSC-PAs are followed by an average local attention block (ALAB). The structure of ALAB is shown in Fig. 5 (b). The ALAB is learned from the multi-way attention block (MWAB) in VLESR [14]. VLESR is inspired by the convolutional attention block, which computes the channel attention map using both global average pooling and global max pooling. Channel attention can be traced back to squeeze and excitation inception, and the authors only used global average pooling. We combined local information with global average pooling to create a simple but effective structure.

Let $F_{GAP}(\cdot)$ denotes a global average pooling operation. Assume that the input to the ALAB is $h_N \in \mathfrak{R}^{H\times W\times C}$ (H, W, and C are the height, width, and number of channels, respectively), and the output of the ALAB is:

$$h_{out} = F_{ALAB}(h_N) = h_N \otimes F_{sig}(F_{1\times1}(F_{1\times1}(h_N + F_{GAP}(h_N)))) \tag{9}$$

where $F_{1\times 1}(\cdot)$ is the point-wise convolution layer, $F_{sig}(\cdot)$ is the sigmoid function.

3.3 Feature Fusion and Reconstruction

Relatively new feature fusion schemes for SR are generally based on a path aggregation network, which integrates multi-scale features and achieves good results in instance segmentation [20]. However, Chan et al. found that feature maps with a resolution smaller than that of the input image play an unessential role in the image restoration task [5], which means that the encoder structure is unsuitable for SR tasks. Existing papers often extract the outputs of various residual groups as multi-scale features.

We tried some methods in a lightweight structure with fewer channels, such as equal proportional concatenation (EPC), unequal proportional concatenation (UPC), and paired fusion concatenation (PFC), but the effect decreased. For the EPC, we directly concatenate all features and extract the final fused feature L_{fusE} using a point-wise convolution. The features of being concatenated ($L_1, L_2, ..., L_M$) refer to the outputs of the M lightweight residual attention groups (LRAGs). We believe deeper features are more critical for UPC, so we use a ratio of ($1/M$: $2/M$: ...: 1) to select features. Then, the final fused feature L_{fusU} is obtained using a point-wise convolution after the select features are concatenated. The frequency grouping fusion block (FGFB) in VLESR [14] adopts the PFC. We remove the MWAB in FGFB as the PFC and employ it to obtain the final fused feature L_{fusP}.

The training stage often manifests as fast convergence in the early stage and slow convergence in the later stage. We speculate that because the concatenation operation requires practical information, the backbone network has reserved some performance to output complete information that subsequent networks cannot use. Thus, we further explored the relationship between the number of channels and feature fusion in the experiment. The concatenation operation usually needs to extract the global features from all the blocks [33], which increases the computational complexity. In addition, it prevents memory from being released, increasing read and write consumption. At the same time, we use a small number of channels, so we choose to output the final features directly through the backbone network.

The reconstruction module consists of P high-resolution convolutional blocks (HR-Convs), as shown in Fig. 1. The nearest neighbor interpolation method is used to enlarge the resolution of the picture, and n_i is the output of the i-th nearest neighbor interpolation. We referred to the general design of PAN [34] and used convolutional layers to shrink the channel after up-sampling. However, we removed pixel attention, saving computation and improving performance. The HR-Conv contains two 3×3 convolution layers. For each $1 \leq i \leq P$, the output of the i-th HR-Conv function can be expressed as:

$$r_i = F_{HR-Conv}^i(n_i) = F_{3\times 3}(F_{3\times 3}(n_i)) \tag{10}$$

where $F_{3\times 3}(\cdot)$ denotes a convolution layer with a 3×3 kernel.

4   Experiments

4.1   Experimental Settings

The DIV2K dataset [28] was used for training and validation for a fair comparison with the state-of-the-art methods. We selected the top ten images in the validation dataset for validation and marked them as DIV2K-10. Similar to other researches, five standard benchmark datasets, Set5, Set14, B100, Urban100, and Manga109, are used for testing. The widely used Peak signal to noise ratio (PSNR) and structural similarity index (SSIM) on the luminance (Y) channel [32] are used as the evaluation metrics.

The L1 loss function and the ADAM optimizer ($\beta 1 = 0.9$, $\beta 2 = 0.999$) were used. The training mini-batch size was set to 32, and the size of the output HR image patch was set to 192×192. The cosine annealing learning scheme is adopted, and the cosine period is 250k iterations. The initial learning rate was set to 1×10^{-3}. The number of the feature channels was 40. The number of LRAGs was 4. Furthermore, the number of the LSC-PAs was 5. For $\times 2$ and $\times 3$ SR, the number of HR-Conv is 1. For $\times 4$ SR, the number of HR-Conv is 2. The model was implemented using an NVIDIA 3080 GPU.

4.2   Ablation Analysis

We designed three comparative experiments to evaluate the effects of LLAN. We used LLAN as the primary network and explored its performance by changing components. We compared the performance in PSNR, the number of parameters, and GFLOPs for all methods. GFLOPs refer to the number of multiplication and addition operations, which is computed by assuming that the resolution of the HR image is 720p. The results are the mean values of PSNR calculated by 328 images on five benchmark datasets.

Experiments on LSC-PA. We conducted three comparative experiments. For the first experiment, we used SC-PAs as residual blocks [34] instead of LSC-PAs in the feature extraction section, and this model that has 16 SC-PA was called SC-PA16; In the second experiment, we reduced the number of LSC-PA to 16, and the model was called LSC-PA16; The third experiment was the LLAN model contained 20 LSC-PA, which was called LSC-PA20 (i.e., LLAN). The PSNR results, the number of parameters, and GFLOPs of $\times 4$ SR for different models on the five benchmark datasets are shown in Table 1. It could be seen that our LSC-PA outperformed the SC-PA in the case of a slight increase in the number of parameters and GFLOPs. Under the condition of significantly reducing the number of parameters and GFLOPs, the effect of our LSC-PA decreased slightly.

Experiments on ALAB. We conducted four comparative experiments. In the first experiment, the LLAN model did not contain the ALAB, and this model was called ALAB_0. For the second experiment, we used MWAB [14] instead of ALAB at the tail of the residual group in the feature extraction section, and this model was called MWAB; In the third experiment, we employed the lower branch of MWAB which adopts the idea of local attention, and the model was called LOCB; The fourth experiment was the LLAN model contained 4 ALAB, which was called ALAB (i.e., LLAN). The PSNR results, the number of parameters, and GFLOPs of $\times 4$ SR for different models on the five

Table 1. The PSNRs, the number of parameters, and GFLOPs of × 4 SR for three models

	PSNR	Parameters (K)	FLOPs (G)
SC-PA16	28.9291	272.37	27.6
LSC-PA16	28.9200	228.53	24.9
LSC-PA20	28.9360	275.25	27.6

benchmark datasets are shown in Table 2. It could be seen that the attention mechanism is effective, and our ALAB outperformed the MWAB.

Table 2. The PSNRs, the number of parameters, and GFLOPs of × 4 SR for four models

	PSNR	Parameters (K)	FLOPs (G)
ALAB_0	28.9269	274.28	27.5
MWAB	28.9233	277.21	27.6
LOCB	28.9307	275.09	27.6
ALAB	28.9360	275.25	27.6

Experiments on Reconstruction. We conducted two comparative experiments. For the first experiment, the reconstruction part of the LLAN contained the pixel attention [34], and this model was called HR-PixA; In the second experiment, the LLAN model did not include the pixel attention, which was called HR-Conv (i.e., LLAN). The PSNR results, the number of parameters, and GFLOPs of × 4 SR for different models on the five benchmark datasets are shown in Table 3. It could be seen that removing PixA not only reduces the number of parameters and GFLOPs but also improves the SR effect.

Table 3. The PSNRs, the number of parameters, and GFLOPs of × 4 SR for two models

	PSNR	Parameters (K)	FLOPs (G)
HR-PixA	28.9279	276.50	28.3
HR-Conv	28.9360	275.25	27.6

4.3 Experiments on Feature Fusion

Due to the numerous experiments in this section, we adopted a faster verification scheme. The training mini-batch size was set to 16. The initial learning rate was set to 3×10^{-3} and halved at every 200 epochs. We record the results in 5×10^5 iterations. Firstly, we designed three models for LLAN that include equal proportion concatenation, unequal proportion concatenation, and paired fusion concatenation, called EPC, UPC, and PFC, respectively. The last experiment was the LLAN model, which did not contain feature fusion, called FF_0 (i.e., LLAN). The PSNRs of × 4 SR for different models on the DIV2K-10 dataset are shown in Fig. 6. It could be seen that the feature fusion mechanism did not have a positive effect when the number of channels is 40. Besides, compared to EPC and PFC, UPC has better performance.

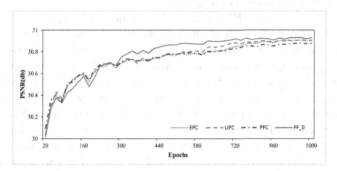

Fig. 6. The PSNRs of the × 4 SR for different models on DIV2K-10 validation dataset

As mentioned earlier, we explored the relationship between feature fusion and the number of channels. We conducted sixteen comparative experiments and increased the number of channels to 48, 56, and 64, respectively. The PSNR results of × 4 SR for different models on the DIV2K-10 dataset are shown in Table 4. It could be seen that when the number of channels is 48, the effect of LLAN increases significantly. Besides, the effect of UPC surpasses that of LLAN when the number of channels is 64. Moreover, both EPC and PFC perform poorly when applied to LLAN. When the number of channels is 48, the PFC deteriorates significantly. When the number of channels is 64, the EPC deteriorates significantly.

Table 4. The PSNR of × 4 SR for different models on the DIV2K-10 validation dataset.

Channel	EPC	UPC	PFC	FF_0
	PSNR	PSNR	PSNR	PSNR
40	30.90644	30.90981	30.87978	30.93195
48	30.96589	30.95166	30.19415	30.98349
56	30.74455	30.99519	30.96557	30.99761
64	29.98725	31.00973	30.98960	31.00615

4.4   Comparison with State-of-the-Art Methods

We conducted a comparison with other 14 state-of-the-art methods, which included SRCNN [9], FSRCNN [10], DRCN [16], VDSR [17], DRRN [26], MemNet [27], SRMDNF [31], IDN [15], AWSRN-S [29], SRFBN-S [18], LAPAR-B [19], A2F-SD [30], PAN [34] and VLESR [14]. The PSNR/SSIM results for $\times$ 2, $\times$ 3, and $\times$ 4 SR on the three standard test datasets are shown in Table 5, and the number of parameters and GFLOPs for each state-of-the-art method are also included. The PSNR/SSIM results for all methods are obtained from Gao et al. [14].

As seen in Table 5, our PSNR/SSIM results significantly outperformed the state-of-the-art methods. The computational effort of our model is smaller than PAN, and the computational effort of our model is not significantly different from VLESR. Moreover, the number of parameters of our model is considerably smaller than that of VLESR. The computational effort and the number of parameters in IDN are close to two times that of our model. For $\times$ 2 SR, the maximum SSIM difference reached 0.0035. For $\times$ 3 SR, the maximum SSIM difference was 0.0065. For $\times$ 4 SR, the maximum SSIM difference was 0.0091. PSNR directly uses image pixels for statistical analysis. SSIM is a structural similarity indicator, which is more in line with human visual characteristics. Our SSIM performance is better on almost all datasets and all scales. The PSNR performance is slightly inferior but also excellent.

Table 5. The PSNRs, the number of parameters, and GFLOPs of $\times$ 2, $\times$ 3, $\times$ 4 SR for different models, with the best and second-best results marked in bold and italic, respectively.

Scale	Model	Parameters (K)	FLOPs (G)	Set5 PSNR/SSIM	Set14 PSNR/SSIM	B100 PSNR/SSIM
$\times$ 2	SRCNN	57	52.7	36.66/0.9524	32.42/0.9063	31.36/0.8879
	FSRCNN	12	6.0	37.00/0.9558	32.63/0.9088	31.53/0.8920
	VDSR	665	612.6	37.53/0.9587	33.03/0.9124	31.90/0.8960
	DRCN	1774	17974.0	37.63/0.9588	33.04/0.9118	31.85/0.8942
	DRRN	297	6797.0	37.74/0.9591	33.23/0.9136	32.05/0.8973
	MemNet	677	2662.4	37.78/0.9597	33.28/0.9142	32.08/0.8978
	SRMDNF	1513	347.7	37.79/0.9600	33.32/0.9150	32.05/0.8980
	IDN	553	127.7	37.83/0.9600	33.30/0.9148	32.08/0.8985
	AWSRN-S	397	91.2	37.75/0.9596	33.31/0.9151	32.00/0.8974
	SRFBN-S	282	–	37.78/0.9597	33.35/0.9156	32.00/0.8970
	LAPAR-B	250	85.0	37.87/0.9600	33.39/0.9162	32.10/0.8987
	A^{2}F-SD	313	71.2	37.91/0.9602	33.45/0.9164	32.08/0.8986
	PAN	261	70.5	37.99/0.9603	33.53/0.9174	32.14/0.8992
	VLESR	311	71.2	*38.01/0.9605*	**33.58**/*0.9177*	*32.16/0.8993*
	LLAN	265	70.2	**38.02/0.9610**	*33.55*/**0.9183**	**32.19/0.9004**
$\times$ 3	SRCNN	57	52.7	32.75/0.9090	29.28/0.8209	28.41/0.7863
	FSRCNN	12	4.6	33.16/0.9104	29.43/0.8242	28.53/0.7910

(continued)

Table 5. (*continued*)

Scale	Model	Parameters (K)	FLOPs (G)	Set5 PSNR/SSIM	Set14 PSNR/SSIM	B100 PSNR/SSIM
	VDSR	665	612.6	33.66/0.9213	29.77/0.8314	28.82/0.7976
	DRCN	1774	17974.0	33.82/0.9226	29.76/0.8311	28.80/0.7963
	DRRN	297	6797.0	34.03/0.9244	29.96/0.8349	28.95/0.8004
	MemNet	677	2662.4	34.09/0.9248	30.00/0.8350	28.96/0.8001
	SRMDNF	1530	156.3	34.12/0.9250	30.04/0.8370	28.97/0.8030
	IDN	553	57.0	34.11/0.9253	29.99/0.8354	28.95/0.8013
	AWSRN-S	477	48.6	34.02/0.9240	30.09/0.8376	28.92/0.8009
	SRFBN-S	376	–	34.20/0.9255	30.10/0.8372	28.96/0.8010
	LAPAR-B	276	61.0	34.20/0.9256	30.17/0.8387	29.03/0.8032
	A^2F-SD	316	31.9	34.23/0.9259	30.22/0.8395	29.01/0.8028
	PAN	261	39.0	34.30/0.9266	30.30/*0.8416*	29.06/0.8042
	VLESR	319	32.9	**34.40**/0.9272	*30.34*/0.8415	*29.08/0.8043*
	LLAN	265	38.6	*34.35*/**0.9278**	**30.36/0.8419**	**29.12/0.8054**
× 4	SRCNN	57	52.7	30.48/0.8628	27.49/0.7503	26.90/0.7101
	FSRCNN	12	4.6	30.71/0.8657	27.59/0.7535	26.98/0.7150
	VDSR	665	612.6	31.35/0.8838	28.01/0.7674	27.29/0.7251
	DRCN	1774	17974.0	31.53/0.8854	28.02/0.7670	27.23/0.7233
	DRRN	297	6797.0	31.68/0.8888	28.21/0.7720	27.38/0.7284
	MemNet	677	2662.4	31.74/0.8893	28.26/0.7723	27.40/0.7281
	SRMDNF	1555	89.3	31.96/0.8930	28.35/0.7770	27.49/0.7340
	IDN	553	32.3	31.82/0.8903	28.25/0.7730	27.41/0.7297
	AWSRN-S	588	33.7	31.77/0.8893	28.35/0.7761	27.41/0.7304
	SRFBN-S	483	–	31.98/0.8923	28.45/0.7779	27.44/0.7313
	LAPAR-B	313	53.0	31.94/0.8917	28.46/0.7784	27.52/0.7335
	A^2F-SD	320	18.2	32.06/0.8928	28.47/0.7790	27.48/**0.7373**
	PAN	272	28.2	32.06/0.8939	*28.56/0.7813*	27.55/0.7352
	VLESR	331	19.5	**32.17**/*0.8945*	28.55/0.7802	*27.55*/0.7345
	LLAN	275	27.6	*32.10*/**0.8949**	**28.60/0.7821**	**27.59**/*0.7368*

4.5 Subjective Visual Comparison

In this experiment, we have selected some representative methods [3] to explore the characteristics of the SISR algorithms. The × 4 SR images reconstructed from our method and other state-of-the-art methods are shown in Fig. 7. Our results were better than previous methods and dependable. For classical convolutional and residual networks, the results were over-blurred and distorted. The results may appear clearer for the GAN-based networks, but there are many hallucinating edges that lose authenticity. Taking image026 from the Urban100 dataset as an example, while other methods have ambiguous results or incorrect directions, our results were close to the original HR image. The results of the × 4 SR for image219090 in the B100 dataset are similar, and our method is better for handling railing and hull markings.

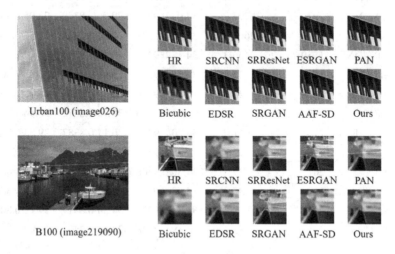

Fig. 7. Visual comparison for upscaling factor × 4.

5   Conclusions

We proposed a lightweight local attention network (LLAN) in this paper. The lightweight self-calibrated residual blocks with pixel attention (LSC-PA) that have the stronger expressive ability can better propagate and fuse features. The average local attention block (ALAB) can combine global features and accelerate the network. We carried out extensive experiments using five benchmark datasets. The experimental results demonstrated that our method outperforms other compared state-of-the-art techniques, and our approach provides a better balance between the complexity and performance of the SISR methods. Furthermore, we explored the performance of LLAN with more channels and the relationship between channel and feature fusion. Experiments have shown that when the number of channels is large enough, our method LLAN performs better, and unequal proportional concatenation can have a positive effect. The code is available at https://github.com/fchaner/LLAN.

References

1. Ahn, N., Kang, B., Sohn, K. A.: Fast, accurate, and lightweight super-resolution with cascading residual network. In: Proceedings of the European Conference on Computer Vision (ECCV), pp. 252–268. Munich, Germany (2018)

2. Ayazoglu, M.: Extremely lightweight quantization robust real-time single-image super resolution for mobile devices. In: Proceedings of the IEEE/CVF Conference on Computer Vision and Pattern Recognition, pp. 2472–2479. Virtual (2021)
3. Behjati, P.: Towards Efficient and Robust Convolutional Neural Networks for Single Image Super-Resolution. Universitat Autònoma de Barcelona, Spain (2022)
4. Cammarasana, S.: Real-time Ultrasound Signals Processing: Denoising and Super-resolution. Diss. University of Genoa, Italy (2023)
5. Chan, K.C., Xu, X., Wang, X., Gu, J., Loy, C.C.: GLEAN: Generative latent bank for image super-resolution and beyond. IEEE Trans. Pattern Anal. Mach. Intell.Intell. **45**(3), 3154–3168 (2022)
6. Chan, K.C.K.: Image and video super-resolution in the wild. Nanyang Technological University, Singapore (2022)
7. Chen, H., et al.: Real-world single image super-resolution: a brief review. Inf. Fusion **79**, 124–145 (2022)
8. Chu, X., Zhang, B., Ma, H., Xu, R., Li, Q.: Fast, accurate and lightweight super-resolution with neural architecture search. In 25th International Conference on Pattern Recognition (ICPR), pp. 59–64. Milan, Italy (2021)
9. Dong, C., Loy, C.C., He, K., Tang, X.: Learning a deep convolutional network for image super-resolution. In: Proceedings of the European Conference on Computer Vision (ECCV), Part IV 13, pp. 184–199. Zurich, Switzerland (2014)
10. Dong, C., Loy, C.C., He, K., Tang, X.: Image super-resolution using deep convolutional networks. IEEE Trans. Pattern Anal. Mach. Intell.Intell. **38**(2), 295–307 (2015)
11. Esmaeilzehi, A., Ahmad, M.O., Swamy, M.N.S.: SRNSSI: a deep light-weight network for single image super resolution using spatial and spectral information. IEEE Trans. Comput. Imaging **7**, 409–421 (2021)
12. Fuoli, D.: Advances in efficient super-resolution and enhancement of images and videos. Diss. ETH Zurich (2023)
13. Gao, Q., Zhao, Y., Li, G., Tong, T.: Image super-resolution using knowledge distillation. In: Asian Conference on Computer Vision, Part II, pp. 527–541. Perth, Australia (2019)
14. Gao, D., Zhou, D.: A very lightweight and efficient image super-resolution network. Expert Syst. Appl. **213**, 118898 (2023)
15. Hui, Z., Wang, X., Gao, X.: Fast and accurate single image super-resolution via information distillation network. In Proceedings of the IEEE Conference on Computer Vision and Pattern Recognition, pp. 723–731. Salt Lake City Utah (2018)
16. Kim, J., Lee, J.K., Lee, K.M.: Deeply-recursive convolutional network for image super-resolution. In Proceedings of the IEEE Conference on Computer Vision and Pattern Recognition, pp. 1637–1645. Las Vegas Nevada (2016)
17. Kim, J., Lee, J.K., Lee, K.M.: Accurate image super-resolution using very deep convolutional networks. In Proceedings of the IEEE Conference on Computer Vision and Pattern Recognition, pp. 1646–1654. Las Vegas Nevada (2016)
18. Li, Z., Yang, J., Liu, Z., Yang, X., Jeon, G., Wu, W.: Feedback network for image super-resolution. In Proceedings of the IEEE/CVF Conference on Computer Vision and Pattern Recognition, pp. 3867–3876. Long Beach California (2019)
19. Li, W., Li, J., Li, J., Huang, Z., Zhou, D.: A lightweight multi-scale channel attention network for image super-resolution. Neurocomputing **456**, 327–337 (2021)
20. Liu, S., Qi, L., Qin, H., Shi, J., Jia, J.: Path aggregation network for instance segmentation. In Proceedings of the IEEE Conference on Computer Vision and Pattern Recognition, pp. 8759–8768. Salt Lake City Utah (2018)
21. Luo, X., Xie, Y., Zhang, Y., Qu, Y., Li, C., Fu, Y.: Latticenet: towards lightweight image super-resolution with lattice block. In: Computer Vision–ECCV 2020: 16th European Conference, Proceedings, Part XXII 16, pp. 272–289. Glasgow, UK (2020)

22. Luo, J., Zhao, L., Zhu, L., Tao, W.: Multi-scale receptive field fusion network for lightweight image super-resolution. Neurocomputing **493**, 314–326 (2022)
23. Qin, J., Liu, F., Liu, K., Jeon, G., Yang, X.: Lightweight hierarchical residual feature fusion network for single-image super-resolution. Neurocomputing **478**, 104–123 (2022)
24. Rashidi, A.: Advanced super-resolution techniques for high quality scanned images. Diss. Université de Bordeaux, France (2022)
25. Song, Z., Zhong, B.: A lightweight local-global attention network for single image super-resolution. In: Proceedings of the Asian Conference on Computer Vision, pp. 4395–4410. Macau SAR, China (2022)
26. Tai, Y., Yang, J., Liu, X.: Image super-resolution via deep recursive residual network. In Proceedings of the IEEE Conference on Computer Vision and Pattern Recognition, pp. 3147–3155. Honolulu Hawaii (2017)
27. Tai, Y., Yang, J., Liu, X., Xu, C.: Memnet: A persistent memory network for image restoration. In Proceedings of the IEEE international conference on computer vision, pp. 4539–4547. Venice Italy (2017)
28. Timofte, R., Agustsson, E., Van Gool, L., Yang, M.H., Zhang, L.: Ntire 2017 challenge on single image super-resolution: Methods and results. In Proceedings of the IEEE Conference on Computer Vision and Pattern Recognition Workshops, pp. 1110–1121. Honolulu Hawaii (2017)
29. Wang, C., Li, Z., Shi, J.: Lightweight image super-resolution with adaptive weighted learning network. arXiv preprint, arXiv:1904.02358 (2019)
30. Wang, X., Wang, Q., Zhao, Y., Yan, J., Fan, L., Chen, L.: Lightweight single-image super-resolution network with attentive auxiliary feature learning. In: Proceedings of the Asian Conference on Computer Vision, Virtual (2020)
31. Zhang, K., Zuo, W., Zhang, L.: Learning a single convolutional super-resolution network for multiple degradations. In Proceedings of the IEEE conference on computer vision and pattern recognition, pp. 3262–3271. Salt Lake City Utah (2018)
32. Zhang, Y., Tian, Y., Kong, Y., Zhong, B., Fu, Y.: Residual dense network for image super-resolution. In Proceedings of the IEEE Conference on Computer Vision and Pattern Recognition, pp. 2472–2481. Salt Lake City Utah (2018)
33. Zhang, R., Isola, P., Efros, A. A., Shechtman, E., Wang, O.: The unreasonable effectiveness of deep features as a perceptual metric. In: Proceedings of the IEEE Conference on Computer Vision and Pattern Recognition, pp. 586–595. Salt Lake City Utah (2018)
34. Zhao, H., Kong, X., He, J., Qiao, Y., Dong, C.: Efficient image super-resolution using pixel attention. In: Computer Vision–ECCV 2020 Workshops, Part III 16, pp. 56–72. Glasgow, UK (2020)
35. Zou, W., Ye, T., Zheng, W., Zhang, Y., Chen, L., Wu, Y.: Self-calibrated efficient transformer for lightweight super-resolution. In: Proceedings of the IEEE/CVF Conference on Computer Vision and Pattern Recognition, pp. 930–939. New Orleans Louisiana (2022)

Domain Adaptation for Speaker Verification Based on Self-supervised Learning with Adversarial Training

Qiulin Li, Junhao Qiang, and Qun Yang[✉]

Nanjing University of Aeronautics and Astronautics, Nanjing 210023, China
qun.yang@nuaa.edu.cn

Abstract. Speaker verification models trained on a single domain have difficulty keeping performance on new domain data. Adversarial training maps different domain data to the same subspace to handle this problem. However, adversarial training only uses domain labels on the target domain and does not mine its speaker information. To improve the domain adaptation performance for speaker verification, we propose a joint training strategy for adversarial training and self-supervised learning. In our method, adversarial training adapts knowledge from the source domain to the target domain, while self-supervised learning obtains speech representations from unlabeled utterances. Further, our self-supervised learning only uses positive pairs to avoid false negative samples. The proposed joint training strategy enables adversarial training to guide self-supervised learning to focus on speaker verification tasks. Experiments show our proposed method outperforms other domain adaptation methods.

Keywords: speaker verification · domain adaptation · self-supervised learning · adversarial training

1 Introduction

Speaker verification is the task of determining whether the input audio is a registered claimed identity. Many speaker verification models use the supervised framework and perform well on labeled datasets [1–3]. However, supervised frameworks require a large number of labeled data for training, whereas getting such labels on a new domain is costly. Moreover, these models show significant performance degradation when testing with new domain data.

To bridge this gap, domain adaptation techniques have been proposed to enable models trained on well-labeled source domain data and then adapt its good performance to a new target domain. Commonly used domain adaptation methods extract domain invariant information by adversarial training [4–8]. This method reduces the domain distribution differences by projecting the

---

Q. Li and J. Qiang—Authors contributed equally to this research.

© The Author(s), under exclusive license to Springer Nature Switzerland AG 2024
S. Rudinac et al. (Eds.): MMM 2024, LNCS 14554, pp. 385–395, 2024.
https://doi.org/10.1007/978-3-031-53305-1_29

data distributions of different domains into the same feature subspace. The model adapts the knowledge from the source domain to the target domain when extracting the target domain representations from the same feature subspace. In [9,10], the researchers propose a variational domain adversarial learning method. They introduce variational autoencoders (VAEs) in adversarial training to adapt domain-invariant speaker representations to the PLDA backend. However, they only use the domain labels and do not mine the unlabeled information of the target domain.

Recently, self-supervised learning has shown significant adaptive performance. Originating in computer vision, the self-supervised learning method can mine the latent information of data without the corresponding labels [11,12]. The main idea for contrastive self-supervised learning is to learn useful representations of the samples by reducing the distance between augmented samples (positive pairs) and increasing the distance between different samples (negative pairs). There are already some works exploring contrastive self-supervised learning methods in domain adaptation for speaker verification [13,14]. However, since the self-supervised pre-training model requires fine-tuning labeled data on downstream tasks, it is hard to achieve good performance without labeled data of the target domain. Some people try to address this problem by using domain adaptation training strategy based on self-supervised learning [15]. They use self-supervised learning to obtain speech information from target domain data while using speaker classification to get speaker information from source domain data. However, contrastive self-supervised learning requires a large number of negative samples, and the model may treat the utterances from the same speaker as different speakers (false negative samples), which interferes with training.

In this paper, we propose a new domain adaptation strategy for speaker verification. We take both adversarial training and self-supervised learning into consideration as well as use a new joint training strategy. We use adversarial training to adapt knowledge from the source domain to the target domain while using contrastive self-supervised learning to learn discriminative speech representations from the similarity of positive samples. Inspired by [16], our contrastive self-supervised learning does not use negative pairs. We propose a joint training strategy to combine adversarial training and self-supervised learning. It enables adversarial training to guide self-supervised learning to focus on more speaker information by using speaker-supervised classification in the source domain. In summary, our method uses adversarial training to guide self-supervised learning without negative samples. We conduct several experiments, the results show that our method improves the adaptation performance of speaker verification models.

2   Related Works

Self-supervised learning constructs task-independent supervised information from unlabeled data and uses auxiliary tasks to get task-independent representations. Therefore, it usually needs labeled data to fine-tune to improve the performance in downstream tasks. In the contrastive self-supervised learning

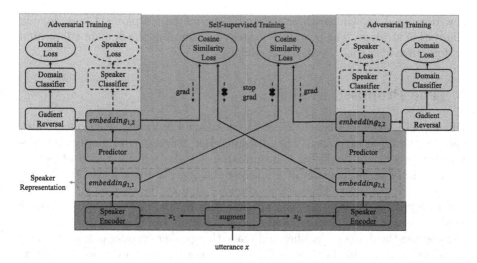

Fig. 1. Overview of the proposed model. The model consists of three components, which are adversarial training, self-supervised training, and speaker encoder. We feed both source and target domain utterances into the model to calculate domain loss and cosine similarity loss. The speaker loss is only calculated for source domain utterances because just the utterances of source domain have the speaker labels. This means speaker classifier performs classification only on the representations of source domain.

method, the auxiliary task usually reduces the distance between positive samples and increases the distance between negative samples. However, contrastive self-supervised learning requires a large number of negative samples for working, such as SimCLR [11] and MoCo [12]. These self-supervised learning methods with negative samples also suffer from the same problem of false negative samples.

Among the domain adaptations for speaker verification, the domain adversarial neural network (DANN) can obtain domain-invariant speaker representations. In [17], the authors propose a Collaborative Adversarial Network (CAN). They focus on both domain-invariant and domain-specific features. But for target domain data, adversarial training only uses its labels for adaptation and does not consider its speech information. In [15], authors propose a self-supervised domain adaptation method (SSDA). They perform speaker classification to get speaker information from the source domain while using self-supervised learning to obtain speech information from the target domain. However, their method does not consider inter-domain distribution differences, which means that the model does not adapt the source domain knowledge well to the target domain. The closest to our method is [18,19]. The authors propose a joint Augmentation Adversarial Training (AAT) and Angular Prototype Loss (AP) method. They use adversarial training to make the speaker representations invariant to the channel environment when data is augmented. However, they only focus on channel variability and still need fine-tuning to get good performance when applying to new domain data. Unlike these methods, our method explores a new domain adaptation strategy. The proposed strategy can improve the domain adaptation performance of speaker verification models.

3 The Proposed Model

Our model is shown in Fig. 1. We start our discription from the encoder, which encodes the input utterance. The self-supervised training is primarily a predictor that provides a nonlinear transformation for the self-supervised objective. Adversarial training consists of a speaker classifier, a domain classifier, and a gradient reversal layer. The speaker classifier enables the speaker encoder output to represent the speaker, i.e., speaker embedding. The domain classifier is used to discriminate the domain of representations. The gradient reversal layer can confuse the domain classifier and make the speaker encoder output domain-invariant features. In the proposed joint training strategy, we place the adversarial training after the predictor of the self-supervised learning to make the adversarial training guide the self-supervised learning.

3.1 Self-supervised Learning Without Negative Samples

To obtain discriminative speech representations, we should ensure that the representations extracted from an utterance are consistent and similar even after different encodings. Therefore, we use self-supervised learning to encourage the non-overlapping segments of an utterance, before and after the nonlinear transformation, to be close to each other. This goal uses only positive pairs instead of both positive and negative pairs. A positive pair can be prepared by selecting two non-overlapping segments from the same input utterance and augmenting them. It can be denoted as x_i ($i = 1\,or\,2$). For the positive pair, the segment selection and augmentation method is important in self-supervised learning. They enable positive pairs to have different speech content and background noise but semantically belong to the same speaker, thus intra-class distance for the speaker can be calculated. The positive pair goes through the speaker encoder E to get the representations $e_{i,1}$ ($i = 1\,or\,2$), which is used to represent the speaker during the test. The representations $e_{i,1}$ go through the predictor p to obtain the nonlinear representations $e_{i,2}$ ($i = 1\,or\,2$). These processes can be expressed as follows:

$$e_{1,1}, e_{2,1} = E(x_1, x_2) \qquad (1)$$

$$e_{1,2}, e_{2,2} = p(e_{1,1}, e_{2,1}) \qquad (2)$$

where x_i denotes the i-th augmented segment of the utterance x and $e_{i,j}$ denotes the j-th transformation of the i-th augmented segment. In order to mine various speech information by self-supervised learning, the source and target domain utterances can be considered as complete unlabeled data to apply the self-supervised loss without negative pairs.

As mentioned before, the objective of self-supervised learning without negative pairs is that one transformed representation of a positive pair should match its other untransformed representation. Therefore, we match $e_{1,1}$ and $e_{2,2}$ using the following negative cosine loss:

$$D(e_{1,1}, e_{2,2}) = -\frac{e_{1,1}}{\|e_{1,1}\|_2} \frac{e_{2,2}}{\|e_{2,2}\|_2} \tag{3}$$

where $\|.\|_2$ is the l2-norm. In Fig. 1, $e_{1,1}$ is the representation without the transformation, and $e_{2,2}$ is the representation after the nonlinear transform. The loss function encourages the speaker representations of positive pairs to be close to each other before and after the transformation. Thus, the speaker encoder tends to learn similar information, i.e., speaker information, from different augmented segments of an utterance to minimize the negative cosine loss.

Since the self-supervised framework has two speaker encoder branches, we can exchange the position of the positive pair and calculate the distance again. Therefore, our self-supervised loss without negative pairs is as follows:

$$L_{self} = \frac{1}{2}D(e_{1,1}, e_{2,2}) + \frac{1}{2}D(e_{2,1}, e_{1,2}) \tag{4}$$

To avoid the error of the speaker encoder outputting the same representations for different speaker utterances, we introduce the gradient stop operation of siamese networks [16]. In Fig. 1, we apply the stop gradient to a branch of the speaker encoder. Thus, the above equation can be expressed in more detail as:

$$L_{self} = \frac{1}{2}D(e_{1,1}, e_{2,2}.stop) + \frac{1}{2}D(e_{2,1}, e_{1,2}.stop) \tag{5}$$

which means that the predictor receives the gradient from $e_{1,2}$ and $e_{2,2}$, and the speaker encoder receives the gradient only from the predictor.

3.2   Adversarial Training and Speaker Classification

The goal of the adversarial training is to make the domain classifier unable to discriminate which domain the speaker represents. Thus, the speaker encoder will find a shared subspace of the source and target domains. Its output representations are domain-invariant, and the source representations will calculate the speaker classification loss.

Different from adversarial training, the goal of the gradient descent process is minimizing the domain classification loss, so the domain classifier can correctly classify the domain of the utterance by its representation. To simplify the training process, we introduce the gradient reversal layer. It does nothing in the forward process but multiplies the gradient by negative parameters in the backward process. Therefore, the model update is performed in the opposite direction, i.e., confusing the domain classifier, after the gradient passes through the gradient reversal layer. We insert the gradient reversal layer between the predictor p and the domain classifier C, and put the speaker classifier S after the predictor p. We calculate the speaker loss and domain loss as follows:

$$L_{spk} = \frac{1}{2}L_S(S(e_{1,2})) + \frac{1}{2}L_S(S(e_{2,2})) \tag{6}$$

$$L_d = \frac{1}{2}L_C(C(GRL(e_{1,2}))) + \frac{1}{2}L_C(C(GRL(e_{2,2}))) \tag{7}$$

where L_S denotes the cross-entropy loss and L_C denotes the binary cross-entropy loss.

3.3 Joint Optimization Strategy

Although self-supervised learning can obtain speaker information from unlabeled data, the performance on downstream tasks still benefits from fine-tuning with labeled data. However, domain adaptation tasks expect to keep performance on unlabeled data of the target domain. Thus we propose to introduce adversarial training that further improves the domain adaptation performance of the model. The speaker-supervised classification of the source domain in adversarial training can guide the self-supervised learning optimization toward speaker verification. Meanwhile, adversarial training can adapt the knowledge from the source domain to the target domain by the shared subspace of the source and target domains. Therefore, we normalize the speaker loss and self-supervised loss and then optimize them together with the domain loss. We define the total loss as follows:

$$L_{total} = \frac{1}{2}L_{spk} + \frac{1}{2}L_{self} + L_d \tag{8}$$

where the source domain data are involved in the loss calculation of L_{spk}, L_{self}, and L_d, while the target domain data is involved in the loss calculation of L_{self} and L_d. The training objective is to output domain-invariant speaker representations with better speaker discrimination.

4 Experimental Setup

4.1 Dataset

To evaluate our proposed method in the domain adaptation, our experiments use two corpora, namely VoxCeleb1&2 [20,21] and CN-Celeb1&2 [22,23], to construct language mismatch. The mainly English corpus, VoxCeleb1 or VoxCeleb2, are taken as source domains in the experiments, and they are denoted as Vox1 and Vox2, respectively. There are 1221 speakers in Vox1 and 5994 speakers in Vox2. In addition, we select the speech genre from the Chinese corpus CN-Celeb1&2 as the target domain and denote it as CN-s, which has 306 speakers. The whole CN-Celeb1 corpus has 800 speakers, which we denote as CN1.

4.2   Data Preprocessing

Since the CN-Celeb corpus is mainly short audio, we need to perform additional processing. We first select utterances from CN-Celeb1&2 in the speech genre. After that, we concatenate and cut the utterances into 5-second segments according to the speaker. After processing, the English training set has 1221 speakers with 148,641 utterances, and the Chinese training set has 306 speakers with 42,942 utterances. For all utterances, we randomly select two non-overlapping 2-second segments and perform independent data augmentation. We extracted 40-dimensional Filterbank features for all utterances with a 25 ms window and 10 ms frame shift.

Data augmentation is usually used to increase the amount of data and channel variability. Experimental results prove that model performance benefits from the amount and variability of data, especially for self-supervised learning [24]. In this paper, we use two general augmentation methods, namely additive noise and room impulse response (RIR) simulation. For additive noise, we use speech, music, and noise utterances from the MUSAN corpus [25] with signal-to-noise ratios randomly chosen from the ranges $\{13, 20\}$, $\{5, 15\}$, and $\{0, 15\}$, respectively. For the RIR simulations, we perform convolution with pre-computed RIR filters [26].

4.3   System Configuration

We use Fast ResNet-34 as the speaker encoder [27]. It has a quarter of the number of channels of the original ResNet-34, and has an output size of 512. Since the number of source domain utterances is much larger than the target domain, we train only the source domain utterances at the beginning of the epoch and feed the source and target domain utterances together into the speaker encoder at the end of the epoch. We use stochastic gradient descent as the optimizer, and the learning rate has a cosine decay schedule. We train 100 epochs in total using a batch size of 200.

5   Results

5.1   Ablation Study

In this section, we evaluate the impact of adversarial training and self-supervised learning. The Equal Error Rate (EER) on the target domain CN-s test set is shown in Table 1. The first case is the baseline, i.e., without adversarial training and self-supervised learning. The second case uses only adversarial training, and the third case uses only self-supervised learning. Our method is the fourth case, that uses both adversarial learning and self-supervised learning. We test self-supervised learning after pre-training on the target domain.

From Table 1, we can find when tested on the target domain after training the speaker classifier on the source domain, the EER is 13.24%. Adding adversarial training improves the model performance, which has an EER of 12.06%. The

EER with self-supervised learning is 18.30%, which is significantly worse than other methods. The main reason is that the self-supervised learning pre-trains with a small number of utterances, and we cannot fine-tune the model on the unlabeled data of the target domain. Obviously, our method obtains the best performance by using jointly optimizing adversarial training and self-supervised learning. The results show that self-supervised learning and adversarial training do not conflict. In contrast, the speaker-supervised classification in adversarial training guides self-supervised learning, while self-supervised learning mines the target domain speech information to help adversarial training. Further, the joint optimization strategy leads to better domain adaptation performance of the speaker verification model. We conclude that our method of jointly optimal adversarial training and self-supervised learning can improve the domain adaptation performance of the speaker verification model.

Table 1. Performance of adversarial training and self-supervised learning in the ablation study using the Vox1 as the source domain and the CN-s as the target domain.

adversarial	self-supervised	EER(%)
–	–	13.24
✓	–	12.06
–	✓	18.30
✓	✓	**10.53**

5.2 Performance Evaluation of Our Method with Other Domain Adaptation Methods

The domain adaptation results are shown in Table 2. Among them, AAT is a self-supervised learning method using augmented adversarial training and angular prototype loss to pre-train the model [19]. Both SSDA and SSDA-Joint are domain adaptation methods based on self-supervised learning that explore the impact of serial and parallel training on speaker classification and self-supervised learning, respectively [15]. CAN uses adversarial training and improves performance by focusing on domain-invariant and domain-specific information [17]. In addition, we show the result that the self-supervised objective with both positive and negative samples (SimCLR) applied in our training strategy.

As a self-supervised learning method that cannot fine-tune the model on the target domain, the AAT has an EER of 14.54%. As an adversarial training method that does not mine the speech information in the target domain, the CAN has an EER of 11.79%. The EER of our proposed joint training strategy is 10.53%, which is better than these methods. In addition, the SSDA and SSDA-Joint methods avoid fine-tuning by speaker classification and obtain the best EER of 10.20%. However, the source domain they use is Vox2, which contains

Table 2. Comparison with previous methods. Ours (SimCLR): applying SimCLR to our training strategy, Ours: applying self-supervised without negative samples to our training strategy

Train Data	Method	Loss	EER(%)
Vox2	AAT	AP + AAT	14.54
Vox2 + CN1 [15]	SSDA	Contrastive	20.72
		Triplet	13.66
		ProtoLoss	13.72
		GE2E	13.34
	SSDA-Joint	Contrastive	20.48
		Triplet	11.27
		ProtoLoss	**10.20**
		GE2E	10.24
Vox1 + CN-s	CAN	Softmax	11.79
	Ours(SimCLR)	InfoNCE	13.43
	Ours	Cosine	**10.53**

a significantly larger number of utterances compared to the Vox1 dataset we use. Therefore, we get competitive performance with fewer data. Furthermore, applying our proposed strategy, the EER with SimCLR self-supervised objective is 13.43%. Thus the relative improvement of self-supervised learning without negative samples is 21.59% compared to that with negative samples. In summary, our proposed method improves the domain adaptation performance of speaker verification models.

6 Conclusions

This paper proposes a self-supervised learning method with adversarial training in domain adaptation for speaker verification. The self-supervised learning uses only positive pairs to avoid false samples during training. Meanwhile, adversarial training adapts source domain knowledge to the target domain through a shared subspace between the source and target domains. The speaker-supervised classification of the source domain in adversarial training can guide self-supervised learning to consider speaker verification tasks. Therefore, the proposed joint training strategy gets better domain-invariant speaker representations for domain adaptation. The experiments show that our method eventually achieves an EER of 10.53%, which is close to the optimal performance (EER of 10.2%) trained with large amounts of source domain data. It can be concluded that our proposed method obtains better speaker representations in the target domain.

References

1. Snyder, D., Garcia-Romero, D., Sell, G., Povey, D., Khudanpur, S.: X-vectors: Robust DNN embeddings for speaker recognition. In: ICASSP, pp. 5329–5333 (2018). https://doi.org/10.1109/ICASSP.2018.8461375

2. Desplanques, B., Thienpondt, J., Demuynck, K.: ECAPA-TDNN: emphasized channel attention, propagation and aggregation in TDNN based speaker verification. In: Meng, H., Xu, B., Zheng, T.F. (eds.) INTERSPEECH, pp. 3830–3834. ISCA (2020)

3. India, M., Safari, P., Hernando, J.: Double multi-head attention for speaker verification. In: ICASSP, pp. 6144–6148 (2021). https://doi.org/10.1109/ICASSP39728.2021.9414877

4. Ganin, Y., Lempitsky, V.: Unsupervised domain adaptation by backpropagation. In: International Conference on Machine Learning, pp. 1180–1189. PMLR (2015)

5. Wang, Q., Rao, W., Guo, P., Xie, L.: Adversarial training for multi-domain speaker recognition. In: 12th International Symposium on Chinese Spoken Language Processing, pp. 1–5. IEEE (2021)

6. Wang, Q., Rao, W., Sun, S., Xie, L., Chng, E.S., Li, H.: Unsupervised domain adaptation via domain adversarial training for speaker recognition. In: ICASSP, pp. 4889–4893 (2018). https://doi.org/10.1109/ICASSP.2018.8461423

7. Wei, G., Lan, C., Zeng, W., Zhang, Z., Chen, Z.: ToAlign: task-oriented alignment for unsupervised domain adaptation. Adv. Neural. Inf. Process. Syst. **34**, 13834–13846 (2021)

8. Chen, Z., Wang, S., Qian, Y.: Adversarial domain adaptation for speaker verification using partially shared network. In: INTERSPEECH, pp. 3017–3021 (2020)

9. Tu, Y., Mak, M.W., Chien, J.T.: Variational domain adversarial learning for speaker verification. In: Interspeech (2019)

10. Tu, Y., Mak, M.W.: Information maximized variational domain adversarial learning for speaker verification. In: ICASSP 2020–2020 IEEE International Conference on Acoustics, Speech and Signal Processing (ICASSP), pp. 6449–6453 (2020). https://doi.org/10.1109/ICASSP40776.2020.9053735

11. Chen, T., Kornblith, S., Norouzi, M., Hinton, G.: A simple framework for contrastive learning of visual representations. In: International Conference on Machine Learning, pp. 1597–1607. PMLR (2020)

12. He, K., Fan, H., Wu, Y., Xie, S., Girshick, R.: Momentum contrast for unsupervised visual representation learning. In: Proceedings of the IEEE/CVF Conference on Computer Vision and Pattern Recognition, pp. 9729–9738 (2020)

13. Zhang, H., Zou, Y., Wang, H.: Contrastive self-supervised learning for text-independent speaker verification. In: ICASSP 2021–2021 IEEE International Conference on Acoustics, Speech and Signal Processing (ICASSP), pp. 6713–6717. IEEE (2021)

14. Xia, W., Zhang, C., Weng, C., Yu, M., Yu, D.: Self-supervised text-independent speaker verification using prototypical momentum contrastive learning. In: ICASSP 2021–2021 IEEE International Conference on Acoustics, Speech and Signal Processing (ICASSP), pp. 6723–6727. IEEE (2021)

15. Chen, Z., Wang, S., Qian, Y.: Self-supervised learning based domain adaptation for robust speaker verification. In: ICASSP 2021–2021 IEEE International Conference on Acoustics, Speech and Signal Processing (ICASSP), pp. 5834–5838. IEEE (2021)

16. Chen, X., He, K.: Exploring simple siamese representation learning. In: Proceedings of the IEEE/CVF Conference on Computer Vision and Pattern Recognition, pp. 15750–15758 (2021)

17. Zhang, W., Ouyang, W., Li, W., Xu, D.: Collaborative and adversarial network for unsupervised domain adaptation. In: 2018 IEEE/CVF Conference on Computer Vision and Pattern Recognition, pp. 3801–3809 (2018). https://doi.org/10.1109/CVPR.2018.00400

18. Huh, J., Heo, H.S., Kang, J., Watanabe, S., Chung, J.S.: Augmentation adversarial training for unsupervised speaker recognition. In: Workshop on Self-Supervised Learning for Speech and Audio Processing, NeurIPS (2020)

19. Kang, J., Huh, J., Heo, H.S., Chung, J.S.: Augmentation adversarial training for self-supervised speaker representation learning. IEEE J. Sel. Top. Sig. Process. **16**(6), 1253–1262 (2022). https://doi.org/10.1109/JSTSP.2022.3200915

20. Nagrani, A., Chung, J.S., Zisserman, A.: Voxceleb: a large-scale speaker identification dataset. In: INTERSPEECH (2017)

21. Chung, J.S., Nagrani, A., Zisserman, A.: Voxceleb2: deep speaker recognition. In: INTERSPEECH (2018)

22. Fan, Y., et al.: Cn-celeb: a challenging Chinese speaker recognition dataset. In: ICASSP 2020–2020 IEEE International Conference on Acoustics, Speech and Signal Processing (ICASSP), pp. 7604–7608. IEEE (2020)

23. Li, L., Liu, R., Kang, J., Fan, Y., Cui, H., Cai, Y., Vipperla, R., Zheng, T.F., Wang, D.: Cn-celeb: multi-genre speaker recognition. Speech Commun. **137**, 77–91 (2022)

24. Sang, M., Li, H., Liu, F., Arnold, A.O., Wan, L.: Self-supervised speaker verification with simple siamese network and self-supervised regularization. In: ICASSP 2022–2022 IEEE International Conference on Acoustics, Speech and Signal Processing (ICASSP), pp. 6127–6131. IEEE (2022)

25. Snyder, D., Chen, G., Povey, D.: Musan: A music, speech, and noise corpus. arXiv preprint arXiv:1510.08484 (2015)

26. Ko, T., Peddinti, V., Povey, D., Seltzer, M.L., Khudanpur, S.: A study on data augmentation of reverberant speech for robust speech recognition. In: 2017 IEEE International Conference on Acoustics, Speech and Signal Processing (ICASSP), pp. 5220–5224 (2017). https://doi.org/10.1109/ICASSP.2017.7953152

27. Chung, J.S., et al.: In defence of metric learning for speaker recognition. arXiv preprint arXiv:2003.11982 (2020)

Quality Scalable Video Coding Based on Neural Representation

Qian Cao[1], Dongdong Zhang[1(✉)], and Chengyu Sun[2]

[1] Department of Computer Science and Technology, Tongji University, Shanghai,
China
2230788@tongji.edu.cn, ddzhang@tongji.edu.cn
[2] Shanghai Key Laboratory of Urban Renewal and Spatial Optimization Technology,
Tongji University, Shanghai, China
cy.sun@tongji.edu.cn

Abstract. Neural Representation for Videos (NeRV) encodes each video
into a network, providing a promising solution to video compression.
However, existing NeRV methods are limited to representing single-
quality videos with fixed-size models. To accommodate varying quality
requirements, NeRV methods need multiple separate networks with different sizes, resulting in additional training and storage costs. To address
this, we propose a Quality Scalable Video Coding method based on Neural Representation, in which a hierarchical network consisting of a base
layer (BL) and several enhancement layers (ELs) represents the same
video with coarse-to-fine qualities. As the smallest subnetwork, the BL
represents basic content. The larger subnetworks can be formed by gradually adding the ELs which capture residuals between the lower-quality
reconstructed frames and original ones. Since the larger subnetworks
share the parameters of the smaller ones, our method saves 40% of storage space. In addition, our structural design and training strategy enable
each subnetwork to outperform the baseline on average +0.29 PSNR.

Keywords: Implicit Neural Representation · Video compression ·
Quality Scalable coding

1 Introduction

Traditional video compression approaches such as H.264 [1], HEVC [2] rely on
manually-designed modules, such as motion estimation and discrete cosine transform (DCT). With the success of deep learning, learning-based codecs [3,4]
replace handcraft modules with neural networks and achieve improved rate-
distortion performance. Recently, Implicit Neural Representation (INR) has
received increasing attention in the signal compression tasks such as image [5–7]
and video [8–16], due to its simpler pipelines and faster decoding speed.

Neural Representation for Videos (NeRV) methods [10–16] train a neural
network to represent a video, thus encoding the video into the network weights.
The encoding is the process of training a neural network to overfit video frames,
and the decoding is a simple forward propagation operation. The network itself

© The Author(s), under exclusive license to Springer Nature Switzerland AG 2024
S. Rudinac et al. (Eds.): MMM 2024, LNCS 14554, pp. 396–409, 2024.
https://doi.org/10.1007/978-3-031-53305-1_30

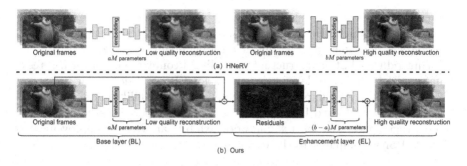

Fig. 1. An example to represent a video at two different quality levels: HNeRV requires two separate networks with different sizes (aM and bM parameters), while our methods achieves this with just one network with totaling bM parameters. The network comprises a BL that can independently decode the lower-quality video and a EL that enhances the quality by adding residual details to the BL.

serves as the video stream, and the model size dictates the trade-off between rate and distortion. Larger networks exhibit greater reconstruction accuracy at the expense of storage space and transmission bandwidth.

The original videos often need to be transmitted in different qualities adapted to terminal device and network conditions. For NeRV methods, this requires multiple sizes of the networks to represent the video at different quality levels. Since these networks of different sizes represent the same video content, there is information redundancy among them. However, existing methods do not consider this, and each network is trained and stored independently, resulting in significant costs. If the larger network can directly leverage the knowledge learned by the smaller ones instead of starting training from scratch, it can clearly converge faster and save storage space by sharing parameters. Scalable Video Coding (SVC) schemes [17,18] encapsulate videos with varying quality in a hierarchical video stream that can be partially decoded to provide flexible quality. Motivated by this, we propose using a single network that can be separated into executable subnetworks with different sizes to represent a video in flexible quality. The small subnetwork can operate independently or as part of larger ones. Therefore, larger subnetworks can leverage information already learned by smaller ones.

In this paper, we propose a quality scalable video coding method based on neural representation (QSVCNR). Building upon HNeRV [15], we design a hierarchical network consisting of a base layer (BL) and one or more enhancement layers (ELs). An example is illustrated in Fig. 1. The hierarchical network enables a coarse-to-fine reconstruction and naturally decomposes the representation into a basic component and progressive levels of details, separately preserved in different layers. As the smallest subnetwork, BL can independently reconstruct a coarser video. Larger subnetworks are constructed by gradually introducing ELs which fit the residuals between the lower-quality reconstructed frames and original ones. We employ per-layer supervision to train the entire network end-to-end. The main contributions are as follows:

- We propose a quality scalable video coding method based on neural representation, in which a hierarchical network consisting of a BL and several ELs represents the same video with coarse-to-fine qualities.
- We introduce additional structures including a context encoder for BL, residual encoders and inter-layer connections for ELs, and design an end-to-end training strategy, to improve the representation capability of our method.
- On the bunny [19] and UVG [20] datasets, Our proposed method saves 40% of storage space, and each subnetwork exhibits an average improvement of +0.29 PSNR compared to the baseline.

2   Related Work

Implicit Neural Representation (INR) [21–23] parameterize the signal as a function approximated by a neural network, which map reference coordinates to their respective signal values. For instance, an image can be represented as function $f(x, y) = (R, G, B)$, where (x, y) are the coordinates of a specific point. This concept has been applied to compression tasks such as image [5–7] and videos [8,9]. However, these multi-layer perceptron (MLP) based INRs are limited to pixel-wise representation. They predict a single point at a time, resulting in an unacceptably slow coding speed when applied to video representation.

Neural Representation for Videos (NeRV) is specifically designed for video, including two branches: Index-based and Hybrid-based. Index-based methods view video as an implicit function $y = f(t)$ where y is the tth frame of video. NeRV [10] first proposes image-wise representation with convolutional layers, achieving better reconstruction quality and greatly improving efficiency. ENeRV [11] Separate the representation into temporal and spatial branches. NVP [12] uses learnable positional features. D-NeRV [13] introducing temporal reasoning and motion information. However, Index-based methods only focus on location information but ignore the visual information. Hybrid-based method CNeRV [14] proposes a hybrid video neural representation with content-adaptive embeddings. HNeRV [15] proposes that auto-encoder is also a hybrid neural representation. The encoder extracts content-adaptive embedding, while the decoder and embeddings are viewed as neural representations. DNeRV [16] introduces difference between frames as motion information. Whereas, existing methods are limited to representing a single video with a fixed-size neural network, which lacks flexibility in practical applications. We adopt the framework proposed in [15] as the baseline and propose a scalable coding method.

Video Compression algorithms such as H.264 [1], HEVC [2] achieve efficient compression performance. The learning-based codecs [3,4] use neural networks to replace certain components but still follow the traditional pipeline. These codecs explicitly encodes videos as latent codes of motion and residual. NeRV methods represent videos as neural networks, and then employ model compression techniques like model pruning and weight quantization to reduce the model size further. It provides a novel pipeline for video compression, achieving fast decoding and satisfactory reconstruction quality with a simpler structure.

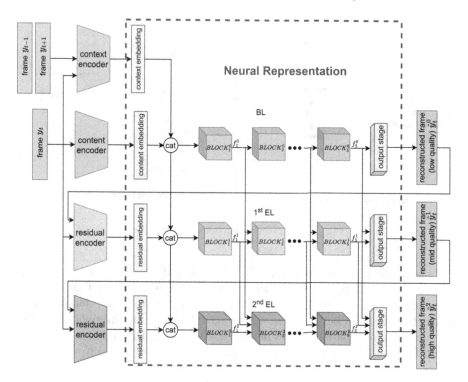

Fig. 2. The pipeline of quality scalable video coding method based on neural representation. Model architecture comprises 3 layers: BL (green part), 1st EL (yellow part) and 2nd EL (red part). The purple part is the implicit neural representation for video. (Color figure online)

Scalable Video Coding (SVC) is an important extension for video compression, e.g., SVC [17] for H.264 and SHVC [18] for HEVC. SVC obtains multilayer streams (BL and ELs) through one encoding to accommodate different conditions. The BL can be independently decoded, and the ELs can be appended to the BL to enhance the output quality. Typical scalability modes include temporal, spatial, and quality scalability. In the NeRV method, the network itself is the video stream. Inspired by multi-scale implicit neural representation [24,25], we propose a hierarchical network for implementing quality scalable coding.

3 Methodology

3.1 Overview

Figure 2 illustrates the overview of our Quality Scalable Video Coding method based on Neural Representation. We follow the hybrid philosophy proposed in [15] that treats the embedding and decoder of the auto-encoder as the INR

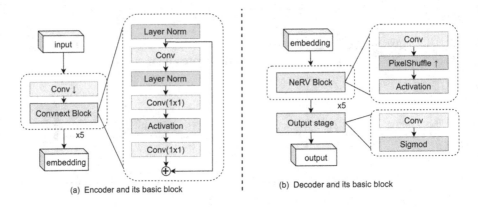

Fig. 3. Structure of encoder and decoder

for videos. Our proposed network is hierarchical and consists of a BL and several ELs. BL learns basic low-frequency information from current frame content and neighboring contextual information. ELs learn high-frequency detailed information from residue between the reconstructed frames and original ones. The modular structure allows for the flexible selection of a BL and any number of ELs to form subnetworks with different sizes, thus achieving quality scalability.

3.2 The Base Layer

The base layer (BL) is the initial layer and can function as an independent neural representation for coarse reconstruction.

The encoder comprises two components: the content encoder and the context encoder. The content encoder extracts the essential content-adaptive embedding $e^{content}$ from the current frame y_t. However, the baseline methods [15] overlook the contextual dependency on adjacent frames, which plays a significant role in video tasks. We draw inspiration from [16] and introduce a context encoder. The context encoder captures short-term context by considering neighboring frames y_{t-1}, y_{t+1}, allowing for an explicit utilization of the temporal correlations.

$$e^{content} = CONTENT\_ENC(y_t)$$
$$e^{context} = CONTEXT\_ENC(y_t, y_{t-1}, y_{t+1}) \tag{1}$$

Each encoder consists of five stages, including a downsampling convolutional layer and a ConvNext block [26], as illustrated in Fig. 3(a). All encoders share the same structure, with variations only in the inputs. This enables all encoders generate the embeddings of the same size, which can be merged through direct concatenation. The dimensions of the embeddings are flexible and determined by the downsampling stride and embedding channels specified in the hyperparameters. We maintain consistency with the baseline approach. For 1920×960 video, the downsampling strides are set as [5, 4, 4, 3, 2], providing the embedding in size of $16 \times 4 \times 2$.

The content embedding $e^{content}$ and context embedding $e^{context}$ are combined as inputs and pass through the decoder. Finally, the output stage maps the final output features into pixel domain and gets the reconstructed frame $\hat{y}_t^0$ as follows:

$$f_1^0 = BLOCK_j^0(e^{content}, e^{context})$$
$$f_j^0 = BLOCK_j^0(f_{j-1}^0) \tag{2}$$
$$\hat{y}_t^0 = Sigmoid(Conv(f_5^0))$$

The decoder includes five blocks, denoted as $BLOCK_1^0$ through $BLOCK_5^0$, and f_j^0 represents the output intermediate feature of the jth block. We utilize NeRV block [10] as the fundamental unit of decoder, as shown in Fig. 3(b). Among these, only the convolutional layer contains trainable parameters, and the pixel shuffle layer serves as upsampling method. To ensure a consistent size between the reconstructed frames and original ones, the upsampling strides are the same as the downsampling strides applied in the encoder.

3.3 The Enhancement Layers

The Enhancement Layers (ELs) builds on the BL to capture progressive levels of details, enabling higher-quality reconstruction.

It is inadequate for ELs to share a single fixed-size embedding with the BL. The ELs primarily learn high-frequency details, but the low-frequency part of embedding leads to poor fitting of the high-frequency information. It is important for each layer's embedding to align with its representation content, and the embedding size should adapt to the changing subnetwork sizes. The high-frequency details that previous layers fail to reconstruct are exactly the residuals. Therefore, we introduce a residual encoder for each EL to extract the residual embedding r^i from the reconstructed frame of the previous layer $\hat{y}_t^{i-1}$ and the original frame y_t. The structure of the residual encoder is the same as the encoders of BL described in the previous section.

$$r^i = RESIDUAL\_ENC(\hat{y}_t^{i-1}, y_t) \tag{3}$$

Regarding the decoder, a straightforward approach is to apply five blocks, similar to the base layer (BL), and aggregate the final features from different layers only in the output stage. In addition, we introduce Inter-layer connections at each block to effectively utilize the information preserved in lower layers, as shown in Fig. 2. Specifically, embeddings and intermediate features from previous layers are included as additional input to each block of EL as follows:

$$f_1^i = Block_j^i(e^{content}, e^{context}, r^1, \cdots r^i)$$
$$f_j^i = Block_j^i(f_{j-1}^0, \cdots f_{j-1}^i) \tag{4}$$
$$\hat{y}_t^i = Sigmoid(Conv(f_5^0, \cdots f_5^i))$$

where, $Block_j^i$ is the jth block of the ith EL, and f_j^i denotes its output intermediate feature. The reconstructed frame, $\hat{y}_t^i$ is the output of the subnetwork consisting of the BL and i ELs.

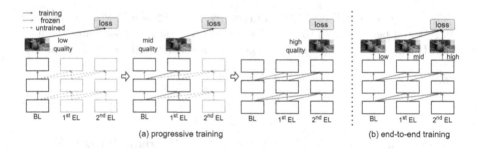

Fig. 4. Training strategy

While the Inter-layer connections require BL to accommodate the representation tasks of higher layers, thereby impacting its performance, ELs can effectively leverage the knowledge acquired from previous layers through this structure to better capture high-frequency residuals. Notably, connections are unidirectional, maintaining the independence of lower layers. Consequently, even without higher ELs, the preceding layers can still serve as a standalone subnetwork representing lower-quality frames. This modular structure allows the BL and ELs to form subnetworks with different sizes, thus providing flexible quality scalability.

3.4   End-to-End Multi-task Training Strategy

There are two training strategies: progressive training [24,27] and end-to-end training. In progressive training, BL is trained first, and then ELs are trained layer by layer with the weights of previous layers frozen, as illustrated in Fig. 4(a). This approach maintains the performance of the previous layers as the training progresses. However, it only considers the current layer, ignoring that each layer is part of larger subnetworks. As a result, this strategy may fail to achieve the global optimal between multiple subnetworks.

Each subnetwork shares specific layers and has similar objective to reconstruct the same frames at different quality. We adopt a multi-task learning strategy that enables adaptive balancing of the objectives across subnetworks. Unlike training each layer sequentially, our strategy trains the entire network end-to-end, as depicted in Fig. 4(b). Each layer is supervised with an individual loss, and these losses are weighted and summed to obtain the total loss. Higher weights are assigned to ELs to emphasize the capture of high-frequency details. The total loss function is expressed in Eq. 5.

$$Loss = \sum_{i=1}^{l} \lambda_i MSE(\hat{y}_t^i, y_t) \tag{5}$$

where l is the total number of layers, MSE is Mean squared error, $\hat{y}_t^i$ represents the reconstructed frame of the ith layer, and y_t represents the original frame, λ_i denotes the weight for the loss of the ith layer.

4 Experiments

4.1 Datasets and Settings

Datasets. Experiments are conducting on the bunny [19] and UVG [20] datasets. The bunny is a video consisting of 132 frames with a 1280×720 resolution. The UVG dataset contains seven videos, in a total of 3900 frames, with a 1920×1080 resolution. To ensure consistency with previous approaches, we apply center-cropping to the videos to match the size of the embeddings. Specifically, we center crop the bunny to 1280×640 and UVG videos to 1920×960.

Settings. The stride list for encoder and decoder is set as [5, 4, 3, 2, 2] for the UVG and [5, 4, 2, 2, 2] for the bunny. We vary the number of channels of the NeRV blocks to obtain representation models of specific sizes. During training, we use the Adam optimizer [28] with cosine learning rate decay. The maximum learning rate is $1e-3$, with 10% of epochs dedicated to warm-up. The batch size is 1. The reconstruction quality is assessed using the peak signal-to-noise ratio (PSNR), while video compression performance is evaluated by measuring the bits per pixel (Bpp) required for encoding the video. All experiments are performed in the PyTorch framework on an RTX 3090 GPU.

Table 1. Video reconstruction on bunny and UVG.

Model	bunny	UVG							
		beauty	bosph	bee	jockey	ready	shake	yach	avg.
NeRV 0.5M	25.77	30.53	28.90	32.05	26.48	20.71	28.41	24.89	27.42
NeRV 1.5M	29.20	32.00	31.09	36.28	28.95	22.79	31.57	26.35	29.86
NeRV 3M	32.67	32.88	33.22	38.44	31.03	24.73	33.52	27.73	31.65
NeRV avg	29.21	31.80	31.07	35.59	28.82	22.74	31.16	26.32	29.64
ENeRV 0.5M	27.07	31.16	29.68	36.10	25.84	20.56	30.99	25.30	28.51
ENeRV 1.5M	31.01	**33.25**	31.11	37.68	27.59	22.36	33.37	26.00	30.19
ENeRV 3M	35.41	**34.06**	33.94	38.59	29.52	24.34	35.30	27.74	31.92
ENeRV avg	31.16	32.82	31.57	37.45	27.65	22.42	33.22	26.34	30.21
HNeRV 0.5M	31.98	31.69	30.49	36.79	26.83	21.02	32.44	25.94	29.31
HNeRV 1.5M	35.57	33.06	**33.06**	38.65	29.79	23.66	34.06	27.85	31.44
HNeRV 3M	37.43	33.58	**34.73**	38.96	32.04	25.74	34.57	29.26	32.69
HNeRV avg	34.99	32.77	**32.76**	38.13	29.55	23.47	33.69	27.68	31.14
QSVCNR(2/3 training) 0.5M	32.07	31.70	30.50	37.10	27.65	21.47	32.30	25.96	29.53
QSVCNR(2/3 training) 1.5M	35.97	32.90	32.83	38.54	30.28	24.02	33.99	27.83	31.48
QSVCNR(2/3 training) 3M	38.02	33.46	34.17	39.12	32.04	25.76	34.97	28.98	32.64
QSVCNR(2/3 training) avg	35.35	32.69	32.50	38.25	29.99	23.75	33.75	27.59	31.22
QSVCNR(full training) 0.5M	**32.38**	**32.02**	**30.60**	37.21	**27.90**	**21.68**	32.44	**26.05**	**29.70**
QSVCNR(full training) 1.5M	**36.25**	33.20	32.99	**38.69**	**30.59**	**24.30**	34.14	**28.00**	**31.70**
QSVCNR(full training) 3M	**38.29**	33.71	34.41	**39.23**	**32.45**	26.04	35.21	29.26	**32.90**
QSVCNR(full training) avg	**35.64**	**32.98**	32.67	**38.38**	**30.31**	24.01	**33.93**	**27.77**	**31.43**

4.2   Main Results

We compare our method QSVCNR with baseline method HNeRV [15] and other implicit methods NeRV [10], ENeRV [11]. Our three-layer network consists of the BL with 0.5M parameters, the first EL with 1M parameters, and the second EL with 1.5M parameters. The network is trained end-to-end for 900 epochs. For other methods [10,11,15], we scale channel width to get three models with sizes of 0.5M, 1.5M and 3M. Each is trained for 300 epochs, totaling 900 epochs. The reconstruction performance is compared on the bunny and UVG datasets, as shown in Table 1. Remarkably, our proposed method, QSVCNR, achieved comparable or even superior performance within just two-thirds of the total epochs. After completing the full training, an average improvement of 0.29 PSNR on UVG dataset is achieved, specifically, +0.39 PSNR for 0.5M model, +0.26 PSNR for 1.5M model, and +0.21 PSNR for 3M model.

Notably, our method only requires storing a single 3M model, while other methods require storing three models, totaling 5M parameters. This saves 40% of storage space. The information represented in models with different sizes is similar, but training three models separately fails to utilize their redundancy and correlation. In our approach, each subnetwork shares lower layers and avoids learning the low-frequency information repeatedly. Hence, within the same total number of training epochs, each layer undergoes more extensive training. Additionally, the hierarchical coarse-to-fine framework and residual embedding enhance the ability of the ELs to capture intricate high-frequency details.

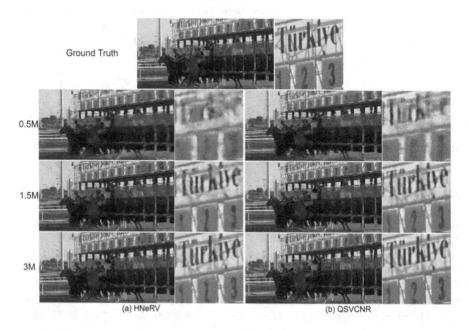

Fig. 5. Video reconstruction visualization. With the same parameters, ours reconstructs videos with better details.

For example, 'Ready' in Fig. 5 illustrates the visualization result of the video reconstruction. At the same memory budget, the improved performance in reconstructing details is evident. Specifically, the numbers and letters in the images exhibit enhanced visual quality.

We apply 8-bit quantization to both the model and the embedding for video compression. Finally, lossless entropy coding is applied. To assess the compression performance, we compare our method QSVCNR with other NeRV methods [10,11,15], as well as traditional methods like H.264 and HEVC [1,2], along with the state-of-the-art SVC method, SHVC [18]. As shown in Fig. 6, QSVCNR exhibits superior performance compared to HNeRV [10](+over 0.3 PSNR) and HEVC [2]. It is worth noting that our experiments are all conducted at small bpp, showing its potential in high-rate video compression.

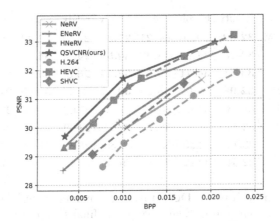

Fig. 6. PSNR vs. BPP on UVG dataset.

4.3   Ablation

Context Encoder. Despite the challenge of modeling long-time contexts with auto-encoder framework, the inclusion of neighbouring frames as short-term context can enhance the reconstruction performance. To demonstrate this, we integrate a context encoder into the baseline method and present the results in Table 2. The incorporation of the context encoder leads to improvements of 0.16 PSNR for the 1.5M model and 0.18 PSNR for the 3M model.

Table 2. Context Encoder ablation on UVG

Model	beauty	bosph	bee	jockey	ready	shake	yach	avg.
HNeRV 1.5M	33.06	33.06	38.65	29.79	23.66	34.06	27.85	31.44
HNeRV 3M	33.58	34.73	38.96	32.04	25.74	34.57	**29.26**	32.69
HNeRV 1.5M +context	**33.11**	**33.14**	**38.68**	**30.24**	**24.12**	**34.10**	**27.96**	**31.62**
HNeRV 3M +context	**33.68**	**34.74**	**39.23**	**32.13**	**25.94**	**35.01**	29.25	**32.85**

Residual Encoder. The residual encoder makes the embedding of each EL consistent with its representation content and helps ELs better fit high-frequency detailed information. To evaluate its effectiveness, we conduct an experiment where we remove the residual encoder from the ELs and allow all layers to share the content and context embedding as input. To ensure a fair comparison, we increased the channels of the content embedding to three times, maintaining the total embedding size unchanged. The results are presented in Table 3. The model without the residual encoder shows an evident decrease over 0.5 PSNR. Each layer represents content of a different frequency, but sharing a common embedding mixes all frequency components together, resulting in poor reconstruction. It is necessary to introduce residual encoders and embeddings for ELs.

Table 3. Residual Encoder ablation on UVG

Model	beauty	bosph	bee	jockey	ready	shake	yach	avg.
QSVCNR 0.5M	**32.02**	**30.60**	**37.21**	**27.90**	**21.68**	**32.44**	**26.05**	**29.70**
QSVCNR 1.5M	**33.20**	**32.99**	**38.69**	**30.59**	**24.30**	**34.14**	**28.00**	**31.70**
QSVCNR 3M	**33.71**	**34.41**	**39.23**	**32.45**	**26.04**	**35.21**	**29.26**	**32.90**
QSVCNR avg	**32.98**	**32.67**	**38.38**	**30.31**	**24.01**	**33.93**	**27.77**	**31.43**
-resiual 0.5M	31.63	30.11	36.54	27.27	21.00	32.23	25.67	29.20
-resiual 1.5M	32.85	32.46	38.35	29.64	23.39	33.99	27.30	31.15
-resiual 3M	33.45	33.80	38.94	31.30	25.02	34.92	28.60	32.29
-resiual avg	32.64	32.12	37.94	29.40	23.13	33.71	27.21	30.88

Inter-layer connection in Decoder. We conduct experiments to remove Inter-layer connections and only aggregate the final feature of layers at output stage. We scale the channels of each block, ensuring that every subnetwork maintains the same total of parameters as before. The results are presented in Table 4. By removing Inter-layer sharing, the BL can focus on representing its specific part, resulting in improved performance. However, the ELs tend to exhibit lower performance. Overall, Inter-layer sharing enables ELs to learn higher-frequency details by leveraging the knowledge gained from the lower layers, leading to better overall performance.

Table 4. Inter-layer connection ablation on UVG

Model	beauty	bosph	bee	jockey	ready	shake	yach	avg.
QSVCNR 0.5M	32.02	30.60	37.21	27.90	21.68	32.44	26.05	29.70
QSVCNR 1.5M	**33.20**	**32.99**	38.69	**30.59**	**24.30**	34.14	28.00	**31.70**
QSVCNR 3M	**33.71**	**34.41**	**39.23**	32.45	**26.04**	35.21	29.26	**32.90**
QSVCNR avg	**32.98**	**32.67**	38.38	**30.31**	**24.01**	33.93	27.77	**31.43**
-interlayer 0.5M	**32.04**	**30.63**	**37.25**	**27.94**	**21.73**	**32.48**	**26.15**	**29.74**
-interlayer 1.5M	33.17	32.95	**38.71**	30.52	24.01	34.11	**28.01**	31.64
-interlayer 3M	33.67	34.39	39.22	32.30	25.90	35.18	**29.28**	32.84
-interlayer avg	32.95	32.65	**38.39**	30.25	23.88	33.92	**27.81**	31.40

Training Strategies. In progressive learning, each layer is individually trained for 300 epochs, and then its weights are frozen before training the next layer. We compare the performance of a 3-layer model trained using progressive training to our end-to-end trained model. The results presented in Table 5 indicate that freezing a portion of the weights harms the representational capability. The end-to-end training of all layers with multi-objective performs better.

Table 5. Training Strategies ablation on UVG

Model	beauty	bosph	bee	jockey	ready	shake	yach	avg.
QSVCNR 0.5M	**32.02**	**30.60**	**37.21**	**27.90**	**21.68**	**32.44**	**26.05**	**29.70**
QSVCNR 1.5M	**33.20**	**32.99**	**38.69**	**30.59**	**24.30**	**34.14**	**28.00**	**31.70**
QSVCNR 3M	**33.71**	**34.41**	**39.23**	**32.45**	**26.04**	**35.21**	**29.26**	**32.90**
QSVCNR avg	**32.98**	**32.67**	**38.38**	**30.31**	**24.01**	**33.93**	**27.77**	**31.43**
progress 0.5M	31.78	30.56	36.69	27.35	21.49	32.62	25.68	29.45
progress 1.5M	32.73	32.25	38.10	29.15	23.01	33.90	27.06	30.88
progress 3M	33.22	33.40	38.66	30.49	24.16	34.65	27.94	31.78
progress avg	32.57	32.07	37.81	28.99	22.88	33.72	26.89	30.70

The experiment confirms that the end-to-end training is preferable. However, in some specific cases where a new EL needs to be added to an already trained network, the progressive training method can serve as an alternative that does not require retraining the entire network.

5 Conclusion

This paper proposes a quality scalable video coding method based on neural representation, in which a hierarchical network comprising a BL and several ELs represents the same video with different qualities simultaneously. Without retraining multiple individual networks, these layers can be gradually combined to form subnetworks with varying sizes. The larger subnetworks share the parameters of small ones to save storage space. In addition, the context encoder improves performance by exploiting temporal redundancy, while the residual encoders and Inter-layer connections effectively enhance the ELs' capability to learn high-frequency details. The entire network can be trained end-to-end, allowing for the dynamic balancing of optimization objectives across different layers. As a result, these subnetworks outperform individually trained networks of the same size.

References

1. Wiegand, T., Sullivan, G.J., Bjontegaard, G., Luthra, A.: Overview of the h. 264/avc video coding standard. IEEE Trans. Circuits Syst. Video Technol. **13**(7), 560–576 (2003)

2. Sullivan, G.J., Ohm, J.R., Han, W.J., Wiegand, T.: Overview of the high efficiency video coding (HEVC) standard. IEEE Trans. Circuits Syst. Video Technol. **22**(12), 1649–1668 (2012)
3. Lu, G., Ouyang, W., Xu, D., Zhang, X., Cai, C., Gao, Z.: DVC: an end-to-end deep video compression framework. In: Proceedings of the IEEE/CVF Conference on Computer Vision and Pattern Recognition, pp. 11006–11015 (2019)
4. Li, J., Li, B., Lu, Y.: Deep contextual video compression. Adv. Neural. Inf. Process. Syst. **34**, 18114–18125 (2021)
5. Dupont, E., Goliński, A., Alizadeh, M., Teh, Y.W., Doucet, A.: COIN: compression with implicit neural representations. arXiv preprint arXiv:2103.03123 (2021)
6. Dupont, E., Loya, H., Alizadeh, M., Goliński, A., Teh, Y.W., Doucet, A.: COIN++: neural compression across modalities. arXiv preprint arXiv:2201.12904 (2022)
7. Strümpler, Y., Postels, J., Yang, R., Gool, L.V., Tombari, F.: Implicit neural representations for image compression. In: European Conference on Computer Vision, pp. 74–91. Springer, Cham (2022). https://doi.org/10.1007/978-3-031-19809-0_5
8. Zhang, Y., van Rozendaal, T., Brehmer, J., Nagel, M., Cohen, T.: Implicit neural video compression. arXiv preprint arXiv:2112.11312 (2021)
9. Rho, D., Cho, J., Ko, J.H., Park, E.: Neural residual flow fields for efficient video representations. In: Proceedings of the Asian Conference on Computer Vision, pp. 3447–3463 (2022)
10. Chen, H., He, B., Wang, H., Ren, Y., Lim, S.N., Shrivastava, A.: NeRV: neural representations for videos. Adv. Neural. Inf. Process. Syst. **34**, 21557–21568 (2021)
11. Li, Z., Wang, M., Pi, H., Xu, K., Mei, J., Liu, Y.: E-NeRV: expedite neural video representation with disentangled spatial-temporal context. In: European Conference on Computer Vision, pp. 267–284. Springer, Cham (2022). https://doi.org/10.1007/978-3-031-19833-5_16
12. Kim, S., Yu, S., Lee, J., Shin, J.: Scalable neural video representations with learnable positional features. Adv. Neural. Inf. Process. Syst. **35**, 12718–12731 (2022)
13. He, B., et al.: Towards scalable neural representation for diverse videos. In: Proceedings of the IEEE/CVF Conference on Computer Vision and Pattern Recognition, pp. 6132–6142 (2023)
14. Chen, H., Gwilliam, M., He, B., Lim, S.N., Shrivastava, A.: CNeRV: content-adaptive neural representation for visual data. arXiv preprint arXiv:2211.10421 (2022)
15. Chen, H., Gwilliam, M., Lim, S.N., Shrivastava, A.: HNeRV: a hybrid neural representation for videos. In: Proceedings of the IEEE/CVF Conference on Computer Vision and Pattern Recognition, pp. 10270–10279 (2023)
16. Zhao, Q., Asif, M.S., Ma, Z.: DNeRV: modeling inherent dynamics via difference neural representation for videos. In: Proceedings of the IEEE/CVF Conference on Computer Vision and Pattern Recognition, pp. 2031–2040 (2023)
17. Schwarz, H., Marpe, D., Wiegand, T.: Overview of the scalable video coding extension of the h. 264/avc standard. IEEE Trans. Circuits Syst. Video Technol. **17**(9), 1103–1120 (2007)
18. Boyce, J.M., Ye, Y., Chen, J., Ramasubramonian, A.K.: Overview of SHVC: scalable extensions of the high efficiency video coding standard. IEEE Trans. Circuits Syst. Video Technol. **26**(1), 20–34 (2015)
19. Big buck bunny. http://bbb3d.renderfarming.net/download.html
20. Mercat, A., Viitanen, M., Vanne, J.: UVG dataset: 50/120fps 4k sequences for video codec analysis and development. In: Proceedings of the 11th ACM Multimedia Systems Conference, pp. 297–302 (2020)

21. Mildenhall, B., Srinivasan, P.P., Tancik, M., Barron, J.T., Ramamoorthi, R., Ng, R.: NeRF: representing scenes as neural radiance fields for view synthesis. Commun. ACM **65**(1), 99–106 (2021)

22. Sitzmann, V., Martel, J., Bergman, A., Lindell, D., Wetzstein, G.: Implicit neural representations with periodic activation functions. Adv. Neural. Inf. Process. Syst. **33**, 7462–7473 (2020)

23. Chen, Y., Liu, S., Wang, X.: Learning continuous image representation with local implicit image function. In: Proceedings of the IEEE/CVF Conference on Computer Vision and Pattern Recognition, pp. 8628–8638 (2021)

24. Cho, J., Nam, S., Rho, D., Ko, J.H., Park, E.: Streamable neural fields. In: European Conference on Computer Vision, pp. 595–612. Springer, Cham (2022). https://doi.org/10.1007/978-3-031-20044-1_34

25. Landgraf, Z., Hornung, A.S., Cabral, R.S.: PINs: progressive implicit networks for multi-scale neural representations. arXiv preprint arXiv:2202.04713 (2022)

26. Liu, Z., Mao, H., Wu, C.Y., Feichtenhofer, C., Darrell, T., Xie, S.: A convnet for the 2020s. In: Proceedings of the IEEE/CVF Conference on Computer Vision and Pattern Recognition, pp. 11976–11986 (2022)

27. Rusu, A.A., et al.: Progressive neural networks. arXiv preprint arXiv:1606.04671 (2016)

28. Kingma, D.P., Ba, J.: Adam: a method for stochastic optimization. arXiv preprint arXiv:1412.6980 (2014)

Hierarchical Bi-directional Temporal Context Mining for Improved Video Compression

Zijian Lin[ID] and Jianping Luo[(✉)][ID]

Guangdong Key Laboratory of Intelligent Information Processing, Shenzhen Key
Laboratory of Media Security and Guangdong Laboratory of Artificial Intelligence
and Digital Economy (SZ), Shenzhen University, Shenzhen, China
linzijian2021@email.szu.edu.cn, ljp@szu.edu.cn

Abstract. Bidirectional video compression leverages information from
both past and future frames to assist in compressing video frames. In this
paper, we propose a novel multi-scale bidirectional context-aware adaptive contextual video compression framework. This framework extracts
bidirectional contextual information across multiple scales and dynamically adjusts video frame encoding based on the temporal interval differences between bidirectional frames. Additionally, we introduce a bidirectional encoding and decoding scheme, which adopts a "Close To
One" access order. Experimental results demonstrate that our proposed
method outperforms traditional video coding standard H.265/HEVC-
HM, as well as advanced deep learning-based video coding frameworks
like DCVC-TCM and B-CANF.

Keywords: Bidirectional video compression · Multi-scale bidirectional
context-aware adaptive · "Close To One" bidirectional encoding and
decoding order

1 Introduction

Currently, video transmission on the internet accounts for a substantial portion
of data traffic, making enhanced video compression efficiency particularly crucial
for efficient data transfer. With the emergence of convolutional neural networks,
deep learning image processing techniques have been continuously advancing.
Benefiting from this progress, research on learning-based video compression is
also evolving. In the past, learning-based video compression primarily involved
coding the residual of video frames. With the introduction of DCVC, the research
direction has shifted towards utilizing contextual information to assist video
frame encoding. Numerous experiments have demonstrated the superiority of
contextual encoding over direct residual coding. Both methods, whether encoding residuals directly or utilizing contextual information, heavily depend on the
accuracy of motion vectors estimated between video frames. Hence, many studies

ⓒ The Author(s), under exclusive license to Springer Nature Switzerland AG 2024
S. Rudinac et al. (Eds.): MMM 2024, LNCS 14554, pp. 410–421, 2024.
https://doi.org/10.1007/978-3-031-53305-1_31

employ techniques such as frame interpolation, optical flow, multiscale processing, and feature domain transformation to enhance the accuracy of motion estimation. However, most of these methods use unidirectional reference frames, and there is relatively limited research on hierarchical bidirectional B-frame encoding, despite the advantage of utilizing past and future information. Therefore, this study proposes a new framework that considers both past and future information, extracts bidirectional contextual information at multiple scales, and employs an adaptive inter-frame time interval encoding model to enhance video compression efficiency using contextual information. Our contributions can be summarized as follows:

- We propose a novel bidirectional encoding-decoding order that significantly improves bidirectional encoding-decoding effectiveness.
- We introduce a new framework that extracts contextual information in a bidirectional multiscale manner and adaptively utilizes contextual information for inter-frame compression.
- Through extensive testing on various datasets, we demonstrate that our encoding framework outperforms existing advanced video compression methods, such as H.265/HEVC-HM [1], DCVC-TCM [19], and B-CANF [6], in terms of PSNR and MS-SSIM.

2   Related Work

At present, most research on lossy learned image compression is based on the groundbreaking image compression framework proposed by the Balle team [2]. This framework replaces traditional quantization with uniform noise and estimates the required bitrate for image coding using the entropy of this noise. They introduced hyperpriors into the entropy model [3], which can be considered as edge information of the image.

In the field of lossy learning-based video compression, DVC [15] stands as a pioneering achievement. This method adopted the aforementioned lossy image compression model to compress motion and residual information, using optical flow for inter-frame motion extraction. Subsequent studies have improved upon DVC. M-LVC [13] optimized motion estimation and image reconstruction by utilizing previously decoded multi-frame information. Scale introduces multiscale spatial flows, capable of recovering higher-quality video frames in scenarios with significant motion compared to traditional optical flow methods. On the other hand, FVC [11] transformed video frames from the pixel space to the feature space, computing the offset map and using deformable convolution in the feature space for motion compensation.

However, these methods rely on compressing residual information. DCVC [12] is the first conditional coding-based framework, utilizing contextual information to aid video frame encoding, replacing traditional residual compression methods, with empirical evidence suggesting improved performance. The team's improved DCVC-TCM [19] extracted contextual information from a fusion of features across different scales, resulting in higher compression ratios. CANF-VC [9] is inspired by ANFIC [8] and utilized the conditional augmented normalizing flow (CANF) as a guiding mechanism for the conditional entropy model. This enables the conditional guide encoder to learn the conditional distribution for inter-frame encoding.

It's worth noting that, in comparison, B-frame coding achieves higher compression efficiency by utilizing both past and future contextual information [6,21,23]. Wu et al. [21] employed frame interpolation to generate B-frames, but due to without motion encoding, leading to suboptimal reconstructed frame quality. Yılmaz et al. [23] utilized optical flow to extract bidirectional reference frame information and performed residual coding. B-CANF [6] presented a pioneering bidirectional encoding framework that employed bidirectional conditionally enhanced normalized flow (ANF) for B-frame encoding. Moreover, B-CANF introduced simulated B* frames for P-frame encoding, thereby concurrently achieving the effect of frame type-adaptive coding. While B-frame encoding indeed offers bitrate optimization, bidirectional video encoding still lacks sufficient attention.

3   Proposed Method

3.1   Overview

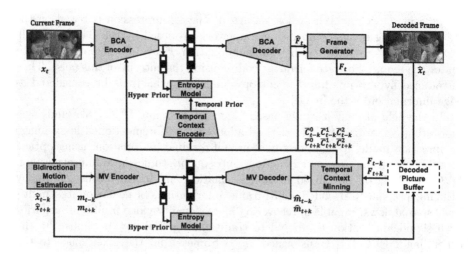

Fig. 1. Overall architecture diagram of our proposed bidirectional video compression framework. The orange modules are used only during the encoding process, while the blue modules are utilized during both the encoding and decoding processes. (Color figure online)

Given the current frame X_t, to acquire bidirectional information for assisting the current frame encoding, we employ reference frames in both the forward and

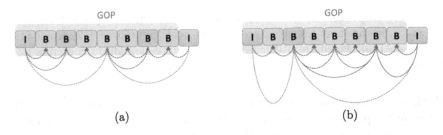

Fig. 2. "Equal Time Intervals" (a) and "Close To One" (b) B-Frame Encoding-Decoding Orders.

backward directions. In a video sequence GOP=8, we designate the first frame of each GOP as an I-frame. In prior methods, the 1st B-frame was positioned equidistantly from two consecutive I-frames. However, this approach could lead to inaccurate motion estimation when the interval between the reference frame and the current frame is significant. We adjusted the encoding-decoding order, as illustrated in the diagram. When the interval is substantial, we adopt the'Close To One' approach, where we prioritize proximity to one of the reference frames to enhance the accuracy of motion compensation in that direction, while the other direction assists in refining the accuracy of contextual information. We present a multi-scale bidirectional contextual video compression framework, with the entire structure illustrated in the diagram. Instead of manually designed modules, we employ trainable neural network modules for various functionalities, including motion estimation, compression, compensation, and residual compression, achieving analogous outcomes. Furthermore, deep context learning for video encoding has demonstrated superiority over residual coding [12]. In our framework, we begin by extracting multi-scale bidirectional contextual information between video frames. This information is then utilized to support the contextual encoder in compressing video frames. Further details are outlined in the following sections (Figs. 1 and 2).

3.2   Multi-scale Bidirectional Contextual Information Prediction

In order to acquire more precise contextual information to assist the contextual encoder-decoder in encoding the current frame, we propose a structure called Multi-scale Bidirectional Contextual Information Prediction. This structure utilizes bidirectional optical flow to generate bidirectional multi-scale contextual information. The details of each module are elaborated as follows:

Motion Estimation. Given the current frame X_t and reference frames X_{t-k} and X_{t-k} in both forward and backward directions, motion estimation aims to estimate the motion vector flow between the reference frames and the current frame, thereby eliminating temporal redundancy in video frame compression. In this approach, we employ Spynet [18], a pre-trained spatial pyramid optical flow network, to extract bidirectional optical flows m_{t-k} and m_{t+k}. The pyramid structure with a large receptive field helps in handling large-scale motion.

Motion Compression. After extracting the bidirectional optical flows m_{t-k} and m_{t+k}, we leverage the structure of mean scale hyperprior [17] provided by CompressAI [4], which are utilized in image compression [3], to compress m_{t-k} and m_{t+k} and reconstruct $\hat{m}_{t-k}$ and $\hat{m}_{t-k}$.

Motion Compensation. Previous methods typically generate prediction frames in the pixel space using optical flows and reference frames through warping. However, due to limitations in the number of feature channels, this approach tends to lose many details. In our approach, we extract the reference frames into a feature space with a channel dimension of M = 64 and warp it in the feature domain according to the optical flow to obtain finer predicted features.

Multi-scale Contextual Mining. Due to the limitations of a single scale, models may struggle to learn diverse information effectively. The outcomes of many previous methods indicate that information extracted at different scales is richer and can enhance model performance. Therefore, our approach also incorporates multi-scale contextual information extraction for bidirectional prediction features. Optical flows and reference features F_{t-k} and F_{t+k} are downsampled and warped at three different scales. At larger scales, channel information primarily focuses on extracting texture and color details from video frames, while at smaller scales, channel information primarily captures information related to significant motion within video frames. To achieve feature fusion across different scales, a refinement network is employed to upsample prediction features from smaller scales and fuse them with prediction features from larger scales, resulting in a more refined output.

3.3   Bidirectional Contextual Adaptive Encoder-Decoder

Previous video coding methods have mostly employed residual compression techniques. However, due to the fact that the entropy of conditional coding is less than or equal to that of residual coding $(H(x_t - \tilde{x}_t) \geq H(x_t|\tilde{x}_t))$, utilizing residual coding methods is not optimal. Recently, in DCVC [12], a method was proposed to utilize acquired contextual information to assist the encoder-decoder in reducing encoding bitrate. Here, given our utilization of bidirectional reference frames, we introduce a bidirectional contextual adaptive encoder-decoder.

Adaptive Context Selection. Considering the potential disparity in time intervals within bidirectional reference frames, the degree of reliance on these frames during encoding and decoding may vary. Before encoding video frames, we incorporate a CA attention module [10] to enable the network to adaptively assign weights to the two reference frames for precise auxiliary information extraction. The Channel Attention (CA) mechanism decomposes channel attention into two parallel 1D feature encoding processes along the x and y directions, effectively integrating spatial coordinate information into the generated attention maps.

Bidirectional Contextual Encoder-Decoder. To achieve higher compression ratios, we hierarchically input the current frame X_t and the contextual information $\bar{C}_{t-k}^n$, $\bar{C}_{t+k}^n$ (n = 0,1,2) into the contextual encoder. As described in DCVC [12], the contextual encoder can autonomously allow the network to learn conditions and facilitate intra-frame compression. Through the contextual encoder, we ultimately obtain a latent code y_t, which is quantized and subsequently decoded by the contextual decoder, producing decoded features. These features are cached for use in subsequent motion compensation, aiding the generation of more accurate motion features. Finally, the reconstructed features are decoded into decoded frames using a frame generator.

Entropy Model. We employ the image compression method [17] from the CompressAI library for bitrate estimation of the context-encoded features. To better approximate the Laplacian distribution of the encoded features, similar to DCVC [12], we provide hyperpriors and the bidirectional temporal prior obtained through the temporal prior contextual encoder to the entropy model. While the autoregressive entropy model can yield certain performance improvements, its serial nature leads to a significant reduction in encoding and decoding speed. Thus, akin to DCVC-TCM [19], we refrain from using this method. Our practical encoding and decoding follow the same method as implemented in the CompressAI library.

3.4 Loss Function

The joint training loss function for the entire network is defined as:

$$L = \lambda D + R = \lambda d(\hat{x}, x) + R_m + R_f \tag{1}$$

Here, $d(\hat{x}, x)$ represents the distortion between the reconstructed frames and the current frame, evaluated by MSE (mean-square-error) or 1-MS-SSIM [14]. For PSNR and MS-SSIM, we set λ values to 256, 512, 1024, 2048, and 8, 16, 32, 64 respectively. R_m represents the bitrate required for compressing quantized motion vectors, and R_f represents the bitrate required for compressing frame features. We employ a progressive training approach to train the entire network.

4 Experiments

4.1 Experimental Setup

Training Data. We utilize the Vimeo-90k [22] septuplet dataset as our training dataset. Each video sequence consists of 7 frames. These frames are cropped randomly to dimensions of 256 × 256.

Testing Data. We conduct testing on the widely used UVG [16], and HEVC B (1080P), C (480P), and D (240P) test sets [20]. The HEVC dataset consists of 13 sequences, including Classe B, C, and D. The UVG dataset contains 7 1080p sequences.

Implementation Details. We train a total of 8 models, each evaluated using the PSNR and MS-SSIM loss metrics. We employ the AdamW optimizer with a batch size of 4. In the training dataset, the 1st and 7th frames are encoded as I-frames and used as bidirectional reference frames for subsequent B-frames. A multi-stage training strategy was adopted. Initially, the MV encoder-decoder is trained with a single B-frame. Subsequently, the MV encoder-decoder is frozen while the remaining models continued training. Finally, the entire model is jointly trained, gradually increasing the number of B-frames trained to 3 and 5 frames. Ultimately, a "cascaded" loss function based on 5 B-frames from a video sequence is employed to fine-tune the entire network.

4.2   Experimental Results

Baseline Methods. Our traditional Baseline Method employs the HM codec, with the low delay P configuration for comparison. For unidirectional predictive learning-based video coding, we evaluate DCVC [12] and DCVC-TCM [19]. We configure the intra period of the unidirectional encoding-decoding methods as 32, which optimally utilizes their P-frame encoding-decoding advantages. Additionally, we conduct comparisons with B-CANF, an advanced bidirectional predictive learning-based video compression framework, we use its recommended settings, the intra-period set to 32 and GOP size set to 16 for all the test sequences. To facilitate a more comprehensive inter-frame encoding comparison, we employ the Cheng2020Anchor [7] deep image compression method from the CompressAI library to encode I-frames. However, we exclud the autoregressive module from Cheng2020Anchor, as while it does offer performance gains, it significantly impacts the encoding and decoding speed.

Table 1. BD-rate for PSNR is HM.

	BD-rate (%) PSNR				BD-rate (%) MS-SSIM			
	HEVC-B	HEVC-C	HEVC-D	UVG	HEVC-B	HEVC-C	HEVC-D	UVG
HM	0.0	0.0	0.0	0.0	0.0	0.0	0.0	0.0
DCVC	53.6	76.0	61.1	54.4	−8.1	−16.2	−31.8	6.9
DCVC-TCM	−6.6	13.8	−0.9	−16.5	−45.3	−48.0	−57.2	−26.5
B-CANF	5.3	18.3	−2.7	1.8	−28.4	−38.7	−52.1	−12.7
ours	−11.3	−13.6	−33.7	−19.0	−51.0	−57.2	−71.0	**−25.2**

Results. Table 1 illustrates the BD-rate [5] results across various PSNR and MS-SSIM metrics on different test datasets, with HM as the anchor. Our approach outperforms other methods across all metrics and test datasets. Specifically, in the HEVC B, C, D, and UVG test sets, our method achieves bitrate savings of 11.3%, 13.6%, 33.7%, and 19.0% in PSNR, and 51.0%, 57.2%, 71.0%, and 25.2% in MS-SSIM. Remarkably, DCVC-TCM and B-CANF even underperform HM on certain datasets and metrics. Figures 4 and 3 depicts the RDO curves for various test datasets with respect to PSNR and MS-SSIM metrics. It is evident from the

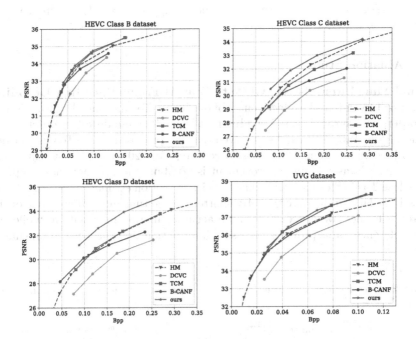

Fig. 3. Rate-Distortion (RD) curves on the HEVC and UVG datasets. The distortion is measured by MS-MSSSIM.

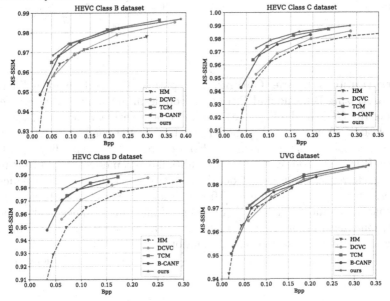

Fig. 4. Rate-Distortion (RD) curves on the HEVC and UVG datasets. The distortion is measured by PSNR.

graphs that our method exhibits minor enhancements compared to DCVC-TCM on the HEVC-B and UVG test sets, while it significantly outperforms DCVC-TCM on the HEVC C, D test sets.

4.3   Ablation Study

"Equal Time Intervals" vs "Close to One" B-Frame Encoding-Decoding Orders. To validate the effectiveness of our 'Close To One' encoding and decoding order, we use the 'Equal Time Intervals' encoding and decoding order as the anchor. From Table 2, it can be observed that our proposed 'Close To One' encoding and decoding order achieves approximately 7%–20% bitrate savings across various test datasets. This improvement is attributed to the fact that adopting the 'Close To One' order avoids inaccuracies in motion estimation resulting from excessively large frame intervals. By shortening the time interval with one of the reference frames, we can extract more accurate motion information, while the other reference frame provides necessary information that the closer frame might lack.

Table 2. Effectiveness of "Close To One" B-Frame encoding-decoding orders of our scheme.

	BD-rate (%) PSNR				BD-rate (%) MS-SSIM			
	HEVC-B	HEVC-C	HEVC-D	UVG	HEVC-B	HEVC-C	HEVC-D	UVG
Equal Time Intervals	0.0	0.0	0.0	0.0	0.0	0.0	0.0	0.0
Close To One	−17.3	−18.9	−20.4	−16.0	−6.2	−7.1	−17.5	−4.3

Bi-directional B-Frame Coding. To contrast the performance difference between utilizing B-frame and P-frame encoding in our model architecture, we configure the GOP of the P-frame encoding model with a value of 32, while the GOP of the B-frame encoding model is set to 8. We conduct tests using the same test set of 96 frames for each video sequence. Due to the disparate GOP settings, B-frame encoding employs 10 additional I-frames compared to P-frame encoding. However, as observed from the results presented in the Table 3, despite the increased number of I-frames, B-frame encoding exhibits a certain degree of performance enhancement over P-frame encoding. This enhancement is particularly pronounced in lower resolutions, which can be attributed to the inherent inaccuracy of motion estimation information in low-quality video frames. The utilization of bidirectional reference frames helps to mitigate this deficiency.

Table 3. Effectiveness of B-frame coding of our scheme.

codec frame type	BD-rate (%)			
	HEVC-B	HEVC-C	HEVC-D	UVG
P frame(GOP = 32)	0.0	0.0	0.0	0.0
B frame(GOP = 8)	−4.0	−26.9	−33.5	−1.5

Multi-scale Architecture. To validate the effectiveness of the multi-scale structure in our approach, we separately train a model without the multi-scale contextual module. In this model, the context encoder only employs bidirectional single-scale contextual assistance for encoding. As shown in Table 4, the incorporation of the multi-scale contextual structure leads to bitrate savings of 24.1%, 12.7%, 8.7%, and 9.0% on the HEVC B, C, D, and UVG test sets, respectively. This improvement is attributed to the model's ability to extract context information at various scales, providing richer contextual assistance to the video frame compression process.

Table 4. Effectiveness of multi-scale architecture of our scheme.

	BD-rate (%)			
	HEVC-B	HEVC-C	HEVC-D	UVG
w/o multi-scale architecture	0.0	0.0	0.0	0.0
w/ multi-scale architecture	−24.1	−12.7	−8.7	−9.0

5   Conclusion

In this work, we propose a novel video compression framework that leverages multi-scale bidirectional contextual information extraction and adapts coding based on frame intervals. Our framework incorporates the 'Close To One' bidirectional B-frame encoding and decoding order to enhance the accuracy of bidirectional contextual information. Experimental results demonstrate that our approach outperforms HM, DCVC-TCM, and B-CANF in terms of bitrate savings.

References

1. HM-16.20. https://vcgit.hhi.fraunhofer.de/jvet/HM/-/tree/HM-1. Accessed 27 June 2023
2. Ballé, J., Laparra, V., Simoncelli, E.P.: End-to-end optimized image compression. arXiv preprint arXiv:1611.01704 (2016)
3. Ballé, J., Minnen, D., Singh, S., Hwang, S.J., Johnston, N.: Variational image compression with a scale hyperprior. arXiv preprint arXiv:1802.01436 (2018)

420   Z. Lin and J. Luo

4. Bégaint, J., Racapé, F., Feltman, S., Pushparaja, A.: Compressai: a pytorch library and evaluation platform for end-to-end compression research. arXiv preprint arXiv:2011.03029 (2020)
5. Bjontegaard, G.: Calculation of average PSNR differences between RD-curves. ITU SG16 Doc. VCEG-M33 (2001)
6. Chen, M.J., Chen, Y.H., Peng, W.H.: B-canf: adaptive b-frame coding with conditional augmented normalizing flows. IEEE Trans. Circuits Syst. Video Technol. (2023)
7. Cheng, Z., Sun, H., Takeuchi, M., Katto, J.: Learned image compression with discretized gaussian mixture likelihoods and attention modules. In: Proceedings of the IEEE/CVF Conference on Computer Vision and Pattern Recognition, pp. 7939–7948 (2020)
8. Ho, Y.H., Chan, C.C., Peng, W.H., Hang, H.M., Domański, M.: Anfic: image compression using augmented normalizing flows. IEEE Open J. Circuits Syst. 2, 613–626 (2021)
9. Ho, Y.-H., Chang, C.-P., Chen, P.-Y., Gnutti, A., Peng, W.-H.: CANF-VC: conditional augmented normalizing flows for video compression. In: Avidan, S., Brostow, G., Cissé, M., Farinella, G.M., Hassner, T. (eds.) ECCV 2022. LNCS, vol. 13676, pp. 207–223. Springer, Cham (2022). https://doi.org/10.1007/978-3-031-19787-1_12
10. Hou, Q., Zhou, D., Feng, J.: Coordinate attention for efficient mobile network design. In: Proceedings of the IEEE/CVF Conference on Computer Vision and Pattern Recognition, pp. 13713–13722 (2021)
11. Hu, Z., Lu, G., Xu, D.: FVC: a new framework towards deep video compression in feature space. In: Proceedings of the IEEE/CVF Conference on Computer Vision and Pattern Recognition, pp. 1502–1511 (2021)
12. Li, J., Li, B., Lu, Y.: Deep contextual video compression. Adv. Neural. Inf. Process. Syst. 34, 18114–18125 (2021)
13. Lin, J., Liu, D., Li, H., Wu, F.: M-LVC: multiple frames prediction for learned video compression. In: Proceedings of the IEEE/CVF Conference on Computer Vision and Pattern Recognition, pp. 3546–3554 (2020)
14. Loshchilov, I., Hutter, F.: Decoupled weight decay regularization. arXiv preprint arXiv:1711.05101 (2017)
15. Lu, G., Ouyang, W., Xu, D., Zhang, X., Cai, C., Gao, Z.: DVC: an end-to-end deep video compression framework. In: Proceedings of the IEEE/CVF Conference on Computer Vision and Pattern Recognition, pp. 11006–11015 (2019)
16. Mercat, A., Viitanen, M., Vanne, J.: UVG dataset: 50/120fps 4k sequences for video codec analysis and development. In: Proceedings of the 11th ACM Multimedia Systems Conference, pp. 297–302 (2020)
17. Minnen, D., Ballé, J., Toderici, G.D.: Joint autoregressive and hierarchical priors for learned image compression. Adv. Neural Inf. Process. Syst. 31 (2018)
18. Ranjan, A., Black, M.J.: Optical flow estimation using a spatial pyramid network. In: Proceedings of the IEEE Conference on Computer Vision and Pattern Recognition, pp. 4161–4170 (2017)
19. Sheng, X., Li, J., Li, B., Li, L., Liu, D., Lu, Y.: Temporal context mining for learned video compression. IEEE Trans. Multim. (2022)
20. Sullivan, G.J., Ohm, J.R., Han, W.J., Wiegand, T.: Overview of the high efficiency video coding (hevc) standard. IEEE Trans. Circuits Syst. Video Technol. 22(12), 1649–1668 (2012)

21. Wu, C.Y., Singhal, N., Krahenbuhl, P.: Video compression through image interpolation. In: Proceedings of the European Conference on Computer Vision (ECCV), pp. 416–431 (2018)
22. Xue, T., Chen, B., Wu, J., Wei, D., Freeman, W.T.: Video enhancement with task-oriented flow. Int. J. Comput. Vision **127**, 1106–1125 (2019)
23. Yılmaz, M.A., Tekalp, A.M.: End-to-end rate-distortion optimized learned hierarchical bi-directional video compression. IEEE Trans. Image Process. **31**, 974–983 (2021)

MAMixer: Multivariate Time Series Forecasting via Multi-axis Mixing

Yongyu Liu, Guoliang Lin, Hanjiang Lai, and Yan Pan[(⊠)]

School of Computer Science and Engineering, Sun Yat-sen University,
Guangzhou, China
{liuyy236,lingliang}@mail2.sysu.edu.cn,
{laihanj3,panyan5}@mail.sysu.edu.cn

Abstract. Sensor data, such as traffic flow monitoring data, constitutes a type of multimedia data. Forecasting sensor data holds significant potential for decision-making. And we can explore its patterns using
time series forecasting methods. In the past few years, Transformer-based
models have gained popularity in multivariate time series forecasting
due to their ability to capture long-range temporal dependencies. Transformer models have quadratic time complexity in the self-attention mechanism, which hinders their efficiency in handling long-term sequences.
The recently proposed PatchTST model has addressed this limitation by
dividing sequences into a series of patches and using patch-wise tokens as
input, which can dramatically reduce the computational complexity in
Transformers. However, PatchTST adopts a channel-independent approach, which ignores the cross-channel interaction within multivariate
time-series data. In this work, we propose Multi-Axis Mixer(MAMixer),
in which we use three layers to capture the spatial, global and local
temporal interactions within multivariate time-series data: a Spatial-
Interaction Layer to capture the interactions among different channels,
a Global-Temporal-Interaction Layer to capture the interactions among
different patches of data, and a Local-Temporal-Interaction Layer to capture the interactions within a patch. We conduct extensive experiments
on various real-world datasets for time-series forecasting. The experimental results demonstrate that the proposed method has superior performance gains over several state-of-the-art methods for time-series forecasting.

Keywords: Multivariate time series forecasting · Multi-layer
perceptron · Sensor data

1 Introduction

In recent years, the ubiquitous employment of sensor technology has resulted in
quantities of sensor data across diverse domains, including traffic flow monitoring [15], energy management systems [9], and weather analysis [6]. Sensor data
prediction has attracted much interest because it can provide invaluable insights

© The Author(s), under exclusive license to Springer Nature Switzerland AG 2024
S. Rudinac et al. (Eds.): MMM 2024, LNCS 14554, pp. 422–435, 2024.
https://doi.org/10.1007/978-3-031-53305-1_32

for decision-making and planning. As a type of multimedia data [5, 21], most sensor data is inherently time series data because it is collected at specific time intervals, capturing sequential observations of a physical phenomenon, system behavior, or environmental conditions over time. Forecasting sensor data with time series methods can capture its underlying trends and relationships.

Multivariate time series forecasting refers to the task of predicting future values of multiple variables over time. As Transformer has demonstrated remarkable success in various applications, such as natural language processing [10], computer vision [12], and speech recognition [11], a surge of Transformer-based models for long-term multivariate time series forecasting tasks have emerged, including Informer [29], Autoformer [26], FEDformer [30], and Pyraformer [18]. These models leverage the attention mechanism [23] or its variants to capture the pair-wise temporal dependencies among time points.

Despite the superior performance of Transformer-based models, there still exist some problems. First, although positional encoding techniques preserve some ordering information, the self-attention mechanism still inevitably leads to the loss of temporal information due to its inherent permutation-invariant nature [27]. Second, the learning of point-wise correlations in Transformers may not truly have interactions at the semantic level. Actually, unlike textual data where words inherently carry semantic meaning, numerical data lacks such inherent semantic context. To solve this issue, PatchTST [19] segments time points into subseries-level patches and applies Transformer architecture. However, it should be noted that PatchTST employs a pure channel-independent approach which does not explicitly model the cross-channel relevance. In real-world scenarios, inter-dependencies among channels are frequently observed and provide valuable information. Consequently, it is crucial to model cross-channel correlations explicitly and effectively. Besides, PatchTST still employs the computationally expensive multi-head self-attention mechanism.

Recently, with the emerge of MLP-Mixers [20], an all-MLP architecture for vision, a series of MLP-based models have been proposed in the computer vision domain [3, 17]. These models are based entirely on multi-layer perceptrons (MLPs) and do not apply computationally intense self-attention mechanism, which have attained competitive performance. For time series forecasting, MLP architecture can preserve the temporal ordering of the input time series and retain semantic information as much as possible [27]. [27] provides extensive empirical validation with DLinear, a simple linear model, which surpasses the performance of the majority of Transformer-based models.

In this work, we propose a novel MLP-based multi-axis mixer architecture for multivariate time series forecasting, named Multi-Axis Mixer(MAMixer). By adopting a patch-based approach akin to PatchTST, MAMixer preserves semantic information. MAMixer aims to capture spatial, global, and local temporal interactions effectively, which is implemented by three layers: a Spatial-Interaction Layer to capture the interactions among different channels, a Global-Temporal-Interaction Layer to capture the interactions among different patches of data, and a Local-Temporal-Interaction Layer to capture the interactions within a patch. The main contributions of this work are as follows:

- We introduce MAMixer, a novel MLP-based architecture designed for multivariate time series forecasting. The framework extracts rich feature information from multiple axes.
- We design an effective block to capture spatial, global, and local temporal interactions separately, and integrate information from all three dimensions to enhance the learning ability of the model.
- Extensive experiments on seven datasets demonstrate that MAMixer achieves competitive performance in multivariate time series forecasting tasks.

2   Related Work

2.1   Transformer-Based Models for Time Series Forecasting

The Transformer architecture [23], known for its ability to capture long-range dependencies, has achieved significant success in diverse domains. While this success has led to a surge of research exploring the application of Transformer-based models in time series forecasting, the inflexibility to support long input sequence caused by the quadratic time and memory complexity is the main bottleneck. Informer [29] addresses the issues by introducing ProbSparse self-attention to reduce complexity and a DMS forecasting strategy. Autoformer [26] introduces a trend and seasonal decomposition architecture with an Auto-Correlation mechanism for data disentanglement and frequency domain transformation. Pyraformer [18] proposes a pyramidal attention module with inter-scale and intra-scale connections with linear complexity. FEDformer [30] adopts a frequency enhanced structure to attain linear complexity. In order to capture comprehensive semantic information, PatchTST [19] and Crossformer [28] use subseries-level patches as input tokens for Transformers, rather than individual time steps. PatchTST adopts channel-independent approach, while Crossformer employs channel-mixing approach.

2.2   MLP-Based Models for Computer Vision

MLP-Mixer [20] applies a simple token-mixing MLP to replace self-attention in ViT [4], composing an all-MLP architecture. It utilizes two types of MLP layers, one for combining the feature of image patches and the other for mixing spatial information. The MLP-Mixer achieves competitive scores on image classification benchmarks. gMLP [17] introduces a spatial gating unit on visual tokens instead of self-attention mechanism. ASMLP [16] pays more attention to local feature interaction by using axial shifted channels of the feature maps. CycleMLP [3] can handle variable image size and provide linear complexity with image size using local windows. Hire-MLP [7] mixes tokens within a local region and across local regions. MAXIM [22] adopts a multi-axis approach that captures both local and global interactions in parallel. DynaMixer [25] dynamically generates mixing matrices by leveraging the contents of all tokens to be mixed, resorting to dynamic information fusion.

2.3   MLP-Based Models for Time Series Forecasting

Recently, the potential of MLPs has also been explored in time series domain. DLinear [27] decomposes the time series into trend and seasonality components and demonstrates the effectiveness of learning a linear mapping from context to horizon, pointing out the limitations of sub-quadratic approximations in the self-attention mechanism. The recently proposed work TSMixer [24] leverages MLP modules to explore various inherent time series characteristics.

3   Methodology

Here we introduce Multi-Axis Mixer(MAMixer), our novel framework for time series forecasting, designed to capture spatial, global, and local temporal interactions. First, we process data with normalization and patching. Second, we use MAMixer Block to capture the interactions. Finally, we predict the future values with linear projection. The overall architecture of MAMixer is shown in Fig. 1(a).

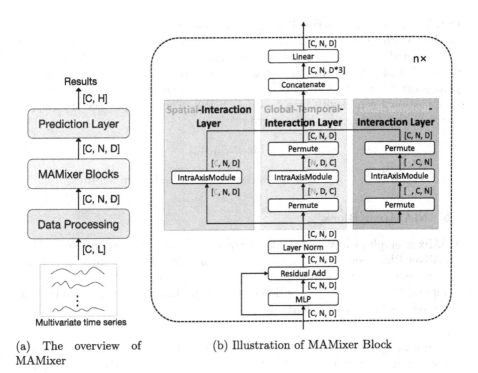

(a)   The   overview   of MAMixer

(b) Illustration of MAMixer Block

Fig. 1. Architecture of the proposed MAMixer

3.1 Preliminary

The multivariate time series is denoted as a feature tensor $\mathcal{X} = [\mathbf{x}_1, \mathbf{x}_2, ..., \mathbf{x}_L] \in \mathbb{R}^{L \times C}$, where L is the look-back window size, C is the number of input sequence channels, and $\mathbf{x}_t \in \mathbb{R}^C$ represents the features across all channels at timestamp t, with $1 \leq t \leq L$. Given the historical L steps, the task involves predicting the subsequent H time steps, with the goal of learning a mapping f:

$$f : [\mathbf{x}_{t-L+1}, ..., \mathbf{x}_t] \to [\mathbf{x}_{t+1}, ..., \mathbf{x}_{t+H}], \tag{1}$$

where $\mathbf{x}_{t+k}$ represents the ground truth value at timestamp $t + k$, $1 \leq k \leq H$.

3.2 Data Processing

Instance Normalization. We employ the recently proposed technique reversible instance normalization (RevIN) [13] to solve the time series forecasting problems against distribution shift, which first normalizes the input sequences before patching and finally denormalizes the model output sequences.

Patching. Input time series $\mathcal{X}$ is segmented into overlapping or non-overlapping patches. With the patch length denoted as P and the stride as S, the tensor $\mathcal{X}$ is segmented into a sequence of patches $\mathcal{X}_p \in \mathbb{R}^{C \times N \times P}$, where N represents the number of patches and is calculated as $N = \lfloor (L - P)/S \rfloor + 2$. The patching process can reduce the number of input tokens to decrease the memory usage and computational complexity by a factor of S [19]. Then the patches are mapped to the hidden features by a linear layer:

$$\mathbf{E} = \text{Linear}_{\text{tail}}(\mathcal{X}_p), \tag{2}$$

where $\mathbf{E} \in \mathbb{R}^{C \times N \times D}$ and D denotes the dimension of hidden features.

3.3 MAMixer Block

MAMixer employs a stack of MAMixer Blocks with skip connections. Each MAMixer Block consists of five layers: one Normalizing Layer, three Interaction Layers(Spatial-Interaction Layer, Global-Temporal-Interaction Layer, Local-Temporal-Interaction Layer) and one Integration Layer. We now introduce the five layers in the following parts.

Normalizing Layer. First, we normalize the data to stabilize the training procedure [2]. Given $\mathbf{E} \in \mathbb{R}^{C \times N \times D}$ as the input features, the input are first normalized as follows:

$$\mathbf{Z} = \text{LN}(\text{MLP}(\mathbf{E}) + \mathbf{E}), \tag{3}$$

where LN represents Layer Normalization [2] and $\mathbf{Z} \in \mathbb{R}^{C \times N \times D}$, MLP consists of two consecutive linear layers with activation layer in the middle.

Interaction Layers. Subsequently, we use Interaction Layers to capture interactions for three axes: spatial axis whose length is the number of channels C, global-temporal axis whose length is the number of patches N, and local-temporal axis whose length is patch feature size D. We feed the features $\mathbf{Z}_1 = \mathbf{Z} = [\mathbf{z}_{1,1}, \mathbf{z}_{1,2}, ..., \mathbf{z}_{1,C}] \in \mathbb{R}^{C \times N \times D}$ into the Spatial-Interaction Layer. For the Global-Temporal-Interaction Layer, the dimensions of $\mathbf{Z}$ are rearranged to yield the input tensor $\mathbf{Z}_2 = [\mathbf{z}_{2,1}, \mathbf{z}_{2,2}, ..., \mathbf{z}_{2,N}] \in \mathbb{R}^{N \times D \times C}$. Regarding the Local-Temporal-Interaction Layer, the tensor $\mathbf{Z}$ is permuted to generate the input tensor $\mathbf{Z}_3 = [\mathbf{z}_{3,1}, \mathbf{z}_{3,2}, ..., \mathbf{z}_{3,D}] \in \mathbb{R}^{D \times C \times N}$.

We use the Intra-Axis Module, which we will explain in Sect. 3.4, to implement interaction:

$$\widetilde{\mathbf{Z}}_1 = \text{IntraAxisModule}(\mathbf{Z}_1), \tag{4}$$

$$\widetilde{\mathbf{Z}}_2 = \text{IntraAxisModule}(\mathbf{Z}_2), \tag{5}$$

$$\widetilde{\mathbf{Z}}_3 = \text{IntraAxisModule}(\mathbf{Z}_3). \tag{6}$$

Then $\widetilde{\mathbf{Z}}_2$ and $\widetilde{\mathbf{Z}}_3$ are transformed into the same shape as $\widetilde{\mathbf{Z}}_1 \in \mathbb{R}^{C \times N \times D}$, denoted as $\widetilde{\mathbf{Z}}'_2 \in \mathbb{R}^{C \times N \times D}$ and $\widetilde{\mathbf{Z}}'_3 \in \mathbb{R}^{C \times N \times D}$.

Integration Layer. In order to mix temporal and spatial information, the outputs of three Interaction Layers are integrated as depicted in Fig. 1(b). Mathematically, this inter-axis interaction is expressed by the following equation:

$$\widetilde{\mathbf{E}} = \text{Agg}([\widetilde{\mathbf{Z}}_1, \widetilde{\mathbf{Z}}'_2, \widetilde{\mathbf{Z}}'_3]), \tag{7}$$

where Agg represents the aggregation function, which can be realized using a linear layer. The result of this process, denoted as $\widetilde{\mathbf{E}} \in \mathbb{R}^{C \times N \times D}$, constructs informative features by combining spatial and temporal information, contributing to the better performance of the MAMixer model.

The procedure for the MAMixer Block is summarized in Algorithm 1.

3.4 Intra-axis Module

The primary objective of the Intra-Axis Module is to capture correlations within the same axis and enhance the representational power of features. In vision domain, Squeeze-and-Excitation(SE) module [8] can recalibrate channel-wise feature responses by learning appropriate channel weights. The SE module consists of two main operations: squeeze and excitation. The features are first squeezed to aggregate information across all spatial locations for each channel. In the excitation operation, generated attention weights adaptively reweight the importance of different channels. SE module can help the network focus more on important channels and suppress less relevant ones.

We adopt the SE module as the Intra-Axis Module, aiming to construct informative features by explicitly modelling inter-dependencies within a single axis. The procedures for spatial interaction, global temporal interaction, and local temporal interaction are the same expect for the axis we focus on.

428   Y. Liu et al.

Algorithm 1: PyTorch-style Pseudo-code for MAMixer Block

```
# initializaiton
Agg_out = nn.Linear(D * 3, D)

# code in forward
def MAMixerBlock(self, E):
  B, C, N, D = E.shape
  Z = LN(MLP(E)+E)
  # Spatial interaction
  Z_C = Z
  Z_C = IntraAxisModule(Z_C)
  # Global temporal interaction
  Z_N = rearrange(Z, 'B, C, N, D -> B, N, D, C')
  Z_N = IntraAxisModule(Z_N)
  Z_N = rearrange(Z_N, 'B, N, D, C -> B, C, N, D'))
  # Local temporal interaction
  Z_D = rearrange(Z, 'B, C, N, D -> B, D, C, N')
  Z_D = IntraAxisModule(Z_D)
  Z_D = rearrange(Z_D, 'B, D, C, N -> B, C, N, D'))
  # Aggregation
  out = Cat(Z_C, Z_N, Z_D)
  return Agg_out(out)
```

To illustrate, let's take the Spatial-Interaction Layer as an example. Given the input $\mathbf{Z}_1 = [\mathbf{z}_{1,1}, \mathbf{z}_{1,2}, ..., \mathbf{z}_{1,C}] \in \mathbb{R}^{C \times N \times D}$, the feature maps are first squeezed and transformed into an axis descriptor $\mathbf{M}_1$. The axis-wise mean value $\mathbf{M}_1 = [m_{1,1}, ..., m_{1,C}] \in \mathbb{R}^C$ is computed by averaging the values of the last two dimensions:

$$m_{1,k} = \frac{1}{N \times D} \sum_{i=1}^{N} \sum_{j=1}^{D} \mathbf{z}_{1,k}(i,j), \tag{8}$$

where $1 \leq k \leq C$ and $m_{1,k}$ serves as the aggregation of temporal information.

Then a simple gating mechanism is introduced, utilizing a sigmoid activation to capture intra-axis dependencies:

$$\mathbf{S}_1 = \sigma(\mathbf{W}_2 \delta(\mathbf{W}_1 \mathbf{M}_1)), \tag{9}$$

where $\delta(\cdot)$ denotes the ReLU [1] function, $\sigma(\cdot)$ represents the sigmoid activation, $\mathbf{S}_1 = [s_{1,1}, s_{1,2}, ..., s_{1,C}] \in \mathbb{R}^C$, and $\mathbf{W}_1$ and $\mathbf{W}_2$ are the learnable parameters.

The final output of the Spatial-Interaction Layer is obtained by rescaling $\mathbf{Z}_1$ with the activations $\mathbf{S}_1$:

$$\tilde{\mathbf{z}}_{1,k} = s_{1,k}\mathbf{z}_{1,k}, \tag{10}$$

where $\tilde{\mathbf{Z}}_1 = [\tilde{\mathbf{z}}_{1,1}, \tilde{\mathbf{z}}_{1,2}, ..., \tilde{\mathbf{z}}_{1,C}]$ and Eq. (9) refers to channel-wise multiplication between the scalar $s_{1,k}$ and the feature map $\mathbf{z}_{1,k}$, where $1 \leq k \leq C$.

The procedures for the Global-Temporal-Interaction Layer and the Local-Temporal-Interaction Layer are the same as the Spatial-Interaction Layer except for the axis of interest. For example, the axis-wise mean value $\mathbf{M}_2$ for $\mathbf{Z}_2$ in Global-Temporal-Interaction Layer is calculated as:

$$m_{2,k} = \frac{1}{C \times D} \sum_{i=1}^{C} \sum_{j=1}^{D} \mathbf{z}_{2,k}(i,j), \tag{11}$$

where $1 \leq k \leq N$.

3.5 Prediction Layer

The prediction layer generates the predicted results $[\hat{\mathbf{x}}_{t+1}, ..., \hat{\mathbf{x}}_{t+H}]$. This prediction layer is constructed by introducing a simple linear transformation applied to the flattened hidden features across all patches.

The mathematical representation of this process can be summarized as follows:

$$[\hat{\mathbf{x}}_{t+1}, ..., \hat{\mathbf{x}}_{t+H}] = \text{Linear}_{\text{head}}(\text{Flatten}(\widetilde{\mathbf{E}})). \tag{12}$$

4 Experiments

We assess the effectiveness of our proposed MAMixer across seven publicly available datasets. In order to show the enhanced performance of MAMixer, we compare it with six benchmark models. Additionally, we perform ablation experiments on the three Interaction Layers to explore their importance. To show the computational efficiency of MAMixer, we also conduct a computational analysis.

Table 1. The statistics of popular datasets for benchmark.

Datasets	Weather	Electricity	Traffic	ETTh1	ETTh2	ETTm1	ETTm2
Channels	21	321	862	7	7	7	7
Timesteps	52696	26304	17544	17420	17420	17420	17420
Granularity	10 min	1 h	1 h	1 h	1 h	5 min	5 min

4.1 Datasets

For our empirical analysis, we conduct experiments on a diverse set of seven real-world time series datasets. These datasets have been widely adopted as benchmark datasets and are publicly accessible via [27]. The statistics of the datasets are shown in Table 1.

The specific datasets are described as follows:

- **Weather Dataset:** This dataset comprises 21 distinct weather indicators, including attributes like air temperature and humidity. Data points are recorded at 10-min intervals throughout the year 2020 in Germany.

Table 2. Multivariate time series forecasting results. The prediction lengths are {96, 192, 336, 720}. The best results are in **bold** and the second best are underlined.

Models		MAMixer		PatchTST/64		DLinear		FEDformer		Pyraformer		Autoformer		Informer	
Metric		MSE	MAE	MSE	MAE	MSE	MAE	MSE	MAE	MSE	MAE	MSE	MAE	MSE	MAE
Weather	96	**0.142**	**0.191**	<u>0.149</u>	<u>0.198</u>	0.176	0.237	0.238	0.314	0.896	0.556	0.249	0.329	0.354	0.405
	192	**0.187**	**0.236**	<u>0.194</u>	<u>0.241</u>	0.220	0.282	0.275	0.329	0.622	0.624	0.325	0.370	0.419	0.434
	336	**0.239**	**0.276**	<u>0.245</u>	<u>0.282</u>	0.265	0.319	0.339	0.377	0.739	0.753	0.351	0.391	0.583	0.543
	720	**0.301**	**0.325**	<u>0.314</u>	<u>0.334</u>	0.323	0.362	0.389	0.409	1.004	0.934	0.415	0.426	0.916	0.705
Electricity	96	**0.127**	**0.222**	<u>0.129</u>	**0.222**	0.140	<u>0.237</u>	0.186	0.302	0.386	0.449	0.196	0.313	0.304	0.393
	192	**0.146**	**0.239**	<u>0.147</u>	<u>0.240</u>	0.153	0.249	0.197	0.311	0.386	0.443	0.211	0.324	0.327	0.417
	336	**0.161**	**0.255**	<u>0.163</u>	<u>0.259</u>	0.169	0.267	0.213	0.328	0.378	0.443	0.214	0.327	0.333	0.422
	720	**0.194**	**0.284**	<u>0.197</u>	<u>0.290</u>	0.203	0.301	0.233	0.344	0.376	0.445	0.236	0.342	0.351	0.427
Traffic	96	**0.354**	<u>0.251</u>	<u>0.360</u>	**0.249**	0.410	0.282	0.576	0.359	2.085	0.468	0.597	0.371	0.733	0.410
	192	<u>0.381</u>	**0.265**	**0.379**	**0.256**	0.423	0.287	0.610	0.380	0.867	0.467	0.607	0.382	0.777	0.435
	336	<u>0.395</u>	**0.269**	**0.392**	**0.264**	0.436	0.296	0.608	0.375	0.869	0.469	0.623	0.387	0.776	0.434
	720	**0.426**	**0.285**	<u>0.432</u>	<u>0.286</u>	0.466	0.315	0.621	0.375	0.881	0.473	0.639	0.395	0.827	0.466
ETTh1	96	**0.367**	**0.395**	<u>0.370</u>	0.400	0.375	<u>0.399</u>	0.376	0.415	0.664	0.612	0.435	0.446	0.941	0.769
	192	**0.405**	<u>0.418</u>	<u>0.413</u>	0.429	**0.405**	**0.416**	0.423	0.446	0.790	0.681	0.456	0.457	1.007	0.786
	336	<u>0.423</u>	**0.432**	**0.422**	<u>0.440</u>	0.439	0.443	0.444	0.462	0.891	0.738	0.486	0.487	1.038	0.784
	720	**0.443**	**0.457**	<u>0.447</u>	<u>0.468</u>	0.472	0.490	0.469	0.492	0.963	0.782	0.515	0.517	1.144	0.857
ETTh2	96	**0.271**	**0.335**	<u>0.274</u>	<u>0.337</u>	0.289	0.353	0.332	0.374	0.645	0.597	0.332	0.368	1.549	0.952
	192	**0.333**	**0.377**	<u>0.341</u>	<u>0.382</u>	0.383	0.418	0.407	0.446	0.788	0.683	0.426	0.434	3.792	1.542
	336	**0.327**	<u>0.385</u>	<u>0.329</u>	**0.384**	0.448	0.465	0.400	0.447	0.907	0.747	0.477	0.479	4.215	1.642
	720	**0.377**	**0.422**	<u>0.379</u>	**0.422**	0.605	0.551	0.412	<u>0.469</u>	0.963	0.783	0.453	0.490	3.656	1.619
ETTm1	96	**0.291**	<u>0.345</u>	<u>0.293</u>	0.346	0.299	**0.343**	0.326	0.390	0.543	0.510	0.510	0.492	0.626	0.560
	192	0.336	0.372	**0.333**	<u>0.370</u>	<u>0.335</u>	**0.365**	0.365	0.415	0.557	0.537	0.514	0.495	0.725	0.619
	336	**0.365**	<u>0.387</u>	<u>0.369</u>	0.392	<u>0.369</u>	**0.386**	0.392	0.425	0.754	0.655	0.510	0.492	1.005	0.741
	720	0.426	**0.418**	0.416	<u>0.420</u>	<u>0.425</u>	0.421	0.446	0.458	0.908	0.724	0.527	0.493	1.133	0.845
ETTm2	96	**0.164**	**0.254**	<u>0.166</u>	<u>0.256</u>	0.167	0.260	0.180	0.271	0.435	0.507	0.205	0.293	0.355	0.462
	192	0.227	**0.296**	**0.223**	**0.296**	<u>0.224</u>	<u>0.303</u>	0.252	0.318	0.730	0.673	0.278	0.336	0.595	0.586
	336	**0.269**	**0.327**	<u>0.274</u>	<u>0.329</u>	0.281	0.342	0.324	0.364	1.201	0.845	0.343	0.379	1.270	0.871
	720	**0.353**	**0.382**	<u>0.362</u>	<u>0.385</u>	0.397	0.421	0.410	0.420	3.625	1.451	0.414	0.419	3.001	1.267

- **Electricity Dataset:** Representing the electricity consumption patterns, this dataset contains hourly consumption records from 321 customers spanning the period between 2012 and 2014.
- **Traffic Dataset:** Focusing on road occupancy rates, this dataset encompasses hourly data collected by sensors installed along San Francisco freeways during the years 2015 and 2016.
- **ETT Datasets:** The Electricity Transformer Temperature (ETT) dataset consists of two distinct subsets, each capturing different temporal granularities. The first subset, ETTh, comprises hourly-level data, while the second subset, ETTm, consists of data recorded at 15-min intervals. Both subsets contain seven relevant oil temperature and load features of electricity transformers. The data spans from July 2016 to July 2018.

Table 3. Ablation results. The prediction lengths are $\{96, 192, 336, 720\}$. The best results are in **bold** and the second best are <u>underlined</u>.

Models		MAMixer		w/o spatial		w/o global		w/o local	
Metric		MSE	MAE	MSE	MAE	MSE	MAE	MSE	MAE
Weather	96	**0.142**	**0.191**	<u>0.148</u>	0.197	**0.142**	<u>0.192</u>	0.142	0.191
	192	<u>0.187</u>	<u>0.236</u>	0.190	0.237	**0.185**	**0.234**	0.185	0.234
	336	0.239	<u>0.276</u>	0.239	0.277	**0.235**	**0.273**	<u>0.236</u>	0.273
	720	**0.301**	**0.325**	0.310	0.330	<u>0.303</u>	<u>0.328</u>	0.311	0.335
Electricity	96	**0.127**	**0.222**	0.130	0.225	0.129	<u>0.223</u>	<u>0.128</u>	<u>0.223</u>
	192	**0.146**	<u>0.239</u>	<u>0.147</u>	0.240	**0.146**	<u>0.239</u>	0.146	0.238
	336	**0.161**	**0.255**	0.164	0.259	**0.161**	**0.255**	<u>0.162</u>	<u>0.256</u>
	720	**0.194**	**0.284**	0.203	0.291	<u>0.197</u>	<u>0.288</u>	0.201	0.291
ETTh1	96	**0.367**	**0.395**	0.367	0.395	0.367	0.395	0.367	0.395
	192	**0.405**	**0.418**	<u>0.406</u>	<u>0.419</u>	<u>0.406</u>	0.420	<u>0.406</u>	0.420
	336	**0.423**	<u>0.432</u>	**0.423**	0.429	<u>0.424</u>	0.429	0424	0.429
	720	**0.443**	**0.457**	<u>0.445</u>	<u>0.462</u>	<u>0.445</u>	<u>0.462</u>	<u>0.445</u>	<u>0.462</u>
ETTh2	96	**0.271**	**0.335**	0.273	<u>0.336</u>	<u>0.272</u>	<u>0.336</u>	<u>0.272</u>	<u>0.336</u>
	192	**0.333**	**0.377**	<u>0.337</u>	<u>0.379</u>	<u>0.337</u>	0.380	<u>0.337</u>	0.380
	336	**0.327**	<u>0.385</u>	<u>0.328</u>	<u>0.385</u>	**0.327**	**0.384**	**0.327**	**0.384**
	720	**0.377**	**0.422**	<u>0.380</u>	0.425	<u>0.380</u>	<u>0.424</u>	<u>0.380</u>	<u>0.424</u>
ETTm2	96	**0.164**	**0.254**	<u>0.165</u>	<u>0.255</u>	0.180	0.265	<u>0.165</u>	<u>0.255</u>
	192	<u>0.227</u>	<u>0.296</u>	**0.223**	**0.295**	**0.223**	**0.295**	**0.223**	**0.295**
	336	**0.269**	0.327	0.272	<u>0.326</u>	<u>0.270</u>	**0.325**	<u>0.270</u>	**0.325**
	720	**0.353**	**0.382**	0.361	<u>0.384</u>	<u>0.360</u>	<u>0.384</u>	<u>0.360</u>	<u>0.384</u>

4.2 Baselines and Experimental Setup

In our comparative analysis, we choose the SOTA Transformer-based models, including **Informer** [29], **Autoformer** [26], **Pyraformer** [18], **FEDformer** [30], **PatchTST** [19], and a recent MLP-based model **DLinear** [27].

In order to maintain consistency across evaluations, all models are subjected to the same experimental configuration with prediction length $H \in \{96, 192, 337, 720\}$. To establish a reliable baseline, we adopt the baseline results from [19], resulting in a look-back window of $L = 512$ for PatchTST, $L = 336$ for DLinear, and $L = 96$ for other Transformer-based models.

Following the methodology of PatchTST, our proposed MAMixer model is instantiated with a look-back window of $L = 512$, a latent space dimensionality of $D = 128$, a patch length of $P = 16$, and a stride size of $S = 8$. The training process encompasses a total of 100 epochs, and the number of MAMixer Blocks was chosen from the set $\{1, 2\}$ via grid search. Adam [14] optimizer is used for training.

To quantitatively assess the performance, we employ two fundamental metrics: the Mean Squared Error (MSE) and the Mean Absolute Error (MAE).

4.3 Main Results

We present the results of our experiments on multivariate time series forecasting, as shown in Table 2. Notably, our MAMixer demonstrates comparable performance. In the Weather dataset, MAMixer exhibits an averaged improvement of 3.7% compared to PatchTST, 11.7% compared to DLinear, 30.0% compared to FEDformer, and 35.1% compared to Autofomer, as measured by the MSE metric. In the Electricity dataset, MAMixer achieves a 1.3% averaged MSE improvement and a 1.1% averaged MAE improvement compared to the most advanced model, PatchTST. Considering dataset size, for small datasets with a lot of noise [24], such as the ETT datasets, MAMxier achieves comparable performance, indicating that MAMixer possesses robustness in effectively handling noisy data. For larger datasets where previous models have performed well, such as the Traffic dataset, MAMixer still yields a 0.4% averaged MSE improvement. The limited improvement could be attributed to the relatively large number of channels in this dataset, making it difficult to effectively capture the complex correlation between channels.

4.4 Ablation Study

In this section, we conduct extensive ablation studies to validate the effectiveness of the Spatial-Interaction Layer, Global-Temporal-Interaction Layer, and Local-Temporal-Interaction Layer. The evaluations are performed on the Weather, Electricity, ETTh1, ETTh2, and ETTm2 datasets. We construct three distinct variants of our MAMixer architecture:

- *w/o* **spatial:** This variant removes the Spatial-Interaction Layer.
- *w/o* **global:** This variant eliminates the Global-Temporal-Interaction Layer.
- *w/o* **local:** This variant excludes the Local-Temporal-Interaction Layer.

The results are comprehensively depicted in Table 3. In summary, all of these interaction components yield positive contributions to the model's performance. It's noteworthy that even when one layer is removed, the remaining interaction layers still demonstrate their effectiveness. Notably, the Spatial-Interaction Layer emerges as a paramount factor, which means that cross-channel relevance acts as a major bottleneck of multivariate time series forecasting. Additionally, temporal interactions exhibit notable importance, as real-world data frequently exhibit hourly and daily periodic patterns. It can be observed that all three Interaction Layers contribute to effectively predicting time series with a length of 720.

4.5 Model Analysis

While PatchTST has managed to mitigate computation and memory usage by patching, its computational burden remains relatively high due to the utilization

Table 4. Comparison of computational efficiency.

Dataset	Models	MACs	Parameter	Epoch time	Memory
Weather	PatchTST	70.64G	1.19M	41.77 s	12124M
	MAMixer	30.83G	1.03M	12.12 s	3954M
Electricity	PatchTST	134.98G	1.19M	291.54 s	21618M
	MAMixer	50.91G	1.23M	75.77 s	5916M

of multi-head self-attention mechanisms. In Table 4, MAMixer exhibits significant computational improvements over PatchTST. To make a comprehensive analysis, we consider the following key metrics: (a) multiply-add cumulative operations on the entire data per epoch (MACs), (b) number of model parameters (Parameter), (c) running time per epoch (Epoch time) and (d) memory occupation during the training process (Memory). To ensure fairness in comparison, both MAMixer and PatchTST models are trained on a single GPU node with identical hardware configurations, using a non-distributed setup for result reporting. Notably, MAMixer, a MLP-based model, effectively reduces MACs, Epoch time and Memory. It's worth noting that inter-channel interactions contribute to an increase in the number of parameters. Specifically, the number of parameter of MAMixer tends to be higher when applied to the Electricity dataset, which encompasses 321 channels. However, in the case of the Weather dataset with 21 channels, MAMixer maintains fewer parameters. Despite the increase in parameters, MAMixer demonstrates significant reductions in MACs, Epoch time, and Memory. This implies that MAMixer achieves comparable performance to PatchTST while considerably reducing training time and memory consumption.

5 Conclusion

In this paper, we introduce MAMixer, a novel framework for multivariate time series forecasting. By incorporating multi-axis interactions, MAMixer effectively captures spatial and temporal dependencies. Experiments across diverse datasets show MAMixer achieves competitive accuracy. Ablation studies confirm the importance of spatial, global, and local temporal interactions. Furthermore, MAMixer exhibits efficiency improvements, including reducing computational complexity, training time, and memory usage compared to benchmark models. MAMixer offers a comprehensive solution for accurate and efficient multivariate time series forecasting. Its ability to achieve a balance between accuracy and efficiency makes it a promising candidate for various forecasting tasks.

Acknowledgements. This work was supported by Guangdong Basic and Applied Basic Research Foundation (2023A1515011400, 2021A1515012172), National Science Foundation of China (61772567, U1811262).

References

1. Agarap, A.F.: Deep learning using rectified linear units (ReLU). arXiv preprint arXiv:1803.08375 (2018)
2. Ba, J.L., Kiros, J.R., Hinton, G.E.: Layer normalization. arXiv preprint arXiv:1607.06450 (2016)
3. Chen, S., Xie, E., Ge, C., Chen, R., Liang, D., Luo, P.: CycleMLP: a MLP-like architecture for dense visual predictions. IEEE Trans. Pattern Anal. Mach. Intell. (2023)
4. Dosovitskiy, A., et al.: An image is worth 16 × 16 words: transformers for image recognition at scale. arXiv preprint arXiv:2010.11929 (2020)
5. Giannakeris, P., et al.: Fusion of multimodal sensor data for effective human action recognition in the service of medical platforms. In: Lokoč, J., et al. (eds.) MMM 2021. LNCS, vol. 12573, pp. 367–378. Springer, Cham (2021). https://doi.org/10.1007/978-3-030-67835-7_31
6. Grover, A., Kapoor, A., Horvitz, E.: A deep hybrid model for weather forecasting. In: Proceedings of the 21th ACM SIGKDD International Conference on Knowledge Discovery and Data Mining, pp. 379–386 (2015)
7. Guo, J., et al.: Hire-MLP: vision MLP via hierarchical rearrangement. In: Proceedings of the IEEE/CVF Conference on Computer Vision and Pattern Recognition, pp. 826–836 (2022)
8. Hu, J., Shen, L., Sun, G.: Squeeze-and-excitation networks. In: Proceedings of the IEEE Conference on Computer Vision and Pattern Recognition, pp. 7132–7141 (2018)
9. Hu, Y.C.: Electricity consumption prediction using a neural-network-based grey forecasting approach. J. Oper. Res. Soc. **68**, 1259–1264 (2017)
10. Kalyan, K.S., Rajasekharan, A., Sangeetha, S.: AMMUS: a survey of transformer-based pretrained models in natural language processing. arXiv preprint arXiv:2108.05542 (2021)
11. Karita, S., et al.: A comparative study on transformer vs RNN in speech applications. In: 2019 IEEE Automatic Speech Recognition and Understanding Workshop (ASRU), pp. 449–456. IEEE (2019)
12. Khan, S., Naseer, M., Hayat, M., Zamir, S.W., Khan, F.S., Shah, M.: Transformers in vision: a survey. ACM Comput. Surv. (CSUR) **54**(10s), 1–41 (2022)
13. Kim, T., Kim, J., Tae, Y., Park, C., Choi, J.H., Choo, J.: Reversible instance normalization for accurate time-series forecasting against distribution shift. In: International Conference on Learning Representations (2021)
14. Kinga, D., Adam, J.B., et al.: A method for stochastic optimization. In: International Conference on Learning Representations (ICLR), San Diego, California, vol. 5, p. 6 (2015)
15. Li, M., Zhu, Z.: Spatial-temporal fusion graph neural networks for traffic flow forecasting. In: Proceedings of the AAAI Conference on Artificial Intelligence, vol. 35, pp. 4189–4196 (2021)
16. Lian, D., Yu, Z., Sun, X., Gao, S.: AS-MLP: an axial shifted MLP architecture for vision. arXiv preprint arXiv:2107.08391 (2021)
17. Liu, H., Dai, Z., So, D., Le, Q.V.: Pay attention to MLPs. In: Advances in Neural Information Processing Systems, vol. 34, pp. 9204–9215 (2021)
18. Liu, S., et al.: Pyraformer: low-complexity pyramidal attention for long-range time series modeling and forecasting. In: International Conference on Learning Representations (2021)

19. Nie, Y., Nguyen, N.H., Sinthong, P., Kalagnanam, J.: A time series is worth 64 words: long-term forecasting with transformers. In: The Eleventh International Conference on Learning Representations (2022)
20. Tolstikhin, I.O., et al.: MLP-mixer: an all-MLP architecture for vision. In: Advances in Neural Information Processing Systems, vol. 34, pp. 24261–24272 (2021)
21. Tsanousa, A., Chatzimichail, A., Meditskos, G., Vrochidis, S., Kompatsiaris, I.: Model-based and class-based fusion of multisensor data. In: Ro, Y.M., et al. (eds.) MMM 2020. LNCS, vol. 11962, pp. 614–625. Springer, Cham (2020). https://doi.org/10.1007/978-3-030-37734-2_50
22. Tu, Z., et al.: Maxim: multi-axis MLP for image processing. In: Proceedings of the IEEE/CVF Conference on Computer Vision and Pattern Recognition, pp. 5769–5780 (2022)
23. Vaswani, A., et al.: Attention is all you need. In: Advances in Neural Information Processing Systems, vol. 30 (2017)
24. Vijay, E., Jati, A., Nguyen, N., Sinthong, G., Kalagnanam, J.: TSMixer: lightweight MLP-mixer model for multivariate time series forecasting. In: ACM SIGKDD International Conference on Knowledge Discovery and Data Mining (2023)
25. Wang, Z., Jiang, W., Zhu, Y.M., Yuan, L., Song, Y., Liu, W.: DynaMixer: a vision MLP architecture with dynamic mixing. In: International Conference on Machine Learning, pp. 22691–22701. PMLR (2022)
26. Wu, H., Xu, J., Wang, J., Long, M.: Autoformer: decomposition transformers with auto-correlation for long-term series forecasting. In: Advances in Neural Information Processing Systems, vol. 34, pp. 22419–22430 (2021)
27. Zeng, A., Chen, M., Zhang, L., Xu, Q.: Are transformers effective for time series forecasting? In: Proceedings of the AAAI Conference on Artificial Intelligence, vol. 37, pp. 11121–11128 (2023)
28. Zhang, Y., Yan, J.: Crossformer: transformer utilizing cross-dimension dependency for multivariate time series forecasting. In: The Eleventh International Conference on Learning Representations (2022)
29. Zhou, H., et al.: Informer: beyond efficient transformer for long sequence time-series forecasting. In: Proceedings of the AAAI Conference on Artificial Intelligence, vol. 35, pp. 11106–11115 (2021)
30. Zhou, T., Ma, Z., Wen, Q., Wang, X., Sun, L., Jin, R.: FEDformer: frequency enhanced decomposed transformer for long-term series forecasting. In: International Conference on Machine Learning, pp. 27268–27286. PMLR (2022)

A Custom GAN-Based Robust Algorithm for Medical Image Watermarking

Kun Zhang[1,2], Chunling Gao[3], and Shuangyuan Yang[1(✉)]

[1] School of Informatics, Xiamen University, Xiamen, China
yangshuangyuan@xmu.edu.cn
[2] National Institute for Data Science in Health and Medicine, Xiamen University, Xiamen, China
[3] Army 73rd Group Military Hospital of PLA, Xiamen, China

Abstract. In recent years, digital healthcare and deep learning have seen substantial advancements, with medical imaging emerging as a crucial component. Despite this progress, the security of these images remains a significant challenge, particularly during network transmission where they are at risk of unauthorized access and tampering. To mitigate these risks, we introduce a novel and robust watermarking algorithm, rooted in advanced deep learning technologies and tailored specifically for medical images. Employing a custom Generative Adversarial Network (GAN), our innovative approach embeds QR codes—encoding confidential, patient-specific data—directly into medical images. Rigorous experimental evaluations confirm the resilience of our solution against a wide array of adversarial attacks and various image distortions, achieving an exceptional average Peak Signal-to-Noise Ratio (PSNR) of 37.12 and an extraction accuracy above 99%. Our algorithm not only enhances the security and integrity of medical images but also fortifies the protection of patient privacy. Importantly, this work fills a research gap by applying Generative Adversarial Networks (GANs) to the domain of end-to-end medical image watermarking.

Keywords: Deep learning · Medical image · Watermarking · Generative adversarial networks (GANs)

1 Introduction

As digital healthcare technologies continue to advance at a rapid pace, the transition from paper-based to digital medical records and images has become increasingly prevalent. While this digital transformation streamlines information exchange processes, it simultaneously raises new challenges in the realm of data security and privacy protection. To address these concerns, a myriad of security solutions have been introduced, including but not limited to, identity verification protocols, data encryption techniques, and digital watermarking methods [2,6,20]. Among these approaches, digital watermarking stands out as a particularly effective means of securing medical images without sacrificing their

© The Author(s), under exclusive license to Springer Nature Switzerland AG 2024
S. Rudinac et al. (Eds.): MMM 2024, LNCS 14554, pp. 436–447, 2024.
https://doi.org/10.1007/978-3-031-53305-1_33

quality. Utilizing advanced algorithms, this technique covertly embeds supplementary information—such as hospital identifiers or patient-specific data—into various domains within the image. It is widely considered the optimal strategy for protecting sensitive medical data in insecure transmission environments [4]. The watermark embedding process is schematically presented in Fig. 1.

While conventional watermarking techniques, based on transformation and hybrid domains, have achieved notable successes [9], the burgeoning complexity and increasing frequency of medical image exchange, spurred by advancements in multimedia technologies, impose new challenges on these traditional, manually designed methods. Deep learning, a highly effective machine learning paradigm capable of automatically distilling high-level and abstract features from data, has emerged as a powerful tool across multiple medical application scenarios. These include, but are not limited to, medical image segmentation, classification, and computer-aided diagnosis [18].

Given this backdrop, this paper introduces a end-to-end deep learning framework centered on Generative Adversarial Networks (GANs) [8]. This framework securely integrates QR codes, containing confidential patient information, into medical images, while enabling precise extraction of the embedded watermark. This ensures both the security and integrity of the data throughout its transmission. Relative to existing solutions, the proposed method demonstrates superior efficacy. The key contributions of the work are:

- Propose a custom GANs algorithm, thereby filling a research void in applying Generative Adversarial Networks (GANs) to the field of medical image watermarking, particularly with respect to end-to-end watermarking algorithms.
- Compared to alternative approaches, the feature extraction method utilizing Convolutional Neural Networks (CNNs) boasts lower computational complexity and enhanced extraction performance.
- The development of a high-performing, CNN-based GAN model for watermark embedding, leveraging the connectivity mechanisms of DenseNet [10].
- The introduction of a custom-designed attack network, named "AttackNet", aimed at bolstering the model's robustness against prevalent image attacks.

The remainder of the paper is structured as follows: Sect. 2 revisits relevant work on digital watermarking in medical images; Sect. 3 provides a detailed description of the proposed watermarking technique; Sect. 4 covers experimental setups and results; Sect. 5 concludes this study.

2 Related Work

Medical image watermarking can be primarily classified into two domains: spatial domain and transform domain watermarking techniques [9].

2.1 Spatial Domain Techniques

Within the realm of spatial domain watermarking, information is directly embedded into the host image's pixel matrix. Methods typically leverage the least

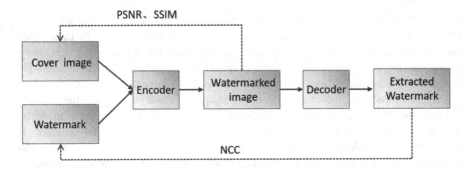

Fig. 1. Watermarking Process

significant bits (LSBs) of the pixel values to ensure minimal degradation of image fidelity. Despite their computational efficiency and elevated watermarking capacity, spatial domain techniques are notably susceptible to issues such as lossy compression and additive noise [19]. Representative algorithms in this category include Least Significant Bit (LSB), Local Binary Pattern (LBP), and Histogram modification. Singh et al. proposed an enhanced LSB technique to safeguard medical data, specifically addressing identity authentication challenges in telemedicine applications [17]. Liu et al. introduced a robust, discriminative zero-watermarking mechanism, founded on Complete Local Binary Pattern (CLBP) operators, to augment the discernibility and robustness of medical imagery [15]. Kelkar et al. employed conventional histogram shifting to augment both the capacity and the concealment of watermarking by segmenting medical images and selecting non-intrusive regions for watermark insertion [12].

2.2 Transform Domain Techniques

In the transform domain, images undergo predefined mathematical transformations, such as the Discrete Fourier Transform (DFT), Discrete Cosine Transform (DCT), Discrete Wavelet Transform (DWT), or Singular Value Decomposition (SVD), prior to the embedding of watermark information into the transformed coefficients. Transform domain watermarking offers enhanced robustness but sacrifices embedding capacity and ease of implementation compared to spatial domain methods. Cedillo-Hernandez et al. designed a watermarking scheme that embeds the watermark into the mid-frequency coefficients of the DFT domain, thereby sidestepping the necessity for separating medical images from Electronic Patient Record (EPR) data [5]. Yang et al. amalgamated perceptual hashing with zero-watermarking technologies and edge detection to introduce a Zernike-DCT-based robust watermarking algorithm [23]. Kahlessenane et al. proposed an integrative, blind watermarking mechanism that embeds electronic medical records into computed tomography scans via a discrete wavelet transform and a saw-tooth scanning topology [11]. Zermi et al. employed DWT decomposition to create a comprehensive watermarking strategy that integrates watermarking into various singular sub-band matrices, thereby incorporating hospital-specific and patient-specific metadata into medical images [24].

Emerging hybrid domain techniques fuse multiple watermarking domains to offer enhanced robustness and embedding efficiency, albeit at the cost of increased computational overhead [1,3,22]. The aforementioned conventional watermarking technique, while transparent and interpretable, often necessitate manual feature selection, posing a challenge in optimizing between embedding capacity, robustness, and perceptual invisibility. To surmount this limitation, Gong et al. introduced a deep-learning-based robust zero-watermarking approach, employing low-frequency DCT features as labels and using a Residual-DenseNet for watermark generation. Enhanced the robustness of Conventional DCT-based Watermarking Algorithms for Medical Imagery in the Face of Geometric Attacks [7]. Nonetheless, the reliance on Manually Selected Low-Frequency DCT Features limits the ability to fully capture the intrinsic characteristics of the images, suggesting potential for further enhancement in watermark embedding performance.

Against this backdrop, we present a novel end-to-end watermarking framework based on Generative Adversarial Networks (GANs). This approach securely embeds QR codes containing sensitive patient data into medical images. Subsequently, a bespoke adversarial network is utilized to subject the watermarked images to a gamut of attack scenarios, thereby robustly assessing and enhancing their resilience. Our methodology not only ensures precise watermarking retrieval from medical images exposed to diverse forms of perturbations but also contributes to the ongoing development of medical image watermarking techniques.

3 Methods

This paper proposes a novel end-to-end algorithm for robust watermarking of medical images using deep learning. Utilizing Generative Adversarial Networks (GANs) for feature learning, our method bypasses manual feature extraction, unlike traditional techniques. We further employ DenseNet connectivity for watermark stealth and introduce a custom Attack Network (AttackNet) for resilience. The overarching framework of the algorithm is depicted in Fig. 2.

3.1   Propounded Watermarking Algorithm

In this section, we delineate our novel algorithmic framework, termed "GDA". The architecture synergistically incorporates Generative Adversarial Networks (GANs), Dense Convolutional Networks (DenseNet), and a bespoke Attack Network (AttackNet).

As depicted in Fig. 2, the GDA framework utilizes cover medical images along with randomly generated noise as inputs for the generator module (G). These inputs are processed through a meticulously engineered Encoder, which is a convolutional neural network that capitalizes on the robust feature extraction capabilities of DenseNet. The Encoder effectively outputs medical images embedded

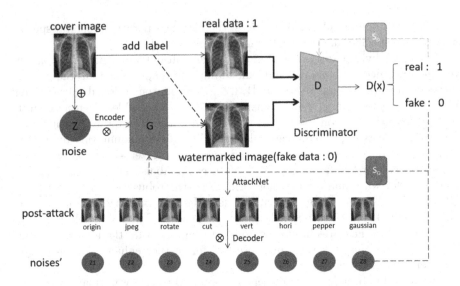

Fig. 2. The overarching framework of the proposed algorithm

with watermarks, thereby fulfilling the objective of efficient watermark embedding.

The algorithm interprets watermarking as an unsupervised learning task, assigning labels "1" for original and "0" for watermarked images. The generator aims to deceive a discriminator (D), which in turn evaluates the watermarked images' quality.

To bolster the robustness of the watermarking, an auxiliary Attack Network (AttackNet) is introduced during training. It subjects the watermarked images to various attacks, as outlined in Table 1, producing compromised images that feed into a Decoder model for watermark extraction.

As depicted in Fig. 3, the architecture starts with a $1 \times 3 \times 512 \times 512$ medical image, which is transformed and combined with upsampled noise to form a $1 \times 17 \times 512 \times 512$ tensor. This feeds into a Densely Connected Encoder with subsequent layers including convolution, ReLU, and Batch Normalization. Post-Encoder, the watermarked image undergoes attacks in the AttackNet, and finally passes through a Decoder model that generates noise images for each attack type.

3.2   Loss Function

The Encoder and Decoder share a single Adam optimizer, while the Discriminator utilizes a separate Adam optimizer. The overall training loss for the proposed watermarking algorithm is defined as per Eq. 1. Which comprises two components: the loss for the Encoder and Decoder, denoted as $Loss_{E-D}$, and the loss for the Discriminator, represented as $Loss_D$.

$$Loss = Loss_{E-D} + Loss_D \qquad (1)$$

Table 1. Types and intensity of attacks in AttackNet.

Type	Intensity
JPEG compress	Q = 90
Rotate	$-2°$ to $2°$
Cut	Randomly cropping 5×512 pixel blocks
Vertical	Translate 5 pixels vertically
Horizontal	Translate 5 pixels horizontally
Salt & Pepper	0.001
Gaussian	mean = 0.01, std = 0.002

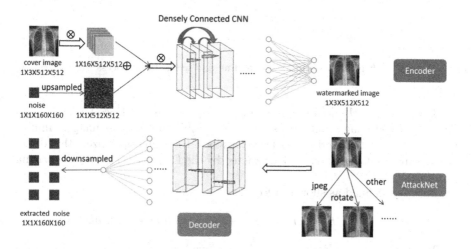

Fig. 3. The structure of GDA

The $Loss_{E-D}$ is defined according to Eq. 2.

$$Loss_{E-D} = \lambda MSE(x, y) + \beta \sum_{0}^{n} \text{BCE\_Logits}(w, wi) \qquad (2)$$

Wherein x represents the cover image, y represents the watermarked image, w represents the original watermark, while wi denotes the watermark extracted after being subjected to the i-th type of attack. λ and β are hyperparameters. BCE_Logits is a loss function that combines the Sigmoid activation function and Binary Cross-Entropy loss. The Sigmoid activation function maps any real number to between 0 and 1, while Binary Cross-Entropy is used to measure the 'distance' between the model's predictions and the true labels, as defined in Eq. 5. MSE stands for Mean Squared Error, which is used to measure the difference between two images or signals, defined as follows.

$$MSE(x, x') = \frac{1}{HW} \sum_{i=1}^{H} \sum_{j=1}^{W} (x(i,j) - x'(i,j))^2 \qquad (3)$$

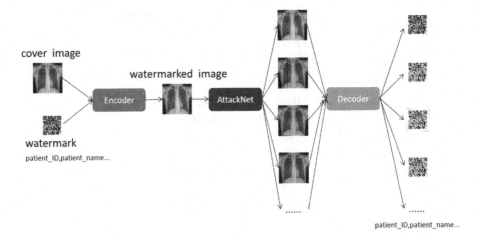

cover image

watermarked image

watermark
patient_ID,patient_name...

patient_ID,patient_name...

Fig. 4. Watermark embedding and extraction process

The design of the loss function, $Loss_{E-D}$, enables the Encoder to generate a watermarked image that is as similar as possible to the cover image. Simultaneously, since the Encoder and Decoder share the same optimizer, this design also ensures that the watermark extracted by the Decoder is close to the original watermark.

$Loss_D$ is defined as shown in Eq. 4.

$$Loss_D = BCE(cover, label\_1) + BCE(watermarked, label\_0) \qquad (4)$$

The Binary Cross-Entropy Loss is employed for binary classification tasks, defined as in Eq. 5. y is the model's predicted probability, and $\hat{y}$ represents the actual binary target value, which is typically either 0 or 1. BCE serves to discriminate between authentic cover images, labeled as '1' and watermarked images, labeled as '0'. During training, the objective is to minimize $Loss_D$ enhance the discriminator's ability to correctly identify cover images as being near '1' and watermarked images as being close to '0'. This, in turn, indirectly bolsters the Encoder's capacity within the generator to produce watermarked images that closely resemble the cover images, thereby improving watermark embedding capabilities.

$$BCE(y, \hat{y}) = -(y \cdot \log(\hat{y}) + (1 - y) \cdot \log(1 - \hat{y})) \qquad (5)$$

3.3   Watermark Embedding and Extraction

We introduce a test scenario to illustrate the embedding and extraction process of the proposed watermarking algorithm. The overall procedure is depicted roughly in Fig. 4. We embed a QR code containing sensitive information such as patient IDs and names into a cover medical image used for testing, using an Encoder.

Subsequently, the watermarked image is fed into AttackNet, which generates a series of post-attack watermarked images. Ultimately, through the Decoder, we successfully extract various post-attack QR codes. We have achieved a 100% extraction accuracy of the embedded patient privacy data across all types of attacks.

4 Experiment

4.1 Experimental Setup and Results

We introduce a medical image watermarking scheme that has been rigorously evaluated and validated across a broad range of medical imaging modalities, including X-rays, CT scans, and MRI images. For our experiments, we utilize medical images with dimensions of 512×512 pixels as the cover images. These are watermarked with custom-generated 160×160 pixel QR codes that embed patient-sensitive data. To quantitatively assess the imperceptibility of the watermark and the visual similarity between the original and watermarked medical images, we employ metrics such as Peak Signal-to-Noise Ratio (PSNR) and Structural Similarity Index (SSIM). NCC refers to Normalized Cross-Correlation, a metric employed to quantify the similarity between the watermark to be embedded and the extracted watermark. The definition is provided in Eq. 6. The value of the Normalized Cross-Correlation (NCC) approaching 1 indicates greater similarity between the two images or signals, with the minimum value being 0.

$$NCC(f,g) = \frac{\sum_{x,y}[f(x,y) \cdot g(x,y)]}{\sqrt{\sum_{x,y} f(x,y)^2 \cdot \sum_{x,y} g(x,y)^2}} \tag{6}$$

The value of PSNR quantifies the amount of noise introduced into the cover image during the watermarking process. The PSNR value decreases as the image distortion increases. Its definition is given in Eq. 7.

$$PSNR = 20 \cdot \log_{10}\left(\frac{MAX_I}{\sqrt{MSE}}\right) \tag{7}$$

wherein MAX_I represents the maximum possible pixel value for the image.

SSIM (Structural Similarity Index) is a metric that measures the similarity between two images in a way that is closer to human perception. It takes values in the range of $[-1, 1]$, with a maximum value of 1 representing identical images. It is defined as Eq. 8.

$$SSIM(x,y) = \frac{(2\mu_x\mu_y + c_1)(2\sigma_{xy} + c_2)}{(\mu_x^2 + \mu_y^2 + c_1)(\sigma_x^2 + \sigma_y^2 + c_2)} \tag{8}$$

The training workflow is delineated in Fig. 3. Our training dataset comprises 5,126 X-ray medical images [13]. During training, the number of epochs is set to

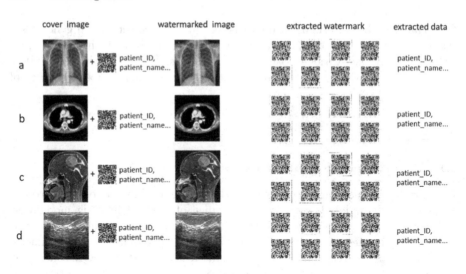

Fig. 5. Experimental results on X-ray, CT, MRI and US

40. Both the Encoder and the Decoder employ a shared Adam optimizer, as does the Discriminator, with all having a learning rate of 0.0003. The hyperparameter λ is set at 700. The attack weights β_1, β_2 ..., β_8 are specified as 0.7, 0.18, 0.02, 0.02, 0.02, 0.02, 0.02, 0.02, respectively. Here, β_1 represents the weighting factor for watermark loss in the absence of any attacks, whereas β_2 through β_8 correspond to the various types of attack weights as outlined in Table 1. Additionally, we have evaluated the watermark embedding performance of our proposed algorithm on other types of medical images, as shown in Fig. 5.

As illustrated in Fig. 5, panels a, b, c, and d represent four different types of medical images: X-ray, CT, MRI, and US, respectively. Table 2 enumerates the performance metrics—PSNR, SSIM, ACC, NCC and the NCC after various attacks—that our algorithm achieves on these medical images. Wherein "ACC" represents the accuracy rate of obtaining patient privacy data from the extracted QR code, which is more crucial than NCC in practical applications. Notably, the PSNR consistently exceeds 36.5, the SSIM surpasses 0.97, the NCC values are all above 0.98, and most importantly, the value of Acc is consistently 1, indicating the complete and accurate extraction of patient privacy data. These results convincingly demonstrate the robust and effective watermarking capabilities of our proposed algorithm across multiple types of medical images.

4.2 Ablation Experiment

In this section, we conduct an ablation experiment to evaluate the importance of each component in our proposed GDA model. Specifically, we will examine the effect of Densely Connected and AttackNet. The results of the ablation study

Table 2. Experimental results on X-ray, CT, MRI, and US, respectively.

	PSNR	SSIM	ACC	NCC	JPEG	Rotate	Cut	Verti	Hori	S & P	Gaussian
a	37.23	0.985	1	0.989	0.955	0.980	0.985	0.983	0.986	0.986	0.988
b	36.51	0.971	1	0.983	0.921	0.957	0.961	0.959	0.960	0.961	0.962
c	36.79	0.973	1	0.982	0.937	0.965	0.967	0.961	0.968	0.969	0.971
d	36.94	0.974	1	0.981	0.939	0.973	0.978	0.972	0.973	0.979	0.980

are presented in Table 3, where "G" signifies Generative Adversarial Networks (GANs), "D" denotes Densely Connected Networks, and "A" represents Attack-Net.

Table 3. Experimental results of the ablation study.

	PSNR	SSIM	ACC	NCC	JPEG	Rotate	Cut	Verti	Hori	S & P	Gaussian
G	35.10	0.964	1	0.982	0.518	0.956	0.948	0.957	0.958	0.958	0.963
G+D	37.24	0.986	1	0.993	0.635	0.964	0.956	0.959	0.957	0.958	0.959
G+D+A	37.12	0.984	1	0.991	0.951	0.989	0.983	0.985	0.984	0.985	0.986

The ablation study follows the same experimental setup as outlined in Sect. 4.1. Insights gleaned from Tables 3 reveal that upon incorporating Densely Connected features, the PSNR and SSIM metrics improved by 2.14 and 2.2%, respectively. Furthermore, with the subsequent addition of AttackNet, the average NCC across the majority of attacks increased by 3%. Under the most frequently encountered JPEG compression attacks, the NCC exceeded 95%, ensuring the accurate extraction of QR code content, which is highly significant. These ablation study findings underscore the pivotal roles that both Densely Connected Networks and AttackNet play in enhancing the stealth and robustness of the watermark embedding process.

4.3   Comparative Experiment

To better understand the effectiveness of our watermarking algorithm, we compared it with advanced algorithms in medical image watermarking, as shown in Table 4. The results clearly indicate that our method outperforms [14,21,22] in the PSNR and SSIM metrics, confirming its superior stealth in watermark embedding. Likewise, exhibits an average NCC exceeding 0.98 under both salt-and-pepper and Gaussian noise, outperforming [14,16], demonstrating its robustness.

Table 4. Comparative experimental outcomes relative to existing algorithms.

	PSNR	SSIM	NCC/ACC	JPEG(Q = 90)	S & P(0.001)	Gaussian(0,0.001)
[22]	36	0.984	1	0.999	0.994	0.992
[16]	45.42	–	0.998	0.993	0.966	0.934
[14]	34.64	–	–	0.999	0.994	0.959
[21]	35.52	0.988	0.933	0.989	–	–
Proposed	37.03	**0.989**	1	0.952	0.979	**0.992**

5 Conclusion

This paper proposed a novel, GAN-based robust watermarking algorithm tailored for medical images, capable of embedding QR codes that encode confidential patient-specific data. This ensures the security and integrity of medical data during transmission. Distinctively, the propounded algorithm adopts an end-to-end watermarking scheme, diverging from the majority of existing deep-learning-based medical image watermarking algorithms. To further enhance the stealth of watermark embedding and the robustness of watermark extraction, we incorporate a Densely Connected architecture and introduce a custom-designed attack network, termed 'AttackNet'. Benchmarking against contemporary advanced medical image watermarking methods reveals that our algorithm excels in performance metrics, substantiating its superiority.

Acknowledgments. The work was supported by the Natural Science Foundation of Fujian Province of China (No. 2022J01003).

References

1. Anand, A., Singh, A.K.: An improved DWT-SVD domain watermarking for medical information security. Comput. Commun. **152**, 72–80 (2020)
2. Anand, A., Singh, A.K.: Watermarking techniques for medical data authentication: a survey. Multimed. Tools Appl. **80**, 30165–30197 (2021)
3. Balasamy, K., Suganyadevi, S.: A fuzzy based ROI selection for encryption and watermarking in medical image using DWT and SVD. Multimed. Tools Appl. **80**(5), 7167–7186 (2021)
4. Boujerfaoui, S., Riad, R., Douzi, H., Ros, F., Harba, R.: Image watermarking between conventional and learning-based techniques: a literature review. Electronics **12**(1), 74 (2022)
5. Cedillo-Hernandez, M., Garcia-Ugalde, F., Nakano-Miyatake, M., Perez-Meana, H.: Robust watermarking method in DFT domain for effective management of medical imaging. SIViP **9**, 1163–1178 (2015)
6. El-Saadawy, A., El-Sayed, A., ElBery, M., Roushdy, M.I.: Medical images watermarking schemes-a review. Int. J. Intell. Comput. Inf. Sci. **21**(1), 119–131 (2021)
7. Gong, C., Liu, J., Gong, M., Li, J., Bhatti, U.A., Ma, J.: Robust medical zero-watermarking algorithm based on residual-densenet. IET Biometrics **11**(6), 547–556 (2022)

8. Goodfellow, I., et al.: Generative adversarial nets. In: Neural Information Processing Systems (2014)

9. Gull, S., Parah, S.A.: Advances in medical image watermarking: a state of the art review. Multimed. Tools Appl. 1–41 (2023)

10. Huang, G., Liu, Z., Van Der Maaten, L., Weinberger, K.Q.: Densely connected convolutional networks. In: Proceedings of the IEEE Conference on Computer Vision and Pattern Recognition, pp. 4700–4708 (2017)

11. Kahlessenane, F., Khaldi, A., Kafi, R., Euschi, S.: A DWT based watermarking approach for medical image protection. J. Ambient. Intell. Humaniz. Comput. 12(2), 2931–2938 (2021)

12. Kelkar, V., Tuckley, K., Nemade, H., et al.: Novel variants of a histogram shift-based reversible watermarking technique for medical images to improve hiding capacity. J. Healthcare Eng. 2017 (2017)

13. Kermany, D., Zhang, K., Goldbaum, M., et al.: Labeled optical coherence tomography (OCT) and chest X-ray images for classification. Mendeley Data 2(2), 651 (2018)

14. Kumar, C., Singh, A.K., Kumar, P.: Improved wavelet-based image watermarking through SPIHT. Multimed. Tools Appl. 79(15–16), 11069–11082 (2020)

15. Liu, X., Lou, J., Wang, Y., Du, J., Zou, B., Chen, Y.: Discriminative and robust zero-watermarking scheme based on completed local binary pattern for authentication and copyright identification of medical images. In: Medical Imaging 2018: Imaging Informatics for Healthcare, Research, and Applications, vol. 10579, pp. 381–389. SPIE (2018)

16. Mehta, R., Rajpal, N., Vishwakarma, V.P.: A robust and efficient image watermarking scheme based on Lagrangian SVR and lifting wavelet transform. Int. J. Mach. Learn. Cybern. 8, 379–395 (2017)

17. Ogundokun, R.O., Abikoye, O.C.: A safe and secured medical textual information using an improved LSB image steganography. Int. J. Digit. Multimed. Broadcasting 2021, 1–8 (2021)

18. Rguibi, Z., Abdelmajid, H., Zitouni, D.: Deep learning in medical imaging: a review, pp. 131–144 (2022). https://doi.org/10.1201/9781003269793-15

19. Su, Q., et al.: New rapid and robust color image watermarking technique in spatial domain. IEEE Access 7, 30398–30409 (2019). https://doi.org/10.1109/ACCESS.2019.2895062

20. Swaraja, K., Meenakshi, K., Kora, P.: An optimized blind dual medical image watermarking framework for tamper localization and content authentication in secured telemedicine. Biomed. Signal Process. Control 55, 101665 (2020)

21. Thakur, S., Singh, A.K., Ghrera, S.P., Elhoseny, M.: Multi-layer security of medical data through watermarking and chaotic encryption for tele-health applications. Multimed. Tools Appl. 78, 3457–3470 (2019)

22. Vaidya, S.P.: Fingerprint-based robust medical image watermarking in hybrid transform. Vis. Comput. 39(6), 2245–2260 (2023)

23. Yang, C., Li, J., Bhatti, U.A., Liu, J., Ma, J., Huang, M.: Robust zero watermarking algorithm for medical images based on Zernike-DCT. Secur. Commun. Netw. 2021, 1–8 (2021)

24. Zermi, N., Khaldi, A., Kafi, M.R., Kahlessenane, F., Euschi, S.: Robust SVD-based schemes for medical image watermarking. Microprocess. Microsyst. 84, 104134 (2021)

A Detail-Guided Multi-source Fusion Network for Remote Sensing Object Detection

Xiaoting Li, Shouhong Wan$^{(\boxtimes)}$, Hantao Zhang, and Peiquan Jin

University of Science and Technology of China, Hefei, China
{lixiaoting,zhanghantao}@mail.ustc.edu.cn, {wansh,jpq}@ustc.edu.cn

Abstract. Optical and synthetic aperture radar (SAR) remote sensing have established themselves as valuable tools for object detection. Optical images exhibit weather-dependence but offer intricate information, whereas SAR images are weather-independent but may exhibit speckle noise and weaker edge details. Combining them can significantly enhance object detection precision. Nonetheless, existing fusion techniques frequently introduce noise of SAR images into fused features, consequently impinging on the efficacy of object detection. In this study, we present an innovative object detection architecture. It extracts detailed richness maps from optical images, subsequently employing these maps to recalibrate the spatial attention weights assigned to optical and SAR features. This strategic adjustment mitigates the impact of SAR noise on fused features within regions abundant in optical intricacies. Moreover, prevailing public optical-SAR fusion datasets need more meticulous instance-level object annotations, rendering them unsuitable for fulfilling object detection. Thus, we introduce two distinct datasets: OPTSAR, characterized by high registration accuracy, and QXS-PART, offering a counterpart with lower registration accuracy to validate the efficiency and generalization of our method. Both datasets encompass instance-level labels for diverse entities such as ships, aircraft, and storage tanks. Empirical assessments conducted on the OPTSAR and QXS-PART datasets underscore the prowess of our method. It substantiates a marked enhancement in object detection precision across well-aligned and poorly aligned optical-SAR fusion scenarios. Our method notably surpasses the efficacy of single-source object detection methodologies and established fusion approaches in terms of accuracy.

Keywords: Remote sensing object detection · Multi-source fusion · Detail-guided

1 Introduction

Remote sensing platforms utilize a variety of sensors that capture images through different wavelengths and frequencies, such as visible light, infrared, radar, and

Supported by Natural Science Foundation of Anhui Province (Grant No. 2208085MF157).

© The Author(s), under exclusive license to Springer Nature Switzerland AG 2024
S. Rudinac et al. (Eds.): MMM 2024, LNCS 14554, pp. 448–461, 2024.
https://doi.org/10.1007/978-3-031-53305-1_34

microwave. Among these, optical and synthetic aperture radar (SAR) images are commonly used for object detection. While optical images offer high-resolution detail, they are often impacted by weather conditions and cloud cover, which can make it challenging to accurately distinguish the objects in these areas. On the other hand, SAR images are not affected by weather and can capture information that is not visible in optical images. However, they may have lower resolution and more noise, which can make it difficult to discern object details. To enhance the limitations of using singular optical or SAR images, a combination of both can be used to leverage their unique strengths and provide a more comprehensive understanding of the object of interest [5,9,24].

The utilization of algorithms that merge various data sources has gained significant popularity in recent times owing to their capability of improving the precision of downstream tasks. These algorithms can be classified into three categories based on the point of fusion: pixel-level fusion, decision-level fusion, and feature-level fusion. Having a clear understanding of the distinctions between these techniques can assist researchers in identifying the most appropriate approach for their particular requirements and objectives.

Pixel-level fusion algorithms have established themselves as a formidable tool for amalgamating images from disparate sources, resulting in the creation of a fused image possessing a more comprehensive set of information that can be used for downstream tasks. Despite the availability of several techniques, such as autoencoders [7,26], convolutional neural networks, Transformers [14, 18,19], and generative adversarial networks [13,28], integrating optical and SAR images remains a challenge, particularly due to the noise present in SAR images. Furthermore, the absence of emphasis on object features in this approach can make it strenuous to differentiate objects in the fused image.

Decision-level fusion algorithms operate by utilizing data from different sources to obtain downstream task results, which are then integrated to produce a final result. Researchers are striving to enhance the fusion effect of decision-level algorithms, and progress in this area is anticipated to enhance the capabilities of these algorithms [1,27]. However, designing the fusion strategy manually can be a difficult task, and the interaction between information from different sources can be limited, resulting in less effective fusion in comparison to pixel-level and feature-level fusion.

Feature-level fusion algorithms combine the acquired features from different sources and use downstream tasks to supervise feature extraction and fusion, effectively extracting and fusing useful information about objects. Thus, it's the best choice for optical-SAR fusion object detection. Unfortunately, the presence of SAR noise can significantly interfere with the fused features of objects, resulting in reduced performance of object detection. Existing feature-level fusion methods [19,25,29] do not account for this interference, resulting in low accuracy in object detection. Therefore, there is a pressing need to develop improved feature-level fusion methods that account for interference from SAR noise to enhance the accuracy of object detection based on the fusion of SAR and optical images.

The quantitative analysis of optical and SAR image fusion methods demands the availability of a precise and reliable optical-SAR fusion object detection dataset. However, current public datasets such as QXS-SAROPT [6], SARptical [20], SEN1-2 [16], and SEN12MS [15] do not possess instance-level labels, and their registration accuracy is suboptimal, rendering them unsuitable for this purpose. Moreover, these datasets do not adequately support the detection of small objects like ships, aircraft, and storage tanks. Therefore, there is an urgent need to create a new fusion object detection dataset that addresses these limitations.

To address the challenges in SAR and optical fusion object detection, we have made the following contributions.

(1) We construct a detail-guided spatial attention feature fusion object detection network for optical and SAR images. it uses a detail-generating module to obtain an optical detail richness map, which is then used to adjust the spatial attention weights of optical and SAR features in the spatial attention feature fusion module.

(2) In order to showcase the effectiveness and robustness of our approach, we devised two distinct datasets to facilitate optical-SAR fusion object detection. The first dataset, OPTSAR, consists of meticulously aligned images, while the second, QXS-PART, features images with poor alignment. Each dataset boasts instance-level labeling and an abundance of challenging objects, such as ships near the shoreline and storage tanks obscured by clouds.

(3) We have rigorously tested our methodology using the OPTSAR and QXS-PART datasets, and the results are impressive. Our approach significantly minimizes the impact of SAR noise, while improving the accuracy of object detection, even in images that are poorly aligned.

2   Related Work

To enable effective fusion of features, the feature-level fusion algorithms can be broadly categorized into attention-based fusion algorithms, Transformer-based fusion algorithms, and source-specific algorithms

Attention-Based Fusion Algorithms: The attention mechanism has shown great promise in object detection by assigning different weights to different channels or areas. It has inspired several researchers to fuse the features of different sources. For instance, DANet [29] generates mask-guided attention weights using features of different sources and assigns different weights to foreground and background branches using depth information. The MT-DETR [2] model incorporates a confidence fusion module to fuse depth features with optical features using the confidence weighting of the depth features. Similarly, DooDLeNet [4] generates confidence weighting and correlation weighting for different sources. CroFuseNet [23] uses the channel attention module to fuse the optical features and SAR features, which can make full use of the key optical and SAR features. Attention-based fusion methods leverage convolutional neural networks to generate feature weights for different data sources, allowing for better utilization of local information. However, due to the strong noise interference in SAR images and the significant domain difference between SAR and optical images, attention-based fusion algorithms may struggle to ignore noise and achieve optimal performance.

Transformer-Based Fusion Algorithms: The application of Transformer in multi-source fusion methods has proven to be promising due to its capability to obtain the correlation between features. Researchers have achieved successful results by utilizing a Transformer framework to extract and fuse features. Some of these methods rely on exchanging keys K, query Q, and values V or tokens generated by the features of different sources to exchange information from diverse sources [19]. Meanwhile, others use one source to generate the key K and another source to generate the query Q and value V [3,12]. However, these algorithms have their limitations. The internal correlation of a patch cannot be well represented as Transformer divides the image into multiple patches, resulting in the loss of local information. Moreover, the interpretability of Transformer is weak, making it challenging to incorporate prior knowledge from sources to enhance the fusion model.

Source-Specific Algorithms: Some researchers have designed fusion models that take into account the characteristics of data sources and expert knowledge. For instance, APFNet [25] designs five attribute-specific fusion branches to integrate RGB and thermal features under the challenges of thermal crossover, illumination variation, scale variation, occlusion, and fast motion, respectively. LAIIFusion [22] proposes a local illumination perception module with more accurate area and illumination labels to fully perceive the lighting differences in various regions of the image to solve the problem of poor detection performance of objects in low-light environments. However, these methods may not be well-suited for object detection based on the fusion of optical and SAR features due to the unique challenges presented by SAR images, such as high levels of noise. Therefore, there is a need for novel fusion models that can effectively integrate optical and SAR images and address the challenges specific to this domain.

Feature-level fusion methods have been widely used for integrating features from different sources based on the characteristics of the object. However, when it comes to object detection based on optical and synthetic aperture radar (SAR) feature fusion, the presence of speckle noise in SAR images can significantly affect the accuracy of object detection. The noise in SAR images can disturb the features that show the details of objects, making it challenging to differentiate objects in the fused image.

3 Proposed Method

Here, we propose a feature-level fusion object detection network that aims to minimize the effect of SAR noise on detail-rich regions of optical images.

Our network architecture comprises a detail branch and two symmetric branches - the optical branch and SAR branch. These branches are clearly distinguished in Fig. 1 by their respective blue, green, and red color. To effectively capture features from optical and SAR images across different scales, a backbone is utilized in both the optical and SAR branches. To adequately represent

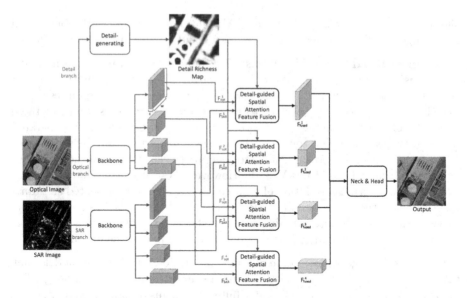

Fig. 1. The framework of proposed detail-guided spatial attention feature fusion network for object detection.

features of varying sizes, we employ $F^i_{opt/SAR/fused}$, where $i \in \{1,2,3,4\}$. The feature size is c×h×w, where c is the number of channels, h represents the height of features, and w represents the width of features. Our detail branch is responsible for generating a detail richness map through the use of optical images. Furthermore, our detail-guided spatial attention feature fusion module combines the features derived from the optical and SAR branches in the guidance of the detail richness map to produce fused features. These features are then integrated into the neck & head module to predict objects.

In the following sections, the detail-generating module embedded in the detail-guided branch is introduced first. Following this, the detail-guided spatial attention feature fusion module is discussed, which effectively combines optical and SAR features in the guidance of the detail richness map. Lastly, the loss function employed in our network is presented.

3.1   Detail-Generating Module

Optical images are known for their ability to capture rich edge structures and detail features of an object, which makes them indispensable in the detection of small objects such as ships, aircraft, and storage tanks. However, SAR features are often marred by excessive noise, which can interfere with fused features, thereby reducing the efficacy of small object detection. To mitigate the impact of SAR noise on fused features in areas with rich optical detail information, we proposed the extraction of detail richness maps from optical images. These maps are then used to guide the fusion of optical and SAR features, resulting in improved detection of small objects.

The detail-generating module is used to get the detail richness map of an optical image. We make i and j represent the horizontal and vertical coordinates of pixel points on the image, respectively. For each point in each channel of an optical image, we calculate the average value of pixels in the surrounding ks×ks region. The calculation formula is as follows:

$$i_{min} = max(0, i - \frac{ks-1}{2}) \tag{1}$$

$$i_{max} = min(weight, i + \frac{ks-1}{2}) \tag{2}$$

$$j_{min} = max(0, j - \frac{ks-1}{2}) \tag{3}$$

$$j_{max} = min(height, j + \frac{ks-1}{2}) \tag{4}$$

$$mean_{i,j}^{k} = mean(x^{k}[i_{min} : i_{max}, j_{min} : j_{max}]) \tag{5}$$

the minimum and maximum horizontal coordinates of the pixels in the ks×ks regions surrounding the pixel at position (i,j) are denoted as i_{min} and i_{max}, respectively, while j_{min} and j_{max} represent the minimum and maximum vertical coordinates of these pixels. Additionally, $mean_{i,j}^{k}$ are defined as the average value of pixels in the ks×ks region of the kth channel on the optical image surrounding the pixel at position (i,j), while x^{k} represents the optical image of the kth channel.

Our next step involves computing the absolute difference between the pixels in the surrounding ks×ks regions of the kth channel's position (i,j) on the optical image and its corresponding $mean_{i,j}^{k}$.

$$abs_{i',j'}^{i,j,k} = | x^{k}[i', j'] - mean_{i,j}^{k} | \tag{6}$$

where,

$$i' \in \{i_{min}, \ldots \ldots, i_{max}\}, j' \in \{j_{min}, \ldots \ldots, j_{max}\} \tag{7}$$

where $x^{k}[i', j']$ means the pixel value at the position (i',j') on the k channel of the optical image.

Finally, we get the average value of $abs_{i',j'}^{i,j,k}$ in the ks×ks region surrounding the pixel at the position (i,j) of three channels. The specific formula is as follows. Where c means the channel number of optical images.

$$ori\_detail_{i,j} = \frac{1}{c \times ks \times ks} \sum_{k=1}^{c} \sum_{i'=i_{min}}^{i_{max}} \sum_{j'=j_{min}}^{j_{max}} abs_{i',j'}^{i,j,k} \tag{8}$$

However, it is observed that the difference between the ori_detail of the region rich in details and that not rich in details is relatively small. Thus, we leverage the value of (9) to amplify this gap. In the formula, $detail_{i,j}$ represents the final value of the detail richness map at the position (i,j), while α denotes the maximum value of pixels that the optical images allow (here, we set α to 255 in our experiments). To achieve the desired results, we carefully tune the super-parameters of the network, denoted by β and γ, which are set to 50 and 2, respectively.

$$detail_{i,j} = \alpha * Sigmoid(\beta * \frac{ori\_detail_{i,j}}{\alpha} - \gamma) \tag{9}$$

3.2 Detail-Guided Spatial Attention Feature Fusion Module

To provide a detailed understanding of the proposed detail-guided spatial attention feature fusion module, we take the generation of F^1_{fused} as an example, which is shown in Fig. 2

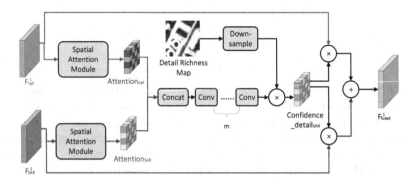

Fig. 2. Detail-guided spatial attention feature fusion module.

To enhance the generalization performance of the fusion model, we incorporate the attention mechanism to adjust the model. Specifically, the optical feature and SAR feature are sent into the spatial attention module to get the spatial attention weights $Attention_{opt}$ and $Attention_{SAR}$, respectively [21]. Two single-channel features are obtained by the max pooling and average pooling at the channel level. Then, we concatenate them and send them to a Conv layer to get $Attention_{opt/SAR}$. The formula is shown in (10):

$$Attention_{source} = Conv(Concat(P_{max}(F^1_{source}), P_{avg}(F^1_{source}))) \tag{10}$$

where P_{max} and P_{avg} represent the max and average pooling, respectively. Source $\in \{opt, SAR\}$.

Then $1\text{-}Attention_{opt}$ and $Attention_{SAR}$ are concatenated and fed into m convolution modules to fuse the attention of two sources, as shown in (11) :

$$Conf_{SAR} = Conv^m(Concat(1 - Attention_{opt}, Attention_{SAR})) \tag{11}$$

Next, the detail richness map is scaled to the size of the feature in the down-sample module, and the final spatial weight of SAR features is obtained by using (12):

$$Conf\_detail_{SAR} = (1 - Down(detail)) * Conf_{SAR} \tag{12}$$

where Down means the down-sample algorithm, and detail represents the detail richness map.

Finally, the fused feature is obtained by using (13):

$$F^1_{fused} = (1 - Conf\_detail_{SAR}) * F^1_{opt} + Conf\_detail_{SAR} * F^1_{SAR} \qquad (13)$$

3.3 Loss Function

The formula of our loss function is shown in (14).

$$Loss = \alpha_{class} * L_{class} + \alpha_{iou} * L_{iou} + \alpha_{dfl} * L_{dfl} \qquad (14)$$

where L_{class} represents the classification loss, L_{iou} represents the IOU loss, and L_{dfl} represents the distribution focal loss [11]. α_{class}, α_{iou}, α_{dfl} represent the weightings of the three loss functions. We set α_{class}=1.0, α_{iou}=2.5, and α_{dfl}=0.5 in our experiments.

4 Datasets and Experiment Settings

4.1 Optical-SAR Fusion Object Detection Datasets

We fabricated two datasets, called OPTSAR and QXS-PART, for optical-SAR fusion object detection.

OPTSAR: To validate the efficacy of our method, we produced a well-aligned optical-SAR fusion object detection dataset OPTSAR. It encompasses 713 pairs of optical-SAR image patches, which have high registration accuracy. The SAR images were procured from TERRASAR-X radar satellite images, while the optical images were sourced from Google Earth images. The SAR images possess a ground resolution of 1m, while the optical images have a resolution of 0.5–1 m. These images cover the ports and airports of Hong Kong and Taiwan. We utilized an advanced registration software called Erdas to register the SAR and optical images, and these images were cropped and annotated after registration. Each image patch is 512-pixel height and 512-pixel weight. We labeled three types of objects: ships, aircraft, and storage tanks, with a total of 13408 objects. We divided the dataset into a training set and a test set, with 500 pairs serving as the training set and 213 pairs serving as the test set.

QXS-PART: Furthermore, to evaluate the generalization of our method, we produced a poorly aligned optical-SAR fusion object detection dataset QXS-PART. The QXS-PART image patches are derived from the publicly available QXS-SAROPT dataset, which uses SAR data from the spotlight mode images of Gaofen-3 with single polarization, provided by China Centre for Resources Satellite Data and Application (CRESDA). The optical images corresponding to the SAR data are obtained from Google Earth. The ground resolution of SAR and optical images is 1m. The images cover the areas of San Diego, Shanghai, and Qingdao. Although the QXS-SAROPT dataset contains numerous images, not all of them contain the objects of interest. To create the QXS-PART dataset, we carefully selected the images from QXS-SAROPT that contain ships, aircraft, or storage tanks and labeled them accordingly. This ensures that the QXS-PART dataset contains only the relevant images needed for our specific use case. The QXS-PART dataset consists of a total of 784 pairs of images, out of which 553 pairs of images are used for training and 231 pairs of images are used for testing. The dataset contains 5000 objects, and each image in the dataset has a size of 256 pixels by 256 pixels.

4.2 Experiment Settings

Our experiment was conducted on an NVIDIA Geforce RTX3090 GPU. The backbone modules and the neck & detect module are derived from Yolov6 3.0. To expedite the training process, we use the pre-trained yolov6m model to initialize their parameters. The total epoch is set to 200, the initial learning rate is 0.01, and the batch size used on the OPTSAR dataset is 16. The batch size used on the QXS-PART dataset is 64.

5   Experimental Comparison and Analysis

5.1   Experimental Results

To assess the performance of our proposed approach, we have compared it with various open-source fusion methods that have been developed in recent years. Among them, DIFNet [8] and DetFusion [17] are pixel-level fusion algorithms based on convolutional neural networks. SwinFusion [14] is a pixel-level fusion method based on Transformer. SDNet [28] and Tardal [13] are pixel-level fusion algorithms based on generative adversarial networks. MT-DETR [2] is an object detection algorithm based on feature-level fusion. In addition to these comparison methods, we also carried out experiments on a single source.

Table 1. Test Experiments on OPTSAR Dataset

Fusion Method	Sources	Detect Model	AP	AP_{50}	AP_{75}	AP_S	AP_M	AP_L
-	SAR	Yolov6 3.0 [10]	12.6	28.7	11.2	16.8	16.7	21.6
-	optical	Yolov6 3.0	43.3	72.3	50.9	41.0	50.1	42.4
DIFNet [8]	SAR&optical	Yolov6 3.0	24.2	53.4	15.4	20.5	40.0	33.5
DetFusion [17]	SAR&optical	Yolov6 3.0	11.0	25.1	6.5	10.5	33.3	26.8
Swinfusion [14]	SAR&optical	Yolov6 3.0	13.8	33.3	5.9	17.5	17.8	35.2
SDNet [28]	SAR&optical	Yolov6 3.0	15.9	33.8	13.7	16.8	22.7	42.9
Tardal [13]	SAR&optical	Yolov6 3.0	17.1	35.7	13.3	18.5	25.1	32.7
MT-DETR [2]	SAR&optical	Deformable DETR [30]	27.4	56.6	19.2	25.1	37.8	13.9
ours	SAR&optical	Yolov6 3.0	46.6	77.1	52.9	43.0	57.7	45.3

In the case of pixel-level fusion methods, we start by generating pixel-level fusion images. These fusion images are then used to train and test the Yolov6 3.0 object detection network. On the other hand, for feature-level fusion methods, we use their respective detection networks to train and test the fusion images.

We use AP, AP_{50}, AP_{75}, AP_S, AP_M and AP_L as the evaluation metrics. AP_{50} and AP_{75} refer to the AP values when the IoU threshold is set to 0.5 and 0.75, respectively, and AP_S, AP_M and AP_L refer to the AP values of objects with less than 32^2 pixels, objects with a pixel count between 32^2 and 96^2, and objects with more than 96^2 pixels respectively.

The performance of the object detection on a single source, as well as the effectiveness of various fusion methods, including our method on the OPTSAR dataset, are presented in Table 1, with the optimal outcomes being highlighted in bold. Notably, the table's second and third rows display the test results achieved by training the Yolov6 3.0 model independently on SAR and optical images, respectively, providing a standard for comparison with the fusion methods. While most fusion methods fail to improve the accuracy of object detection, our method stands out by successfully combining optical and SAR images to improve AP by 3.3.

Table 2. Test Experiments on QXS-PART dataset

Fusion Method	Sources	Detect Model	AP	AP_{50}	AP_{75}	AP_S	AP_M	AP_L
-	SAR	Yolov6 3.0	11.2	22.7	10.5	7.8	22.0	55.5
-	optical	Yolov6 3.0	39.2	69.6	**39.2**	36.0	50.1	50.3
DIFNet	SAR&optical	Yolov6 3.0	16.1	32.6	13.2	14.0	24.5	49.9
DetFusion	SAR&optical	Yolov6 3.0	34.9	60.1	36.3	32.4	45.4	54.6
Swinfusion	SAR&optical	Yolov6 3.0	19.0	37.7	17.6	15.9	31.8	46.2
SDNet	SAR&optical	Yolov6 3.0	33.9	59.3	35.5	30.9	46.4	35.4
Tardal	SAR&optical	Yolov6 3.0	23.1	45.0	20.1	19.8	35.5	64.1
MT-DETR	SAR&optical	Deformable DETR	27.7	50.9	27.6	24.9	37.7	47.4
ours	SAR&optical	Yolov6 3.0	**40.8**	**73.0**	39.1	**36.2**	**54.8**	**69.8**

Table 2 shows the performance of different object detection networks on the QXS-PART dataset. The results indicate that our network also has effective improvement on datasets with low registration accuracy.

5.2 Ablation Studies

We performed ablation studies to confirm the effectiveness of the proposed modules and the values of super-parameters. Table 3 shows the result of the ablation study of the proposed detail-generating module and spatial attention feature fusion module.

Where fusion represents the spatial attention feature fusion module, and detail represents the detail-generating module. We use the concat fusion method to get the result without the detail-generating module and detail-guided spatial

Table 3. Ablation experiment of modules

fusion	detail	AP	AP_{50}	AP_{75}	AP_S	AP_M	AP_L
		43.1	76.3	46.7	40.1	52.9	45.0
✓		44.6	76.9	47.4	**45.0**	50.2	43.4
	✓	45.4	76.1	50.3	41.8	56.6	**50.2**
✓	✓	**46.6**	**77.1**	**52.9**	43.0	**57.7**	45.3

attention feature fusion module. To get the result without the detail-generating module, we remove the detail guidance in the detail-guided spatial attention feature fusion module. In the fourth line of Table 3, we use detail richness map as $Conf\_details_{SAR}$ to test the result without spatial attention feature fusion.

Table 4. Ablation experiment on the kernel size ks

ks	AP	AP_{50}	AP_{75}	AP_S	AP_M	AP_L
71	46.0	75.9	55.2	43.2	56.3	47.9
61	**46.6**	77.1	52.9	43.0	57.7	45.3
51	45.3	76.7	51.7	42.9	56.1	45.7
41	45.5	76.2	51.2	41.8	**57.8**	**55.1**
31	45.5	**77.6**	**56.3**	**44.6**	53.1	38.8
21	45.3	77.2	52.8	43.7	54.6	38.4

What's more, we also carried out the ablation study on the value of kernel size in the detail-generating module. The results are shown in Table 4. We experimented with ks chosen at intervals of 10 from 21 to 71. The result shows that AP is maximum when ks is 61.

The results of the ablation study on the number of Conv in the detail-guided spatial attention feature fusion module are shown in Table 5. When m is set to 2, the AP is the max.

Table 5. Ablation experiment on the Conv count m

m	AP	AP_{50}	AP_{75}	AP_S	AP_M	AP_L
6	45.1	76.5	51.8	42.0	55.6	41.1
5	45.3	**77.2**	48.2	42.3	56.8	54.0
4	46.1	76.9	51.4	42.4	57.4	**57.2**
3	45.2	75.8	48.4	43.4	54.2	56.9
2	**46.6**	77.1	**52.9**	43.0	**57.7**	45.3
1	45.5	76.0	41.8	**43.5**	54.9	49.4

5.3   Visualization

The results obtained from the OPTSAR dataset are visualized in Fig. 3. The green, bluish-violet, and red boxes correspond to the boxes of ships, storage tanks, and aircraft, respectively. The first column presents the ground truth label, while the second and third columns depict the outcomes of the single-source method Yolov6 3.0 on optical and SAR images, respectively. The last column showcases the result of our proposed method. It is noticeable that the optical single-source method fails to detect objects under clouds, while the SAR single-source method often loses objects due to noise. Other methods appear to struggle with error detection and missing detection in non-cloud-covered areas. However, our proposed method displays good performance in both cloud-covered and non-cloud-covered areas.

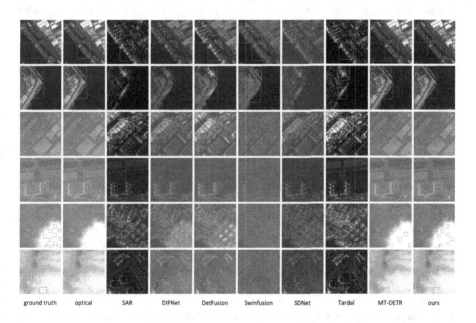

Fig. 3. Results visualization on the OPTSAR dataset.

6   Conclusion

In this paper, we propose an innovative object detection network built upon the principle of detail-guided spatial attention feature fusion, leveraging both optical and SAR remote sensing images. It introduces a novel detail richness map to quantify the intricate information within optical images. This map effectively governs the spatial attention allocation not only for optical features but also for SAR features. Furthermore, we present two distinct datasets designed specifically for optical-SAR fusion object detection: OPTSAR, a meticulously aligned dataset, and QXSPART, a dataset notable for its inherent alignment challenges. These datasets serve the crucial purpose of rigorously assessing the model's capacity for generalization and robust performance. Empirical results underscore the superiority of our model compared to alternative fusion techniques and single-source object detection methods. Notably, our model achieves enhanced object detection accuracy across both well-aligned and poorly-aligned datasets.

References

1. Cheng, Y., Cai, R., Li, Z., Zhao, X., Huang, K.: Locality-sensitive deconvolution networks with gated fusion for RGB-D indoor semantic segmentation. In: IEEE Conference on Computer Vision & Pattern Recognition (2017)
2. Chu, S.Y., Lee, M.S.: MT-DETR: robust end-to-end multimodal detection with confidence fusion. In: Proceedings of the IEEE/CVF Winter Conference on Applications of Computer Vision, pp. 5252–5261 (2023)
3. Chudasama, V., Kar, P., Gudmalwar, A., Shah, N., Wasnik, P., Onoe, N.: M2FNet: multi-modal fusion network for emotion recognition in conversation. In: Proceedings of the IEEE/CVF Conference on Computer Vision and Pattern Recognition, pp. 4652–4661 (2022)
4. Frigo, O., Martin-Gaffe, L., Wacongne, C.: DooDLeNet: Double deepLab enhanced feature fusion for thermal-color semantic segmentation. In: Proceedings of the IEEE/CVF Conference on Computer Vision and Pattern Recognition, pp. 3021–3029 (2022)
5. Fu, S., Xu, F., Jin, Y.Q.: Reciprocal translation between SAR and optical remote sensing images with cascaded-residual adversarial networks. Sci. China Inf. Sci. **64**, 1–15 (2021)
6. Huang, M., et al.: The QXS-SAROPT dataset for deep learning in SAR-optical data fusion (2021)
7. Li, H., Wu, X.J.: DenseFuse: a fusion approach to infrared and visible images. IEEE Trans. Image Process. **28**, 2614–2623 (2018)
8. Jung, H., Kim, Y., Jang, H., Ha, N., Sohn, K.: Unsupervised deep image fusion with structure tensor representations. IEEE Trans. Image Process. **29**(99), 3845–3858 (2020)
9. Kulkarni, S.C., Rege, P.P.: Pixel level fusion techniques for SAR and optical images: a review. Inf. Fusion **59**, 13–29 (2020)
10. Li, C., et al.: Yolov6 v3. 0: A full-scale reloading. arXiv preprint arXiv:2301.05586 (2023)
11. Li, X., et al.: Generalized focal loss: learning qualified and distributed bounding boxes for dense object detection. In: Advances in Neural Information Processing Systems vol. 33, pp. 21002–21012 (2020)
12. Li, Y., et al.: DeepFusion: lidar-camera deep fusion for multi-modal 3D object detection (2022)
13. Liu, J., et al.: Target-aware dual adversarial learning and a multi-scenario multi-modality benchmark to fuse infrared and visible for object detection. In: Proceedings of the IEEE/CVF Conference on Computer Vision and Pattern Recognition, pp. 5802–5811 (2022)

14. Ma, J., Tang, L., Fan, F., Huang, J., Mei, X., Ma, Y.: SwinFusion: cross-domain long-range learning for general image fusion via Swin transformer. IEEE/CAA J. Automatica Sinica **9**(7), 1200–1217 (2022)

15. Schmitt, M., Hughes, L.H., Qiu, C., Zhu, X.X.: SEN12MS - a curated dataset of georeferenced multi-spectral sentinel-1/2 imagery for deep learning and data fusion (2019)

16. Schmitt, M., Hughes, L.H., Zhu, X.X.: The sen1-2 dataset for deep learning in SAR-optical data fusion (2018)

17. Sun, Y., Cao, B., Zhu, P., Hu, Q.: DetFusion: a detection-driven infrared and visible image fusion network. In: Proceedings of the 30th ACM International Conference on Multimedia, pp. 4003–4011 (2022)

18. Vaswani, A., et al.: Attention is all you need. arXiv (2017)

19. Wang, Y., Chen, X., Cao, L., Huang, W., Sun, F., Wang, Y.: Multimodal token fusion for vision transformers. In: Proceedings of the IEEE/CVF Conference on Computer Vision and Pattern Recognition, pp. 12186–12195 (2022)

20. Wang, Y., Zhu, X.X., Zeisl, B., Pollefeys, M.: Fusing meter-resolution 4-D InSAR point clouds and optical images for semantic urban infrastructure monitoring. IEEE Trans. Geosci. Remote Sens. 1–13 (2017)

21. Woo, S., Park, J., Lee, J.Y., Kweon, I.S.: CBAM: convolutional block attention module. In: Proceedings of the European Conference on Computer Vision (ECCV), pp. 3–19 (2018)

22. Wu, J., Shen, T., Wang, Q., Tao, Z., Zeng, K., Song, J.: Local adaptive illumination-driven input-level fusion for infrared and visible object detection. Remote Sens. **15**(3), 660 (2023)

23. Wu, W., Guo, S., Shao, Z., Li, D.: CroFuseNet: a semantic segmentation network for urban impervious surface extraction based on cross fusion of optical and SAR images. IEEE J. Sel. Top. Appl. Earth Observations Remote Sens. **16**, 2573–2588 (2023)

24. Xia, Y., Zhang, H., Zhang, L., Fan, Z.: Cloud removal of optical remote sensing imagery with multitemporal SAR-optical data using X-Mtgan. In: IGARSS 2019–2019 IEEE International Geoscience and Remote Sensing Symposium, pp. 3396–3399. IEEE (2019)

25. Xiao, Y., Yang, M., Li, C., Liu, L., Tang, J.: Attribute-based progressive fusion network for RGBT tracking. In: Proceedings of the AAAI Conference on Artificial Intelligence, vol. 36, pp. 2831–2838 (2022)

26. Xu, H., Wang, X., Ma, J.: DRF: disentangled representation for visible and infrared image fusion. IEEE Trans. Instrum. Meas. **70**, 1–13 (2021)

27. Yao, Y., Mihalcea, R.: Modality-specific learning rates for effective multimodal additive late-fusion. In: Findings of the Association for Computational Linguistics: ACL 2022, pp. 1824–1834 (2022)

28. Zhang, H., Ma, J.: SDNet: a versatile squeeze-and-decomposition network for real-time image fusion. Int. J. Comput. Vis. **129**, 2761–2785 (2021)

29. Zhao, X., Zhang, L., Pang, Y., Lu, H., Zhang, L.: A single stream network for robust and real-time RGB-D salient object detection (2020)

30. Zhu, X., Su, W., Lu, L., Li, B., Wang, X., Dai, J.: Deformable DETR: deformable transformers for end-to-end object detection. arXiv preprint arXiv:2010.04159 (2020)

A Secure and Fair Federated Learning Protocol Under the Universal Composability Framework

Li Qiuxian[1]📧, Zhou Quanxing[1]([📧])📧, and Ding Hongfa[2]📧

[1] College of Big Data Engineering, Kaili University, Kaili 556011, China
qiuxianLL@163.com
[2] School of Information, Guizhou University of Finance and Economics,
Guiyang 550025, China

Abstract. Federated learning is a paradigm of distributed machine learning that enables multiple participants to collaboratively train a global model while preserving data privacy and locality. However, federated learning faces the dual challenges of data privacy and model fairness. Balancing these requirements while achieving efficient and robust learning outcomes remains a pressing issue. In this paper, based on the Universal Composable framework, we introduce an ideal federated learning functionality F_{FSFL} and a fair and secure federated learning real-world protocol π_{FSFL}. Our protocol integrates differential privacy and fairness incentive mechanisms, safeguarding client data privacy and countering potential threats from dishonest or malicious clients to model fairness. We demonstrate that our protocol can securely simulate the ideal functionality F_{FSFL} under the semi-honest model and resist passive attacks from polynomial-time adversaries. We further conduct a series of experiments on three widely used datasets (MNIST, CIFAR-10, and Fashion-MNIST) to validate the efficacy and robustness of our protocol under varying noise levels and malicious client ratios. Experimental results reveal that compared to other federated learning protocols, our method ensures data privacy and model fairness while delivering performance on par with, if not better than, baseline protocols.

Keywords: Federated Learning · Universal Composability Framework · Differential Privacy · Fairness Incentive Mechanism · Polynomial-time Adversary

Supported by organizations the National Natural Science Foundation of China (62002080), Guizhou Province Science and Technology Plan Project for 2023 (Guizhou Province Science Foundation - General [2023] No. 440), Project for Improving the Quality of Universities in Municipalities and Provinces (Ministry Office Issued [2022] No. 10-32), Natural Science Research Project of Guizhou Provincial Department of Education (Guizhou Education Union KY [2020] No. 179, No. 180, [2021] No. 140), Major Special Project Plan of Science and Technology in Guizhou Province (20183001), School-level Project of Kaili University (2022YB08), and School-level Research Project of Guizhou University of Finance and Economics (2020XYB02).

© The Author(s), under exclusive license to Springer Nature Switzerland AG 2024
S. Rudinac et al. (Eds.): MMM 2024, LNCS 14554, pp. 462–474, 2024.
https://doi.org/10.1007/978-3-031-53305-1_35

1   Introduction

In the digital age, various entities and organizations are producing and collecting vast amounts of data, which hold immense value in the fields of machine learning and artificial intelligence [1]. However, due to privacy and regulatory constraints, this data often cannot be directly shared or aggregated. Federated learning, as a distributed machine learning paradigm, offers a solution. It allows multiple clients to train models locally using their data and then send the model parameters to a server for aggregation [2]. This approach leverages the diversity and richness of data while ensuring data privacy and security.

However, federated learning also faces a series of challenges. Among them, communication efficiency, model fairness, and client honesty are the three core issues of utmost concern to researchers and practitioners [3–5]. Frequent communication not only increases network costs but also poses risks in terms of security and privacy. The fairness of the model and the honesty of the client directly affect the stability and reliability of the federated learning system.

To address these challenges, past research has mainly focused on individual aspects, such as improving communication efficiency or ensuring model fairness. For instance, Ro et al. proposed a federated averaging algorithm to reduce the communication overhead between clients and servers in federated learning [6]. Ji et al. further studied how to select clients to participate in training to enhance efficiency under limited bandwidth and a large number of clients [7]. Moreover, Woo et al., considering communication efficiency, introduced a fairness-based multi-AP coordination method used in Wi-Fi 7 for federated learning [8]. Liu et al. explored the integration of multi-task intelligent scheduling with cross-device federated learning [9]. Arouj et al. discussed energy-aware federated learning on battery-powered clients [10]. Rosero et al. studied the application of federated learning in time series prediction [11]. Lim et al. conducted a comprehensive survey on federated learning in mobile edge networks [12].

On the other hand, to balance the contributions and benefits of each client, some research has focused on model fairness. For example, Lyu et al. introduced a collaborative fairness federated learning framework that uses reputation to ensure participants converge to different models, achieving fairness and accuracy [13]. Du et al. further developed a fairness-aware agnostic federated learning framework that uses a kernel reweighting function to allocate reweighting values to each training sample, achieving high accuracy and fairness guarantees on unknown test data [14]. Ezzeldin et al. proposed FairFed, a novel fair federated learning algorithm that calculates the global model through a fair aggregation method, maintaining utility under different data heterogeneities [15]. Shejwalkar and Houmansadr demonstrated the vulnerability of federated learning to model poisoning attacks and introduced a new model poisoning attack framework [16]. Javed et al. proposed an integrated approach that combines blockchain technology and federated learning in vehicular networks to enhance security and privacy [17]. Liu et al. explored the application of federated learning in fintech, especially on how to protect privacy in personal identity information data analysis [18].

Furthermore, the issue of client honesty has garnered widespread attention, as malicious or dishonest clients might exploit protocol vulnerabilities to attack or disrupt. For instance, Zhao et al. introduced a novel privacy-enhanced federated learning framework that achieves local differential privacy protection through client self-sampling and data perturbation mechanisms [19]. Cao et al. further proposed FLCert, an integrated federated learning framework that has provable security against poisoning attacks from a bounded number of malicious clients [20]. Li et al. introduced a composite step synchronization protocol to coordinate the training process within and between super nodes, showing strong robustness to data heterogeneity [21]. Zhang et al. studied how to select non-IID data federated learning clients in mobile edge computing [22]. Luo et al. introduced an adaptive client sampling algorithm that handles system and statistical heterogeneity to minimize wall-clock convergence time [23]. Ferrag et al. provided a new comprehensive real dataset for network security in IoT and IIoT applications suitable for centralized and federated learning [24]. Finally, Rehman et al. introduced TrustFed, a framework for fair and trustworthy cross-device federated learning in IIoT based on blockchain, which detects model poisoning attacks, achieves fair training settings, and maintains the reputation of participating devices [25].

In this paper, to address the aforementioned challenges, we propose a novel federated learning protocol that, under a universally composable framework [26], balances data privacy, model fairness, and client honesty. Our protocol combines differential privacy and fairness incentive mechanisms, offering multifaceted protection of user privacy, ensuring model fairness, and effectively defending against malicious attacks. We also employ simulator techniques to prove that our protocol is secure against polynomial-time adversaries. To validate our approach, we conducted experiments on three widely-used datasets, and the results demonstrate that our protocol exhibits superior and robust performance across various scenarios.

2   Preliminary Knowledge

2.1   Universal Composability Framework

The Universal Composability (UC) Framework is a general model used for proving the security of cryptographic protocols. It ensures that a protocol remains secure even when arbitrarily combined with other protocols. Security is defined in terms of protocol emulation: a protocol is deemed secure if no external environment can distinguish its behavior from that of an ideal functionality. The ideal functionality is an abstract construct that describes the desired behavior of the protocol and is executed by a trusted third party. Mathematically, the UC framework is articulated as follows:

Given two protocols π and F, if there exists a simulator S such that for any environment Z, the view produced by executing π and S is indistinguishable from the view produced when executing F, then π is said to UC-emulate F. This implies that π can perfectly simulate F without revealing any additional information. If this can be proven, then it is assured that π is secure in any context.

2.2 Differential Privacy

Differential Privacy (DP) is a technique designed specifically to protect individual data privacy within a dataset. The core objective is to ensure that the results of data analysis are not significantly affected by the presence or absence of a single data point. To achieve this, differential privacy introduces specific random noise to obscure query outputs. As a result, even if a query result becomes publicly available, it is not possible to determine whether a particular data point exists in the dataset, thereby ensuring individual data privacy. For the precise implementation of differential privacy, it is essential to first determine the sensitivity of a query in order to decide the amount of noise that should be added. Sensitivity is defined as the maximum possible range of output differences for a query function f, when applied to any two adjacent datasets D and D' (which differ by only one data record). Mathematically, this is represented as:

$$\Delta f = \max_{D,D'} \|f(D) - f(D')\| \tag{1}$$

Once the query's sensitivity is known, the amount of noise required to meet a specific ϵ-differential privacy standard can be determined. A commonly used noise distribution for this purpose is the Laplace distribution, whose probability density function is:

$$p(x) = \frac{1}{2b} e^{-\frac{|x|}{b}} \tag{2}$$

Here, parameter b represents the scale of the distribution. If b is chosen to be $\frac{\Delta f}{\epsilon}$, then the noise from this Laplace distribution satisfies ϵ-differential privacy. This means that by adding this Laplace noise to the output of query function f, the query result will meet the ϵ-differential privacy standard.

3 Ideal Function

This section elaborates on the ideal function F_{FSFL} for federated learning, serving as an abstract representation of the process and objectives of federated learning. The function involves multiple participants, including a set of clients C, a server S, and an optional trusted third party T. Each client C_i owns its dataset D_i, consisting of pairs $(x_{i,j}, y_{i,j})$, where $x_{i,j}$ is a d-dimensional feature vector, and $y_{i,j}$ is the corresponding label. These data may come from different distributions, representing unique data characteristics of each client. The dataset is further divided into a training set $D_{i,\text{train}}$ and a test set $D_{i,\text{test}}$. The server is responsible for aggregating the model parameters from various clients and updating the global model. The trusted third party can provide encryption for communication between clients and the server, preventing data leakage during transmission.

3.1   Initialization Phase

The global model parameter w_0 is randomly initialized to avoid any initial bias. Additionally, fairness and incentive parameters α and β are initialized to determine the size of rewards and penalties. Moreover, a round counter r is set to 0 to track the training progress.

3.2   Client Selection Phase

A subset C_r is either randomly or criterion-based selected from the client set C to participate in the current round of training. This ensures sufficient client participation in each round and coverage of different data distributions.

3.3   Client Update Phase

Each selected client C_i performs the following operations: Initially, it trains on its local dataset D_i with the global model parameter w_0 and generates model updates. These updates are then sent to F_{FSFL}, allowing the server to understand the model improvements. The server does not directly access the actual data of the client, thereby maintaining data privacy. Subsequently, the client calculates its loss function value L_i locally using its training set $D_{i,\text{train}}$ and global model parameter w_0, evaluates its model performance, and sends this loss value L_i to F_{FSFL} to inform the server about its model error.

3.4   Server-Side Aggregation Phase

In this phase, F_{FSFL} sends the global model parameter w_0 to S, allowing the server to understand the status of the global model. Then, S aggregates the model parameters w_i from all the selected clients to update the global model parameter w_0, making the global model reflect the data from all clients. Following this, S calculates the global loss function value L_0 using the training sets $D_{i,\text{train}}$ from all selected clients and the global model parameter w_0, evaluates the global model's performance, and sends this global loss value L_0 to F_{FSFL} to let all clients know the error of the global model.

3.5   Fairness and Incentive Mechanism Phase

Each selected client C_i receives a utility U_i to incentivize its participation in federated learning and maintain honesty. This utility consists of a reward R_i and a penalty P_i, representing C_i's contribution to model utility and threat to model security, respectively. The reward R_i is calculated based on the global loss function value L_0 and the client's loss function value L_i, reflecting the improvement in C_i's model performance:

$$R_i = \alpha \times (L_0 - L_i) \tag{3}$$

The penalty P_i is calculated based on the global model parameter w_0 and the client's model parameter w_i, reflecting the deviation in C_i's model state:

$$P_i = \beta \times \|w_0 - w_i\|_2^2 \tag{4}$$

The utility U_i is the difference between the reward and the penalty, reflecting C_i's overall benefit:

$$U_i = R_i - P_i \tag{5}$$

F_{FSFL} sends U_i to C_i, allowing each client to know its utility.

3.6 Termination Condition Phase

The final phase is to evaluate whether the model has converged. The ideal function checks the model's convergence state based on predetermined conditions. If it has, the training terminates; otherwise, it continues to the next round. The conditions for convergence can be as follows:

- The model's convergence error is measured using the Mean Squared Error (MSE), defined as:

$$\text{MSE} = \frac{1}{n} \sum_{i=1}^{n} (L_i - L_0)^2 \tag{6}$$

  If MSE is less than a certain threshold ϵ, the model is considered converged.
- The model's convergence accuracy is measured using the accuracy (ACC), defined as:

$$\text{ACC} = \frac{1}{n} \sum_{i=1}^{n} I(y_{i,\text{test}} = f_i(x_{i,\text{test}})) \tag{7}$$

  If ACC exceeds a certain threshold η, the model is considered converged.
- A maximum round threshold can be preset, for instance, $R_{\max}$. If the model's training rounds r reach $R_{\max}$, even if the model hasn't fully converged, the training will terminate to avoid over-training or ineffective training.

4 Fair and Secure Federated Learning Protocol

The federated learning process confronts dual challenges of data privacy and model fairness. To address these issues, this section introduces a robust fair and secure federated learning protocol π_{FSFL}. The comprehensive description of this protocol is as follows:

4.1 Protocol's Input and Output

Given the challenges posed by real-world network environments, such as data loss, data tampering, and data breaches, it's imperative that the protocol's input and output adapt to these variations. Specifically, the protocol takes as input the set of clients C, a server S, a trusted third party T, global model parameters w_0, fairness and incentive parameters α and β, and differential privacy parameters ϵ and δ. Upon execution, the protocol aims to produce the updated global model parameters w_0' and the utility U_i for each client, accompanied by reward or penalty information.

4.2   Client Model Update

At the heart of the protocol lies the client model update. Each client C_i utilizes its local dataset D_i and the prevailing global model parameters w_0 for training. This is typically achieved through iterative gradient descent or other optimization algorithms. Post-training, each client calculates the gradient difference g_i between its local model parameters w_i and the global model parameters w_0, where J represents the loss function. To preserve data privacy, a Laplace noise is added to the gradient g_i to adhere to differential privacy. This noise is sampled from a Laplace distribution scaled inversely with ϵ and δ. The formula for this is:

$$g_i' = g_i + \text{Laplace}\left(\frac{\Delta J}{\epsilon}\right) \tag{8}$$

where ΔJ represents the sensitivity of the loss function.

4.3   Server Model Aggregation

Upon receiving the noise-perturbed gradients g_i' from all clients, the server undertakes model aggregation. The server uses these gradients to update the global model parameters. The specific update formula is:

$$w_0' = w_0 - \eta \sum_{i \in C} g_i' \tag{9}$$

where η denotes the learning rate.

4.4   Fairness and Incentive Mechanism

Following model aggregation, the protocol ensures model fairness and motivates honest client participation. The server calculates the loss for each client and, based on this, assigns rewards or penalties. For each client C_i, the reward R_i and penalty P_i are computed as: $R_i = \alpha(L_0 - L_i)$ and $P_i = \beta\|w_0 - w_i\|^2$, where L_0 represents the global loss.

4.5   Detection and Penalty for Dishonest or Malicious Behavior

Lastly, to ensure the security of the federated learning environment, potential dishonest or malicious behaviors need to be detected and penalized. For those clients with a negative utility, a trusted third party T is introduced for validation. Clients failing this validation are likely indulging in dishonest or malicious behaviors. Preset penalties, such as reducing the client's weight, restricting its participation in the next round, or entirely removing it from the system, can be imposed on such clients.

5 Security Proof

This section proves the security of our proposed Fair and Secure Federated Learning Protocol π_{FSFL}. Specifically, it ensures that the protocol can securely simulate the ideal functionality F_{FSFL} when confronted with a polynomial-time adversary A. To establish this security proof, we first clarify the adversary model and the cryptographic assumptions adopted. Next, we design a simulator S to mimic the behavior of adversary A in an ideal scenario. Finally, through reduction techniques, we argue that the observations of the adversary in both real and ideal scenarios are indistinguishable, ensuring the security of the protocol.

Theorem 1. *If the Fair and Secure Federated Learning Protocol π_{FSFL} ensures secure data transmission in a semi-honest model, then it can resist passive attacks from a polynomial-time adversary, thus ensuring its security.*

Proof. We assume a static, semi-honest adversary model. This implies that the adversary chooses some clients to control at the onset of the protocol, always adheres to the protocol, but attempts to glean additional information from the execution. The adversary can also control any portion of the client set C, but not more than half. Additionally, the cryptographic primitives employed (like differential privacy and zero-knowledge proofs) are assumed to be secure against polynomial-time adversaries.

Simulating Client Computations: Whenever a client controlled by A computes, the simulator S interacts with the ideal functionality F_{FSFL} to mimic and return the computation. Specifically, S doesn't rely on any unrealistic information but bases its simulation on its interaction with F_{FSFL}.

Simulating Differential Privacy: Since the client adds differential privacy noise to its gradient, S can mimic the addition of the same noise in the ideal world and return the result. For this, S needs to know the differential privacy parameters ϵ, δ and the sensitivity ΔJ of the loss function. Since these parameters are typically public, S can directly use them.

Simulating Rewards/Penalties: S can simulate the computation of rewards and penalties and return the result. For this, S needs to be aware of the fairness and incentive parameters α, β, the global loss function value L_0, and the loss function values of the clients L_i. These values are typically public, allowing S to use them directly.

Indistinguishability Proof: To prove the indistinguishability between the two worlds, we need to construct a game and analyze its probabilities. We define a game $G(b)$, where $b \in \{0, 1\}$. The process of the game is as follows:

- If $b = 0$, the game operates in the real world, where the environment Z interacts with the adversary A and the protocol π_{FSFL}.
- If $b = 1$, the game operates in the ideal world, where the environment Z interacts with the simulator S and the functionality F_{FSFL}.

Our objective is to prove that for any polynomial-time environment Z, the probability of it outputting different bits in $G(0)$ and $G(1)$ is negligible. That is:

$$|\Pr[G(0) = 1] - \Pr[G(1) = 1]| \leq \text{negl}(n) \tag{10}$$

where n is the security parameter, and negl(n) is a negligible function.

To prove this, we can use a reduction approach. If we assume the existence of a polynomial-time distinguisher D that can effectively differentiate between the two worlds, then by leveraging D, we can construct a polynomial-time algorithm B that breaks a certain cryptographic assumption. This leads to a contradiction, thus proving that the two worlds are indistinguishable.

In conclusion, we can assert that the real protocol π_{FSFL} can securely simulate the ideal functionality F_{FSFL} when facing a polynomial-time adversary A. This indicates that our protocol is secure and can be deployed in practical environments without concerns about data privacy and model fairness.

6   Experiment

6.1   Experimental Environment

To ensure the validity and reproducibility of all experimental results, we conducted all experiments in a controlled computational environment. The hardware setup consists of a computer equipped with an Intel(R) Core(TM) i7-8700 CPU @ 3.70 GHz and 32 GB DDR4 RAM. For tasks requiring intensive graphical computation, we employed an NVIDIA GeForce GTX 1080 Ti GPU. All experiments were run on a machine with a 1TB SSD, operating on Ubuntu 20.04 LTS.

On the software side, our primary programming language was Python 3.8. The deep learning models were constructed and trained using the PyTorch 1.8 framework. Our experiments utilized three datasets: the MNIST dataset provided by Yann LeCun, the CIFAR-10 dataset containing a diverse set of small images across 10 categories, and the FashionMNIST dataset, a more challenging variant of MNIST comprising images from 10 distinct fashion categories.

6.2   Experimental Results

In this section, we systematically validated the effectiveness of the "Secure and Fair Federated Learning Protocol in the Universal Composable Framework" through a series of experiments. We employed three widely-used datasets: MNIST, CIFAR-10, and FashionMNIST, and based our experiments on a simple multi-layer perceptron (MLP) model. We assessed the impact of different noise levels on model convergence and accuracy and enhanced data privacy and model fairness using differential privacy and fairness incentive mechanisms.

Figures 1 shows the accuracy evolution of the model over 20 training epochs under different noise levels across the three datasets. The results revealed that, in the absence of noise interference, each model could quickly converge and achieve a satisfactory accuracy level. However, as the noise level increased, the performance of the model began to face challenges. Especially on the CIFAR-10 dataset, a mere 0.1 noise level led to a decrease in accuracy by about 6% points.

These initial experimental results clarified that, in a noise-free environment, our federated learning protocol can offer superior performance across multiple datasets and models. However, in a noisy environment, while our protocol

(a) Model convergence on the MNIST dataset. (b) Model convergence on the Fashion MNIST dataset. (c) Model convergence on the CIFAR-10 dataset.

Fig. 1. Model convergence under different noise levels on various datasets.

can still maintain data privacy and model fairness, the learning outcome might degrade.

To further assess the robustness of our protocol, especially in the face of malicious clients, we designed two additional experiments. Figures 2 depict the performance changes of the model over 20 training epochs on the three datasets when confronted with malicious clients of different proportions. Clearly, when malicious clients are present, the performance of the model without the incentive mechanism drastically deteriorates. However, with the introduction of the incentive mechanism, the model can significantly resist these malicious attacks, especially as the proportion of malicious clients increases.

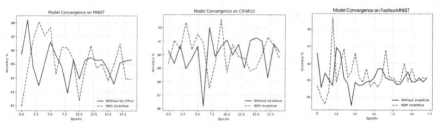

(a) Model accuracy on the MNIST dataset. (b) Model accuracy on the CIFAR-10 dataset. (c) Model accuracy on the FashionMNIST dataset.

Fig. 2. Model accuracy comparison on various datasets (with/without incentive mechanisms).

Figure 3 further compares the performance of the model on the FashionM-NIST dataset when faced with 30% malicious clients employing random category interference. The results are consistent with previous findings, indicating that the incentive mechanism can significantly enhance the model's robustness in malicious environments.

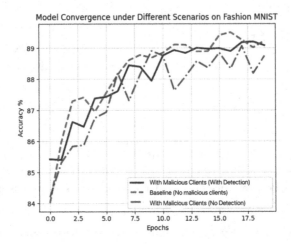

Fig. 3. Model convergence comparison under the influence of malicious clients and their detection.

Finally, our experiments also involved a comparative evaluation of four different protocols, namely: the basic protocol, differential privacy, the protocol we proposed, and federated averaging. This experiment focused on the impact of each protocol on model performance. As shown in Fig. 4, as expected, while differential privacy enhances data privacy protection, it also results in a sacrifice in model performance. However, it can be observed that our protocol, while ensuring data privacy and fairness, is comparable to the basic protocol and even surpasses it in certain scenarios.

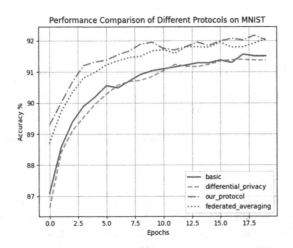

Fig. 4. Performance comparison of different protocols.

In conclusion, this series of experiments provided compelling evidence that our proposed federated learning protocol exhibits superior and robust performance across multiple scenarios.

7 Conclusion

In this study, we designed a federated learning protocol under the universal composability framework, integrating differential privacy and fairness incentive mechanisms to ensure data privacy and combat dishonest or malicious interference. Through theoretical analysis and experimental validation, we demonstrated that the protocol can securely simulate the ideal functionality in a semi-honest model, defend against passive adversaries within polynomial time, and exhibit outstanding performance and robustness across various noise levels and proportions of malicious clients. Despite the significant advancements we achieved, further research directions remain. For instance, our current protocol primarily targets static, semi-honest adversary models, and in the future, we plan to extend to more complex adversary settings. Moreover, while our foundational model predominantly focuses on simple multilayer perceptrons, we intend to explore more intricate deep learning models as base models in future endeavors.

References

1. Ahmed, S., Miskon, S.: IoT driven resiliency with artificial intelligence, machine learning and analytics for digital transformation. In: 2020 International Conference on Decision Aid Sciences and Application (DASA), pp. 1205–1208. IEEE (2020)
2. Hosseinalipour, S., Brinton, C.G., Aggarwal, V., et al.: From federated to fog learning: distributed machine learning over heterogeneous wireless networks. IEEE Commun. Mag. **58**(12), 41–47 (2020)
3. Konečný, J., McMahan, H.B., Yu, F.X., et al.: Federated learning: strategies for improving communication efficiency. arXiv preprint arXiv:1610.05492 (2016)
4. Li, T., Sanjabi, M., Beirami, A., et al.: Fair resource allocation in federated learning. arXiv preprint arXiv:1905.10497 (2019)
5. Dorner, F.E., Konstantinov, N., Pashaliev, G., et al.: Incentivizing honesty among competitors in collaborative learning and optimization]. arXiv preprint arXiv:2305.16272 (2023)
6. Ro, J., Chen, M., Mathews, R., et al.: Communication-efficient agnostic federated averaging (2021). https://doi.org/10.48550/arXiv.2104.02748
7. Ji, Y., Kou, Z., Zhong, X., et al.: Client selection and bandwidth allocation for federated learning: an online optimization perspective (2022). https://doi.org/10.48550/arXiv.2205.04709
8. Woo, G., Kim, H., Park, S., You, C., Park, H.: Fairness-based multi-AP coordination using federated learning in Wi-Fi 7. Sensors **22**(24), 9776 (2022). https://doi.org/10.3390/s22249776
9. Liu, J., et al.: Multi-job intelligent scheduling with cross-device federated learning. IEEE Trans. Parallel Distrib. Syst. **34**(2), 535–551 (2023). https://doi.org/10.1109/TPDS.2022.3224941

10. Arouj, A., Abdelmoniem, A.M.: Towards energy-aware federated learning on battery-powered clients. arXiv e-prints (2022). https://doi.org/10.48550/arXiv. 2208.04505
11. Llasag Rosero, R., Silva, C., Ribeiro, B.: Forecasting functional time series using federated learning. In: Iliadis, L., Maglogiannis, I., Alonso, S., Jayne, C., Pimenidis, E. (eds.) EANN 2023. CCIS, vol. 1826, pp. 491–504. Springer, Cham (2023). https://doi.org/10.1007/978-3-031-34204-2_40
12. Lim, W.Y.B., et al.: Federated learning in mobile edge networks: a comprehensive survey. IEEE Commun. Surv. Tutorials **22**(3), 2031–2063 (2020). https://doi.org/ 10.1109/COMST.2020.2986024
13. Lyu, L., Xu, X., Wang, Q., et al.: Collaborative fairness in federated learning, pp. 189–204. Privacy and Incentive, Federated Learning (2020)
14. Du, W., Xu, D., Wu, X., et al.: Fairness-aware agnostic federated learning. Proceedings (2020). https://doi.org/10.48550/arXiv.2010.05057
15. Ezzeldin, Y.H., Yan, S., He, C., et al.: Fairfed: enabling group fairness in federated learning. In: Proceedings of the AAAI Conference on Artificial Intelligence, pp. 37. no. 6, pp. 7494–7502 (2023)
16. Shejwalkar, V., Houmansadr, A.: Manipulating the byzantine: optimizing model poisoning attacks and defenses for federated learning. In: Proceedings 2021 Network and Distributed System Security Symposium (2021). https://doi.org/10. 14722/NDSS.2021.24498
17. Javed, A.R., Hassan, M.A., Shahzad, F., et al.: Integration of blockchain technology and federated learning in vehicular (IoT) networks: a comprehensive survey. Sensors **22**(12), 4394 (2022). https://doi.org/10.3390/s22124394
18. Liu, J., He, X., Sun, R., et al.: Privacy-preserving data sharing scheme with FL via MPC in financial permissioned blockchain. In: ICC 2021-IEEE International Conference on Communications, pp. 1–6. IEEE (2021). https://doi.org/10.1109/ ICC42927.2021.9500868
19. Zhao, J., et al.: Privacy-enhanced federated learning: a restrictively self-sampled and data-perturbed local differential privacy method. Electronics **11**(23), 4007 (2022). https://doi.org/10.3390/electronics11234007
20. Cao, X., Zhang, Z., Jia, J., et al.: Flcert: provably secure federated learning against poisoning attacks. IEEE Trans. Inf. Forensics Secur. **17**, 3691–3705 (2022)
21. Li, Z., He, Y., Yu, H., et al.: Data heterogeneity-robust federated learning via group client selection in industrial IoT. IEEE Internet Things J. **9**(18), 17844–17857 (2022)
22. Zhang, W., Wang, X., Zhou, P., et al.: Client selection for federated learning with non-IID data in mobile edge computing. IEEE Access **9**, 24462–24474 (2021)
23. Luo, B., Xiao, W., Wang, S., et al.: Tackling system and statistical heterogeneity for federated learning with adaptive client sampling. In: IEEE INFOCOM 2022-IEEE Conference on Computer Communications, pp. 1739–1748. IEEE (2022)
24. Ferrag, M.A., Friha, O., Hamouda, D., et al.: Edge-IIoTset: a new comprehensive realistic cyber security dataset of IoT and IIoT applications for centralized and federated learning. IEEE Access **10**, 40281–40306 (2022)
25. ur Rehman, M.H, Dirir, A.M., Salah, K., et al.: TrustFed: a framework for fair and trustworthy cross-device federated learning in IIoT. IEEE Trans. Ind. Inform. **17**(12), 8485–8494 (2021)
26. Cheng, Z., Jiang, Y., Huang, X., et al.: Universal interactive verification framework for federated learning protocol. In: Proceedings of the 2021 10th International Conference on Networks, Communication and Computing, pp. 108–113 (2021)

Bi-directional Interaction and Dense Aggregation Network for RGB-D Salient Object Detection

Kang Yi[1], Haoran Tang[2], Hongyu Bai[1], Yinjie Wang[1], Jing Xu[1(✉)],
and Ping Li[2]

[1] College of Artificial Intelligence, Nankai University, Tianjin, China
{kong_yi,2013479,2120210424}@mail.nankai.edu.cn, xujing@nankai.edu.cn
[2] Department of Computing, The Hong Kong Polytechnic University,
Kowloon, Hong Kong
haoran.tang@connect.polyu.hk, p.li@polyu.edu.hk

Abstract. RGB-D salient object detection (SOD) which aims to detect the prominent regions in figures has attracted much attention recently. It jointly models the RGB and depth information. However, existing methods explore cross-modality information from RGB images and depth maps without considering the potential coupling correlation between them. This may lead to insufficient information learning of these two modalities and even bring conflict due to their de-coupled representations. Thus, in this paper, we propose a novel framework called **Bi**-directional **I**nteraction and **D**ense **A**ggregation **Net**work (BIDANet) for RGB-D salient object detection. Firstly, we carefully design the depth-guided enhancement (DGE) and RGB-induced style transfer (RST) to allow the depth map and RGB image to learn information from each other through the bi-directional interaction network. Secondly, we adopt an adaptive cross-modal fusion (ACF) to flexibly integrate these learned multi-modal features. Last, we propose a dense aggregation network (DAN) to effectively aggregate cross-stage outcomes and generate accurate saliency prediction. Extensive experiments on 5 widely-used datasets demonstrate that our proposed BIDANet achieves superior performance compared with 14 state-of-the-art methods.

Keywords: RGB-D · Salient Object Detection · Bi-directional Interaction · Dense Aggregation · Multi-modal Learning

1 Introduction

Salient object detection (SOD) that visually locates the most outstanding parts or regions of images [29] has become one of the fundamental tasks in computer vision due to its extendability. SOD could help us understand the high-level semantic information of images and thus remarkable progress of this area has been made in recent years from traditional computer vision technology to

© The Author(s), under exclusive license to Springer Nature Switzerland AG 2024
S. Rudinac et al. (Eds.): MMM 2024, LNCS 14554, pp. 475–489, 2024.
https://doi.org/10.1007/978-3-031-53305-1_36

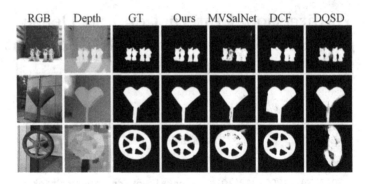

Fig. 1. Comparison examples of our proposed BIDANet and other representative methods, including MVSalNet [39], DCF [13], and DQSD [4].

multi-modal learning. Despite the success of the SOD study, its performance may degrade when being applied to complex real-world scenarios. 3D images, which provide high-quality depth structural information, could effectively help to improve SOD and protect its performance [13,39], making it a new area called RGB-D SOD task. This is gaining fast-growing interest nowadays and a large amount of studies have been devoted to it by utilizing RGB-D image pairs [36].

However, there are still some thorny problems that need to be solved. First is how to utilize and integrate the multi-model information derived from RGB and depth modalities in a general framework. The regular RGB images with three channels present an explicit description of how the objects are displayed in their way while in real-world scenarios, the structural depth map that contains sufficient spatial and geometric information could serve as supplementary clues for these RGB images. This is owing to its capability of capturing multi-dimension layout information and further attacking light and color changes. Some of the existing RGB-D SOD methods have explored interaction paradigms for cross-modality fusion between RGB and depth [2,11]. Nevertheless, due to the poor hardware devices and low-precision sensors, there exists noise in depth maps during the acquisition process, which leads to unsatisfactory results [15]. Hence, it is necessary and crucial to design an appropriate framework to accurately encode the depth maps as well as the RGB images for RGB-D SOD task.

The second problem is the deep potential correlation between RGB and depth information. Most existing works model the two kinds of modalities independently without considering whether they would have implicit impacts on each other [8,13]. In other words, they rarely investigate the modality gap problem caused by imbalanced semantic distribution between the RGB images and depth maps. However, RGB information would indicate the formalization of structure in the depth map while the depth map would deliver rich contrast information from RGB images.

Thus, separately modeling the two kinds of modalities may learn incomplete semantic information which would even destroy the performance of RGB-D SOD task. Moreover, exploring the potential correlation between them could help us understand why RGB-D utilization presents superior performance than single RGB adoption for the SOD task.

Therefore, to solve the above problems, we propose a novel framework called **Bi**-directional **I**nteraction and **D**ense **A**ggregation **Net**work (BIDANet) for RGB-D salient object detection. Specifically, we design the depth-guided enhancement (DGE) and RGB-induced style transfer (RST) through a bi-directional coupling interaction network. DGE allows low-level depth maps with richer structure information to guide the high-level RGB features while RST lets higher-level RGB representations with advanced semantic information guide the low-level depth features. Then, with the help of DGE and RST, we adopt an adaptive cross-modal fusion (ACF) to flexibly integrate these multi-modal features extracted from different stages. Next, to generate accurate saliency prediction, we propose a dense aggregation network (DAN) to effectively aggregate cross-stage outcomes from the ACF module. This further ensures obtaining a unified framework of BIDANet through joint training. We have shown comparison examples of BIDANet and other methods in Fig. 1.

In summary, the main contributions of this paper are as follows:

1. We propose an end-to-end framework BIDANet for RGB-D SOD by efficiently exploring the potential correlation between RGB image and depth map and adaptively decoding cross-stage features to generate final prediction.
2. We carefully design depth-guided enhancement and RGB-induced style transfer to allow the depth map and RGB image to learn information from each other in a bi-directional coupling interactive paradigm.
3. We construct the dense aggregation network to aggregate cross-stage outcomes from the adaptive cross-modal fusions by a top-down pathway, aiming to overcome the insufficient decoding problem for prediction.
4. We conduct extensive experiments on 5 public datasets and compare our BIDANet with 14 state-of-the-art methods. The results demonstrate our proposed model achieves better performance.

2 Related Work

2.1 RGB Salient Object Detection

Previous methods focus on the adoption of prior knowledge to predict saliency regions from RGB images. Despite their capabilities of feature extraction, deep learning-based methods have been introduced into the SOD task and achieved significant progress [29]. Yao et al. [35] designed a boundary information progressive guidance network to detect the salient regions and boundary features simultaneously. Liu et al. [19] explored the potential of pooling techniques and introduced a classic U-shape architecture to progressively fuse the coarse-level semantic information with the fine-level features. Tu et al. [28] proposed an edge-guided feature learning strategy that extracts hierarchical global and local information to incorporate non-local features, and maintains the clear edge structure of salient objects. Li et al. [18] developed a cross-layer feature pyramid network, which dynamically aggregates and distributes multi-scale features for information communication across multiple layers. Zhang et al. [37] proposed a hybrid priors SOD method, which fuses and optimizes the feature maps extracted from region contrast, background prior, depth prior and surface orientation prior.

2.2   RGB-D Salient Object Detection

Nonetheless, they are limited to insufficient representation of single RGB modality, leading to suboptimal performance of the SOD task. The RGB-D SOD methods which design both handcrafted features and fusion strategies to integrate RGB images and depth maps have been devoted to this area [25]. They inherited advanced extraction and learning capability of CNN and have achieved remarkable progress in complex real-world scenarios. Mao et al. [22] proposed a novel cross-modality fusion and progressive integration Network to address saliency prediction on stereoscopic 3D images. Fu et al. [10] adopted a Siamese network to extract deep hierarchical features from both input modalities simultaneously, and introduced a densely cooperative fusion module for complementary feature discovery. Sun et al. [27] presented a depth-sensitive RGB feature modeling scheme, which utilizes the depth decomposition and the depth-sensitive attention module for effective multi-modal feature fusion. Yang et al. [34] introduced a bi-directional progressive guidance framework to extract and enhance the unimodal features with the aid of another modality data. Jin et al. [16] designed an inverted bottleneck fusion strategy to effectively capture the cross-modal complementary information meanwhile accelerate elementwise operations for lightweight RGB-D SOD. For exploring the relationships of inter-pixel and intra-pixel, and long-range cross-modal interactions, Wu et al. [32] proposed a Transformer fusion and pixel-level contrastive learning method for RGB-D SOD.

However, there is still one significant shortcoming. Most of them explicitly ignore the potential coupling correlation between the RGB image and depth map, which may brings inconsistent representations of the cross modalities. This motivates our work in this paper.

3   Method

In this section, we first briefly introduce the overall framework of our BIDANet. Then we provide details of each modules in BIDANet, such as depth-guided enhancement, RGB-induced style transfer, adaptive cross-modal fusion and dense aggregation network. Finally, we leverage our joint loss function.

3.1   Overview

The overall framework of BIDANet is shown in Fig. 2. Given a RGB image with depth map, we first capture their deep representations and potential coupling correlation by two interactive modules, depth-guided enhancement and RGB-induced style transfer. Secondly, we adopt a cross-modal fusion strategy to

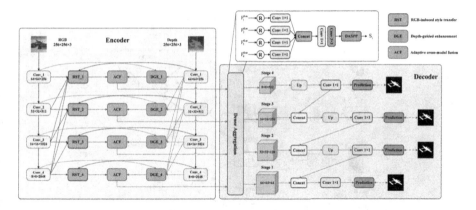

Fig. 2. Overall framework of our BIDANet. The bi-directional interactive encoder contains depth-guided enhancement (DGE), RGB-induced style transfer (RST) and adaptive cross-modal fusion strategy (ACF). The dense aggregation decoder (DAN) adopts top-down paradigm to make predictions.

adaptively integrate these multi-modal features over different extraction stages. Next, all the outcomes are send to our decoder which contains dense aggregation network to generate accurate predictions in a top-down way.

3.2 Bi-directional Interactive Encoder

Depth Guidance Enhancement. Compared with single RGB images, depth maps could provide structural and textural information, which could strengthen the representation of RGB images and improve the performance of saliency detection [27]. Thus, we design a depth guidance enhancement that utilizes low-level depth cues to enhance the high-level RGB representation because low-level depth features have more detailed information. DGE is shown in the top part of Fig. 3.

Specifically, we first obtain different RGB features and the depth features at stacked backbones. Then take the i-th level DGE as an example. It accepts all the depth features at lower levels to guide the representations of the i-th RGB features, which is shown as follows:

$$
\begin{aligned}
F_i^{DGE} = {} & Conv_{1\times1}(SA(Conv_{1\times1}(Conv_{1\times1}(F_i^{Depth})))) \otimes F_i^{RGB}) \\
& \oplus Conv_{1\times1}(SA(Conv_{3\times3}(Conv_{1\times1}(F_{i-1}^{Depth})))) \otimes F_i^{RGB}) \\
& \oplus Conv_{1\times1}(SA(Conv_{5\times5}(Conv_{1\times1}(F_{i-2}^{Depth})))) \otimes F_i^{RGB}) \\
& \oplus Conv_{1\times1}(SA(Conv_{7\times7}(Conv_{1\times1}(F_{i-3}^{Depth})))) \otimes F_i^{RGB}) \qquad (1)
\end{aligned}
$$

where $Conv$ is the convolutional layer, and $\{1 \times 1, 3 \times 3, 5 \times 5, 7 \times 7\}$ is the kernel size of convolutional layer. SA is spatial attention [31] to generate the depth-based attentional mask for RGB information. By DGE, the structural information from the depth map would be injected into the extracted RGB feature.

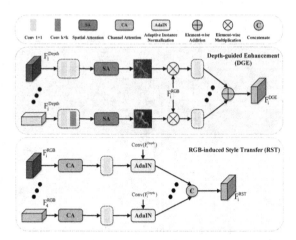

Fig. 3. Illustration of two key components: Depth Guidance Enhancement and RGB-induced Style Transfer. The DGE utilizes the lower-level depth features to guide the higher-level RBG features while the RST uses the higher-level RGB features to transfer the style of the lower-level depth features.

RGB-Induced Style Transfer. Since we aim to investigate the potential correlation between RGB images and depth maps, it is also necessary to utilize the RGB feature to strengthen the representation of depth information. The RGB-induced Style Transfer module is designed to use high-level RGB features to guide the extraction of depth features, which is shown in the bottom part of Fig. 3. Thus, the learned depth information would be transferred from single structural features to both structural and color features, revealing the correlation between depth map and RGB image and further helping the SOD task [5]. RST is defined as follows:

$$\begin{aligned}
F_i^{RST} = Cat[&AdaIN(Conv_{1\times1}(Conv_{1\times1}(F_i^{Depth})), f_i^{RGB}); \\
&AdaIN(Conv_{3\times3}(Conv_{1\times1}(F_i^{Depth})), f_{i+1}^{RGB}); \\
&AdaIN(Conv_{5\times5}(Conv_{1\times1}(F_i^{Depth})), f_{i+2}^{RGB}); \\
&AdaIN(Conv_{7\times7}(Conv_{1\times1}(F_i^{Depth})), f_{i+3}^{RGB})]
\end{aligned} \tag{2}$$

where $Cat(\cdot)$ denotes the concatenation operation and we use the Adaptive Instance Normalization $(AdaIN)$ [12] to realize the style transfer over different-level depth features. Moreover, to extract information from RGB more accurately, we process the RGB features as:

$$f_i^{RGB} = Conv_{1\times1}(CA(F_i^{RGB})) \tag{3}$$

where CA is the channel attention [31].

Adaptive Cross-Modal Fusion. So far, we have obtained the multi-level cross-modal RGB and depth features where they are both aware of the information from each other in a bi-directional process. Thus, we are able to ensure the consistency between them for combination. We first re-size the results of all stages:

$$f_i^{DGE} = Conv_{1\times1}(F_i^{DGE}), \ f_i^{RST} = Conv_{1\times1}(F_i^{RST}) \tag{4}$$

Then, we adaptively compute the weight vectors for the RGB feature and depth feature by *Softmax* and fuse them together:

$$(w_i^{DGE}, w_i^{RST}) = softmax(FC(split(FC(GAP(f_i^{DGE} \oplus f_i^{RST}))))) \tag{5}$$

$$F_i^{fuse} = f_i^{DGE} \otimes w_i^{DGE} \oplus f_i^{RST} \otimes w_i^{RST} \tag{6}$$

where FC and GAP are fully-connected layer and global average pooling. $\oplus$ and $\otimes$ are element-wise addition and product.

3.3 Dense Aggregation Network for Prediction

In our previous encoder, we capture the potential interactive correlation between RGB images and depth maps to guide more accurate representations of them and then fuse them adaptively to utilize them together. Here, we propose the Dense Aggregation Network for decoding the SOD task.

We re-size the shapes of all levels into the same shape and concatenate them:

$$f_i^{cat} = Cat[Conv_{1\times1}(Resize(F_1^{fuse})); Conv_{1\times1}(Resize(F_2^{fuse}));$$
$$Conv_{1\times1}(Resize(F_3^{fuse})); Conv_{1\times1}(Resize(F_4^{fuse}))] \tag{7}$$

where 1×1 convolution is also used for reducing channels. For the SOD task, if we directly use the unsampled fused features, it may face excessive noise problems [21]. Therefore, we carefully consider a de-noising and protective way to overcome this shortcoming and propose a top-down pathway to integrate outcomes from different levels. Before this, we apply the DASPP [33] to process multi-scale contextual information:

$$S_i = DASPP(Conv_{3\times3}(Conv_{1\times1}(f_i^{cat}))) \tag{8}$$

Finally, our top-down decoding for prediction is defined as follows:

$$f_4^{out} = Conv_{1\times1}(Up_{\times2}(S_4)), \tag{9}$$
$$f_3^{out} = Conv_{1\times1}(Up_{\times2}(Cat[S_3; f_4^{out}])), \tag{10}$$
$$f_2^{out} = Conv_{1\times1}(Up_{\times2}(Cat[S_2; f_3^{out}])), \tag{11}$$
$$f_1^{out} = Conv_{1\times1}(Cat[S_1; f_2^{out}]) \tag{12}$$

where we also add deep supervisions to the outputs F_i^{out} ($i \in \{1, 2, 3, 4\}$) for speeding up the convergence of the dense aggregation decoder. In our work, we only output the prediction of f_1^{out} while other three predictions are omitted in the test phase.

3.4 Loss Function

For each stage in the decoder, we adopt hybrid loss [26] to preserve sharp boundaries. It contains binary cross-entropy (BCE) loss, SSIM loss, and IoU loss, which could be defined as follows:

$$\mathcal{L}_i^{hybrid} = \mathcal{L}_i^{BCE} + \mathcal{L}_i^{SSIM} + \mathcal{L}_i^{IoU}, i \in \{1, 2, 3, 4\}. \tag{13}$$

where BCE loss measures the pixel-level difference between the predicted saliency map and ground truth, SSIM loss evaluates the structural similarity and IoU loss captures fine structures in image level.

As a result, the total loss of the whole framework can be denoted as:

$$\mathcal{L}_{total} = \sum_{i=1}^{4} \mathcal{L}_i^{hybrid}(P_i, G). \tag{14}$$

Note that all the predicted saliency maps and the ground truths have the same resolution as the original input images.

4 Experiments

In this section, we first introduce the experimental setup and implementation details. Then, we compare the proposed BIDANet with 14 state-of-the-art methods over 5 benchmark datasets and analyze the results. The ablation studies are also conducted to verify the effectiveness of each component.

4.1 Datasets and Evaluation Metrics

Datasets. To demonstrate the generalization capability of BIDANet, we conduct experiments on five public datasets. **DUT** [14] contains 1,200 images captured by Lytro Illum camera. **NJUD** [17] is composed of 1,985 image pairs from 3D movies and photographs. **STERE** [23] collects 1,000 pairs of binocular images from the Internet. **SSD** [42] is a small-scale dataset with only 80 image pairs. **SIP** [8] involves 929 images with high quality depth maps. It covers many challenging scenes of salient objects and persons. Following the same settings as previous works [9,16], we use a combined set of three datasets to train the proposed model, including 700 images from NLPR [24], 800 pairs from DUT, and 1,485 samples from NJUD. The remaining images with corresponding depth maps are used for test.

Evaluation Metrics. We adopt four widely-used metrics to evaluate the performance of BIDANet, including mean absolute error (MAE) [3], S-measure (S_α) [6], max F-measure (F_β) [1], and max E-measure (E_ξ) [7]. MAE calculates the mean absolute difference in pixel level between the predicted saliency map and the ground truth. The S-measure is designed to evaluate the structural similarities at object level and region level. Both F-measure and E-measure first

Table 1. Quantitative results compared with 14 state-of-the-art methods for RGB-D SOD task. "-" means that the results are unavailable since the authors did not release them. ↑ (↓) indicates the larger (lower) result would be the better. The best and the second best performance are bold and underlined, respectively.

Model	SIP [8] $E_\xi\uparrow$	MAE↓	DUT [14] $E_\xi\uparrow$	MAE↓	NJUD [17] $E_\xi\uparrow$	MAE↓	STERE [23] $E_\xi\uparrow$	MAE↓	SSD [42] $E_\xi\uparrow$	MAE↓
Ours	**0.933**	**0.042**	**0.958**	**0.031**	**0.946**	**0.034**	**0.941**	**0.040**	0.913	**0.042**
M2RNet(23PR) [9]	0.921	<u>0.049</u>	0.935	0.042	0.904	0.049	0.929	<u>0.042</u>	-	-
HINet(23PR) [2]	0.899	0.066	-	-	<u>0.945</u>	0.039	0.933	0.049	<u>0.916</u>	0.049
DMRA(22TIP) [14]	0.906	0.060	0.941	0.036	0.920	<u>0.037</u>	0.921	0.044	-	-
CCAFNet(22TMM) [41]	0.915	0.054	0.941	0.036	0.920	<u>0.037</u>	0.921	0.044	-	-
DRSD(22TMM) [38]	-	-	0.902	0.072	0.927	0.050	0.933	0.046	**0.917**	0.049
MoADNet(22TCSVT) [16]	0.911	0.058	0.945	0.033	0.929	0.042	0.931	0.043	0.900	0.057
MMNet(22TCSVT) [11]	0.871	0.080	<u>0.951</u>	<u>0.032</u>	0.922	0.038	0.916	0.046	0.912	<u>0.047</u>
DQSD(21TIP) [4]	0.900	0.065	0.889	0.073	0.912	0.051	0.911	0.052	0.890	0.053
DRLF(21TIP) [30]	0.891	0.071	0.870	0.080	0.901	0.055	0.915	0.050	0.879	0.066
JLDCF(21TPAMI) [10]	<u>0.923</u>	0.050	0.938	0.043	0.935	0.041	0.937	**0.040**	-	-
IRFRNet(21TNNLS) [40]	0.921	0.054	<u>0.951</u>	0.035	<u>0.945</u>	0.040	**0.941**	0.044	0.910	0.053
A2dele(20CVPR) [25]	0.890	0.070	0.928	0.043	0.916	0.051	0.928	0.045	0.862	0.070
S2MA(20CVPR) [20]	0.919	0.058	0.937	0.043	0.930	0.053	0.932	0.051	0.909	0.052
D3Net(20TNNLS) [8]	0.909	0.063	0.889	0.071	0.939	0.046	<u>0.938</u>	0.046	0.910	0.058

Model	SIP [8] $S_\alpha\uparrow$	$F_\beta\uparrow$	DUT [14] $S_\alpha\uparrow$	$F_\beta\uparrow$	NJUD [17] $S_\alpha\uparrow$	$F_\beta\uparrow$	STERE [23] $S_\alpha\uparrow$	$F_\beta\uparrow$	SSD [42] $S_\alpha\uparrow$	$F_\beta\uparrow$
Ours	**0.892**	**0.915**	**0.926**	**0.940**	**0.915**	**0.927**	<u>0.900</u>	0.911	**0.875**	**0.884**
M2RNet(23PR) [9]	<u>0.882</u>	0.902	0.903	0.925	<u>0.910</u>	<u>0.922</u>	0.899	<u>0.913</u>	-	-
HINet(23PR) [2]	0.856	0.880	-	-	**0.915**	0.914	0.892	0.883	0.865	0.852
DMRA(22TIP) [14]	0.852	0.863	0.905	0.915	<u>0.910</u>	0.911	0.891	0.887	-	-
CCAFNet(22TMM) [41]	0.877	0.881	0.905	0.915	<u>0.910</u>	0.911	0.891	0.887	-	-
DRSD(22TMM) [38]	-	-	0.864	0.853	0.886	0.876	0.899	0.887	0.861	0.832
MoADNet(22TCSVT) [16]	0.865	0.890	0.907	0.920	0.901	0.907	0.896	0.901	0.854	0.863
MMNet(22TCSVT) [11]	0.824	0.860	<u>0.920</u>	<u>0.939</u>	<u>0.910</u>	0.918	0.884	0.896	<u>0.871</u>	0.872
DQSD(21TIP) [4]	0.863	0.890	0.844	0.859	0.898	0.910	0.891	0.900	0.868	<u>0.877</u>
DRLF(21TIP) [30]	0.850	0.868	0.825	0.851	0.886	0.883	0.888	0.878	0.834	0.859
JLDCF(21TPAMI) [10]	0.881	<u>0.905</u>	0.905	0.924	0.902	0.912	**0.903**	0.914	-	-
IRFRNet(21TNNLS) [40]	0.879	0.881	0.919	0.924	0.909	0.908	0.897	0.893	0.864	0.841
A2dele(20CVPR) [25]	0.829	0.834	0.885	0.891	0.871	0.874	0.879	0.880	0.803	0.777
S2MA(20CVPR) [20]	0.872	0.877	0.903	0.901	0.894	0.889	0.890	0.882	0.868	0.848
D3Net(20TNNLS) [8]	0.860	0.861	0.850	0.842	0.900	0.900	0.899	0.891	0.857	0.834

obtain the binary saliency maps by varying the thresholds. The max F-measure computes the harmonic mean of average precision and average recall in multiple thresholds, while E-measure utilizes global image-level and local pixel-level information to measure the converted binary maps.

4.2 Implementation Details

We conduct all experiments with an NVIDIA GeForce RTX 3080Ti GPU on Pytorch. The backbones of both RGB and depth branches are pre-trained on

| RGB | Depth | GT | Ours | HINet | DMRA | CCAFNet | MoADNet | MMNet | D3Net |

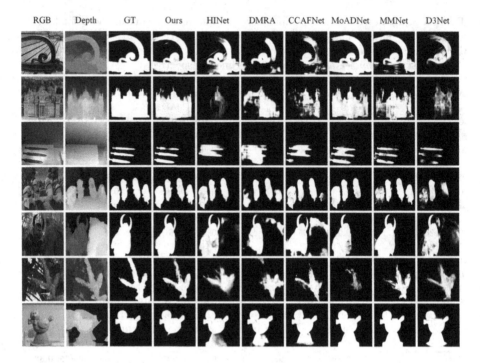

Fig. 4. Visual comparisons of the proposed BIDANet and other state-of-the-art RGB-D SOD methods. Our approach obtains competitive performance in a variety of complex scenarios.

ImageNet. The other layers are initialized to the default settings in Pytorch. During the training and testing, we simply copy the input depth maps to three channels and resize all images to 256 × 256. Besides, multiple data augmentations, such as random cropping, flipping, rotating, and color enhancement, are used in the training to avoid overfitting. We use the Adam optimizer to train the proposed framework in an end-to-end manner with a batch size of 4 for 100 epochs. The initial learning rate is 5e−5 and then decreases by a decay factor of 0.95. The whole training procedure takes about 3 h for convergence.

4.3   Comparison with State-of-the-Art Methods

In this subsection, we compare the proposed BIDANet with 14 state-of-the-art methods for RGB-D SOD task, including M2RNet [9], HINet [2], DMRA [14], CCAFNet [41], DRSD [38], MoADNet [16], MMNet [11], DQSD [4], DRLF [30], JLDCF [10], IRFRNet [40], A2dele [25], S2MA [20] and D3Net [8]. We would report their performance from published papers if accessible, or calculate the evaluation metrics by the saliency maps provided by the authors. All the metrics are calculated by the official evaluation tools [8].

Quantitative Analysis. The quantitative results of the comparison over five datasets are presented in Table 1. Higher values of S_α, F_β and E_ξ indicate better performance while lower MAE is better. We rank the baselines from top to bottom according to chronological order. It is obvious that compared with the existing methods, our BIDANet achieves the best performance in most cases. We attribute that to the multi-scale bi-directional enhancement and dense aggregation network. Hence, our BIDANet could pay more attention to sufficient and complementary information across different modalities and capture their potential gap, enabling the model to generate better structures.

Qualitative Analysis. To show qualitative comparison, we provide some representative saliency maps predicted by BIDANet and other SOTA methods in Fig. 4. Moreover, these saliency maps are sampled from different scenarios where the SOD task would be more challenging (e.g., fine structures, large objects, small objects, multiple objects, poor-quality depth maps, and low contrast). We can observe that our BIDANet could achieve structural integrity and internal consistency of the salient objects while other methods miss the detailed parts or mistakenly recognize the background as the foreground. Therefore, our method achieves better performance and stability for the SOD task when facing complex scenarios.

4.4 Ablation Study

To further investigate the effectiveness of each component in our BIDANet, we conduct ablation studies on two challenging datasets, NJUD [17], and SIP [8]. The quantitative results are reported in Table 2. Note that we only remove one component at a time and retrain the model.

Table 2. Ablation study on key components: RST, DGE, ACF, and DAN.

Model	NJUD [17]				SIP [8]			
	MAE ↓	S_α ↑	F_β ↑	E_ξ ↑	MAE ↓	S_α ↑	F_β ↑	E_ξ ↑
Ours	**0.034**	**0.915**	**0.927**	**0.946**	**0.042**	**0.892**	**0.915**	**0.933**
w/o RST	0.040	0.903	0.915	0.931	0.049	0.880	0.902	0.920
w/o DGE	0.043	0.901	0.912	0.930	0.051	0.876	0.899	0.918
w/o ACF	0.037	0.911	0.923	0.939	0.047	0.885	0.910	0.926
w/o DAN	0.039	0.906	0.916	0.937	0.049	0.883	0.903	0.925

Effectiveness of the RST. Complementary information between RGB image and depth map is critical in RGB-D SOD. To test how our designed RST performs, we remove all RST modules and directly utilize four stages of depth maps for fusion. From the results, we find that due to the lack of rich semantic information from the RGB for enhancing depth, our method may be affected by severe noise and inaccurate position, misaligning with the true objects.

Effectiveness of the DGE. We then investigate the impact of the proposed DGE. As reported in the second row of Table 2, this variation performs inferior performance compared to the complete BIDANet. Since low-level depth features indicate rich spatial details, utilizing the features of abundant geometric and textural information can enhance the deep representation of high-level RGB features and further improve the performance of saliency detection.

Effectiveness of the ACF. In our BIDANet, ACF plays an important role in fusing cross-modality features. Here, we directly substitute it with two convolutions and a concatenation. As shown in Table 2, we observe that removing ACF would bring significant degradation of performance over the used datasets, which indicates that the proposed ACF could adaptively select critical information in channel dimensions for the downstream task.

Effectiveness of DAN. In this part, we discard the multi-scale aggregation strategy with DASPP modules to demonstrate the effectiveness of the proposed DAN. Compared with the complete BIDANet, this attempt would obtain unsatisfactory performance on the two datasets. The result indicates the importance of the multi-scale aggregation strategy which integrates the cross-level complementary information. Moreover, the DASPP module expands the receptive fields of the decoding features. They both improve the performance of our BIDANet.

5 Conclusion

In this work, we have proposed a end-to-end framework called bi-directional interaction and dense aggregation network for RGB-D salient object detection, which aims to capture the potential correlation and fulfill the gap between RGB images and depth maps. Firstly, we design the depth-guided enhancement for RGB feature and RGB-induced style transfer for the depth feature to allow them to interact with each other. Then, an adaptive cross-modal fusion is adopted to integrate the information from multiple modalities over different stages. Next, we construct a decoder called dense aggregation network which is a top-down pathway to further improve the quality of final SOD prediction. Extensive experiments over 5 public datasets and 14 the-state-of-art methods demonstrate that our proposed BIDANet could achieve better performance for RGB-D SOD.

References

1. Achanta, R., Hemami, S., Estrada, F., Susstrunk, S.: Frequency-tuned salient region detection. In: Proceedings of IEEE Conference on Computer Vision and Pattern Recognition, pp. 1597–1604 (2009)
2. Bi, H., Wu, R., Liu, Z., Zhu, H., Zhang, C., Xiang, T.Z.: Cross-modal hierarchical interaction network for RGB-D salient object detection. Pattern Recogn. **136**, 109194 (2023)
3. Borji, A., Cheng, M.M., Jiang, H., Li, J.: Salient object detection: a benchmark. IEEE Trans. Image Process. **24**(12), 5706–5722 (2015)

4. Chen, C., Wei, J., Peng, C., Qin, H.: Depth-quality-aware salient object detection. IEEE Trans. Image Process. **30**, 2350–2363 (2021)
5. Chen, G., et al.: Modality-induced transfer-fusion network for RGB-D and RGB-T salient object detection. IEEE Trans. Circ. Syst. Video Technol. **33**(4), 1787–1801 (2023)
6. Fan, D.P., Cheng, M.M., Liu, Y., Li, T., Borji, A.: Structure-measure: a new way to evaluate foreground maps. In: Proceedings of IEEE International Conference on Computer Vision, pp. 4558–4567 (2017)
7. Fan, D.P., Gong, C., Cao, Y., Ren, B., Cheng, M.M., Borji, A.: Enhanced-alignment measure for binary foreground map evaluation. In: Proceedings of the International Joint Conference on Artificial Intelligence, pp. 698–704 (2018)
8. Fan, D.P., Lin, Z., Zhang, Z., Zhu, M., Cheng, M.M.: Rethinking RGB-D salient object detection: models, data sets, and large-scale benchmarks. IEEE Trans. Neural Netw. Learn. Syst. **32**(5), 2075–2089 (2021)
9. Fang, X., Jiang, M., Zhu, J., Shao, X., Wang, H.: M2RNet: multi-modal and multi-scale refined network for RGB-D salient object detection. Pattern Recogn. **135**, 109139 (2023)
10. Fu, K., Fan, D.P., Ji, G.P., Zhao, Q., Shen, J., Zhu, C.: Siamese network for RGB-D salient object detection and beyond. IEEE Trans. Pattern Anal. Mach. Intell. **44**(9), 5541–5559 (2022)
11. Gao, W., Liao, G., Ma, S., Li, G., Liang, Y., Lin, W.: Unified information fusion network for multi-modal RGB-D and RGB-T salient object detection. IEEE Trans. Circ. Syst. Video Technol. **32**(4), 2091–2106 (2022)
12. Huang, X., Belongie, S.: Arbitrary style transfer in real-time with adaptive instance normalization. In: Proceedings of IEEE International Conference on Computer Vision, pp. 1501–1510 (2017)
13. Ji, W., et al.: Calibrated RGB-D salient object detection. In: Proceedings of IEEE International Conference on Computer Vision and Pattern Recognition, pp. 9466–9476 (2021)
14. Ji, W., et al.: DMRA: depth-induced multi-scale recurrent attention network for RGB-D saliency detection. IEEE Trans. Image Process. **31**, 2321–2336 (2022)
15. Jin, W.D., Xu, J., Han, Q., Zhang, Y., Cheng, M.M.: CDNet: complementary depth network for RGB-D salient object detection. IEEE Trans. Image Process. **30**, 3376–3390 (2021)
16. Jin, X., Yi, K., Xu, J.: MoADNet: mobile asymmetric dual-stream networks for real-time and lightweight RGB-D salient object detection. IEEE Trans. Circ. Syst. Video Technol. **32**(11), 7632–7645 (2022)
17. Ju, R., Ge, L., Geng, W., Ren, T., Wu, G.: Depth saliency based on anisotropic center-surround difference. In: Proceedings of the International Conference on Image Processing, pp. 1115–1119 (2014)
18. Li, Z., Lang, C., Liew, J.H., Li, Y., Hou, Q., Feng, J.: Cross-layer feature pyramid network for salient object detection. IEEE Trans. Image Process. **30**, 4587–4598 (2021)
19. Liu, J.J., Hou, Q., Liu, Z.A., Cheng, M.M.: PoolNet+: exploring the potential of pooling for salient object detection. IEEE Trans. Pattern Anal. Mach. Intell. **45**(1), 887–904 (2023)
20. Liu, N., Zhang, N., Han, J.: Learning selective self-mutual attention for RGB-D saliency detection. In: Proceedings of IEEE Computer Society Conference on Computer Vision and Pattern Recognition, pp. 13756–13765 (2020)

21. Liu, Z., Wang, Y., Tu, Z., Xiao, Y., Tang, B.: TriTransNet: RGB-D salient object detection with a triplet transformer embedding network. In: Proceedings of the ACM International Conference on Multimedia, pp. 4481–4490 (2021)
22. Mao, Y., Jiang, Q., Cong, R., Gao, W., Shao, F., Kwong, S.: Cross-modality fusion and progressive integration network for saliency prediction on stereoscopic 3D images. IEEE Trans. Multimedia **24**, 2435–2448 (2022)
23. Niu, Y., Geng, Y., Li, X., Liu, F.: Leveraging stereopsis for saliency analysis. In: Proceedings of the IEEE Conference on Computer Vision and Pattern Recognition, pp. 454–461 (2012)
24. Peng, H., Li, B., Xiong, W., Hu, W., Ji, R.: RGBD salient object detection: a benchmark and algorithms. In: Fleet, D., Pajdla, T., Schiele, B., Tuytelaars, T. (eds.) ECCV 2014. LNCS, vol. 8691, pp. 92–109. Springer, Cham (2014). https://doi.org/10.1007/978-3-319-10578-9_7
25. Piao, Y., Rong, Z., Zhang, M., Ren, W., Lu, H.: A2dele: adaptive and attentive depth distiller for efficient RGB-D salient object detection. In: Proceedings of the IEEE Conference on Computer Vision and Pattern Recognition, pp. 9060–9069 (2020)
26. Qin, X., Zhang, Z., Huang, C., Gao, C., Dehghan, M., Jagersand, M.: BASNet: boundary-aware salient object detection. In: Proceedings of the IEEE Conference on Computer Vision and Pattern Recognition, pp. 7471–7481 (2019)
27. Sun, P., Zhang, W., Wang, H., Li, S., Li, X.: Deep RGB-D saliency detection with depth-sensitive attention and automatic multi-modal fusion. In: Proceedings of the IEEE Conference on Computer Vision and Pattern Recognition, pp. 1407–1417 (2021)
28. Tu, Z., Ma, Y., Li, C., Tang, J., Luo, B.: Edge-guided non-local fully convolutional network for salient object detection. IEEE Trans. Circ. Syst. Video Technol. **31**(2), 582–593 (2021)
29. Wang, W., Lai, Q., Fu, H., Shen, J., Ling, H., Yang, R.: Salient object detection in the deep learning era: an in-depth survey. IEEE Trans. Pattern Anal. Mach. Intell. **44**(6), 3239–3259 (2022)
30. Wang, X., Li, S., Chen, C., Fang, Y., Hao, A., Qin, H.: Data-level recombination and lightweight fusion scheme for RGB-D salient object detection. IEEE Trans. Image Process. **30**, 458–471 (2021)
31. Woo, S., Park, J., Lee, J.-Y., Kweon, I.S.: CBAM: convolutional block attention module. In: Ferrari, V., Hebert, M., Sminchisescu, C., Weiss, Y. (eds.) ECCV 2018. LNCS, vol. 11211, pp. 3–19. Springer, Cham (2018). https://doi.org/10.1007/978-3-030-01234-2_1
32. Wu, J., Hao, F., Liang, W., Xu, J.: Transformer fusion and pixel-level contrastive learning for RGB-D salient object detection. IEEE Trans. Multimedia, 1–16 (2023)
33. Yang, M., Yu, K., Zhang, C., Li, Z., Yang, K.: DenseASPP for semantic segmentation in street scenes. In: Proceedings of the IEEE Conference on Computer Vision and Pattern Recognition, pp. 3684–3692 (2018)
34. Yang, Y., Qin, Q., Luo, Y., Liu, Y., Zhang, Q., Han, J.: Bi-directional progressive guidance network for RGB-D salient object detection. IEEE Trans. Circ. Syst. Video Technol. **32**, 5346–5360 (2022)
35. Yao, Z., Wang, L.: Boundary information progressive guidance network for salient object detection. IEEE Trans. Multimedia **24**, 4236–4249 (2022)
36. Yi, K., Zhu, J., Guo, F., Xu, J.: Cross-stage multi-scale interaction network for RGB-D salient object detection. IEEE Sig. Process. Lett. **29**, 2402–2406 (2022)

37. Zhang, J., Wang, X.: Light field salient object detection via hybrid priors. In: Proceedings of the International Conference on Multimedia Modeling, pp. 361–372 (2020)
38. Zhang, Y., et al.: Deep RGB-D saliency detection without depth. IEEE Trans. Multimedia **24**, 755–767 (2022)
39. Zhou, J., Wang, L., Lu, H., Huang, K., Shi, X., Liu, B.: MVSalNet: multi-view augmentation for RGB-D salient object detection. In: Avidan, S., Brostow, G., Cissé, M., Farinella, G.M., Hassner, T. (eds.) Computer Vision, ECCV 2022. LNCS, vol. 13689, pp. 270–287. Springer, Cham (2022). https://doi.org/10.1007/978-3-031-19818-2_16
40. Zhou, W., Guo, Q., Lei, J., Yu, L., Hwang, J.N.: IRFR-Net: interactive recursive feature-reshaping network for detecting salient objects in RGB-D images. IEEE Trans. Neural Netw. Learn. Syst., 1–13 (2021)
41. Zhou, W., Zhu, Y., Lei, J., Wan, J., Yu, L.: CCAFNet: crossflow and cross-scale adaptive fusion network for detecting salient objects in RGB-D images. IEEE Trans. Multimedia **24**, 2192–2204 (2022)
42. Zhu, C., Li, G.: A three-pathway psychobiological framework of salient object detection using stereoscopic technology. In: Proceedings of the IEEE International Conference on Computer Vision Workshops, pp. 3008–3014 (2017)

Face Forgery Detection via Texture and Saliency Enhancement

Sizheng Guo, Haozhe Yang, and Xianming Lin[✉]

Key Laboratory of Multimedia Trusted Perception and Efficient Computing,
Ministry of Education of China, Xiamen University, Xiamen 361005,
People's Republic of China
{szguo,yanghaozhe}@stu.xmu.edu.cn, linxm@xmu.edu.cn

Abstract. In recent years, AI-driven advancements have resulted in increasingly sophisticated face forgery techniques, posing a challenge in distinguishing genuine images from manipulated ones. This presents significant societal and trust-related concerns. Most current methods approach this task as a binary classification problem via CNN, which potentially overlooks subtle forgery cues due to the down-sampling operations. This may lead to inadequate extraction of discriminative features. In this paper, we argue that forgery clues are hidden within facial textures and salient regions. Motivated by this insight, we propose the Texture and Saliency Enhancement Network (TSE-Net). TSE-Net contains two primary components: the Dynamic Texture Enhancement Module (DTEM) and the Salient Region Attention Module (SRAM). DTEM employs the Gray-Level Co-occurrence Matrix to extract facial texture information, while SRAM concentrates on salient regions. Extensive experiments demonstrate that the TSE-Net has superior performance in comparison to competing methods, emphasizing its effectiveness in detecting subtle face forgery cues through enhanced texture and saliency information extraction.

Keywords: Face Forgery Detection · Texture Enhancement · Salient Region Attention

1 Introduction

In recent years, the development of AI technology has led to increasingly realistic face forgery, making it harder to distinguish between authentic and fake images [29]. Furthermore, the cost of generating counterfeit faces has been decreasing, which may result in severe societal and trust-related issues. Consequently, face forgery detection, which aims to accurately discern between genuine and counterfeit facial representations, has become an urgent necessity and has consequently captured the attention of the computer vision community.

The basic face forgery detection methods regard this task as a binary classification problems [19,24–27], and leverage convolutional neural network (CNN) to distinguish the authenticity. However, such a baseline method has two serious

© The Author(s), under exclusive license to Springer Nature Switzerland AG 2024
S. Rudinac et al. (Eds.): MMM 2024, LNCS 14554, pp. 490–502, 2024.
https://doi.org/10.1007/978-3-031-53305-1_37

problems: 1) when facing the lifelike face, some subtle forgery clues might be overlooked due to the downsampling operations in Convolutional Neural Networks (CNNs), leading to insufficient extraction of fake features. 2) CNNs may struggle to identify differences in facial textures between authentic and counterfeit images. 3) CNNs can be prone to overfitting on the training dataset, resulting in a lack of generalization when applied to new or unseen data.

To address these challenges, face forgery detection methods can be broadly classified into three categories: spatial-based, frequency-based, and biological signals-based approaches. Spatial-based detection techniques leverage the ability of CNN models to identify subtle differences between real and fake faces within the spatial domain. Gram-Net [18], for instance, employs global texture features to enhance the robustness and generalization capabilities. Moving beyond the spatial domain, frequency-based methods mainly rely on spectra introduced by GANs and facial image frequency domain features. Biological signals-based approaches, on the other hand, focus on visual artifacts like eyes, teeth, and facial contours. While these methods have achieved considerable success, they often rely on subtle and local properties, such as hair shape and color. Moreover, they tend to overlook

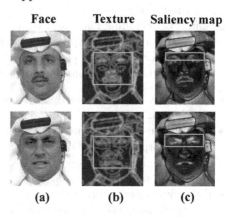

Face   **Texture**   **Saliency map**

(a)   **(b)**   **(c)**

Fig. 1. The first row shows the real faces and the second row shows the fake faces. (a) represents the original image, (b) represents the texture information extracted by GLCM, and (c) represents the saliency map. We find that texture information differs between real faces and forged regions, and saliency maps can roughly distinguish key locations of faces as well as suspected forged regions.

salient regions manipulated by technology, such as the eyes. This oversight results in limited generalization capabilities against unknown synthetic techniques.

In this paper, we argue that there are discernible distinctions in texture and saliency information between real and fake faces, which can help reveal subtle forgery cues, as illustrated in Fig. 1. Inspired by [10,21], we employ the Gray-level Co-occurrence Matrix (GLCM) to analyze texture statistics. GLCM is a potent feature extraction technique capable of capturing intricate patterns and summarizing the frequency of various pixel intensity combinations in an image. Using GLCM, we extract texture features such as entropy and homogeneity to differentiate between genuine and fake images. Unlike previous methods [10,21], we do not merely feed the GLCM into the network; instead, we calculate the entropy information and design a **Dynamic Texture Enhancement Module (DTEM)** to adaptively capture texture details. To further emphasize critical manipulated areas, we compute the saliency map and propose a **Salient Region Attention Module (SRAM)**, which identifies regions that are perceptually more distinct and relevant than others. By concentrating on these regions, we can reduce the complexity of forgery detection and increase detection accuracy. Finally, we add up these features to integrate texture and saliency information. Taking inspiration from the Two-branch Recurrent Network [19], we combine the aforementioned modules to create a Three-branch architecture called **Texture and Saliency Enhancement Network (TSE-Net)**. To demonstrate our method's effectiveness, we conduct extensive experiments on various existing datasets, including FaceForensics++ [24] and CelebDF [16]. The results indicate that our approach outperforms vanilla binary classifier baselines, achieving excellent performance and generalization capabilities.

To sum up, our contributions are as follows:

- We identify that previous methods did not fully utilize texture calculated by GLCM and salient regions in fake faces.
- We introduce the Texture and Saliency Enhancement Network (TSE-Net), which incorporates the Dynamic Texture Enhancement Module (DTEM) and Salient Region Attention Module (SRAM) to utilize texture features and emphasize salient facial areas.
- Extensive experiments demonstrate that our method outperforms the vanilla binary classification baselines and achieves great detection performance and generalization.

2   Related Work

2.1   Face Forgery Detection

Face forgery detection has long been a challenging task in the field of computer vision. Early research primarily relied on hand-crafted features such as inconsistent head poses and eye blinking. However, with the advent of deep learning, convolutional neural networks (CNNs) have gained popularity as an effective approach for detecting face forgery with improved performance [33]. For instance, [19] proposed a Two-branch feature extractor to merge data from both the color and frequency domains, using a multi-scale Laplacian of Gaussian (LoG) operator. Face X-ray [14] is supervised using the forged boundary, while Local Relation [4] measures the similarity between the features of local regions. However, these approaches do not focus on investigating texture discrepancies between real and fake images. To address these issues, our proposed method delves deep into the facial texture for potential clues.

2.2   Gray-Level Co-occurrence Matrix

The Gray-level Co-occurrence Matrix (GLCM) is a widely used feature extraction technique in image processing and computer vision. It characterizes the spatial relationships between pixels in an image by computing the occurrence of pixel pairs with specific relative positions and gray-level values. The GLCM considers the relative position of pixel pairs in different directions and is then normalized to obtain a probability distribution. Various statistical measures, such as contrast and entropy, are calculated from the GLCM and utilized as features. Prior research [21] directly computes co-occurrence matrices on image pixels for each channel and feeds them into convolutional neural networks. While this approach enables the network to learn critical features, these methods do not exploit the statistical features derived from the GLCM, resulting in limited feature information. In contrast, our approach leverages the texture information, such as the entropy derived from GLCM, to obtain more distinct features.

2.3 Saliency Map

A saliency map is a widely used tool in computer vision for highlighting the most salient regions of an image, which can facilitate various tasks, including object recognition. Numerous saliency map generation methods have been proposed, among which using low-level features, such texture and edge information, to calculate the saliency map is a popular approach. In the context of face forgery detection, saliency maps can be used to locate the anomaly area and identify the most critical regions of the face. Due to the computational complexity of some approaches, we adopt a static saliency map [20], which employs image features and statistics to identify the salient regions of the face.

3 Method

3.1 Overall Framework

In this section, we introduce our Texture and Saliency Enhancement Network (TSE-Net) for face forgery detection (seen in Fig. 2). TSE-Net comprises two crucial components: the Dynamic Texture Enhancement Module (DTEM) and the Salient Region Attention Module (SRAM). To extract texture features, the input images are processed by the DTEM, whereas the SRAM extracts saliency features. Specifically, in the DTEM, the input image is initially synthesized using the texture information and then combined with the original image for feature extraction via the backbone network. The dynamic texture enhancement module further refines the extracted texture information. Analogously, in the SRAM, the image generates a saliency map, which is integrated with the original image for feature extraction, and then the Channel Block Attention Module (CBAM) is employed to enhance information extraction from the salient regions. The resulting texture, saliency, and backbone features are subsequently added up and fed into the classification network to distinguish between real and fake faces.

3.2 Dynamic Texture Enhancement Module (DTEM)

For face forgery detection, the subtle differences inside the forged area are of vital importance to discriminate fake face. To comprehensively investigate subtle information in images, we introduce a novel DTEM module. This module comprises a texture generation module and a dynamic texture enhancement module.

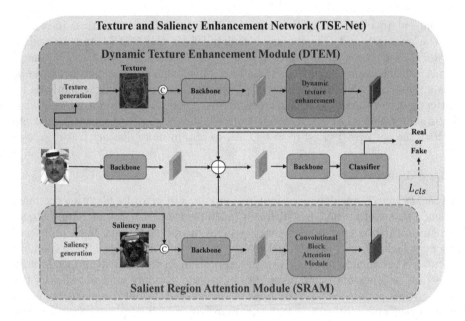

Fig. 2. The overall architecture of the proposed model consists of two key components: Dynamic Texture Enhancement Module (DTEM) and Salient Region Attention Module (SRAM). To be specific, the DTEM dynamically enhance the texture information and the SRAM helps focus on salient region. Then, we add these features up to maintain consistent information. Finally, the features after refining are fed into the network to detect the face is real or fake.

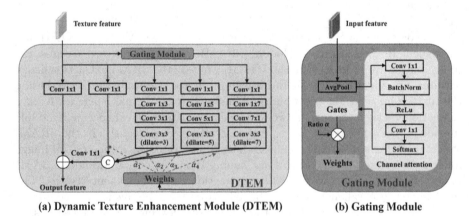

(a) Dynamic Texture Enhancement Module (DTEM)   **(b) Gating Module**

Fig. 3. (a) Dynamic Texture Enhancement Module (DTEM) consists of 4 branches of texture refinement process and a Gating module. (b) The Gating module assign 4 weights to dynamically select and enhance the texture features.

Within the texture generation module, we leverage Gray-level Co-occurrence Matrix to describe the gray-scale relationship between local or overall image pixels and adjacent pixels or those within a specific distance. Specifically, given a gray-scale image M, we extract pairs of adjacent pixels and count pairs of gray pixels in different combinations to calculate the GLCM P. Although P furnishes information about the gray direction, interval, and amplitude of image variation, it is not sufficient to distinguish texture characteristics. Therefore, we derive the *Entropy* from matrix P, which reflects the complexity of the image texture, as:

$$\mathcal{L} = \lambda_1 \mathcal{L}_\tau + \lambda_2 \mathcal{L}_C$$

$$= \lambda_1 \sum_i^N \sum_{j \neq l_i}^C (\hat{W}_j^T \hat{z}_i - \hat{W}_{l_i}^T \hat{z}_i + \gamma)_+ - \lambda_2 \sum_{i=1}^N \log \frac{e^{\hat{W}_{l_i}^T \hat{z}_i}}{\sum_j e^{\hat{W}_j^T \hat{z}_i}} \tag{1}$$

$$\hat{W}_i = \alpha \times \frac{W_i}{\|W_i\|} \tag{2}$$

$$\tilde{W}_m = \alpha \times \frac{(1-\beta)\hat{W}_m + \beta C_m}{\|(1-\beta)\hat{W}_m + \beta C_m\|} \tag{3}$$

$$C_m = \frac{\sum_i^n (1\{l_i = m\}\hat{z}_i)}{\sum_i^n 1\{l_i = m\}} \tag{4}$$

$$\mathcal{L} = \lambda_1 \mathcal{L}_\tau + \lambda_2 \mathcal{L}_C$$

$$= \lambda_1 \sum_i^N \sum_{j \neq l_i}^C (\tilde{W}_j^T \hat{z}_i - \tilde{W}_{l_i}^T \hat{z}_i + \gamma)_+ - \lambda_2 \sum_{i=1}^N \log \frac{e^{\tilde{W}_{l_i}^T \hat{z}_i}}{\sum_j e^{\tilde{W}_j^T \hat{z}_i}} \tag{5}$$

N is the number of gray levels in the image.

To further explore the correlation between these texture features, it is necessary to design various sized receptive fields, which is sensitive to both small and large spatial shifts. In the dynamic texture enhancement module, we adopt TEM [8] as the fundamental texture enhancement method. TEM incorporates four residual branches, represented as $b_i, i \in \{1,2,3,4\}$, each with distinct dilation rates $d \in \{1,3,5,7\}$ as well as a shortcut branch. For every branch b_i, a Conv layer is initially employed to adjust the channel. This is succeeded by two other layers: a $(2i-1) \times (2i-1)$ Conv layer and a 3×3 Conv layer using a specific dilation rate $(2i-1)$ for $i > 1$. These four branches are then merged and the channel size is reduced via a 3×3 Conv operation.

However, the forging methods and areas exist distinction, resulting in limitations in recognizing different forgery feature. To solve the problem, we design a weighting module motivated by [12] for the branches to dynamically select the corresponding forgery feature, as shown in Fig. 3(a). To be specific, the input features will be fed to average pooling and then processed through an attention module, consisting of a Conv-BatchNorm-ReLu, and concluded with a Conv1x1 and Softmax activation, yielding gates that correspond to 4 branches of TEM, as shown in Fig. 3(b). Additionally, we introduced a learnable parameter α, which is employed to dynamically adjust the values of gates, resulting in the derivation of four corresponding weights. These weights enable the dynamic selection and enhancement of texture features, further augmenting the efficacy of the model.

3.3 Salient Region Attention Module (SRAM)

Subtle distinctions exist between fake and real facial regions, and the majority of these regions hold some degree of importance in relation to the entire face, which tends to be damaged in down sampling. To overcome this problem, we propose Salient Region Attention Module (SRAM) which consists of generation of saliency map and an attention module.

For the generation of Saliency map, we utilized Static Saliency Fine Grained technique [20] to generate saliency maps from the original image. This technique calculates on-center and off-center differences of pixels, and computes the intensity submaps using various filter windows over the original gray-scale image, and then sums them pixel by pixel to obtain the saliency map in spatial scale. Besides, we use the center cropping to remove the disturbing saliency information of the background and retain the saliency information of the face as much as possible.

To enable the network to focus on these saliency regions, we integrated the Convolutional Block Attention Module [31] to capture forged clues of significant regions both in the spatial and channel domains. Specifically, when we extract the saliency feature F, CBAM infers a 1D channel attention map Mc of size $C \times 1 \times 1$ and a 2D spatial attention map Ms of size $1 \times H \times W$ sequentially. The attention process can be summarized as follows:

$$F_c = M_c(F) \otimes (F) \tag{6}$$

$$F_s = M_s(F_c) \otimes (F_c) \tag{7}$$

where $\otimes$ denotes element-wise multiplication, F_s denotes the refined feature.

3.4 Loss Function

We use the Cross-entropy as loss function, which is defined as:

$$L_{ce} = -\frac{1}{n} \sum_{i=1}^{n} y_i log(\hat{y_i}) + (1 - y_i)log(1 - \hat{y_i}) \tag{8}$$

where n is the number of images, y_i is the label of the sample, and $\hat{y_i}$ is the prediction of the i-th image.

4   Experiments

4.1   Experimental Setup

Datasets. In order to evaluate the effectiveness of our methodology, we conducted experiments on 2 renowned and challenging datasets. One such dataset is FaceForensics++ [24], which comprises 720 videos for training and 280 videos for validation or testing, and serves as a large-scale forgery face dataset. Face-Forensics++ includes four distinct face synthesis techniques, namely two deep learning-based methods (DeepFakes and NeuralTextures) and two graphics-based methods (Face2Face and FaceSwap). Moreover, the videos in FaceForensics++ exhibit two different levels of video quality, namely high-quality (quantization parameter equal to 23) and low-quality (quantization parameter equal to 40). Another widely-used dataset is CelebDF [16], which includes 590 genuine videos and 5639 counterfeit videos. In CelebDF, the forgery videos are generated using face swap technique for each pair of the 59 subjects.

Evaluation Metrics. We employ the accuracy score (ACC) of the face forgery detection and the area under the receiver operating characteristic curve (AUC) as the fundamental evaluation metrics in our study.

Implementation Details. We adopt EfficientNet-b4 [28] pretrained on the ImageNet as backbones which is commonly employed for this purpose. Our proposed modules are integrated into the output of layer 2 of EfficientNet-b4, as the middle layers capture middle-level features that effectively reveal subtle artifact clues. We resize each input image to 299×299. The network parameters are trained using the Adam optimizer with weight decay of $1e^{-5}$, β_1 of 0.9 and β_2 of 0.999. The initial learning rate is established as 0.001, while the StepLR scheduler with a decay of 5 step-size and a gamma of 0.1 is employed. The batch size for training is set at 32.

4.2   Experimental Results

Intra-testing. We assess the performance of the proposed method on the Face-Forensics++ dataset with respect to two quality settings. Specifically, we investigate the method's efficacy in both high-quality and low-quality settings. The outcomes presented in Table 1 demonstrate that the proposed approach achieves great performance on low-quality settings and is close to attention-based Multi-attentional approach [32] in high-quality settings. In contrast to the Two Branch approach [19], which utilizes the RGB and LoG operator to capture frequency information, our proposed method capitalizes on facial texture by incorporating an enhancement module. Consequently, our method enables the model to attend more effectively to potentially fraudulent regions, resulting in improved performance.

Table 1. The FaceForensics++ dataset was analyzed in terms of ACC and AUC with different image qualities (HQ and LQ), and the results with the highest performance were highlighted in bold.

Table 2. The AUC values obtained from the FF++ (LQ) dataset to evaluate the deepfake classes of FF++ and Celeb-DF datasets were compared. The highest results are highlight in bold.

Methods	ACC (LQ)	AUC (LQ)	ACC (HQ)	AUC (HQ)
MesoNet [1]	70.47	-	83.10	-
Face X-ray [14]	-	61.60	-	87.40
Xception [6]	86.86	89.30	95.73	96.30
Xception-ELA [11]	79.63	82.90	93.86	94.80
Xception-PAF [3]	87.16	90.20	-	-
Two Branch [19]	86.34	86.59	86.34	86.59
MAT [32]	88.69	90.40	**97.60**	**99.29**
SPSL [17]	81.57	82.82	91.50	95.32
RFM [30]	87.06	89.83	95.69	98.79
Freq-SCL [13]	89.00	92.39	96.69	99.28
Ours	**89.19**	**92.82**	96.96	**99.29**

Methods	FF++	Celeb-DF
Two-stream [33]	70.10	53.80
Meso4 [1]	83.00	53.60
FWA [15]	80.10	56.90
DSP-FWA [15]	93.00	64.60
Xception [24]	95.50	65.50
Multi-task [22]	76.30	54.30
Ours	**96.25**	**71.58**

Cross-Testing. In order to further substantiate the generalizability of the proposed method, we carry out cross-dataset evaluations. Specifically, we train our model on the FaceForensics++(LQ) dataset and assess its performance on the Deepfakes class and Celeb-DF datasets. The quantitative results of these evaluations are presented in Table 2, from which it is evident that our approach achieves great performance, particularly in the cross-database setting. Our proposed TSE-Net achieves superior performance compared Xception approach in cross-dataset evaluations, while also exhibiting improvement in intra-domain settings. This enhanced efficacy can be attributed to the capability of our module to guide the backbone network to selectively attend to informative subtle details that are ubiquitous in all forged faces.

Testing on Different Manipulation Types. Additionally, we conduct an evaluation of the proposed method on various face manipulation techniques as outlined in [24]. Specifically, the models are trained and tested solely on low-quality videos generated by a particular face manipulation method. The experimental outcomes are presented in Table 3. Among the four manipulation techniques, videos generated by NT pose an exceptionally difficult challenge. Notably, the performance of our proposed method is particularly impressive in detecting manipulated faces generated by NT, exhibiting an improvement of approximately 3.37% in the accuracy (Acc) score.

4.3 Visualization

To analyze the texture and saliency information, we visually represent the maps generated by the corresponding module. The results are presented in Fig. 4. It

Table 3. The accuracy (Acc) of the proposed method was evaluated on the FaceForensics++ (LQ) dataset, which includes four common manipulation methods, namely DeepFakes (DF), Face2Face (F2F), FaceSwap (FS), and NeuralTextures (NT). The bold results are the best.

Methods	DF	F2F	FS	NT
Steg.Features [9]	67.00	48.00	49.00	56.00
LD-CNN [7]	75.00	56.00	51.00	62.00
Constrained Conv [2]	87.00	82.00	74.00	74.00
CustomPooling CNN [23]	80.00	62.00	59.00	59.00
MesoNet [1]	90.00	83.00	83.00	75.00
Xception [5]	94.27	**91.98**	93.03	76.43
Ours	**96.29**	90.83	**95.41**	**79.80**

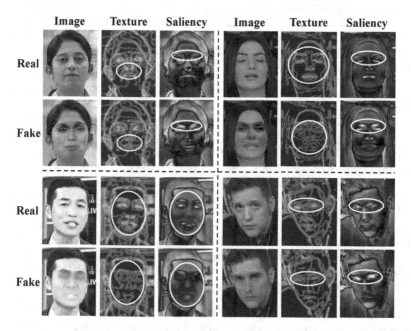

Fig. 4. The visualization of the texture extracted by Entropy derived from GLCM and the saliency map computed by Static Saliency Fine Grained technique respectively. (Color figure online)

is observed that there is a substantial difference in texture information between real and fake faces (seen in the red circle). For instance, under the Deepfakes manipulation, real faces contain more subtle details, while the texture features of fake faces tend to be relatively smooth. Saliency maps enable more accurate identification of the abnormal regions of the face, such as the mouth and eyes, and enable the model to concentrate on areas that may have been manipulated. These visualizations demonstrate the effectiveness of our modules in extracting subtle clues and informative region, which are critical for improving performance.

5 Conclusion

In this paper, we propose a Texture and Saliency Enhancement Network (TSE-Net) for face forgery detection. TSE-Net leverages the texture information of the face and pays attention to the salient region of the face, resulting in an effective and interpretable approach that achieves superior performance in face forgery detection. Experimental results demonstrate the effectiveness of our proposed method.

Acknowledgments. This work was supported by National Key R & D Program of China (No. 2022ZD0118202), the National Science Fund for Distinguished Young Scholars (No. 62025603), the National Natural Science Foundation of China (No. U21B2037, No. U22B2051, No. 62176222, No. 62176223, No. 62176226, No. 62072386, No. 62072387, No. 62072389, No. 62002305 and No. 62272401), and the Natural Science Foundation of Fujian Province of China (No. 2021J01002, No. 2022J06001).

References

1. Afchar, D., Nozick, V., Yamagishi, J., Echizen, I.: MesoNet: a compact facial video forgery detection network. In: 2018 IEEE International Workshop on Information Forensics and Security (WIFS), pp. 1–7. IEEE (2018)
2. Bayar, B., Stamm, M.C.: A deep learning approach to universal image manipulation detection using a new convolutional layer. In: Proceedings of the 4th ACM Workshop on Information Hiding and Multimedia Security, pp. 5–10 (2016)
3. Chen, M., Sedighi, V., Boroumand, M., Fridrich, J.: JPEG-phase-aware convolutional neural network for steganalysis of JPEG images. In: Proceedings of the 5th ACM Workshop on Information Hiding and Multimedia Security, pp. 75–84 (2017)
4. Chen, S., Yao, T., Chen, Y., Ding, S., Li, J., Ji, R.: Local relation learning for face forgery detection. In: Proceedings of the AAAI Conference on Artificial Intelligence, vol. 35, pp. 1081–1088 (2021)
5. Choi, Y., Choi, M., Kim, M., Ha, J.W., Kim, S., Choo, J.: StarGAN: unified generative adversarial networks for multi-domain image-to-image translation. In: Proceedings of the IEEE Conference on Computer Vision and Pattern Recognition, pp. 8789–8797 (2018)
6. Chollet, F.: Xception: deep learning with depthwise separable convolutions. In: Proceedings of the IEEE Conference on Computer Vision and Pattern Recognition, pp. 1251–1258 (2017)
7. Cozzolino, D., Poggi, G., Verdoliva, L.: Recasting residual-based local descriptors as convolutional neural networks: an application to image forgery detection. In: Proceedings of the 5th ACM Workshop on Information Hiding and Multimedia Security, pp. 159–164 (2017)
8. Fan, D.P., Ji, G.P., Cheng, M.M., Shao, L.: Concealed object detection. IEEE Trans. Pattern Anal. Mach. Intell. 44(10), 6024–6042 (2021)

9. Fridrich, J., Kodovsky, J.: Rich models for steganalysis of digital images. IEEE Trans. Inf. Forensics Secur. **7**(3), 868–882 (2012)
10. Goebel, M., Nataraj, L., Nanjundaswamy, T., Mohammed, T.M., Chandrasekaran, S., Manjunath, B.: Detection, attribution and localization of GAN generated images. arXiv preprint arXiv:2007.10466 (2020)
11. Gunawan, T.S., Hanafiah, S.A.M., Kartiwi, M., Ismail, N., Za'bah, N.F., Nordin, A.N.: Development of photo forensics algorithm by detecting photoshop manipulation using error level analysis. Indonesian J. Electr. Eng. Comput. Sci. **7**(1), 131–137 (2017)
12. Han, Y., Huang, G., Song, S., Yang, L., Wang, H., Wang, Y.: Dynamic neural networks: a survey. IEEE Trans. Pattern Anal. Mach. Intell. **44**(11), 7436–7456 (2021)
13. Li, J., Xie, H., Li, J., Wang, Z., Zhang, Y.: Frequency-aware discriminative feature learning supervised by single-center loss for face forgery detection. In: Proceedings of the IEEE/CVF Conference on Computer Vision and Pattern Recognition, pp. 6458–6467 (2021)
14. Li, L., et al.: Face X-ray for more general face forgery detection. In: Proceedings of the IEEE/CVF Conference on Computer Vision and Pattern Recognition, pp. 5001–5010 (2020)
15. Li, Y., Lyu, S.: Exposing DeepFake videos by detecting face warping artifacts. arXiv preprint arXiv:1811.00656 (2018)
16. Li, Y., Yang, X., Sun, P., Qi, H., Lyu, S.: Celeb-DF: a large-scale challenging dataset for DeepFake forensics. In: Proceedings of the IEEE/CVF Conference on Computer Vision and Pattern Recognition, pp. 3207–3216 (2020)
17. Liu, H., et al.: Spatial-phase shallow learning: rethinking face forgery detection in frequency domain. In: Proceedings of the IEEE/CVF Conference on Computer Vision and Pattern Recognition, pp. 772–781 (2021)
18. Liu, Z., Qi, X., Torr, P.H.: Global texture enhancement for fake face detection in the wild. In: Proceedings of the IEEE/CVF Conference on Computer Vision and Pattern Recognition, pp. 8060–8069 (2020)
19. Masi, I., Killekar, A., Mascarenhas, R.M., Gurudatt, S.P., AbdAlmageed, W.: Two-branch recurrent network for isolating Deepfakes in videos. In: Vedaldi, A., Bischof, H., Brox, T., Frahm, J.-M. (eds.) ECCV 2020. LNCS, vol. 12352, pp. 667–684. Springer, Cham (2020). https://doi.org/10.1007/978-3-030-58571-6_39
20. Montabone, S., Soto, A.: Human detection using a mobile platform and novel features derived from a visual saliency mechanism. Image Vis. Comput. **28**(3), 391–402 (2010)
21. Nataraj, L., et al.: Detecting GAN generated fake images using co-occurrence matrices. arXiv preprint arXiv:1903.06836 (2019)
22. Nguyen, H.H., Fang, F., Yamagishi, J., Echizen, I.: Multi-task learning for detecting and segmenting manipulated facial images and videos. In: 2019 IEEE 10th International Conference on Biometrics Theory, Applications and Systems (BTAS), pp. 1–8. IEEE (2019)
23. Rahmouni, N., Nozick, V., Yamagishi, J., Echizen, I.: Distinguishing computer graphics from natural images using convolution neural networks. In: 2017 IEEE Workshop on Information Forensics and Security (WIFS), pp. 1–6. IEEE (2017)
24. Rossler, A., Cozzolino, D., Verdoliva, L., Riess, C., Thies, J., Nießner, M.: Face-Forensics++: learning to detect manipulated facial images. In: Proceedings of the IEEE/CVF International Conference on Computer Vision, pp. 1–11 (2019)

25. Sun, K., et al.: An information theoretic approach for attention-driven face forgery detection. In: Avidan, S., Brostow, G., Cissé, M., Farinella, G.M., Hassner, T. (eds.) European Conference on Computer Vision, vol. 13674, pp. 111–127. Springer, Cham (2022). https://doi.org/10.1007/978-3-031-19781-9_7

26. Sun, K., et al.: Domain general face forgery detection by learning to weight. In: Proceedings of the AAAI Conference on Artificial Intelligence, vol. 35, pp. 2638–2646 (2021)

27. Sun, K., Yao, T., Chen, S., Ding, S., Li, J., Ji, R.: Dual contrastive learning for general face forgery detection. In: Proceedings of the AAAI Conference on Artificial Intelligence, vol. 36, pp. 2316–2324 (2022)

28. Tan, M., Le, Q.: EfficientNet: rethinking model scaling for convolutional neural networks. In: International Conference on Machine Learning, pp. 6105–6114. PMLR (2019)

29. Thies, J., Zollhofer, M., Stamminger, M., Theobalt, C., Nießner, M.: Face2Face: real-time face capture and reenactment of RGB videos. In: Proceedings of the IEEE Conference on Computer Vision and Pattern Recognition, pp. 2387–2395 (2016)

30. Wang, C., Deng, W.: Representative forgery mining for fake face detection. In: Proceedings of the IEEE/CVF Conference on Computer Vision and Pattern Recognition, pp. 14923–14932 (2021)

31. Woo, S., Park, J., Lee, J.-Y., Kweon, I.S.: CBAM: convolutional block attention module. In: Ferrari, V., Hebert, M., Sminchisescu, C., Weiss, Y. (eds.) ECCV 2018. LNCS, vol. 11211, pp. 3–19. Springer, Cham (2018). https://doi.org/10.1007/978-3-030-01234-2_1

32. Zhao, H., Zhou, W., Chen, D., Wei, T., Zhang, W., Yu, N.: Multi-attentional DeepFake detection. In: Proceedings of the IEEE/CVF Conference on Computer Vision and Pattern Recognition, pp. 2185–2194 (2021)

33. Zhou, P., Han, X., Morariu, V.I., Davis, L.S.: Two-stream neural networks for tampered face detection. In: 2017 IEEE Conference on Computer Vision and Pattern Recognition Workshops (CVPRW), pp. 1831–1839. IEEE (2017)

Author Index

A

An, Yukun 172

B

Bai, Hongyu 475
Bai, Ruwen 28

C

Cai, Pingping 216, 230
Cao, Chuqing 300
Cao, Qian 396
Cao, Weipeng 300
Chen, Chi-Yu 1
Chen, Feng 370
Chen, Xin 244
Chen, Zhenru 300
Ching, Pu 1

D

Du, Xiaoqin 159

F

Fang, Zhaoyan 134
Feng, Chengwei 286
Feng, Zunlei 356

G

Gan, Wenjun 201
Gao, Chenghua 28
Gao, Chunling 436
Gao, Guangyu 172
Gao, Neng 80
Guo, Sizheng 490
Guo, Yanrong 244

H

He, XueQin 109
Hong, JiXuan 109
Hong, Richang 244

Hong, Zixuan 300
Hongfa, Ding 462
Hu, Min-Chun 1
Hu, Yue 121, 187
Huang, Chen-Hsiu 94
Huang, Pei-Hsin 1
Huang, Peng 286
Huang, Zihao 28

J

Jiang, Yong 134
Jin, Peiquan 448
Ju, Lunhao 273

K

Kong, Zhe 80

L

Lai, Hanjiang 15, 422
Lan, Xin 314
Lei, Jie 356
Li, Fanzhang 258
Li, Fengfa 28
Li, Min 28
Li, Minghe 286
Li, Ping 475
Li, Qiulin 385
Li, Xiaoting 448
Li, Xuan 356
Li, Yuan 314
Li, Zebin 53
Li, Zituo 273
Liang, Ronghua 356
Liang, Weibo 314
Lin, Guoliang 422
Lin, Xianming 490
Lin, Zijian 410
Liu, Bei 286
Liu, Guihong 314

© The Editor(s) (if applicable) and The Author(s), under exclusive license to Springer Nature Switzerland AG 2024
S. Rudinac et al. (Eds.): MMM 2024, LNCS 14554, pp. 503–505, 2024.
https://doi.org/10.1007/978-3-031-53305-1

Liu, Jiawei   201
Liu, Lisa   216, 230
Liu, Ting   121, 187
Liu, Wei   327
Liu, Yongqi   159
Liu, Yongyu   422
Liu, Yuhan   80
Liu, Zhao   356
Lu, Guangtong   258
Luo, Jianping   53, 410
Lv, Jiancheng   314
Lv, Zhuocheng   172

M
Meng, Bo   28
Ming, Zhong   300
Mudgal, Vaibhav   342

P
Pan, Yan   422
Pan, Yehong   314
Peng, Yan   327
Peng, Yong   121

Q
Qiang, Junhao   385
Qiao, Qian   258
Qin, Yuqi   273
Qiu, Kedi   41
Qiuxian, Li   462
Quanxing, Zhou   462

R
Ren, Junxing   28

S
Shen, Zhengye   258
Shi, Jinyu   147
Shi, Shoudong   41
Smeaton, Alan F.   342
Song, Xin   370
Sun, Chengyu   396
Sun, Jianbin   273
Sweeney, Lorin   342

T
Tang, Haoran   475
Tao, Xi   300
Tian, Qi   244

W
Wan, Shouhong   448
Wang, Chao   327
Wang, Jiahuan   327
Wang, Jian   314
Wang, Qingyang   342
Wang, Shuaiwei   356
Wang, William Y.   216, 230
Wang, Yinjie   475
Wang, Youkai   121, 187
Wu, Ja-Ling   94
Wu, Qiushuo   314
Wu, Wansen   121, 187
Wu, Wenjie   147
Wu, Yong   201

X
Xie, JingJing   109
Xie, Shaorong   327
Xu, Jing   475
Xu, Juan   356
Xu, Kai   187
Xu, Mengying   15
Xu, Zhiwu   300

Y
Yan, Xiao   66
Yang, ChenHui   109
Yang, Haozhe   490
Yang, Kewei   273
Yang, Qun   385
Yang, Shuangyuan   436
Yang, Yuhang   66
Ye, Yongfang   41
Yi, Kang   475
Yin, Jian   15
Yin, Quanjun   187
Yu, Die   134
Yu, Jun   244
Yuan, Tianwen   286

Z
Zha, Zheng-Jun   201
Zhang, Anqi   172
Zhang, Dongdong   396
Zhang, Hantao   448
Zhang, Jian   286
Zhang, Kun   436
Zhang, Sanyuan   66
Zhang, Yifei   80

Zhao, Guozhi   201
Zhao, Tianxiang   41
Zheng, Liang   300
Zheng, Yazi   314

Zhou, Jiashuang   159
Zhou, Yuan   244
Zhu, Liang   370
Zhu, Yangchun   201

THE
by Biomedical Photolithographic

Printed in the United States
by Baker & Taylor Publisher Services